Microwaves:

Theory *and* Application *in* Materials Processing III

Related titles published by the American Ceramic Society:

Phase Equilibria Diagrams, Phase Diagrams for Ceramists, Volume XI: Oxides
Edited by Robert S. Roth
© 1995, ISBN 0-944904-90-4

Environmental Issues and Waste Management Technologies in the Ceramic and Nuclear Industries (Ceramic Transactions Volume 61)
Edited by Vijay Jain and Ron Palmer
© 1995, ISBN 1-57498-004-1

High-Temperature Ceramic-Matrix Composites II: Manufacturing and Materials Development (Ceramic Transactions Volume 58)
Edited by A.G. Evans and R. Naslain
© 1995, ISBN 0-944904-99-8

High-Temperature Ceramic-Matrix Composites I: Design, Durability, and Performance (Ceramic Transactions Volume 57)
Edited by A.G. Evans and R. Naslain
© 1995, ISBN 0-944904-98-X

Advanced Synthesis and Processing of Composites and Advanced Ceramics (Ceramic Transactions Volume 56)
Edited by Kathryn V. Logan
© 1995, ISBN 1-57498-000-9

Sol-Gel Science and Technology (Ceramic Transactions Volume 55)
Edited by Edward J.A. Pope, Sumio Sakka, and Lisa Klein
© 1995, ISBN 0-944904-97-1

Ceramic Processing Science and Technology (Ceramic Transactions Volume 51)
Edited by Hans Hausner, Gary L. Messing, and Shin-ichi Hirano
© 1995, ISBN 0-944904-89-0

Environmental and Waste Management Issues in the Ceramic Industry II (Ceramic Transactions Volume 45)
Edited by Dennis Bickford, Steven Bates, Vijay Jain, and Gary Smith
© 1994, ISBN 0-944904-79-3

Microwaves: Theory and Application in Materials Processing II (Ceramic Transactions Volume 36)
Edited by David E. Clark, Wayne R. Tinga, and Joseph R. Laia Jr.
© 1993, ISBN 0-944904-66-1

Microwaves: Theory and Application in Materials Processing (Ceramic Transactions Volume 21)
Edited by David E. Clark, Frank D. Gac, and Willard H. Sutton
©1991, ISBN 0-944904-43-2

For information on ordering titles published by the American Ceramic Society, or to request a publications catalog, please call 614-890-4700, or write to Customer Service Department, 735 Ceramic Place, Westerville, OH 43081.

Microwaves:

Theory *and* Application *in* Materials Processing III

Edited by

David E. Clark
University of Florida

Diane C. Folz
University of Florida

Steven J. Oda
Consultant

Richard Silberglitt
FM Technologies, Inc.

Volume 59

Published by
The American Ceramic Society
735 Ceramic Place
Westerville, Ohio 43081

Proceedings of the Microwaves: Theory and Application in Materials Processing III Symposium, presented at the 97th Annual Meeting of the American Ceramic Society, held in Cincinnati, May 1–3, 1995.

Library of Congress Cataloging-in-Publication Data

Microwaves : theory and application in materials processing III / edited by David E. Clark . . . [et al.].
p. cm.
Symposium continues series on microwave processing of materials initiated in 1988 at the Materials Research Society Spring Meeting.
Includes index.
ISBN 1-54798-002-5 (alk. paper)
1. Ceramics--Congresses. 2. Microwaves--Industrial applications--Congresses. I. Clark, David E. II. Series.
TP786.M533 1995
666--dc20 95-43601
CIP

For information on ordering titles published by the American Ceramic Society, or to request a publications catalog, please call 614-890-4700.

Printed in the United States of America.

1 2 3 4–99 98 97 96 95

ISSN 1042-1122
ISBN 1-57498-002-5

Contents

Waste Remediation

Processing Equipment

Microwave/Materials Interactions

Dielectric Properties

Process Modeling

Joining

Processing and Thermal Effects

Chemical and Reaction Synthesis

Current Issues and Future Activities

Preface

This symposium continued the series on microwave processing of materials initiated in 1988 at the Materials Research Society Spring Meeting. Symposia have been held annually (except 1989), with the American Ceramic Society and the Materials Research Society taking the lead role on alternating years. In addition to the many excellent technical papers that have been presented and published, these symposia have helped to identify and unify the researchers and equipment manufacturers and, to some extent, the users of the technology. A brief assessment of the last seven years of symposia experiences reveals that:

- Interest and activity in microwave processing are steadily increasing.
- Microwave energy is being used in industrial processing to provide unique benefits that cannot be accomplished using conventional techniques alone.
- Although microwave energy has been used in a large number of industrial processes, case studies of its success are rarely disclosed, especially at open meetings.
- There are many opportunities in both low-temperature (<350°C) and high-temperature processing using microwave energy. However, the larger database, fewer technical barriers, and available equipment will probably lead to more opportunities in low-temperature processing in the near term.
- There is a need for more communication and interaction between industrial users and the rest of the microwave processing community (*i.e.*, researchers equipment manufacturers and suppliers) to identify, develop, and implement applications in niche areas that will improve user productivity.
- There has been significant progress in modeling, understanding basic mechanisms, of microwave/material interactions and equipment design. Similar progress in system integration and controls is needed to foster more successful industrial processes.

It is clear that we need more participation of the users (and potential users) at these symposia. The researchers and equipment manufacturers have demonstrated their ability and willingness to address and solve technical issues that will allow implementation. What is needed is more input from the users who know their markets and can help identify processes where microwave energy can provide technological and/or economic benefits. Therefore, the present symposium was designed to encourage more industrial participation. Researchers, equipment manufacturers, experienced consultants, and users representing nine countries presented 83 papers and participated in an informative panel discussion and lively open forum. Sixty-six papers, as well as summaries of the panel discussion and open forum, are included in these proceedings. A few of the highlights of the meeting include:

- Keynote talks addressing industrial scale-up of microwave manufacturing components, and historical development of microwave energy use and its influence on current processing technologies.
- Fifteen invited papers discussing technologies ranging from mathematical modeling to applications for microwave processing.
- A panel discussion concerning existing microwave systems, the uses and advantages for each, as well as the potential for scaling up to meet the needs of manufacturing operations.
- The willingness of the attendees to engage in frank, open, and collegial discussions.

Although the meeting length was shortened to three days, more papers were accommodated and a strong program was presented. On Wednesday afternoon, 54 attendees remained to hear the last papers! This excellent attendance was due in part to the strong closing session in which industrial applications were presented.

The use of microwave energy in materials processing is still in its infancy. We expect these symposia to continue for many years as this field emerges and matures.

David E. Clark
Diane C. Folz
Steven J. Oda
Richard Silberglitt

Research sponsored by the U.S. Department of Energy, Assistant Secretary for Conservation and Renewable Energy, Office of Industrial Technologies, Advanced Industrial Materials (AIM) Program, under contract DE-AC05-84OR21400 with Martin Marietta Energy Systems, Inc.

Acknowledgments

The editors are grateful for the support provided by various professional societies, government agencies, and companies. We also thank the staff of the American Ceramic Society for their assistance in organizing the facilities for the symposium and expediting the publication of the proceedings.

Of course, the most important constituents of any symposium are the speakers, authors, and session chairs. Thanks to the creativity and dedication of these individuals, the symposium was high-quality and smoothly run, resulting in an important and timely publication.

We extend special thanks to Rebecca Schulz for assistance in organizing the symposium and for managing the collection and compilation of manuscripts.

Microwaves: Theory and Application in Materials Processing III

Sponsored by:

The National Institute of Ceramic Engineers

The Engineering Ceramics Division of
The American Ceramic Society

Financial support provided by:

U.S. Department of Energy—Advanced Industrial Materials Program
Los Alamos National Laboratory
Ontario Hydro
Cober Electronics
National Institute of Ceramic Engineers

Session Chairs:

David E. Clark
University of Florida

Steven J. Oda
Consultant

Ibrahim Balbaa
Ontario Hydro Technologies

John Gerling
AGL, Inc.

David Lewis
IBM, Thomas J. Watson Res. Ctr.

Willard Sutton
United Technologies Res. Ctr.

Monika Willert-Porada
University of Dortmund

Joel Katz
Los Alamos National Lab.

Richard Silberglitt
FM Technologies, Inc.

Wayne Tinga
University of Alberta

D. Lynn Johnson
Northwestern University

J.R. Thomas Jr.
Virginia Polytechnic Institute

Ron M. Hutcheon
AECL–Chalk River Lab.

Martin Barmatz
Jet Propulsion Lab.

Ruth Wroe
EA Technologies, Inc.

Ralph Bruce
U.S. Naval Academy

Jon Binner
University of Nottingham

Mark Janney
Oak Ridge National Lab.

John Booske
University of Wisconsin

T.-Y. Yiin
Jet Propulsion Lab.

Jes Asmussen
Michigan State University

Zak Fathi
Lamba Technologies, Inc.

Terry Tiegs
Oak Ridge National Lab.

Iftikhar Ahmad
FM Technologies, Inc.

Diane C. Folz
University of Florida

Microwave Processing: Steps to Successful Commercialization

MICROWAVE PROCESSING: STEPS TO SUCCESSFUL COMMERCIALIZATION - PANEL DISCUSSIONS

Willard H. Sutton
United Technologies Research Center
East Hartford, CT

INTRODUCTORY REMARKS

A panel of experts, each with many years of experience in the development, marketing and sales of microwave processing equipment, was invited to participate in this symposium. Their task was to discuss the factors, the issues, and the reasons behind the successes or failures during the introduction of commercial microwave processing applications. This topic is particularly timely, since in recent years, microwave processing research and development has been expanding into many new areas, such as ceramics, polymers, composites, and chemicals. The lessons learned from past commercialization experiences can provide important guidelines for future transitioning from new developments in the laboratory to full scale manufacturing processes.

Although the application of microwave energy in industrial and commercial processes has been in progress for over 40 years, the actual number of large scale successes has been very limited. While several reasons for this have been discussed in the past, one major limitation has been the overriding need for close communication between the science and technology, the supplier of microwave equipment, and the customer or user of the microwave energy [1]. This is illustrated in Figure 1, where in the past, gaps in the understanding and application of microwave processing technology needed to be bridged (arrows) by better communication and interaction between the three groups (circles). In this respect, the microwave equipment supplier plays a vital role in the communication link by helping to transition the available technology and understanding into an effective and profitable manufacturing process. This requires teaming and working closely with the customer.

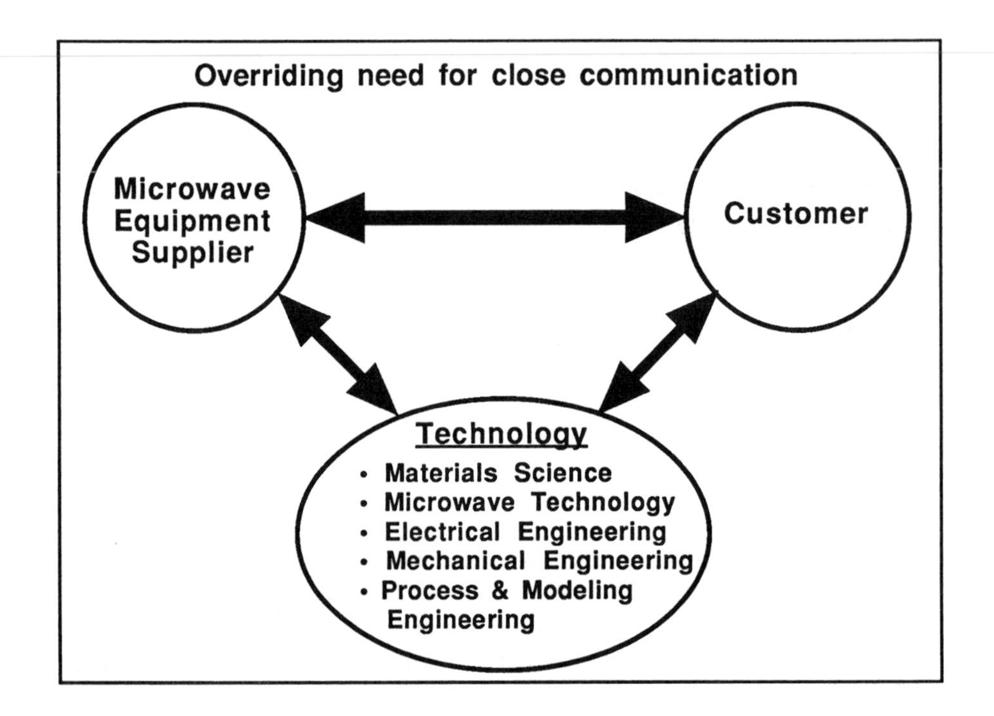

Figure 1. Microwave processing - steps to successful applications.

In the papers that follow, each of the panelists presents case histories and experiences that document the successes or failures during attempts to commercialize microwave processing in various industrial and commercial applications. The reasons for the successes or failures are diverse and varied, and depend on specific customer requirements, on changing economic and global environments, and on many other factors. Technical breakthroughs in microwave processing and economic benefits, such as rapid heating and processing, energy and cost savings, have not provided sufficient reason in themselves to lead directly to successful scaleup and profitable processes and products. The problems of introducing new manufacturing processing methods are described well in the Panel papers, which also led to stimulating discussions during the Panel Session. Perhaps one of the most valuable contributions by the Panel members was a better and more thorough appreciation of the critical problems that can be encountered during the transitioning of microwave processing technology from laboratory developments to the adoption and use by manufacturing processors. Also, key steps to successful applications were highlighted.

In recent years, there have been major improvements in microwave processing technology and understanding, in equipment and controls, in the R & D of new material developments, in pilot plant applications, and in an awareness for better defining new customer requirements. This is illustrated in Figure 2, where the communication gaps are closing (overlapping circles), where the needs and requirements of all the groups involved are better understood mutually, and where the future focus will be on developing a market "pull" for the technology and applications.

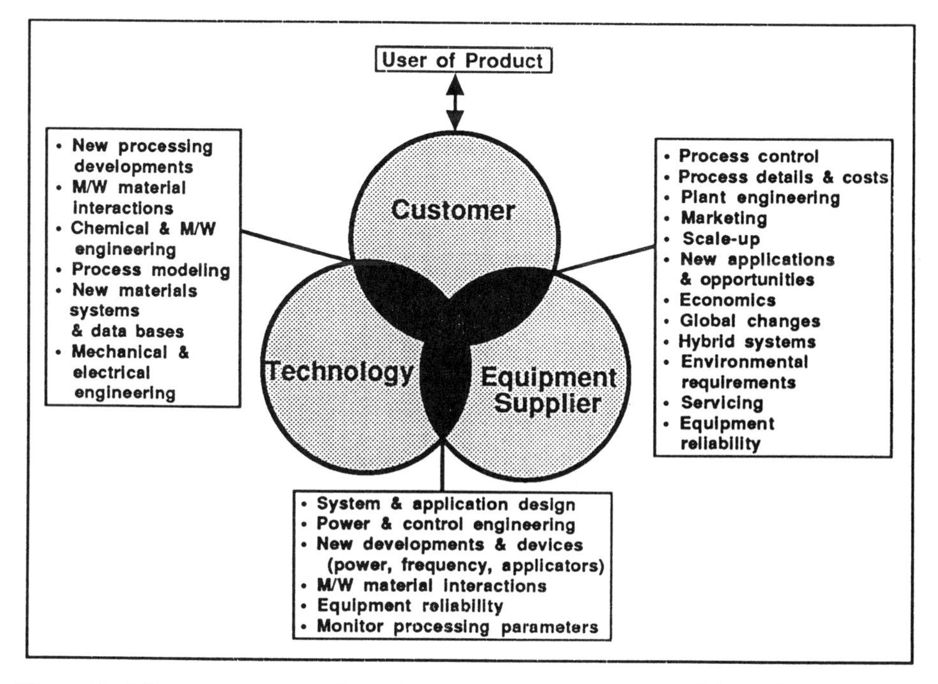

Figure 2. Microwave processing - interacting steps to successful applications.

REFERENCES

1. A.C. Metaxas and R.J. Meredith, **Industrial Microwave Heating**, Peter Peregrinus, Ltd., London, p. 297 (1983).

COMMERCIALIZING MICROWAVE SYSTEMS: PATHS TO SUCCESS OR FAILURE

Robert F. Schiffmann
RF Schiffmann Associates, Inc.
149 West 88 Street
New York, NY 10024

ABSTRACT

Although the application of microwave power to industrial processing began 40 years ago, the actual adoption by processors has been disappointingly small. Approximately 90% of all installed microwave processing systems in the US are devoted to three applications - meat tempering, bacon cooking and rubber vulcanization. Among the reasons are: unrealistic expectations for the microwave process; lack of resources and capital to develop commercial processes; high cost of microwave processing equipment; and competitive conventional processes. In order to improve the opportunities for success the following questions must be addressed: (1) what is the real problem; (2) what are the true economics of the situation; (3) is the effect of the microwave system truly unique; would direct or indirect labor charges be reduced; what is the best combination of the energy sources for the new application?

HISTORICAL PERSPECTIVE

It is now 50 years since Percy Spencer of Raytheon first conceived of a heating device which would use electromagnetic energy to cook foodstuffs (1), and shortly thereafter described the first microwavable food product - popcorn (2), and then a means for conveyorizing materials through a microwave oven (3). It awaited the development of higher power magnetrons and Cryodry's development in 1962 of practical choke systems for large tunnel openings for the use of microwave power for industrial heating to become useful. Predictions for the adoption of microwaves by industry were bullish while most technical people did not forecast a similar rosy future for domestic microwave ovens. After all, industrial heating processes could be revolutionized by this new technology whereas there would be little interest in a cooking appliance which would not brown, crisp or toast foods and would cost hundreds of dollars. How wrong we all were. Table I compares the domestic microwave ovens and industrial heating markets in the US in terms of numbers of units, their total selling price and the number of installed megawatts of output power.

Table I. A Comparison of Domestic Microwave Ovens and Industrial Microwave Heating Systems in the United States

	Domestic Ovens	Industrial Systems
Number of units installed	150 million	<1000
Total Sales Value	$22.5 Billion*	$500 Million **
Installed Megawatts	75,000***	100

(Note the following assumptions were made:)
*$150 per oven
**$5,000/kilowatt installed
***500 watts per oven

Obviously, these numbers are only estimates since most of the industrial data is proprietary; however, they are useful as a measure of success or lack thereof. They indicate that most of the experts were wrong. They never realized how important convenience is to a consumer especially when tied to the great social changes of working parents, and more diversity among families which lead to changes in people's eating habits. Even the food industry did not anticipate the impact of the microwave oven since it did not really begin providing microwavable foods until there was already a 40% saturation level of ovens in our homes.

As for industrial microwaves, the 1960's and 1970's saw a number of major companies join the bandwagon that they thought would take them to success: Litton Industries, Bechtel, Varian, Raytheon, Armour-Cryodry. All invested a great deal of money in personnel, laboratory and pilot plant facilities and conducted hundreds of tests for companies in many industries. Many of the tests were quite successful but few commercial systems resulted from them. Even those which showed great promise, such as potato chip drying, chicken precooking, plywood veneer redry and donut processing did not last (see Table II). That is not to say that they did not work, but for very different reasons each one disappeared to be replaced by other technologies. Some of the reasons are spelled out below.

Table II. Major Early Applications of Microwave Power in Industry

	# of installations
Potato Chip Drying	approx 100
Donut Proofing	24
Donut Frying	6
Chicken Precooking	3

By the 1980's, of the companies mentioned, only Raytheon remained in the business with a small but successful industrial application group. All the other players were small companies with annual sales well under $10 million, and that is the situation which prevails today.

Not only are the number of installations limited but the number of current successful applications is very small as seen in Table III.

Table III. Current Industrial Installations of Microwave Heating

Applications	# Installations	Average Power
Meat Tempering	400	75 kW
Bacon Cooking	25	200 kW
Rubber Vulcanization	<200 (US)	18 kW
	600 (Worldwide)	
Pasta Drying	20	75 kW
Pasteurization/Sterilization	>10 (Europe)	25 kW
Pharmaceutical Drying	>25 (Worldwide)	10 kW
Plasma Generation		3-4 MW total use
Miscellaneous including:		
Wood Curing		
Foundry Core Drying		
Potato Chip Cooking		

Note: Unless otherwise noted, the numbers are for the United States. All numbers are approximate since all data is considered proprietary. Plasma Generation refers to systems used for etching, ashing, and chemical vapor deposition for the semiconductor industry, as well as diamond deposition. While not heating applications, they represent a significant area for the use of microwave technology.

The list is by no means complete - many more single custom systems exist. However, what is apparent is that at least 90% of the numbers of installations and installed power is the result of three major applications: meat tempering, bacon cooking and rubber vulcanization. This is not to say that many other applications are not technical and commercial successes, rather, they have not been adopted to any large extent in their respective industries and many that were installed ended up being shut down, and never perpetuated themselves.

THE ELEMENTS OF SUCCESS

Several authors have outlined the criteria for successful adoption of a microwave processing system. One of the earliest and most quoted is Jeppson (4) and this is summarized in Table IV.

Table IV. Criteria for Successful Adoption of a New Microwave Application

- Product cannot be produced any other way
- Cost is reduced
- Quality is improved
- Yield is higher
- Raw material can be cheaper
- The resulting product is more convenient to the consumer
- Production throughput is increased *

(M.R. Jeppson, 1968)
*Added by this author

On the other hand, Krieger (5) has suggested a different set of criteria as seen in Table V.

Table V. Criteria for Successful Adoption of New Microwave Applications

1. Energy Transfer not heat transfer
    a. High speeds
    b. Short line lengths (less floor space)
    c. Less scrap
    d. Energy conservation because only the material is heated

2. Deeply penetrating heat from molecular agitation
    a. Better quality from more uniform heating
    b. Greater yields

3. Selective heating based on dielectric properties
    a. Unique solutions to problems
    b. Ability to heat within a package

4. An electronic mechanism
    a. Automation
    b. Conntrol and data logging
    c. Environmentally clean and quiet
    d. Quick set up and fast changeover to new products

(B. Krieger, 1993)

Whereas Jeppson's list can be applied to many different technologies, Krieger's is focused particularly upon the benefits provided by microwave heating. However, even if these criteria are met, there may still be other barriers to the adoption or continued success of specific microwave systems. These may become the overriding factors in some cases.

- Timeliness: In order to be adopted, the technology must come at the right time. Microwave donut processing offered major advantages to bakeries but its introduction in 1969 was so early that once other factors occurred it rapidly disappeared. The common appearance of microwave oven in homes has created a familiarity which can be beneficial in this regard, although it may also raise unrealistic expectations for the processor.
- Reluctance to be first: True innovation is frightening - it may lead to failure. Very few users are willing to be the first to adopt a new technology unless there is security in its success which can only result from a great deal of work, and it will still be a gamble. In 1972 Freedman (6) described a survey done for Raytheon relating to the acceptability of microwave heating equipment. The primary answer to the question of the conditions under which the company would purchase microwave equipment was: when the application is proven.
- Need for support technology which may not be available: The absence of ferrite circulators in early microwave systems permitted the destruction of numerous magnetrons and klystrons due to sudden mismatches during

processing. This was a major reason for the failure of chicken processing. Rapid in-package pasteurization of foods is rapid only in how rapidly the food is brought to the pasteurization temperature - the hold and cooling times are the same - perhaps someone can invent a much faster cooling procedure so the overall time is shorter and the process and results will be more attractive.

- Ability to provide a suitable microwavable product: Often, with processed products such as foods, it is necessary to modify the formulas in order to make them microwave compatible. As an example, the sharp decline in conventional dough proofing time of 45 minutes to only 4 minutes in the microwave proofer demanded the development of entirely new science of dough formulation, a task which took over one year. By comparison, the microwave hardware was simple and straight forward. Such research and development is expensive and time consuming and requires corporate dedication which must be founded on a solid cost/benefit analysis.

- Can another technology do the job as well as or better than microwaves? In the case of potato chip drying conventional frying systems were improved and were largely responsible for the disappearance of microwave systems. Sometimes innovation spurs further innovation in a competitive technology.

- Understanding the real benefits: Speed by itself is rarely the single reason to use microwaves. What is needed is a careful analysis of those things which will improve profits - increased throughput, reduction in labor costs, greater yield, etc. Even these must be studied in depth. For example, yield improvements may sound good but is it commercially useful? Improving the yield of chicken when it is sold by the piece may prove not to be beneficial.

- Comparison of the cost of the microwave system to the conventional system: In determining the former it is important to evaluate both capital and operating costs and these should be in terms useful to the customer - cost per hour or per unit of product, for example. The capital cost analysis should account for the time over which the system is to be amortized and the cost of money over that period. Operating cost analysis should include: utility, tube replacement, maintenance, cooling water and floor space plus any other unusual system demands. On the other hand, it is sometimes difficult to assess the cost of a presently operating system as a comparison.

- Effect of the microwave process on the rest of the production system: If a microwave process is capable of tripling the output of a product, for example, then all other parts of the process, up and down stream, will have to operate three times faster. This may lead to an unacceptable increase in capital expenditures.

- Psychological comfort of the operators: This is perhaps no longer as important a factor today as when domestic microwave ovens were rare. However, an operator who is afraid of microwaves can do harm to himself or herself or the system.

- Responsibility for developing the system: This is perhaps the single factor which has led to the adoption of certain microwave technology and not others. The manufacturers of microwave hardware are expert at that and generally not in the users technology. Users tend to know little about microwaves (although this is changing in some industries). So, who does the R&D which may take one or more years and cost a great deal? In the three most successful applications - meat tempering, bacon cooking and rubber vulcanization - microwave hardware manufacturers saw these as areas of potentially large sales and became expert in the technology and marketing of those applications, often relying on outside expertise. In donut processing, it was a bakery mix and machinery manufacturer with its own sales force who identified a significant corporate business opportunity and so developed the technology internally so that it could sell microwave proofers and patented donut mix.

  Conflicting corporate goals may inhibit this sort of effort. A microwave manufacturer developing a process applicable to ceramics wants to sell as many systems to as many ceramics manufacturers as possible. But limited resources may prevent it from doing so on his own. On the other hand, the ceramics manufacturer will generally only support the development of a proprietary technology which prevents the sale of microwave systems to its competitors. Perhaps a third party is required to spearhead the process development - a component manufacturer whose sales to the ceramics industry would be increased by the adoption of a new microwave process using that component. Sometimes these developments are best done with strategic alliances or assistance from the outside.

  Still another factor for the user is the lack of skilled microwave experts. This is a small field. The teaching of microwave power applications at the graduate and undergraduate levels is sparse. So, a company wishing to develop microwave technology may not have the internal resources and will have to go to the outside - to a consultant, to a microwave manufacturer or try to hire an expert - but does the project warrant this latter investment?

  The fundamental question is "What is the business benefit desired from the microwave system?" If it is to sell many microwave systems that would be the goal of the microwave equipment manufacturer and might be the case of a process equipment manufacturer or perhaps a utility or a processor with many plants and/or processing lines, then the application has an opportunity to lead to many installations. Otherwise, it is likely that the systems will all be custom built and will usually be hybrids of microwave-gas oven or microwave-high velocity hot air oven, etc.

- Presence of a "Champion" or "Hero": This is often an individual or small group in a company which forces the development and the adoption of an innovative technology against the desire to maintain the status quo. This can be one of the most important components of innovation.
- Presence of "Linker": This is usually an independent expert knowledgeable in both microwaves and user technologies who can bridge the technology gap between microwave hardware and application. The Linker can speak both technology languages and assists in technology transfer. The Linker may also serve an R&D function sometimes in the form of an independent laboratory.

IMPROVING THE LIKELIHOOD OF SUCCESS

In order to maximize the chances for success and avoid costly failures of microwave processing applications it is important that processors and equipment manufacturers realistically assess the potential applications. Of utmost importance are in-depth examinations with meaningful answers to the following questions: (7).

1. What is the real problem? The most basic of all questions and should be followed by examining why other solutions to the problem have been rejected. The approach should be "Why Microwaves?" rather than "Microwaves!" The cost of microwave equipment is usually high and the investment of time and manpower to develop and install a microwave system can be several times that so the problem must be clearly understood and approached.

2. What are the true economics of the situation? As noted earlier, this is sometimes hard to determine for the old system but can be easily calculated for the microwave equipment. As a rule of thumb, an industrial microwave system usually has an installed cost of $4,000 to $7,000 per kilowatt (which includes the generator, applicator, conveyor and control system). The required amount of microwave power for the process is easily estimated by standard engineering heat balance equations. To illustrate a quick and dirty case for drying: one kilowatt of microwave will dry 2 to 2 1/2 lbs. of water in an hour. So if a drying operation needs to remove 250 lbs. of water in an hour, it will require about 100 kilowatts at a capital cost of $500,000 or more. It is apparent that

capital cost is why this energy source is never used alone to dry huge quantities of product.

Another thing to keep in mind is that in order to be practical the intrinsic value of the material should be high preferably, at least, several dollars a pound and the higher the better. That is why microwave drying of pharmaceuticals, where the major ingredients may be valued at thousands of dollars, is far more practical than food drying.

3. Is the effect of the microwaves in the system truly unique? Microwave tempering and bacon cooking are successful because no other energy source can provide all the unique benefits that microwaves can. But, most often, the proposed systems could be done as well or nearly as well with another cheaper energy source.

4. Would direct and indirect labor changes on a given process be reduced by the addition of microwave power? With ever increasing labor costs and need for improved profits microwaves have a unique opportunity among other energy producing processes. They may make it possible to increase the throughput of a present process with little or no increase in floor space and a net decrease in labor on a pound of product produced basis. Increases of 50% to 100% above the present throughput are within reason. Sometimes there are other significant labor savings. A microwave sausage processing system installed over a decade ago saved 16 hours of cleanup labor daily because of better temperature control and ease of sanitation, plus eliminated the fires that occurred daily in the gas fired grill.

5. What is the best combination of energy sources for the new application? Rarely are microwaves used alone but rather in combination with hot air, steam, infrared or other heating means. This will usually reduce the cost of the system since these are usually far less expensive than microwave energy and the combination is often synergistic.

FINAL THOUGHTS

In a recent issue of Scientific American (8) Mark Weiser stated: "The most profound technologies are those that disappear. They weave themselves into the fabric of everyday life until they are indistinguishable from it." Hot air ovens are a good example - they are used daily in all industries for innumerable heating applications. We no longer say "Wow - a hot air oven!" And yet, that is still the attitude regarding microwave heating. It is still thought of mysterious and magical and, unfortunately, processors expect magical effects which are not likely to occur.

What is necessary for success is sober judgment, a thorough technical understanding and a great deal of laboratory and pilot plant work to determine not only that the process is satisfactory but that all process parameters have been examined. This should be followed by a thorough evaluation of the economic parameters to ensure that there is a positive effect upon the bottom line. Finally, everyone from the potential line operators to the vice presidents of sales and marketing should be interviewed to be sure that no hidden flaws will rear their ugly heads after major investments have been made. Even then, the likelihood of adoption of the microwave process may not be great but if adopted it is likely to be successful.

REFERENCES

1. P.L. Spencer, "Method of Treating Foodstuffs", US Patent 2,495,429 (1950).

2. P.L. Spencer, "Prepared Food Article and Method of Preparing", US Patent 2,480,679 (1949).

3. P.L. Spencer, "Means for Treating Foodstuffs" US Patent 2,605,383 (1952).

4. M.R. Jeppson, "The Evolution of Industrial Microwave Processing in the United States", J. Microwave Power Vol 3(1), pp 29-38(1968).

5. B. Krieger, "Why Use Microwave", Cober Food Processing Seminar (1993).

6. G. Freedman, "The Future of Microwave Power in Industrial Applications", J. Microwave Power, Vol 7(4), pp 353-365 (1972).

7. K. Bedrosian, "The Necessary Ingredients for Successful Microwave Applications in the Food Industry", J. Microwave Power, Vol 8(2), pp 173-178 (1973).

8. M. Weiser, "The Computer in the 21st Century", Scientific American, September, (1991).

COMMERCIALIZATION-STEPS TO SUCCESSFUL APPLICATIONS AND SCALEUP: May 1,1995

Bernard Krieger
Cober Electronics, Inc.
102 Hamilton Avenue
Stamford, CT 06902

ABSTRACT

Successful applications of Microwave Heating Technology result from much more than successful engineering, modeling and pilot demonstration. There must be beneficial fulfillment of a customer's need!

Understanding that need requires thorough knowledge of the industry being served and all of the process and conventional heating alternatives. We hamper ourselves with excellence in physics and chemistry and only superficial experience in ceramic casting and manufacture. The microwave supplier must change his focus if he is to realize commercialization.

INTRODUCTION

The keynote speaker, Robert F. Schiffmann, brought home the reality that the industrial microwave heating industry, which was initiated with such great promise more than thirty years ago, has in reality commercialized relatively few substitutive applications. Tempering of meats, high power production of precooked bacon and the vulcanization of rubber, are most prominent. It should be pointed out however, that the concept of microwave heating has been so intriguing over these years that at the same time it has stimulated many thousands of ideas, equal numbers of laboratory and pilot systems, and even more inventions.

Why have so many innovations resulted in so few commercialized successes? Will the application of microwaves to the processing of ceramics follow the same route? The fact that we are attending this session on steps to successful applications seems to indicate our own concerns.

DISCUSSION

The session chair asked me to address this issue based on my personal experience and from my own perspective. I must tell you frankly that I see the challenges not deriving from Maxwell, but rather from Peter Drucker. Peter Drucker is one of our nation's most prominent management consultants, and one of the creators of "The Marketing Concept." He chaired the Management Science Department at the Graduate School of Business at New York University years ago when I was a student. Drucker explains that the purpose of a business is "to create a customer." This means that commercialization is dependent upon our industry providing something for which a customer will exchange his purchasing power to acquire or not to acquire, considering all of the alternatives which are available to him. Marketing and the creation of customers is not merely selling. It is far deeper and more complicated. It requires very thorough knowledge of the industry to be served, their problems, processes, management culture, competitors and objectives. Commercialization results not nearly from solving technical challenges, but more importantly, from satisfying the wants of customers. In this room we have more than adequate physics, electronics, chemistry and engineering. What we lack is marketing science.

I find that what we are doing with microwave sciences is pushing technology, rather than responding to the pull of the needs of customers. The steps to successful scaleup are therefore largely marketing related.

We desperately need the "science" of creating customers. We continue to believe that if we can get something to work technically, customers will come to us to buy it and therefore, we will have commercialization. Consequently, we develop microwave technology in response to the challenge of heating ceramics, a material with dielectric properties that intrinsically defy heating with microwaves at room temperature. We, therefore, create pre-heaters, reflectors, develop microwave receptive additives, go to higher power levels, special resonant cavities, gyrotrons and even variable frequency high powered traveling wave tubes. Through these techniques we can get ceramics to heat very rapidly with microwaves. Is that enough to get customers to buy it? Obviously not!

Why isn't more effort put into understanding what would really satisfy our customers? If it is ceramics that is our objective, why aren't we in the ceramics plants to determine real needs? Why do we stuff our own staff with microwave scientists and consultants, when we should really be adding ceramics processing technologists?

Perhaps the problem is even larger than that. Maybe we do not even know who the customer is, and we have the wish that if we process faster and better, the customers will find us. We think that if EPRI or DOE, or similar funding agencies, provide money for research that they are the customer. Yet today we talk about commercialization. These agencies are not commercial!

If we continue pushing through technology rather than pulling through customers, I believe that there is little chance to commercialize. The answer is not in the microwave laboratory, it is someplace else; a place very remote from us.

Let us assume for this discussion that the manufacturer of ceramics materials, who is interested in making higher technology parts, is the real customer. (We may very well be wrong - it could be the aerospace manufacturer or the automotive plant who are looking for higher performance engines and components, or even the computer manufacturer who is really looking for smaller size, greater efficiencies and better data processing concepts). The identification of the real customer is a major part of the science of marketing. Let us stick for the moment with the model of the ceramics manufacturer and let us examine what we are offering him through the microwave sciences.

1. Speed: Microwave heating is energy transfer, not just heat transfer. We generate "cold" microwave energy with microwave generators and put them into a microwave oven, which we call an applicator and the ceramic material takes that energy and converts it into heat within itself. Microwave ovens, therefore, are rated in kilowatts and not in temperature setting. The microwave energy excites the material to be heated through the process of dipole rotation or ionic conduction. Since it is a process of energy conversion, rather than heat transfer through conduction or convection, very high speeds can be obtained by the application of incremental amounts of microwave power.

From the users point of view, however, high speed is usually equated with savings in floor space. If a machine can produce product faster, it can do it in shorter space if you look at it on an individual part basis. So, if we can produce sixty parts in an hour with microwaves, and the conventional process required an hour per part, we believe that we have made significant innovation. However, if the conventional production line is longer, or if the customer puts sixty parts in a batch oven for an hour, then production rates are really the same. The real trade off is in terms of real estate. If the customer's plant has absolutely no room to expand, the value of this microwave attribute is therefore high. Conversely, if there is plant space for multiple conventional processes, the value becomes much lower.

2. Penetration: We know that the principles of microwave heating are through molecular activity, therefore the microwave energy penetrates deeply into the material. The poorer the microwave receptivity, the deeper the penetration. This attribute of microwave heating is perhaps the most important feature, since it can result in higher quality products for the ceramic manufacturer. We speak of better grain structure and other properties. These attributes can be distinct advantages, but are they practical? Are they economical? Is this technology an investment that has the flexibility needed in a manufacturing plant, not only for the products being manufactured today, but also for those which will be manufactured tomorrow? Will the customer be willing to add microwave receptive materials to his product for the sake of his microwave process? I doubt that! Probably more than 90% of the manufacturer's cost relates to the cost of his materials. I suspect that less than 10% of the total cost is in processing, and, of course, less than that in heating. He will not add cost per pound, even though he may get higher efficiencies, unless there is very, very dramatic advantage. Can we quantify these advantages? Are they meaningful?

3. Electronic Control: Microwave heating, of course, is an electronic process. As such, it ties into automation and process control. Feedback systems, instrumentation and data logging are all means for achieving high quality repeatable precise production results. Additionally, this form of electronic heating is quiet, not heat intensive and extremely clean and efficient. Microwaves have more of these attributes than alternative means of processing. However, alternatives are by no means without the capability for advanced process control. Will the microwave advantages be worthwhile to the customer if he has to preheat by conventional means, or be involved in changing frequencies along with all of the attendant RFI implications? Additionally, is he willing to pay the high cost for microwave capital equipment and solve the perplexing problem of uniformity of heating, which results from differences in position, shape, materials, etc. within the microwave cavity?

Coming back to Drucker and the marketing concept, if we know that what a customer buys is satisfaction, we must realize that we are competing with all sorts of different alternatives. Let us look at some of the more easily defined competitive alternatives.

1. Other Microwave Companies. The least of our problems is competition from within our own industry. Our problem is not competing with each other; it is creating customers. If anything, that is a mutual objective, because the customers are not now using microwave technology. Therefore, the companies which you may normally think of as competitors, should really be allies at this stage of our marketing science. Indeed, I see that as we sit together in this room as a microwave industry at a microwave symposium.

2. Conventional Heating Alternatives. Consider the competition from direct fire gas heat, convection hot air, conduction, steam, infrared, UV, electron beam or even leaving something out in the sun to dry. Conventional technology, of course, is the enormous competitor. How much do we know about any of these alternatives? Have we considered where their attributes are best used, either alone or in conjunction with microwave? Are we merely trying to justify microwave, or are we trying to justify the wants of the ceramic manufacturer? Do we realize that only if we do the latter, can we commercialize?

3. Non-Heating Alternatives. If we know that what the customer buys is satisfaction, then we must realize that we must compete with are all sorts of different alternatives in the ceramics plant which may take a higher priority for funds in the customer's mind. This can include things which serve much different purposes, such as pressers, parking lots, office furniture, etc. We all know the disappointment that comes when we work hard to achieve a technical concept and find out that in the end there are no funds available from the customer. The funds were obviously spent by the customer on something else that gave him greater satisfaction. Just because we can do it with microwaves, does not mean that he will buy it. The dimension of this aspect of alternative purchase competition is enormous and should impress upon us the importance of being closely tuned in to customer's needs.

CONCLUSIONS

We are here at a microwave symposium today which we prepared for and anticipated for two years. Are we sitting with our customers? I do not think so! We are, however, at a meeting of the American Ceramic Society. Many thousands of people from the ceramics industry who may be the customers that we are looking for, are at this event and are, perhaps, next door, but I doubt that they are in this room. This room is focused on our technology and not on our customers, and yet our customers are the only people that can make commercialization possible. They are next door - do we understand them? Are we talking to them and do we realize that satisfying them is the purpose and essence of our science?

PROCESS COMMERCIALIZATION: DON'T RISK LOSING THE BENEFITS BY IGNORING THE HAZARDS

John F. Gerling
AGL, Inc.
1132 Doker Drive
Modesto, CA 95351

INTRODUCTION

"What do you do about metal TV dinner trays?" Referring to microwave ovens, this question was once asked of appliance manufacturers. They will soon be asked "What do you do about metal zippers and buttons?" These questions may not be directly related to industrial microwave processes, but they are indicative of a major hurdle often faced by process developers. Benefits of new technologies are often promoted without addressing the associated hazards. If suddenly exposed during the first attempts at commercialization, these hazards can quench any hope of commercial success. Hazards are typically addressed during the research phase of development, but their existence must not be forgotten when the technology leaves the laboratory.

HAZARDS TO MAN AND MACHINE

When discussing hazards associated with new technologies, examples which come to mind might be radiation leaks from nuclear reactors or breached o-rings on solid rocket boosters. While these might represent rather extreme cases, the hazards which can cripple a new technology don't necessarily have to be life-threatening. Consider the following examples.

Consumer Products

Although the microwave oven is well accepted by consumers as a major appliance, some of its associated hazards still provide considerable controversy. Both food and non-food products can be subject to uneven or runaway heating, leading to problems ranging from a spoiled dinner to eye injury [1]. Other reported problems are specifically related to the oven itself, such as self-starting and failure to shut off, rather than with the products with which they are used. While the microwave oven itself may not suffer from a poorly conceived product, the product itself may never gain acceptance if similar problems are not addressed and resolved before introduction to the market.

Industrial Processes

The Occupational Safety and Health Administration (OSHA) was created to address the concerns of equipment operators over hazards to which they are exposed. To ensure worker's safety, OSHA has establish stringent guidelines for design and operation of commercial and industrial equipment. However, it is possible for new technologies to introduce new hazards to worker safety which are not covered under the guidelines of agencies such as OSHA, Underwriters Laboratory (UL), Center for Devices and Radiological Health (CDRH) or other regulatory bodies. In such cases, these agencies may have the power to block the introduction of a technology until the hazards are mitigated.

The concern over environmental effects of new processes will certainly have an impact on their commercial potential. In fact, many new processes are finding widespread acceptance *because* of their beneficial effects on the environment. But given the social concerns for protection of the environment, any new process must have at least a neutral, preferably beneficial, environmental impact in order to survive commercially.

Hazards to Equipment and Products

Few industries are more sensitive to the performance of their equipment as the semiconductor manufacturing industry. When the manufacturing of a single part involves hundreds of individual steps and costs hundreds of thousands of dollars, the technology developers and equipment designers go to great lengths to ensure the most reliable process possible. A failure as simple as a water leak can have devastating consequences to both the equipment and the product at a cost many times that of the failed part.

Intangible Hazards

The well-publicized flaw in Intel's Pentium microprocessor is a good example of how *information* can be placed at risk by a piece of equipment. A similarly intangible hazard is to electronic data which can be accessed as the result of flaws in network security technology. While it may not be fair to say that new technology in microprocessors was to blame for the Pentium's flaw, the events relating to the handling of this hazard are relevant to this discussion.

Perceived Hazards

The term "electrophobia" is popularly used to describe a phenomenon of human nature resulting from a lack of technical understanding of electromagnetic energy and how it behaves. Any new technology, when not well understood by its users, can be subject to "false accusations" about effects on humans, the environment, food, or even the process itself.

IDENTIFYING THE HAZARDS

The ultimate responsibility for identifying process hazards and finding solutions for them rests with the scientists and engineers. While these technical professionals typically have an extensive background and understanding of many hazards likely to be encountered, they often times must rely on the input and advise from others in the industry. During the early stages of development, it is extremely important to interface closely with the target market to gain a better understanding of the problems.

The development of a microwave clothes dryer is an excellent example of how close ties with the market, in this case the appliance manufacturers, may have been the key to success. The initial design criteria were partially based on feedback from consumer focus groups who expressed interest in faster drying with less damage to clothing than with a conventional dryer. Also out of these discussions came the need to address certain obvious potential hazards such as metal buttons and zippers, the solutions to which were deceptively simple. However, extensive discussions with appliance manufacturers lead to investigations into hazards such as pencils, bobby pins, and butane cigarette lighters. The solutions to these so called "tramp materials" was not at all straight forward, but had they not been investigated, the failure of the microwave clothes dryer as a consumer product would almost certainly have been assured.

MANAGING THE HAZARDS

While the research and engineering staff are chiefly responsible for identifying and mitigating the hazards associated with the technology under development, the management staff plays a key roll in "managing the expectations" of the market in which the technology will be commercialized. The market will expect perfection unless told otherwise, and is typically unforgiving when its expectations are not met.

Risk/Reward Ratio

Many companies tend to take an extremely conservative approach towards commercialization of a new technology which, in most cases, is driven by the need to maximize the return on their investment. The resulting fear of market rejection leads to efforts to suppress any information about the technology which may appear negative or counterproductive. While this strategy may be successful in getting the process to market initially, it is inevitable that the negative aspects of the technology will be discovered and questions such as "Why were we not told?" will be asked. Depending on the severity of the problem, the reaction could lead to complete rejection of the technology as well as its developer.

It is recognized that some problems or hazards legitimately go unforeseen or undetected before they reach commercialization. In such cases where problems surface after market introduction, the response by the developers is also very critical in deciding their fate.

Plan For The Unexpected

The importance of a market study can never be overestimated. From this important step in the product's development cycle, one can learn what problems may be encountered as well as the level of acceptance in the market. While the concept of a market study traditionally applies to consumer products, it can take almost any form as appropriate for the technology at hand.

The mistake often made is that the development is considered to be complete when the marketing effort begins. In such cases, there is no provision made for returning to the drawing board to make whatever changes are shown to be necessary, and the inertia to change continues to grow.

Conquering The Skeptics

As with almost any novel technology, there are plenty of skeptics whose sole purpose in life may seem to be to thwart the efforts of the pioneering scientist.

Many seem to make their living by exploiting the problems and hazards associated with consumer products, and are particularly successful when allowed to take an offensive position [2]. But whether the users of the new technology are consumers or factory operators, the skeptics will find it much more difficult to generate support for their position if the promoters make a proactive effort to address the relevant issues and provide solutions which are well beyond the users' expectations.

SUMMARY

There are three main points to be considered when developing a new technology. First, the hazards to be addressed should not be limited to those which threaten human health and safety, but should include any form of hazard. Second, the process developer should draw from the knowledge and wisdom of the target market as well as his/her own technical expertise when identifying and mitigating hazards. Finally, attempts to suppress or hide the existence of hazards, even though they may have been properly addressed, will ultimately fail and result in a loss of credibility and, more importantly, commercial success.

REFERENCES

1. Routhier, P., et. Al., "Eye Injury from Microwave Popcorn", New England Journal of Medicine, Nov. 20, 1986.
2. Risman, P.O., "Microwave Food Hazards", Microwave World, Vol. 14, No. 2, Winter 1993, page 21-23.

CONTROLLING & UNDERSTANDING THE VARIABLES: KEY TO COMMERCIALIZING MICROWAVE PROCESSING OF ADVANCED MATERIALS

Richard S. Garard
Lambda Technologies, Inc., 8600 Jersey Court, Suite C, Raleigh, NC 27612

ABSTRACT:

Commercial use of microwave energy for processing advanced materials has been a "promising new development" for over a decade. However, the realization of actual commercial use in most advanced material cases has not yet been achieved. As with any new processing technique, the control and application of process conditions must be reliable, repeatable, and thoroughly understood. This paper will discuss the variables associated with both economic analysis and material properties when determining the potential of microwave processing for a given application. The importance of having a microwave system capable of controlling those variables and distributing the microwave energy uniformly over large volumes within a microwave oven is reviewed. The need for a production equipment supplier to combine materials science expertise with strong microwave engineering background is also discussed with emphasis on ensuring that a good understanding of the material/microwave interaction exists for each specific application.

COMMENTS FOR PANEL DISCUSSION ON COMMERCIALIZATION STATUS:

Advanced materials is an industry that is forecast to realize explosive growth over the next decade. It is also an industry that was forecast to achieve explosive growth over the last decade. However, critical factors assumed in those forecasts have been the ability to significantly reduce the cost and to improve the reliability of manufacturing those advanced materials. By reducing cost and improving process reliability of advanced ceramics, composite materials, etc., extended applications and the ability to reach a broader market base would be viable. While progress is being made in this direction, the forecast potential is still in the future.

Microwave energy is one tool that has been investigated to provide enhanced processing capability. Results to date have been promising on a laboratory scale. However, interest in microwave processing of advanced materials has fluctuated over the past 20 to 30 years since first showing promise. In the early days of applying microwave ovens to food processing, interest was driven by the technology potential for interaction with advanced materials to strictly reduce time and, therefore, cost saving opportunity. Researchers found that coupling microwave energy into most advanced materials was difficult and almost always not repeatable and non uniform. On the economic front, emphasis was placed on energy savings and analysis of this factor did not offer a large enough incentive to continue the investigation when control and reliability problems were persistent. Hence, the early "technology driven" time of the cycle never moved away from fundamental research and these market interest (and funding) declined.

During the mid to late 1980's, the materials industry began to pull in a "market driven" period where very encouraging results on a laboratory scale were being reported and some success of commercial processing was achieved.(eg: rubber industry). Once again, however, momentum turned as the task of converting laboratory results to commercial scale applications for advanced materials proved to have too many hurdles. These hurdles consisted primarily of controllability and uniformity when scaling up. There now appears to be a period where the ball is back in the technology court, as renewed proof of feasibility is again necessary to generate the next level of interest in commercial opportunities. For this period, it will be more important than ever that microwave equipment suppliers and material scientists/manufacturers work closely together in defining the potential applications and then conducting successful process development.

The key to the present market and technology hurdles that stand in the way of ultimate commercialization is understanding and then controlling the variables of microwave processing. Those hurdles and their respective variables can be separated into two categories; economic and technical. The economic variables, while obviously needing to be addressed, are very much application dependent. Factors such as cycle time, energy costs, production throughput, equipment utilization, personnel utilization, space utilization, quality control, reliability and finally, equipment cost, all play a role in the economic assessment of microwave processing. Each of these factors provide a positive return in some fashion when using microwave energy but are based on the specific process and/or application. Therefore, before a good technical research program is launched, it is critical that the microwave system developer and the materials/product manufacturer (working together as a team) define and acknowledge the economic factors and their potential impact prior to selecting a target process application and establishing the goals for successful commercialization.

The technical variables associated with developing the application of microwave energy for advanced materials must also be evaluated prior to selecting a target process. These variables can be defined in two separate, but related areas; 1) material/product properties and 2) microwave system design parameters. The material and product property considerations include the following factors:

- -chemical composition
- -existing process schedule
- -quality & reliability issues
- -mass, size & shape
- -dielectric properties
- -specific heat
- -structural changes during process
- -alternate compositions

Knowledge of the above with regard to their interdependence on one another and with respect to microwave interaction can be evaluated by empirical trials and via modeling techniques prior to determining its potential for further development. With the above input, variables associated with microwave interaction and control of energy within the microwave oven can be addressed.

The factors requiring design and control in this critical area include the following:

-cavity design; for optimum electromagnetic field distribution and interface with existing process conditions & equipment
-coupling efficiency; control of coupling energy via field control and/or selecting the optimum incident frequency for the material or composition.
-process cycle; microwave power requirements & frequency selection

-uniformity; controlling the energy distribution & "hot spots" via material placement, susceptors, and/or controlled frequency sweeping.
-reliability; automated control of reproducible microwave cycles

Lambda Technologies is a relatively new company completely dedicated to commercialize microwave processing for advanced materials and other high-tech process applications. Our approach is to work closely with materials suppliers to understand and then control their process parameters. We have established the ability to analyze a material/product for microwave interaction via determining dielectric properties across a selected frequency range, establishing baseline process conditions via analytical & numerical modeling techniques, and then conduct feasibility trials with broadband microwave frequencies. We attempt to control the process variables by controlling those variables with a technology called variable frequency microwave energy. Once the economic and material property factors are established, our process optimization includes the following considerations, the first two of which(*) are common with conventional microwave technology:

-Microwave cavity design*
-Incident microwave power*
-Incident frequency
-Selective control of incident frequency
-Frequency sweep/ time average uniformity
-Range of sweep (bandwidth control)
-Rate of bandwidth sweeping

While our product development is in the early stage and our applications development limited to a few markets to date, we have demonstrated that by understanding the material related variables and control of the above microwave energy variables-- the major hurdles of commercializing microwave use for advanced materials, such as scale-up, uniformity, repeatability, and dynamic control during processing, can be overcome.

EQUIPMENT SUPPLIER'S ROLE IN PROCESS DEVELOPMENT

Eugene Eves
Ferrite Components, Inc.
24 Flagstone Drive, Hudson, NH 03051

ABSTRACT

The equipment supplier must drive new development with his own time and his own funding. A substantial new facility for high power materials processing and the process of evaluating and selecting candidate process development opportunities are presented.

INTRODUCTION

The industrial equipment supplier's role in new process development extends beyond the need to design, build, and support his individual product. The integrated nature of industrial process development requires a technical program manager to work closely with a marketeer. The process equipment supplier must provide both disciplines from within his company and expect his customer to do the same. Both must constantly assess the effects of inevitable evolution on the cost and viability of the process. Does the process output, the product, continue to meet the cost and quality demands of the customer?

In a process which is microwave intensive, such as cooking, the microwave equipment supplier may take the lead role to integrate front end and down stream processes because his equipment dominates. In a process where microwave energy is used in a novel manner, the microwave supplier will assume the lead position because only he has the knowledge needed to resolve technical trade-offs. In other cases, he may be a vital team player, but not the leader. In any case, he must participate from the conceptual stage through commercialization to make the process successful.

The balance of this discussion will cover the steps and involvement of the equipment supplier. To keep things simple, I will assume the equipment producer builds the tools and not the end product; that he makes cooking ovens, plasma reactors, or furnaces, and not bacon, diamonds, or turbine blades. Some businesses do both very successfully, but their equipment tends to be inaccessible to other users who would become competitors. I will further assume a "partnership" between one supplier and user.

THE PROCESS OF EQUIPMENT DEVELOPMENT

Identification of a real customer need (or wish) is always first. This vital information usually comes from the end product producer (the user), not the process equipment manufacturer (the supplier).

SUBSTANTIATION

Survey & evaluation of existing processes is second. This is the point where the user and the supplier usually meet. The user is unhappy with his current process, or has no process for his new product, or he is just shopping. The supplier must determine at this point:

a) that there is a good microwave process from the technical viewpoint;

b) that the process development will meet the time constraints of the user;

c) that the process will yield a product that the user can sell at a profit;

d) that there is sufficient investment capital to carry the process development;

e) that, for a new product, no other process technology is better;

f) that, for an existing product, a new microwave process is at least twice as good as the existing process.

The market emphasis focuses on the end product while the technical study focuses on the process. The supplier must participate in both. He may require the support of an expert consultant to evaluate the market and competing technologies.

PROTOTYPES

Prototype development proceeds after the survey and evaluation phase. The user and supplier learn about the machinery, test procedures, and scaling tools in their industries. Lab work is done and baseline assumptions are proven or disproven. This stage requires close physical cooperation and, usually, written agreements regarding information disclosure and commercial development. Emphasis must always be placed on scale-up potential. The prototype process will ideally run as it would in the factory with raw material and finished goods being handled in a natural manner and not "frozen" in mid process and shipped from the "real" factory to the prototype line.

Prototype process development requires space, equipment, people, and <u>focus</u> to draw together a robust simulation of the actual process. In addition to strengths in materials selection, mechanical design, and state-of-the-art controls, the microwave supplier must bring special tools to this process.

a) Microwave power generation equipment at the right frequency to the production process.

b) Microwave test and measurement systems.

c) Design capability for special microwave structures and devices.

d) Thermal measurement and imaging equipment peculiar to the microwave environment.

e) Mathematical tools for modelling heating pattern and rates heating in microwave fields.

f) Knowledge of special safety and environment regulation.

g) Access to the other knowledge resources of this field.

Money spent by both parties is best allocated by their natural division. The supplier pays his design and manufacturing costs while the user supplies auxiliary equipment and materials for processing.

PRE-PRODUCTION

Pilot production operation follows prototype runs. A pilot line should run at a minimum of 10% of full production. At this stage, all the scaling and simulation of the prototype is removed. Raw materials are converted to finished, packaged goods. Scrap, yield, and loss are measured as real numbers. Ideally, the final product is placed in the market in the ultimate consumer test. The pilot line is run in close collaboration between the supplier and the user. The location may be the suppliers plant, the users plant or an independent location.

PRODUCTION

In the full production phase, supplier and user roles will usually revert to the standard relationships of vendor and customer in equipment purchase. The partnership remains, however. This supplier and user remain in competition in the end market with other supplier-user partnerships. The supplier must maintain his technical currency and continue development so that his customer using his equipment is the market leader.

SUMMARY

Significant knowledge growth in areas as diverse as material science, electronic computer control systems design, mechanical design, environmental regulation, and many others must be integrated with microwave equipment design. Only in this way will new process potentials be realized as new process systems. Close co-operation between several development partners is needed to bring sufficient technology to bear in the short time required to hit market windows.

It is the equipment suppliers role to provide leadership in development with his time, facilities, and materials.

The following pages illustrate the scope of FCI's resource commitment in its Process Development Center. The figure illustrates the area while the facing table provides a key to the elements of the facility.

One or two major developments are supported at one time. Processes are carried through pre-production in the Development Center. Full production is run in the end user's factory.

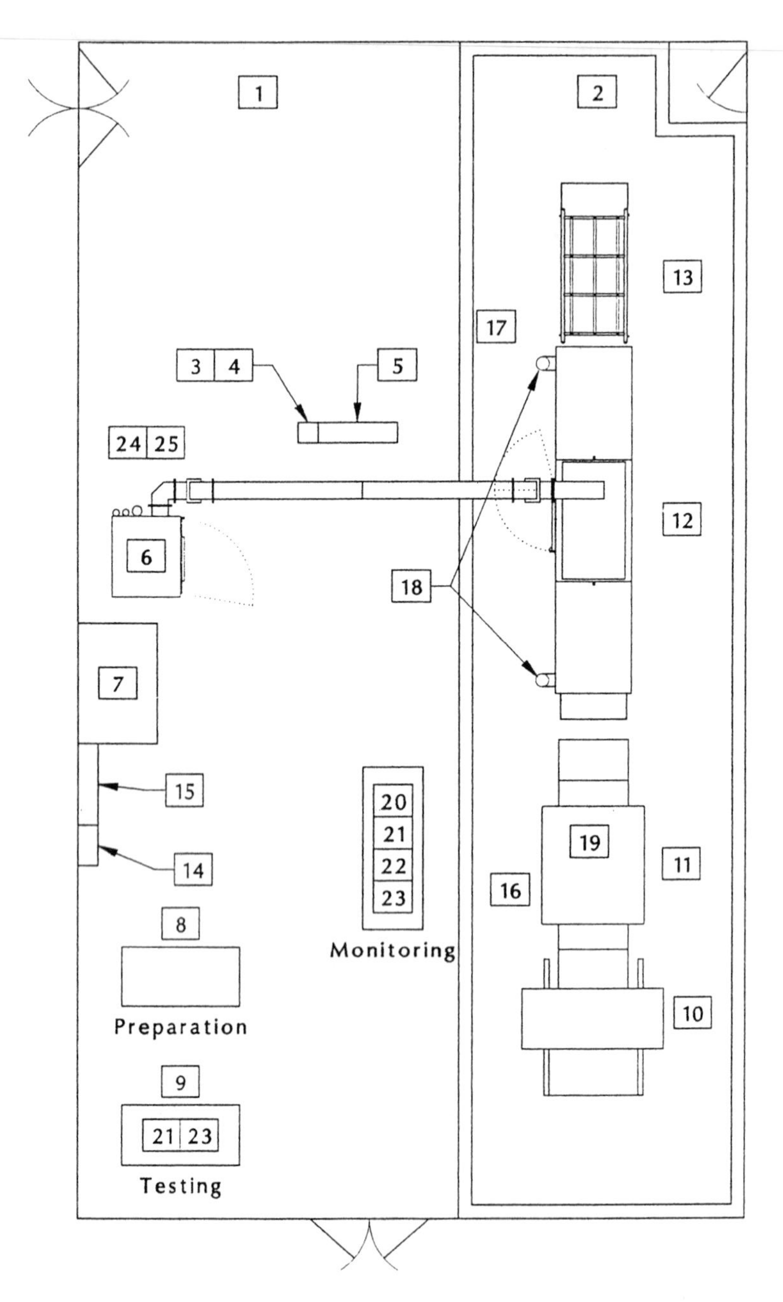

FCI Process Development Center (58' x 34')

FCI PROCESS DEVELOPMENT CENTER

MAJOR FIXED FACILITY

1. Laboratory and test area (20 x 58).
2. Full washdown process area (14 x 58).
3. Communication and data lines.
4. Climate and atmosphere control.
5. Computerized process control center.
6. Microwave generator, 915mhz, 0-75kw.

PROCESS SIMULATION

7. Process material storage, temperature controlled.
8. Process material preparation area.
9. Q.C. and pretest area.
10. Automated material feed.
11. Conveyorized pre-processing.
12. Conveyorized microwave processor.
13. Conveyorized post-processing.

PROCESS TEST FACILITIES

14. Power panel (480V, 208V, 115V).
15. Cooling and process water lines.
16. Natural gas lines.
17. Compressed air lines.
18. Powered exhaust for effluent removal.
19. High temperature stack.

INSTRUMENTATION

20. Data collection computer.
21. Digital thermometers.
22. IR Imaging System.
23. Digital weigh scales.
24. Microwave power safety monitor.
25. Spectrum analyzer and network analyzer.

Manufacturing with Microwaves

COMMERCIALIZATION OF MICROWAVE PROCESSES

Murray J. Kennedy
Richardson Electronics, Ltd..
40W267 Keslinger Road
LaFox, IL 60147

ABSTRACT

Effective commercialization of microwave processing is dependent on the economics of the equipment versus the value of the end product. The current trend by component suppliers is to simplify microwave system design and implementation. This paper outlines three system designs and presents the differences based on application requirements. Case studies will be presented on the curing of resins and epoxies, drying and joining of ceramics, and microwave plasma generation using ceramic based applicators. Optimization of the process by effective equipment design will be stressed. Design parameters including frequency, power supply, output power, and monitoring and feedback circuits will be discussed in relation to system performance and cost. Recent efforts to support the microwave equipment user will be evaluated from a systems design standpoint.

INTRODUCTION

Industry has been successful in identifying specific applications that exploit the advantages of microwave heating (1). In addition to the obvious benefit of time savings, industrial users have discovered applications that improve product quality and yields with the added benefit of small equipment size and lower product cost. Innovative process techniques have led to the use of microwaves for applications that are not possible by conventional industrial methods. In order to move from conception to commercialization, scientist and engineers have sought the assistance of the microwave component and equipment manufacturers. Users requests have gone unanswered until recently. In order to spur growth, the status-quo road blocks encountered in commercialization of microwave processes must be eliminated.

COMMERCIALIZATION TRENDS

There are four steps in any commercialization process. The first step is that of conception. As stated above, industry has been successful in identifying processes that could benefit from the use of microwaves. In the other three commercialization stages requiring feasibility studies (prototyping, and production equipment construction and installation), users are seeking support from the component and equipment manufacturers. This paper will discuss the trends in commercialization of microwave processes. Developments in material testing, component and system development, and computer aided design will be presented. In addition to the above; three innovative system designs will be outlined with emphasis on component integration and use.

FEASIBILITY STUDY

In the first stage of commercialization, an idea is developed that seeks to gain some advantage through the use of microwave processing. After conception, a feasibility study is undertaken. Feasibility studies attempt to answer questions relating to microwavability of the product, potential cost savings, and quality improvements. Researchers also use feasibility studies to gather data for use in the prototyping stage of commercialization. Companies are typically not willing to commit large budgets for feasibility programs where the goal is replacement of an existing technology. This presents the scientist or engineer with the task of proofing the process at a low cost. As a result, initial experiments are often carried out in a conventional home microwave oven. Although this type of testing can be effective, it can often result in abandonment of a program that could otherwise have been successful. Accurate temperature and material property measurements are difficult or impossible when using a conventional home oven. Testing is restricted to multi-mode cavities where power control, applicator design, and operational frequencies essential to optimization of the process are ignored. Cost over-runs can result as multiple experiments are required due to lack of adequate material property information.

To combat these shortcomings, microwave equipment and component manufacturers are reacting to customers needs in several ways. Equipment has been set aside at many facilities for customers use. This allows for industrial equipment testing that replicates final operational conditions. Process specialists work in conjunction with microwave equipment designers to gather data. Through the use of confidentiality agreements, researchers can test a process in industrial equipment without jeopardizing proprietary processes.

The first step in a successful commercialization program utilizes a technique known as cold testing. Cold testing can be used to gather important information on many product properties. Use of a state-of-the-art network analyzer allows for measurement of reflection coefficients and loss factors at various temperatures. By subjecting the sample to a low level microwave signal, accurate material property measurements can be made quickly and at no risk of damage to product or equipment. Single mode cavity measurements can be taken that are essential to effective equipment design.

Results of cold tests can then be used in computer aided design programs to help reduce the number of iterations required in hot testing (full power) samples. The computer programs developed utilize basic electrical and thermodynamic equations to model the process. Tedious and time consuming 3-D programs have been replaced with a simplified program that provides useful system power and product temperature information. The program allows for early identification of process problem areas. Complex processes can be subdivided for ease of analysis by facilitating a composite materials approach that treats preheat, vaporization and curing or drying as separate problems. Although the program does not include the detail of a true 3-D program, it results in a reduction in experimental costs and a substantial reduction in programming time. One microwave equipment manufacturer has developed a computer based system that stores the results of cold testing and uses actual measurements to optimize the process.

PROTOTYPE STAGE

Armed with the information gathered from the feasibility study, the researcher can provide upper level management with the information necessary to proceed to the prototype stage. Prototyping forced those well versed in material science to become microwave equipment experts in the past. Many researchers attempted to build their own equipment in order to reduce cost. In this scenario, schedules were rarely met. Development times were extended as researchers coordinated efforts between component suppliers, tube manufacturers, and waveguide and machine shop houses. Recently, manufacturers of microwave equipment have assisted the researcher in equipment design.

The first step for effective prototyping is equipment specification and layout. Richardson Electronics, Ltd. has developed a microwave CAD program that is user friendly. Users are led through a branching program by answering several questions based on information gathered during the feasibility study. The program offers the user a list of components that will meet their needs. After choosing from the list, the user can construct a system schematic through a click and drag routine. The end result is a personalized design that employs commonality of components while providing system parameters, part and component listings, and prices. In addition, a representational system diagram is provided. A sample output of the Microwave CAD program is presented in Figures 1, 4, and 5. The designer can use the output of this system in the construction stage of prototyping.

In order to further eliminate problems in prototype construction, the microwave industry has developed matched components and microwave power packs. The matched component concept utilizes a "Heath Kit" approach to microwaves. This method was developed with the low power, or modularized system users in mind. For those utilizing tubes with output power levels of 3 kW and below, the matched component concept is ideal. The user can construct a microwave system using components supplied from a single source. The components are developed to work in unison. System diagrams exist that facilitate construction, and the user can build a system without becoming a microwave expert. The matched component concept has further advantages in development time and cost. Commonality of parts supplied by the manufacturer reduces cost. Development time is limited to the applicator which in most cases has been conceptualized during initial feasibility work. Packages are carried in stock and complete systems can be constructed in hours.

For users weary of building their own system, OEM power packs are available. These systems offer the user a complete turnkey microwave system. Customer defined applicators can easily be attached to the power pack. Delivery time is typically short for standard systems. Customizing can be added by specifying circulators, tuners, and directional couplers, or other waveguide components as needs are identified.

Once prototyping is complete, the researchers job is typically over. The design is transferred to production for commercialization and the start of manufacturing. Through the recent developments by component and equipment manufacturers, the journey from conception to commercialization has been shortened.

SYSTEM DESIGNS

Presentation of the following three system designs will help to clarify and demonstrate the design concepts discussed above. Concentration will be on equipment and microwaves as they relate to general processing techniques.

18 kW Epoxy Resin Curing System Operation

The first system is pictured in Figure 1. This system was designed for curing of epoxy resins. The system operates by applying 18kW of microwave power to a 4 X 8 foot mold that is divided into three equal sections. At the beginning of the process, microwave power is applied to the first section of the system using three full-wave doubler power supplies to energize the first three 2kW magnetrons. The microwave properties of the product are altered as curing occurs. Through feasibility testing, it was found that a material property known as the reflection coefficient decreases as the product begins to harden. The reflection coefficient is a ratio of the power reflected to the power absorbed of an incident wave as it travels from one medium to another. In this case, the microwave is traveling from air to the product. The system uses a directional coupler to monitor reflected power. The signal from the directional coupler is fed into the power control circuit. The decrease in the reflection coefficient results in a decrease in the reflected power and a proportional reduction in the signal from the directional coupler. The signal from the directional coupler is fed to a power controller. A prespecified threshold level is used to signal the end of the first stage of the curing process.

At this point, the first three magnetrons are switched off and the second set of magnetrons are energized. As in the first stage, the change in the reflection coefficient is used to trigger the end of the second process step. A similar scheme is then used to activate the third set of magnetrons. Processing is completed as the signal from the third directional coupler falls below the specified threshold level and all tubes are disabled.

Conventional methods of curing epoxy resins result in significant scrap due to product shrinkage. The material is typically cast oversize and trimmed to final dimensions. The microwave process described above facilitates casting to final size. The preferential heating method (one section at a time) reduces shrinkage as material from untreated areas of the mold can flow and fill in areas where shrinkage has occurred. This results in less scrap and additional cost savings as trimming is limited to only one side of the cast part.

By utilizing information relating to the change in the reflection coefficient, an additional product benefit is realized. Normal variance in production batches can be overcome through the use of automated process timing. Timing circuits are replaced with threshold controls that are batch sensitive. This results in a more consistent product as curing times are allowed to vary in order to stabilize hardness levels.

Tests of microwave cured epoxy resins have shown that the material strength of the end product is improved. This has been attributed to the uniformity and speed of the processing. Conventional

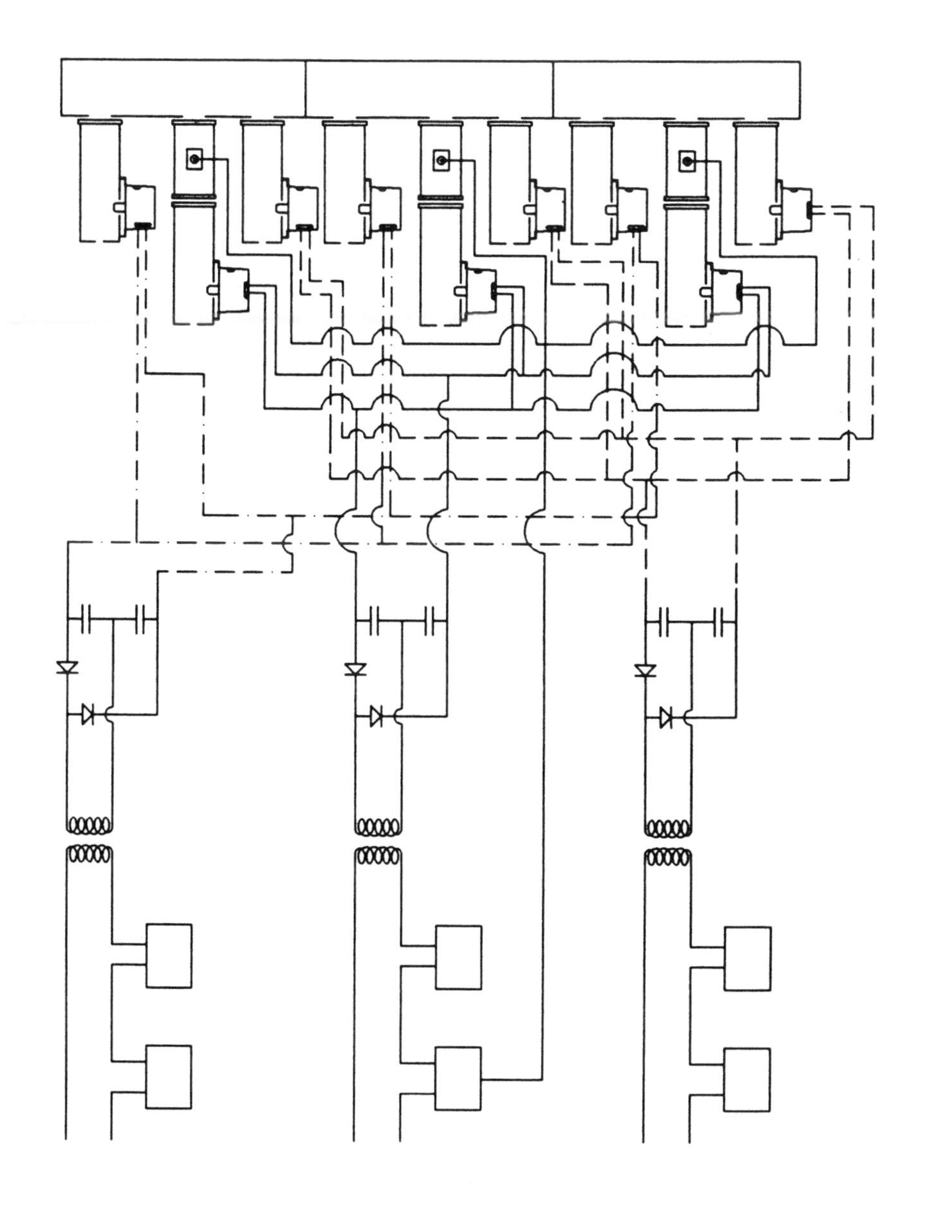

Figure 1. Epoxy resin curing - 18kW.

processing requires that molds sit undisturbed for several hours. This practice requires a large factory area. The molds are typically moved after pouring which results in defects or inclusions in the product that can reduce material strength. The speed of the microwave process eliminates the need for moving the product for storage during curing. It also allows for production in a relatively small factory space.

Although a benefit in process timing, the varying reflection coefficient is detrimental to system integrity and magnetron life. As previously stated, a change in the reflection coefficient is coupled to a change in the reflected power, and consequently the load into which the magnetron operates. In systems where the load conditions can change, a condition called moding can develop in the magnetron. Moding is hazardous to system performance as magnetron current suddenly increases above specified values, and power delivered to the load decreases dramatically. An oscilloscope representation of moding is pictured in Figure 2.

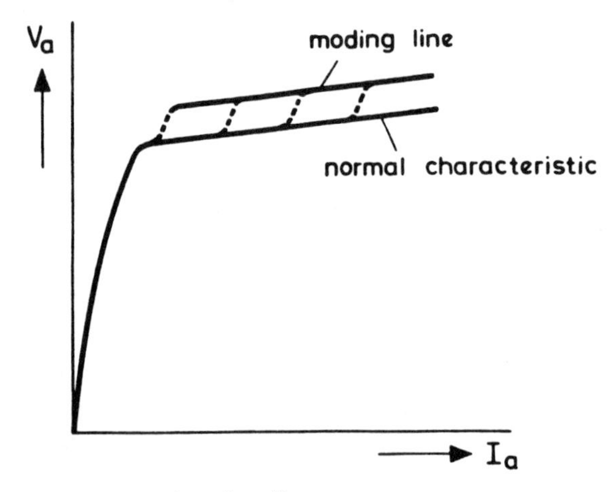

Figure 2. Oscilloscope representation of moding.

Isolators can be employed to protect the magnetron and the product from the effects of varying load conditions and reflected power. Isolators allow power from the magnetron to reach the load while harmful reflected power is channeled to a dummy load. The downside of using isolators in low power magnetron systems is component cost. The waveguide of this system was tuned in order to suppress moding without the added cost of isolators. A polar plot of the waveguide section from the curing system is pictured in Figure 3.

Polar plots provide a representational image of the voltage standing wave ratio and phase relationship of the load as seen by the magnetron. Points 1 and 2 on Figure 3 show the position of the dominate pi mode and the competing pi-1 mode as recorded in the original equipment design. Changing the length of the waveguide by multiples of 1/2 of a guide wavelength maintained the position of the dominate pi mode while moving the pi-1 mode to a less favorable position. This reduces the sensitivity of the system to variations in load conditions and suppresses moding. Waveguide tuning results in improved system reliability at a low system cost.

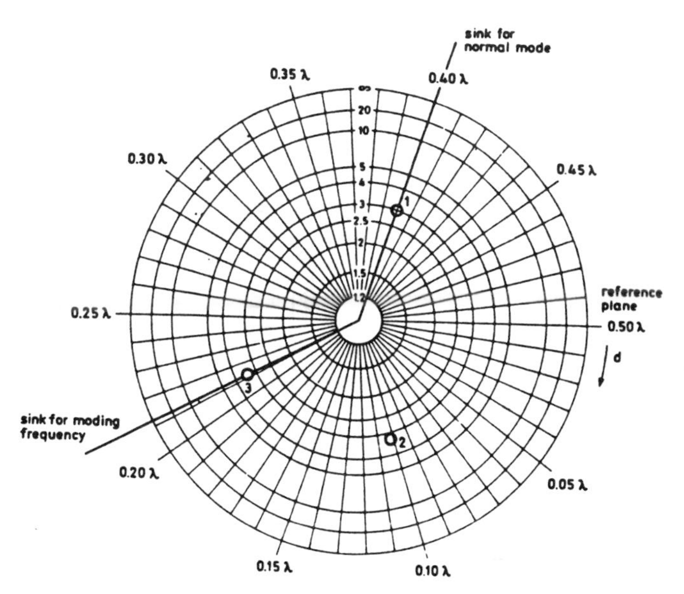

Figure 3. Polar plot showing the dominate pi mode and competing pi-1 mode as recorded in the original equipment.

The use of three power supplies to operate 9 tubes provides a substantial capital cost savings. The microwave power is moved across the product with no mechanical parts. System lifetime is improved by elimination of mechanical components.

Matched system components improve the reliability of the machine. Transformers, capacitors and diodes are designed to enhance performance of the magnetron. A full-wave doubler circuit is utilized to reduce peak current values while preserving tube efficiency at a level of 71%. The transformer leakage inductance is tuned to the circuit capacitance to improve magnetron output power stability. The tuned components act to stabilize magnetron anode current by compensating for normal variations in line voltage. The components utilized are identical to those used in the feasibility and prototype stage, and therefore, design and construction time is reduced.

50 kW Ceramic Drying System

A 50kW system used for ceramic drying is pictured in Figure 4. Dual cavities are used for different stages of the process. The first multi-mode cavity is used for processing of ceramic molds while the second cavity dries the ceramic based final product. The system operates with a single 50kW, 915 MHz. magnetron. The power from the magnetron is channeled to the appropriate cavity through the use of a circulator.

Prior to start-up, both tuners are tuned for maximum reflection in order to prevent microwave energy from reaching the cavities. The magnetron is then energized by a conventional three-phase full-wave rectified supply. Power is channeled in the direction of the circulator arrows toward

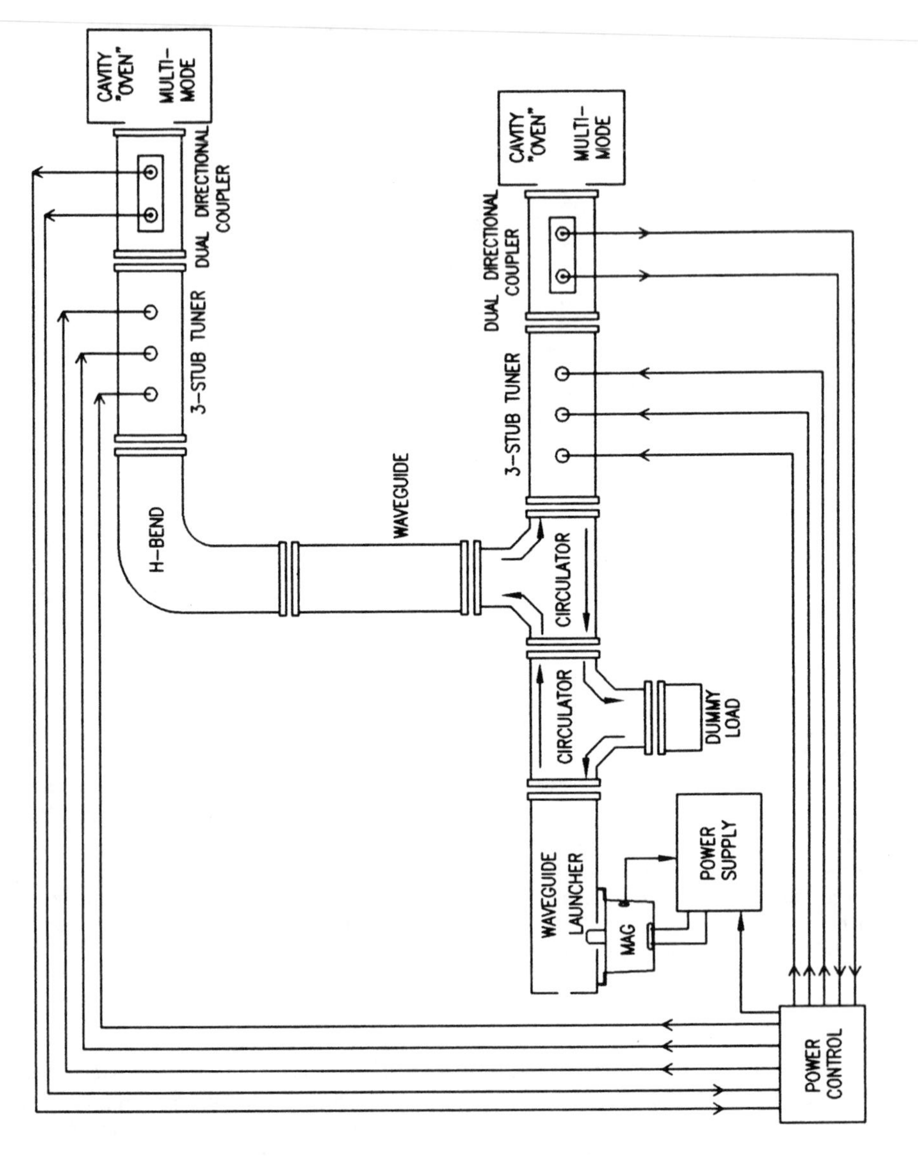

Figure 4. The 50kW dual cavity - single magnetron drying system.

cavity one. The mismatch presented by the tuning stubs reflects the power and prevents it from reaching the first cavity. The reflected power is then channeled toward the second cavity where it is reflected by the second tuner. The energy is finally absorbed by the dummy load of the second circulator. This start-up procedure is utilized to prevent a rapid increase in power to the cavities.

Once the magnetron is energized, the operator can divide the power between the two cavities in an infinite number of ways. The system tuners are automatically tuned to allow a specified amount of power to reach the cavities through adjustments made at the power controller.

The main product benefit realized by the use of microwaves in this system is process speed. The conventional method of drying requires extensive time as both the molds and the product must pass through large ovens or heaters. Use of microwaves allows for drying of the molds and the product simultaneously. Process speeds can be increased ten fold.

Uniformity of heating is also an issue in conventional drying processes. In the system described above, the 915 Mhz signals can penetrate the product which results in uniform heat distribution. This leads to improved strength as the product is dried from the inside out.

High efficiency operation equals a substantial process cost savings. The 915 Mhz magnetron operates at an efficiency of 88%. Additional cost savings are realized in the factory floor space required for the system. Conveyor systems can be replaced with batch drying multi-mode cavities with smaller footprints.

Feedback signals from the directional couplers work in concert with the tuners to balance the power supplied to each of the 2 cavities. A PI controller was programmed to properly adjust the tuners based on signals from the directional couplers and the asked for power of the system. The sequence required to reduce the power in cavity one will be described as a demonstration of the control system. For this example, it will be assumed that a steady state condition exist where both cavities are receiving equal power from the magnetron. The signals from the forward power measuring port of the directional couplers are summed by the PI controller. This "sum" signal gives a representation of the total power delivered to the cavities. As the operator reduces the power control for the first cavity, the signals from the directional couplers are compared to the asked for power of the total system. Since the operator is asking for less power, the magnetron electromagnet power supply current will be increased. Increasing the electromagnet current results in a reduction of the magnetron output power level. The electromagnet current is increased until the signal received from the directional couplers matches the asked for power level for the total system. Once the required output power of the magnetron is achieved, the first cavity tuner will be adjusted until the power delivered to the first cavity is equal to the specified process level. This adjustment results in some power being dumped into the dummy load of the first circulator. The second tuner then adjusts to a matched load condition so that no power is wasted. This scenario can take place in a matter of seconds with no adverse effects to the product.

Feedback control systems as described above would have been impossible in the past. The common use of microprocessor based controllers is what makes the use of a single magnetron, dual cavity system plausible. The single magnetron approach results in a substantial capital cost savings as two cavities can be powered with a single tube and power supply.

30 kW Injection Locked System

Injection locking is a technique used where a high power oscillator will operate with the phase and frequency stability of a lower power source. This results in frequency stability and control that is otherwise unachievable in a magnetron system. The system shown in Figure 5 is a 30 kW plasma generator. The plasma is ignited and maintained with microwave power from the 30 kW source. The 30 kW tube operates into a single-mode cavity. In order to create a uniform plasma, the power delivered to the cavity must be constant. Constant power is delivered to the load by assuring that the frequency of the delivered power matches that of the cavity. By "injecting" a low power microwave signal into the high power magnetron, the frequency and phase of the 30 kW high power tube can be controlled.

There are several characteristics of both the magnetron and the cavity that could cause them to become frequency unstable. Variations in manufacturing techniques can result in a frequency variation of as much as 5 Mhz for magnetrons. Magnetrons are also sensitive to changes in ambient temperature. Load conditions, power supply ripple and current can also effect the output frequency of a magnetron. Likewise, single mode cavities can be affected by changes in ambient temperatures and loads, and both magnetrons and cavities will undergo a frequency shift with operational life. These frequency instabilities can have disastrous effects on system performance.

Typically, systems are designed to overcome these effects by utilization of cavity tuning. Although this will help to maintain matched frequency performance, there are negative side effects. Tuning of the cavity will result in a change in the quality factor. The quality factor is a measure of the ability of the cavity to store energy. As the cavity is tuned to match the variation of the magnetron frequency, the quality factor is changed and so is the process. Changes in the process will affect product quality.

Use of injection locking results in matched frequency operation without the side effects of cavity tuning. The system shown in Figure 5 utilizes the pushing and pulling characteristics of the low power magnetron to control the operational frequency of the high power magnetron. This allows the cavity frequency to be precisely tracked. The frequency of the cavity is monitored by a directional coupler. The cavity frequency is characterized as a dip in the reflected power. This dip is due to the storage of energy or quality factor of the cavity. By adjustment of the load and operational current of the low power magnetron, the output frequency can be changed to match that of the cavity. Through the use of injection locking, the high power magnetron matches that of the low power magnetron.

Injection locking is only warranted in systems where product costs are high. In such systems, process uniformity results in increased product quality and reduced scrap. This system lends itself to applications that are not feasible by other means. Large scale plasma processing is difficult, if not impossible, without such equipment. Therefore, high capital costs are justified. Product produced in these machines can have manufacturing costs as high as $100,000 per batch.

The main attraction of this system is the frequency stability. This is realized by using a low power master tube to control the frequency of the higher power slave tube. As discussed earlier, the frequency of a magnetron will change as the anode current or magnetron load is varied. The amount of change in frequency due to variations in the anode current is called the pushing figure

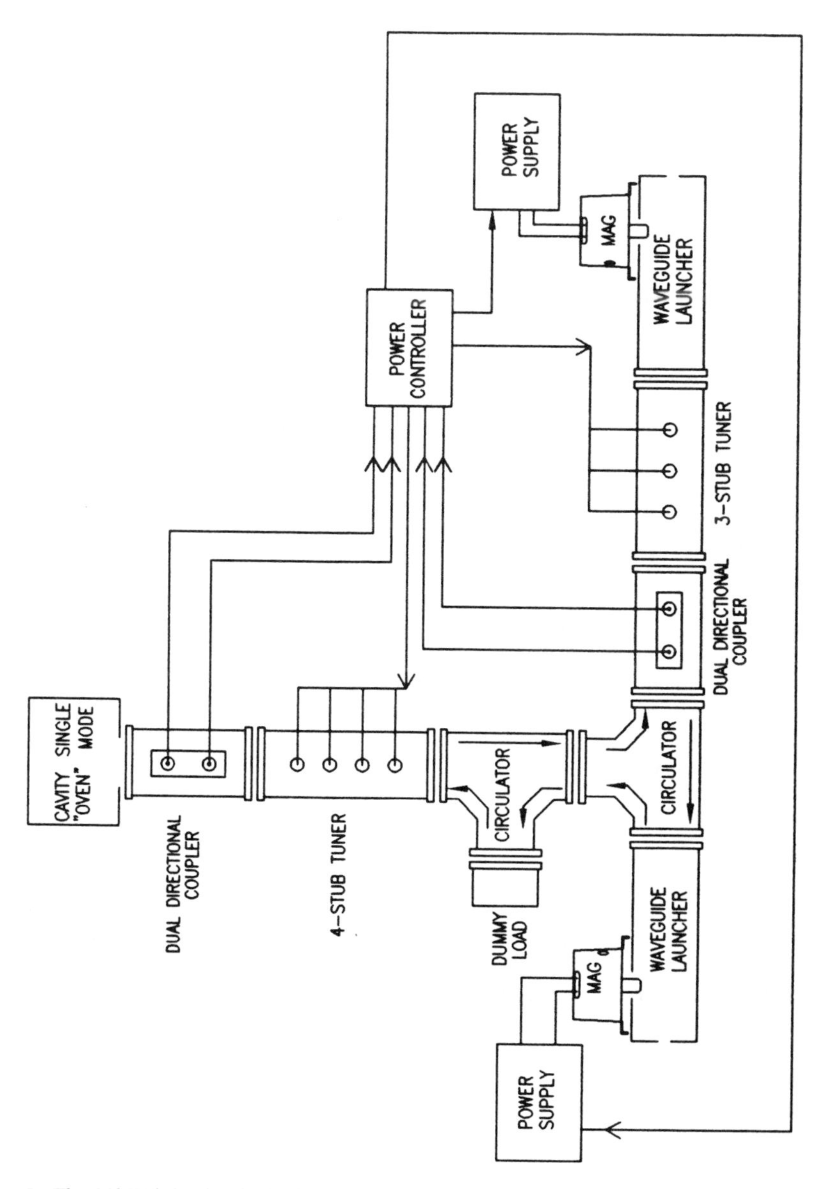

Figure 5. The 30kW injection locked system.

of a magnetron. The change in frequency with varying load conditions is called the pulling figure of the tube. The frequency of the master tube is varied by tuning and load control to match the power of the cavity. Control of the power supply and the tuner is used to push and pull the frequency of the master tube to match the frequency of the cavity. The frequency of the slave tube could be controlled in the same manner, but this would result in a power variation in the cavity and a corresponding process change. The power output from the master tube has no effect on the frequency of the slave tube. The power output of the master tube only affects a parameter referred to as the locking bandwidth. The locking bandwidth is defined as

$$BW = 2F_o/Q_eG^{1/2} \quad (1)$$

Where;
BW = Locking Bandwidth
F_o = Center Frequency of Operation
Q_e = External Q or Quality Factor of the Magnetron
G = Power output of slave tube / Power output of master tube (2).

When the frequency of the magnetron running as a free running oscillator (i.e. no injection locking) differs from that of the master tube by more than the locking bandwidth, the master tube will lose control of the slave tube. The locking bandwidth establishes the limits of frequency control.

In order to stay within the limits of the locking bandwidth, the high power magnetron is operated by a switch mode power supply. The switch mode power supply is ideal for this system. The first advantage of switch mode technology is low ripple, constant current operation. Constant current operation results in increased frequency stability as the frequency pushing of the magnetron is reduced. Another advantage of switch mode operation is the low stored energy resulting from the high frequency operation. If a magnetron arc occurs, the tube is not subjected to high currents that can result from the use of a standard power supply.

Conclusion

The discussion above shows that favorable changes in the microwave industry should spur growth. Equipment and component manufacturers are acting on the request of the end user by providing design support and product testing. These changes, coupled with advancement of technology as a whole, have resulted in the design and commercialization of many new systems. Operation of three such systems was described to show how proper machine design can act to enhance and utilize important material characteristics. If current trends continue, material and microwave specialists can work together to bridge the gap between research and commercialization.

References:

1. J.A. Jolly, "Economics and Energy Utilization Aspects of the Application of Microwaves: A Tutorial Review", Journal of Microwave Power, 11(3), pp. 1-37, 1976.

2. C.A. Walker, "Injection-Locked Magnetrons", Microwaves & RF, March, 1991.

MICROWAVE SINTERING KILOGRAM BATCHES OF SILICON NITRIDE

Prasad S. Apte and William D. MacDonald
Sherritt Inc., Fort Saskatchewan, Alberta, Canada T8L 3W4

ABSTRACT

Microwave heating has been shown to be a very useful process for sintering ceramic materials because it helps overcome several constraints originating from their thermal properties and from furnace design. Thus volumetric heating helps avoid the effects of thermal conductivity, shadowing and convective heat transfer. One of the major limitations is ceramic materials susceptibility to non-uniform heating and thermal runaway which limits the volume of material that can be processed. The challenge is to heat up commercially viable batches of material in a controlled fashion without local overheating. The work done at Sherritt Inc. has been focused on the scale up of the heating process. It has been shown that the uniform hot zone volume can be extended to over 3 litres. The experimental work used silicon nitride as the ceramic material and the uniformity of the heating process was assessed by evaluating the properties of the sintered pieces. The results show that it is possible to sinter over 1800 grams of silicon nitride based samples in about two hours using less than 5 kWhrs/kg.

INTRODUCTION

Microwave processing for sintering ceramic materials was proven to be feasible three decades ago through the work done at General Motors when several spark plugs were sintered simultaneously [1]. This work also showed that the process was economically advantageous in terms of the energy used for sintering. There have been other reports dealing with the energy efficiency of microwave heating which have shown that compared to conventional electric furnaces, microwave furnaces could be up to an order of magnitude lower in terms of energy consumption [2]. This work also showed that the energy efficiency can be obtained only as the batch size was increased. The critical factor in obtaining a larger batch size is the ability to heat up a larger volume of material uniformly. This is a particularly difficult thing to achieve since microwave furnaces are very susceptible to thermal runaway if there is any non uniformity [3]. It is essential that thermal runaway be consistently eliminated for any successful commercial application. Katz [4] gave a general review of microwave sintering of ceramics that summarizes work up to 1992.

There have been several studies dealing with the sintering of silicon nitride. The most notable is the work done on sintering, annealing and reaction bonding at Oak Ride National Laboratory by Janney, Kimrey, Tiegs, Holcombe and others [5,6,7]. Their work has addressed several fundamental issues regarding the sintering of silicon nitride. Physical properties of sintered materials were addressed in this work but only to evaluate the efficiency of microwave heating for larger volumes.

Katz [8] and his coworkers at Los Alamos National Laboratory have reported extensive work on sintering aluminum oxide. Our work was restricted to silicon nitride because here a batch operation, controlled atmosphere and electric heating are features common to both microwave and conventional sintering.

The research and development work at Sherritt Inc. has been directed towards scale up to a "reasonable" size by commercial standards. In this context, reasonable is defined as a batch size that could process the same volume of material as a typical commercial batch furnace over a 24 hour period. Based on discussions with furnace makers this size was determined to be between 1.5 to 2 kg (for silicon nitride) to be processed in about two hours. In terms of a hot zone, this corresponds to a cylindrical zone about 15 cm (6 in.) diameter by 15 cm long.

The heating of ceramic bodies by incident electromagnetic radiation is shown schematically in Fig. 1. In conventional heating the incident flux is in the visible part of the spectrum, penetrates ceramic bodies to a very thin skin depth (a few micrometers) and the interior of the ceramic body heats up mainly by conduction. Those pieces that do not 'see' the radiation directly have to depend on radiation from other ceramics and hence they heat up at different rates than those that do 'see' the radiation. Sintering cycles are generally slow because they have to account for these non-uniform heating conditions.

At a frequency of 2.45 GHz microwaves penetrate most ceramic materials to depths of several centimeters and for many of the commercially produced ceramic bodies this exceeds the dimensions of the pieces. Microwave energy is absorbed throughout the ceramic, based on its dielectric properties and volumetric heating is obtained. The surface of the ceramic bodies is generally slightly cooler because of heat loss from the surface by radiation. The most important factors in determining the sintering process are insulation and the prevention of thermal runaway. If a hot spot is formed, it tends to absorb more microwave energy than its surroundings causing thermal runaway. Microwave heating offers the following advantages:

- the heating can be very rapid
- shadowing is not a factor
- the sintered body has a fine grained microstructure
- surface degradation is minimized.

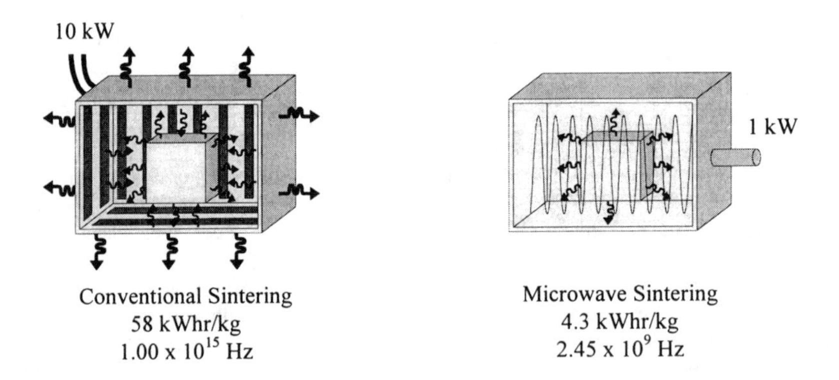

Figure 1: Schematic representation of heating by microwave and visible radiation. The penetration or skin depth of microwaves is orders of magnitude greater than for light. Most of the energy used in conventional high temperature furnaces is lost to the surroundings.

One of the main features of the technology is the use of a powder bed in which the pieces to be sintered are placed. The use of a powder bed for sintering silicon nitride is common practice. In this case, the bed composition has to be altered to make it suitable for use in a microwave field. The powder bed performs several useful tasks in the sintering cycle such as

- •initiate heating from room temperature when silicon nitride itself does not couple well with microwaves,
- •insulates the ceramic pieces once they couple and start heating,
- •enables a switch over from heating the powder bed to heating the ceramic pieces based on differences in bulk density,
- •provides control over the atmosphere, and
- •acts like 'kiln furniture' holding the samples apart.

The experimental work at Sherritt started by developing a process for sintering about 25 grams of silicon nitride, representing a heating volume of about 50 mL. Successive designs and developments progressed from 100 to 1800 gram batches, representing a volume of about 3 litres for the largest batch. Further scale up based on this technology is only dependent on the availability of sufficient microwave power. The uniformity of the heating process was assessed by the uniformity of the properties and the absolute values of the properties, which provide an indication of the advantages of microwave heating. The data reported in this paper addresses some of the processing parameters such as the energy utilization, the effect of cycle time, powder preparation and the properties of the sintered samples.

EXPERIMENTAL

The process described in this paper has been developed for pressureless sintering of silicon nitride. The sintering aids consisted of alumina and yttria. The silicon nitride powder was grade SN-E-10[1], the alumina was grade A-16[2], and the yttrium oxide was Grade C fine[3].

The apparatus used for the experimental work consisted of a 5 kW microwave power supply[4] fed into a three port circulator. Forward and reflected power are measured and reported on the control panel. The power is fed to the applicator through a triple stub tuner which is used as a matching network and enables better utilization of the power. A three-port circulator and a dummy water load ensure that the reflected power reaching the magnetron is small. The microwave power is fed into an applicator consisting of metallic container of a suitable size and shape for the number of pieces to be sintered. It is possible to drill small holes, ~ 5/8"dia., in the applicator to flow in gases or to insert a temperature probe provided no conducting material is inserted through these holes. Microwave power is controlled by a computer which also gathers data from the different temperature probes, the forward and reflected power meters and microwave leak detectors placed around the applicators. The leak detectors are independently linked to the power supply such that if a leak greater than 0.5 milliwatts per square centimetre is detected at any time through the run the power is shut off.

The temperature inside the hot zone is measured by an iridium tipped sapphire probe[5]. The surface temperature of the heating casket is monitored by pyrometry and calibrated with the hot zone temperature. Once a set of process conditions has been developed for a particular product, only the surface temperature is monitored in production. The process is controlled by the power input curve provided the temperature attained falls within a well-defined band.

Silicon nitride, alumina and yttria powders, other process additives, binders, lubricants and other organic aides are mixed in a ball mill using hexane as the milling medium and alumina

[1]Ube Industries, [2]Alcoa, [3]H.C. Starck, [4]Astex A-5000 Supply, [5]Accufibre by Luxtron

balls to facilitate the milling and mixing process. The powders are mixed for 24 hours and the solvent is removed using a rotary evaporator. The dried powder is compacted in a hydraulic press using tungsten carbide tooling at a pressure of 275 MPa (40,000 PSI), and subsequently cold isostatically pressed at 350 MPa (55,000 PSI). The pressed compacts range from 12 to 25 cm in diameter, 5 to 12 cm thick and weigh 3 to 35 g.

The pressed compacts are placed in an alumina crucible and filled with a powder bed. The alumina crucible is insulated by a refractory thermal blanket. The assembly is placed in an applicator large enough to accommodate the number of pieces to be sintered and the power is applied. The actual power cycle varies because the maximum power required to sinter depends on the number of pieces to be sintered as shown in Table 1.

The samples are removed from the insulation blanket and their density is measured to evaluate whether the heating had been adequate and uniform. This cycle is repeated by making appropriate changes to the power rate and the maximum in the power cycle depending on sample density and surface quality. When the power is too high, the parts are "overcooked" as evidenced by silicon droplets accumulated on the part surface. Once a satisfactory density and density distribution is obtained, the cycle is repeated to ensure reproducibility. The data from the power measurements is then analyzed to obtain a sintering program for the particular applicator design and load.

RESULTS AND DISCUSSION

Process Trials

Over the last three years, the scale up of the microwave sintering process has progressed from sintering a few samples at a time to over 500 samples simultaneously. To accomplish this, over 750 sintering runs have been made. A selection of the results from representative runs from each time period in the development of the process is shown in Table 1.

Table 1: Summary of scale up results on microwave sintering of silicon nitride.

No.	No. of Parts	Sample Mass (grams)	Hot Zone Volume (litres)	Max. Power (Watts)	Time (min.)	Specific Energy* (kWhr/kg)	Density (g/cm^3)	Density Range ($\pm g/cm^3$)
1	5	25	0.05	900	70	50.0	3.10	0.10
2	10	50	0.20	900	90	25.0	3.16	0.04
3	49	412	0.75	2400	100	7.9	3.21	0.03
4	104	333	0.90	2250	120	8.7	3.25	0.01
5	560	1790	3.00	5000	120	3.6	3.26	0.007

*Microwave net energy absorbed, as obtained from input minus reflected power.

It can be seen that the efficiency of microwave power utilization and the density improves as the mass of material processed increases. The recent power consumption is consistent with the results of Patterson et al [2] who calculated the net power consumption for alumina to be 3.8 kWhr/kg of alumina. The maximum power required for sintering the 1.8 kg batch was 5 kW, which is the limit of our power supply.

The maximum density obtained represents over 99% of the theoretical density and the variation is less than 0.007 g/cm^3 as shown in Fig. 2, which is comparable to the accuracy of measurement by Archimedes principle. Further densification can be attained by hot isostatic pressing the samples at 1800° C in a nitrogen atmosphere at 206 MPa. However, the maximum density increase was never greater than 0.03g/cm^3. In earlier trials, major deviations in density were observed when the alumina crucible cracked and the protective atmosphere was lost.

Product Evaluation

As part of our scale up process, product quality standards have been developed for each stage, from incoming powder inspection to final quality assurance before shipping. For the rest of this paper, we focus on the recent results of the microwave sintering stage. For each run following sintering, the parts are visually inspected for distortion, cracks, and surface damage. Sintered densities are obtained from preselected samples, including standards, in certain locations within the hot zone. The weight change for these samples are measured as well as the linear shrinkage. At least one sample from each run is polished for metallographic examination and porosity evaluation as per ASTM B276. Hardness measurements are made using Rockwell "A" and Vickers at a 10 kg load. Typical property data is shown in Table 2.

Table 2: Physical properties of microwave sintered silicon nitride

Density	3.26 g/cm^3
Hardness	94.1 ± 0.1 Rockwell A
	16.7 - 17.5 GPa Vickers
Toughness	7.5 MPa m$^{1/2}$
Phase Structure	>95% β - Si_3N_4
Grain Size	Acicular, approximately 0.5 x 5 μm

The data shows that the sintered material has excellent physical properties and a fine grained structure expected from volumetric heating of the ceramics. The range in Vickers hardness values reflect the dependence of that measurement on the applied load. At low loads, ~ 1kg, the material gives high hardness values with considerable scatter. At higher loads, the

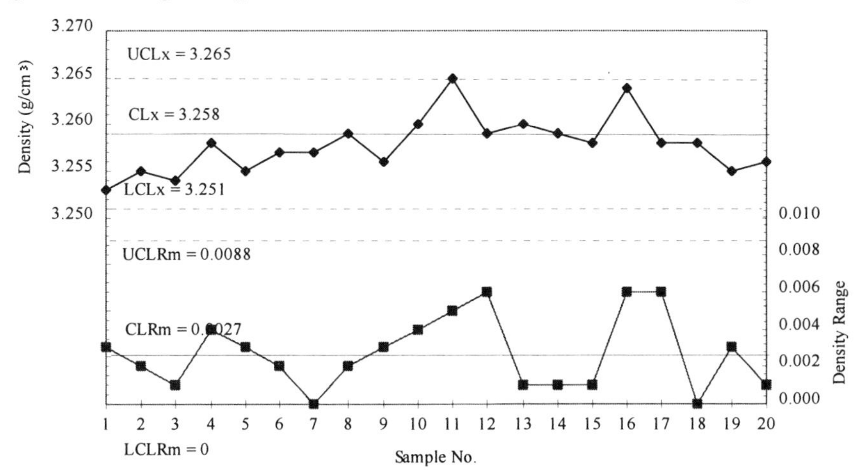

Fig 2: Variation in sintered density on a typical run as shown by an x-Rm chart. Note that the density variation is less than 0.2% and the process is well under statistical control.

hardness and the scatter decrease, as shown in Fig. 3, indicating that a more reliable value is obtained. That this is not a particular characteristic of microwave silicon nitride is indicated by the similar trend for a competitive material.

A fractograph of a sintered sample, shown in Fig. 4, reveals a very fine grain size and an aspect ratio of about 10 on the acicular grains. The lack of surface degradation is shown in Fig. 5 on a CNGA-432 cutting tool insert blank. The Vickers hardness variation from surface to center was ±0.5 GPa. Only a minimum of grinding is required to bring the sample within the cutting tool class G tolerance of ± 0.001". Hardness measurements made on the as sintered surface are typically only 0.2 points lower in Rockwell A hardness than on a ground and polished internal surface. Since it is possible to make sintered silicon nitride with a very thin degraded zone, it appears feasible to make near net shape ceramics by microwave sintering.

CONCLUSION

It is possible to scale up microwave sintering to a heating volume of 3 litres. Further extension of the uniform hot zone is feasible with a larger power supply. The sintered product obtained has uniformly high density and excellent physical. Microwave sintering does provide a fine grained product with reduced surface damaged layer as would be expected from a rapid, volumetric heating process.

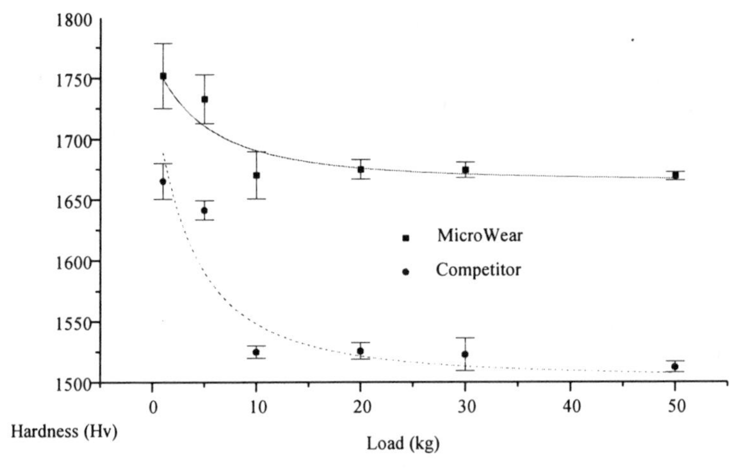

Figure 3: Vickers macrohardness versus indent load on microwave sintered silicon nitride and a competitive conventionally sintered silicon nitride. The error bars, which are plus/minus one standard deviation, decrease with load.

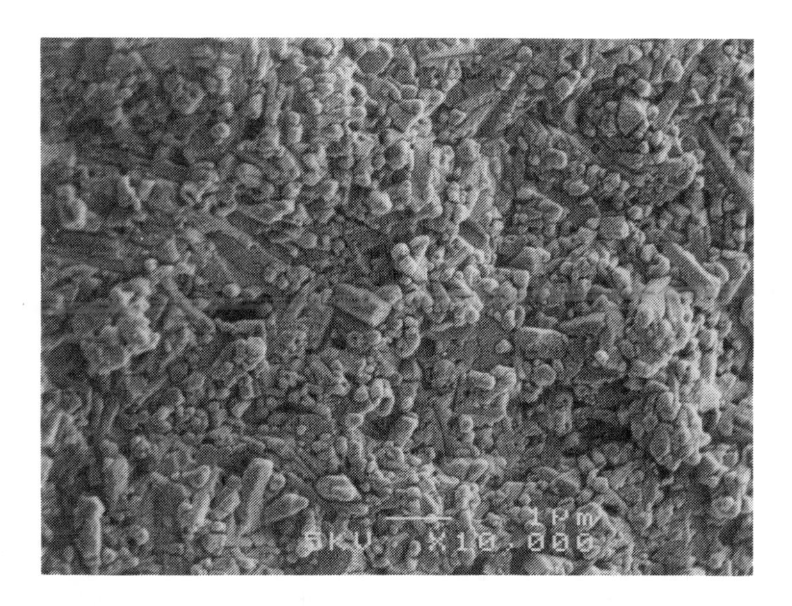

Figure 4: SEM Fractograph of a microwave sintered wear part. The ß-phase needles are embedded in glassy matrix that has a grain size on the order of 0.2 to 0.4 μm. Original 10,000X

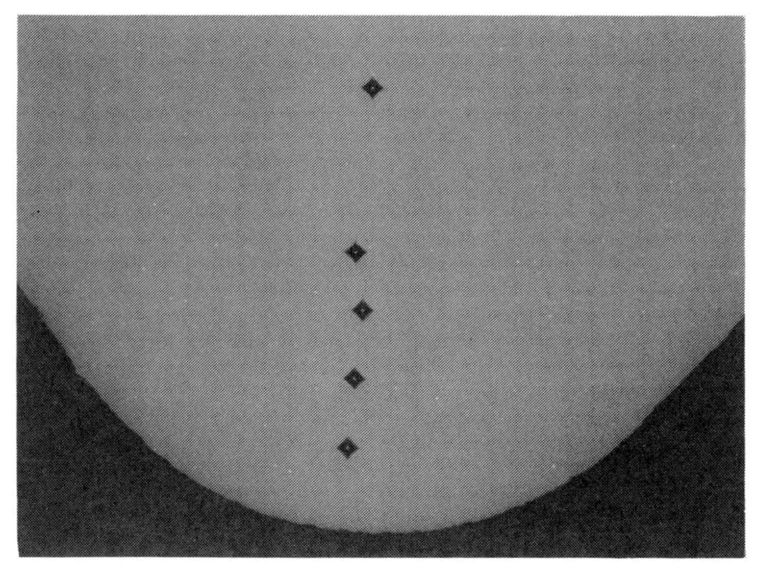

Figure 5: Polished section through the cutting tip of a CNGA-432 style insert blank. The closest microhardness indent is approximately 0.005" from the surface and gave a reading of 17.4 GPa. Original 100X

ACKNOWLEDGMENTS

The authors would like to thank the technologists and co-op students involved in this project for their help. The authors also appreciate the permission of Sherritt, Inc. to publish this work.

REFERENCES

1 Schubring, N.W., "Microwave Sintering of Alumina Spark Plug Insulators", Am. Ceram. Soc., Electron. Div., Grosinger, NY (1983)

2 Patterson, M.C.L., Kimber, R.M., Apte, P.S., and Roy, R., "Batch Process for Microwave Sintering of Si_3N_4" in Microwave Processing of Materials, Mater. Res. Soc. Symp. Proc. (Snyder, W.B. et al eds)., 257-72 (1991)

3 Johnson, D.L, Skamser, D.J., and Spotz, M.S., "Temperature Gradients in Microwave Processing: Boon and Bane", Ceramic Transactions, 36: Microwaves: Theory and Application in Material Processing II (Clark, D.E., Tinga, W.R. and Laia, J.R. eds) 133-46 (1993)

4. Katz, J.D., "Microwave Sintering of Ceramics", Annu. Rev. Mater. Soc. 22, 153-70 (1992)

5 Holcombe et al., "Method for Heat Treating and Sintering Metal Oxides with Microwave Radiation" U.S. Patent #4,880,578 (1989)

6 Tiegs, T.N., Kiggans, J.O. and Kimrey, H.D., "Microwave Processing of Silicon Nitride", Microwave Processing of Materials II, Mater. Res. Soc. Symp. Proc., 189 (Snyder, W.B. et al eds) 267-72 (1991)

7 Janney, M.A. and Kimrey, H.D., "Microwave Sintering of Alumina at 28 GHz", Ceramic Powder Science, II, B, Ceramic Transactions, 1, 919-24 (1988)

8 Katz, J., and Blake, R.D., Am. Ceram. Soc. Bull. 70, 1304-08 (1991)

PARALLAM - NEW STRUCTURAL WOOD COMPOSITE

M.T. Churchland
MacMillan Bloedel Research
4225 Kincaid Street
Burnaby, British Columbia

ABSTRACT

The work presented here represents the accomplishments of many talented people from many disciplines. Many of us have devoted our careers to the Parallam® project and we have much more to do before we can rest and reflect on our success lest they cease to be successes. We are in the middle of the difficult task of converting a technical and marketing home run into a profitable business. By July 1987, MacMillan Bloedel had spent over 40 million dollars on Parallam research and development. In July 1987, MacMillan Bloedel committed a further 97 million in capital. These funds were used to build two commercial plants - one in Vancouver, British Columbia and the other in Colbert, Georgia. A third plant has been built in Buckhannon, West Virginia and will start in the third quarter of 1995.

PRODUCT

Parallam® PSL is made from long veneer strands. A photo of strands is shown in Figure 1. Standard plywood technology is used to produce rotary-peeled veneer, typically 102 inches along the grain, and 1/8 to 1/10 inches thick. The veneer is cut into strips about 1/2 inch in width. Most strands are 102 inches long but strands as short as 24 inches are used. We have experimented successfully with many different species, and are largely producing with Douglas fir in Vancouver and Southern pine in Colbert. The two products have strength properties so similar they have the same code allowable strength values.

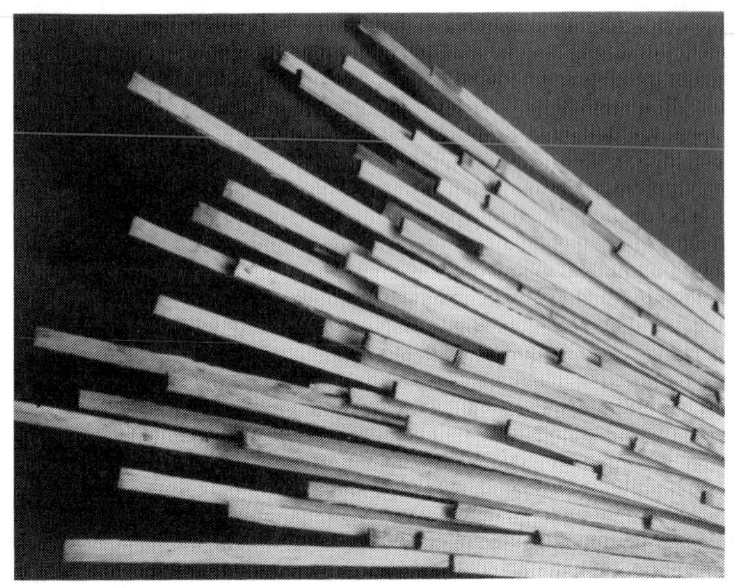

Figure 1. Photograph showing veneer strands used in the fabrication of Parallam®.

Phenol formaldehyde resin is used to bind the strands together. The formulation is similar to that used for exterior grade plywood. It is waterproof, chemically stable and emits effectively no free formaldehyde. Because this resin has been used successfully in the plywood industry for over 40 years, we are sure of its long-term reliability.

Parallam® PSL is assembled in the following steps:

- The resin is applied to all surfaces of the strands and penetrates effectively into lathe checks and cracks.
- The strands are placed into a trough where each strand is oriented parallel to the product axis with more accuracy than other oriented strand products.
- The ends of the strands are randomly distributed along the lay-up.

The mat of parallel glued strands is continuously formed in this way, compressed to a fixed cross-section and heated with microwave energy. The compression produces only about a ten percent over-density compared to the host species. The process steps will be discussed in more detail later.

ATTRIBUTES

What we gain over other products and processes relates to usable strength properties and dimensional stability.

First, the product is very strong. The stranding process inherently weeds out knots, severe grain deviations and weak wood. Such weaknesses cause the strand to break up into short lengths. These are automatically discarded. The mat when formed from the remaining strands has evenly distributed very small strand ends. Each of these ends, although theoretically a weakness, has little impact and the inherent strength of each individual strand is fully utilized.

Second, the product is very consistent. Because a single 3.5 inches by 16 inches beam contains over a thousand strands in any given cross-section, the weakness of a single strand or the placement of a strand end does not significantly affect strength. In the testing of full size beams, the beam-to-beam variability in strength and stiffness is less than 10% (coefficient of variation). Solid wood varies from 25% to 50% (coefficient of variation).

Third, the product is very durable. Because of the waterproof resin and the even microwave heating, the product behaves like solid wood in the presence of water. It is strong even when wet and has minimal swelling, effectively recovering both strength and original dimensions on redrying. In a recent 5-year exterior exposure test conducted by the American Plywood Association, Parallam® PSL retained 100 % of its original strength after conditioning back to initial moisture levels.

Fourth, the product properties are controllable. Because the process steps are controllable, the final product can be engineered to ensure the advertised strength properties are met. The combination of strength, consistency, durability and controllability combine to give a range of usable strength properties unparalleled in the industry.

PRODUCT TESTING

We have continued to evolve Parallam® PSL in subtle ways to improve its strength and durability. In doing so, we have undertaken the most thorough testing program in the history of new wood products. Over 30 different tests of over 50,000 specimens have been done. They include many tests for: bending

(MORE, MOE, shear modulus), compression, tension, shear, cyclic aging, etc.. They also include duration of load, water absorption, dimensional stability, treatability, fire response, etc..

An important part of our philosophy is to test full size beams or sub-assemblies whenever possible (e.g., trusses). Size effects in structural wood products are not well understood by the industry, so full size testing is very important in predicting serviceability. We have tested significant numbers of 40 feet long, 4-inch by 12-inch beams and 4-inch by 24-inch beams. We have also extensively tested parallel cord roof and floor trusses as well as double pitched roof trusses at similar lengths.

CONVERSION OF VENEER TO BILLETS

The part of process that has consumed most of the research effort for the first 10 years is the conversion of dry veneer into billet. The manufacture of Parallam® PSL billets from dry veneer can be discussed under the following headings:

- Veneer clipping (to 0.5-inch wide strands)
- Defect removal (short strand elimination)
- Glue application (all sides covered)
- Continuous mat formation (parallel strand lay-up)
- Continuous pressing
- Continuous microwave heating.

I will focus on the last two points - pressing and microwave heating. They form the heart of the process and the two areas really integrate into one machine.

The essential components of the press and microwave windows are shown in Figure 2. The symmetrically opposed steel belts are continuously pulled between the press platens by the 4-drive drums. The continuous compression action on the mat to form the product comes when the material is drawn, by the motion of the steel belts, into the fixed gap formed by the platens. The loose mat of wood strands needed to make the product is about 3 feet thick. The finished product is about 1 foot thick. The principal reason for compressing is to ensure good contact between strands and therefore, good glue bond quality. The wood density is increased only about 10 percent above the host species.

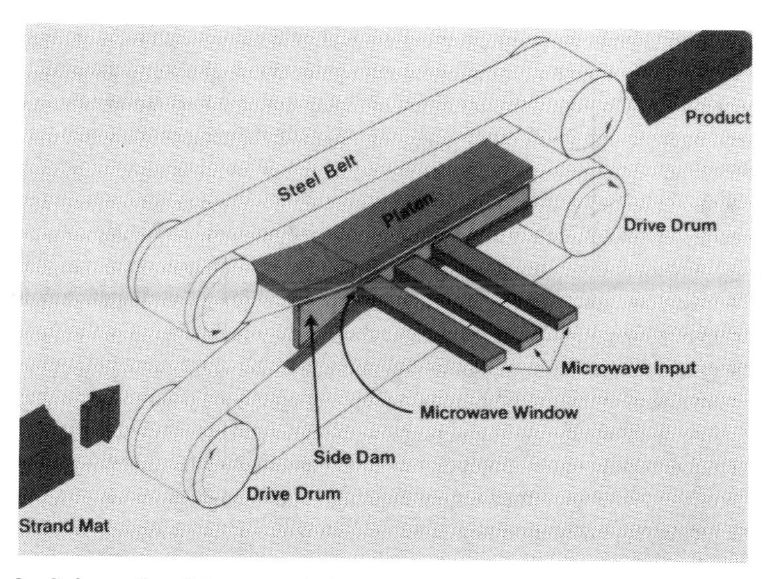

Figure 2. Schematic of the press/microwave set-up .

Two features of our process differ from standard continuous presses:

- Sidedams physically constrain the lay-up.
- Microwave energy is continuously applied.

Figure 2 also shows the sidedams installed in the press. The width of most of our production varies between 12 inches and 16 inches, depending on sidedam positioning. The sidedams maintain full supportive forces on the edge of the product as the press belts continuously draw strands into the press. Even though the sidedams are fixed, compressive forces can be in excess of 300 psi.

Unlike particleboard or oriented strand board processes, edge trim waste is virtually non-existent. Further, because the process continuously forms a beam, we can cut the lengths required by the market without stopping and without end trim losses.

The microwave windows are between the microwave waveguides and the sidedams. In our current press, these windows are made of ceramic. The ceramic, as part of the sidedam, also holds full compressive forces on the edge of the product. The ceramic is transparent to microwave energy and allows microwave power to penetrate into the product.

Microwaves heat materials like wood (which contain water and various ions) effectively. The choices of wood moisture content and of chemical additives such as the phenolic glue affect the penetration of the microwaves into the product. By controlling these variables, we can control the evenness of heating of the compressed mat.

Our use of microwave energy allows us to evenly heat much larger cross-sections (12" by 20") than would be possible with conventional hot pressing. The high frequency microwaves also allow us to use conventional, well-proven, low cost alkaline phenolic resins. These resins are exterior grade and waterproof. Radio frequency systems used, for example, in wood composite preheating do not work well with phenolics because electrical arcing occurs.

Industry standard hot platen presses can cure product up to about 2 inches thick. Thicker products become impractical because wood is a good thermal insulator and press times increase beyond practical limits. In contrast, our microwave system is currently heating a 12-inch thick product in about two minutes. Further, it heats much more evenly than conventional hot presses resulting in more consistent product properties.

SUMMARY OF PRODUCT AND PROCESS EVOLUTION

The evolution of the product and the process have been intertwined inextricably. The process now in use has very high fiber recovery and is highly automated (low labor cost). Because of this, we have low operating costs. The product has high, consistent strength properties and excellent dimensional stability. Because of this, we gain premium sales value. Further, the product and process are protected by 14 internationally patented inventions.

We have exceeded the original concepts of making a composite wood product equivalent in strength to lumber and now manufacture the highest strength wood-based building material available today. Our strength properties exceed those of machine stress related lumber, glulam beams and most laminated veneer lumber.

MICROWAVE ENHANCED SINTERING OF CERAMICS

F.C.R. Wroe and A.T. Rowley
EA Technology, Capenhurst, Chester CH1 6ES, England, UK.

ABSTRACT

It is now well known that microwave dielectric heating can be used to increase the sintering rates and to reduce the sintering times of ceramic materials. However, the nature of the mechanism causing this enhanced sintering is still far from understood, with many workers attributing the effect to a reduction in the activation energy even though there is no real physical basis for this assumption.

Although the mechanism is not understood, many results have indicated that the effect is non thermal in nature, i.e. the enhancement would not be reproduced if conventional heat could be applied in exactly the same way (volumetrically) as microwave heat.

By careful control of the relative proportions of microwave and conventional heating, it has been possible to separate the thermal (heating) effects from the non-thermal effects. This paper discusses the results obtained, and show that they are consistent with recent theories of enhanced diffusion.

INTRODUCTION

The thermal sintering of a ceramic is a complex process, with a number of diffusion mechanisms (volume, grain boundary and surface diffusion) taking place which together produce the required increase in density, and changes in mechanical (and other) properties of the material. In conventional sintering, the dominant force driving the diffusion is due to the capillary stresses which exist within the material associated with the reduction of surface energy. The diffusion of ions or vacancies along a gradient in species concentration is described by Fick's law[1]

$$\mathbf{J} = -D\,\mathrm{grad}(n) \qquad (1)$$

where **J** is the particle flux, n is the species concentration and D is the diffusion coefficient, given by

$$D = \nu_0 a^2 e^{-\frac{\Delta g_m}{kT}} \qquad (2)$$

Here, ν_0 is the characteristic lattice vibration frequency, a is the lattice constant, k is Boltzmann's constant, T is the absolute temperature, and Δg_m is the Gibbs free energy for motion through the lattice (the activation energy). In a given material, the activation energy, and hence the diffusion coefficient, will be different for each diffusion mechanism - surface and grain-boundary diffusion normally have significantly lower activation energies than volume diffusion. The vacancy flux, represented by Fick's law, can be thought to be the product of a mobility term (D) and a driving force term (-grad(n)). If either term is increased, then the diffusion will be enhanced. Other force terms are possible, such as that due to an electric or an electrochemical potential gradient.

When a ceramic material with complex relative permittivity $\varepsilon_r'-\varepsilon_r''j$ is placed in a high frequency electric field, the power per unit volume, P_V, dissipated within it is given by

$$P_V = \omega\varepsilon_0\varepsilon_r''E^2 \tag{3}$$

where ω is the angular frequency of the applied field, ε_0 is the permittivity of free space ($8.854\times10^{-12}Fm^{-1}$), and E is the local electric field strength. The electric field within the material depends on both the applied electric field (due to the microwave applicator) and the real part of the permittivity, ε_r', also known as the dielectric constant. The imaginary part of the permittivity, ε_r'', is often referred to as the dielectric loss factor of the material.

There is now a growing amount of experimental evidence to support the fact that microwave heating of ionic crystalline solids leads to an enhancement of the sintering processes taking place in these materials. Samples processed within a microwave furnace are observed to sinter at a faster rate or at a lower temperature than those processed in a conventional system. Wilson and Kunz[2] demonstrated that partially stabilised zirconia (with 3mol% yttria) could be rapidly sintered using 2.45GHz microwaves with no significant difference in the final grain size. The sintering time was reduced from 7200s to about 600s. Kim and Kim[3] successfully sintered yttria doped zirconia samples (3, 6.6 and 8mol%) in a microwave furnace at very high heating rates (up to 500^oC/min) with near maximum theoretical final densities (>99%). Many authors have incorporated this enhancement into the conventional theories of sintering through the introduction of an effective activation energy for the diffusion processes taking place during sintering (increasing the vacancy mobility through the lattice). Most notably, Janney and Kimrey[4,5] have shown that, at 28GHz, the microwave enhanced densification of high purity alumina proceeds as if the activation energy is reduced from 575kJ/mol to 160kJ/mol. Similarly, the same authors[5] have shown that when stabilised zirconia (with 8mol% yttria) is processed in a 2.45GHz microwave furnace, the sintering curve is shifted by ~150^oC with respect to the conventional sintering curve - equivalent to an apparent reduction in the activation energy for this sintering process. Furthermore, the grain size of the final dense zirconia was finer in the microwave-fired sample. Despite the potential implication for the ceramics industry, there has been little work published which discusses the possible physical mechanisms for such a large reduction in activation energy.

The microwave enhancement of sintering seems to be a genuine non-thermal effect, i.e. even if conventional heat could be applied in the same volumetric way as microwave heating, the same enhancement of the sintering process would not be observed. The microwaves must interact with the ceramic so as to reduce the actual activation energy or to increase the effective driving force experienced by the diffusing species. Bookse et al[8] suggested that the microwaves interact with the crystal lattice in such a way as to render the Boltzmann approximation used in the derivation of the diffusion coefficient (2) invalid. However, in later calculations by the same group, Freeman et al[9] indicated that extremely high microwave electric field strengths would be required for this to occur (~10^7V/m) - much higher than typical field strengths in a microwave applicator (~10^3-10^4V/m). In the same paper, Freeman et al published results of conductivity measurements made on a single crystal of NaCl which indicated that the vacancy mobility is not enhanced by the application of microwaves, but that the diffusion driving force is enhanced. This result is consistent with the calculations of Rybakov and Semenov[10] who show that the driving forces for vacancy motion can be enhanced near a surface or boundary.

When an electric driving force, due to a gradient in the electric potential, is present in addition to those due to capillary stresses, the total particle flux becomes:

$$\mathbf{J} = -D\mathbf{grad}(n) - D\frac{nq}{kT}\mathbf{grad}(V) \tag{4}$$

where q is the electronic charge on the ion or vacancy and V is the local electric potential. The particle flux is enhanced by the presence of an electric field, $\mathbf{E} = -\mathbf{grad}(V)$.

In previous measurements of microwave enhanced sintering, two different furnaces have had to be used, one for the conventional radiant heating experiments and the other for the microwave experiments. The operation of the microwave furnaces has been further complicated by the need for a microwave susceptor (usually zirconia) to provide the initial heating of the sample. The development of a combination (hybrid) furnace removes the need for such a susceptor, and allows the relative amounts of microwave and conventional heating to be varied during processing. Consequently, for the first time, non-thermal effects can be separated from thermal effects by varying the electric field while keeping the heating rate constant during sintering.

EXPERIMENTAL DETAILS

The furnace used for these sintering experiments was specially designed to allow microwave and conventional heating sources to operate simultaneously[11]. The system, shown schematically in Figure 1, is essentially a conventional furnace built within a multi-mode 2.45GHz microwave cavity powered by a 1.2kW industrial standard magnetron. Care must be taken in the design and construction of the radiant heating elements to avoid any unwanted interaction between the two heating sources. Similarly, the dilatometer used to measure sample shrinkage has been modified to avoid its operation being affected by the application of microwaves, and to minimise its effect on the microwave field distribution within the sample space. The thermocouples used in the furnace are positioned to minimise their effect on the microwave field distribution. Two thermocouples are used; one to measure the ambient furnace temperature and one to measure the actual sample temperature. Either thermocouple can be used to control one or both heating sources (microwave and radiant).

All samples used in this work were prepared from Tosoh TZ3Y partially yttria stabilised zirconia (YSZ) powder supplied by Tosoh Corporation. The samples were prepared by wet-milling the powders with 1% polyethylene binder. The powder was then dried, crushed and sieved, prior to uniaxial pressing using a 12mm tool-steel die at a pressure of 96MPa. The binder was then burnt out of the compacts in a muffle furnace using a slow heating cycle. The resulting compacts had a green density of approximately 50-55% of the optimum final sintered density. Approximately 2.5g of powder were used producing a sample with a length of about 6mm.

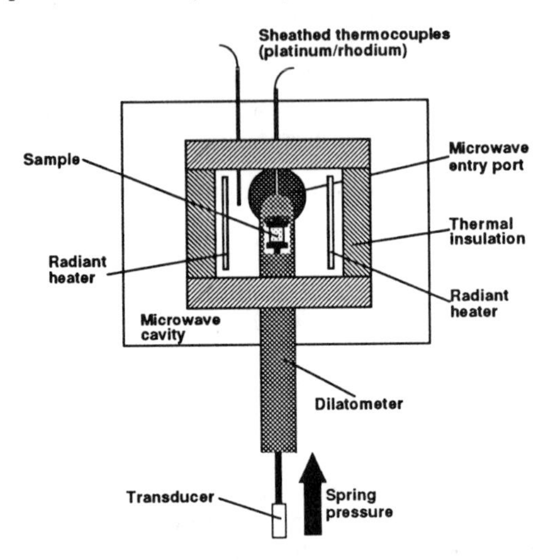

Figure 1: Microwave-assisted furnace used in sintering experiments.

During each sintering experiment, the sample temperature, ambient furnace temperature, dilatometer reading, microwave power and radiant power are all continuously recorded. The sample temperature was controlled by the radiant heating elements, and the microwave power (if used) was kept approximately constant throughout the sintering process. For most experiments, the sample was heated at 10°C/minute to 1500°C, held there for 1 hour and then ramped down at 20°C/minute to room temperature.

Relatively small samples of partially stabilised zirconia were used (~700mm^3) to ensure that any temperature variation throughout the sample was small (<2°C at these heating rates). The difference in the sample and ambient furnace temperatures during equivalent conventional and microwave-assisted experiments can be used as an indication of the actual proportion of power absorbed by the sample. In fact, the microwave power dissipated in the sample is reasonably constant over the temperature range of interest (~800-1400°C).

RESULTS AND DISCUSSION

Figure 2 shows the normalised linear shrinkage, $\Delta l/l_0$, plotted as a function of temperature (l_0 is the original sample length) for conventional and microwave-assisted sintering of partially stabilised zirconia (3mol% yttria). The enhancement of the sintering is clearly demonstrated, the microwave-assisted curve is displaced by ~80°C from the conventional shrinkage curve.

Furthermore, the total shrinkage is greater in the microwave-assisted cases, leading to an increase in the final sample density. At about 1250°C, there is a significant change in gradient in the microwave-assisted curve. Towards the end of the microwave-assisted sintering, although the applied microwave power is still approximately constant, the electric field will be falling due to the increase in the dielectric loss factor, ε_r" with increasing temperature.

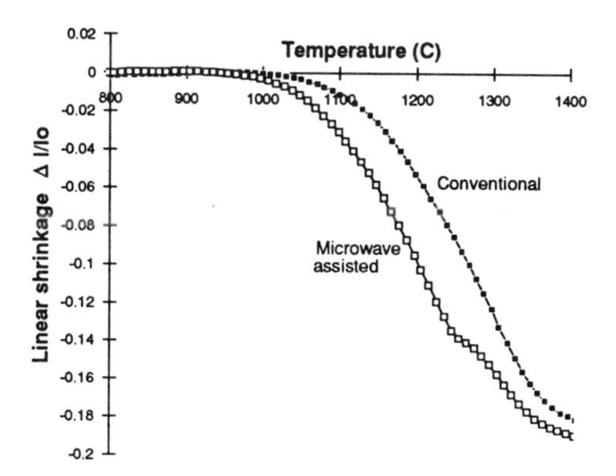

Figure 2: Normalised linear shrinkage of zirconia (3mol% yttria) plotted as a function of temperature for conventional (radiant heat only) and microwave-assisted sintering.

The enhancement is more clearly demonstrated in Figure 3, which plots the rate of shrinkage as a

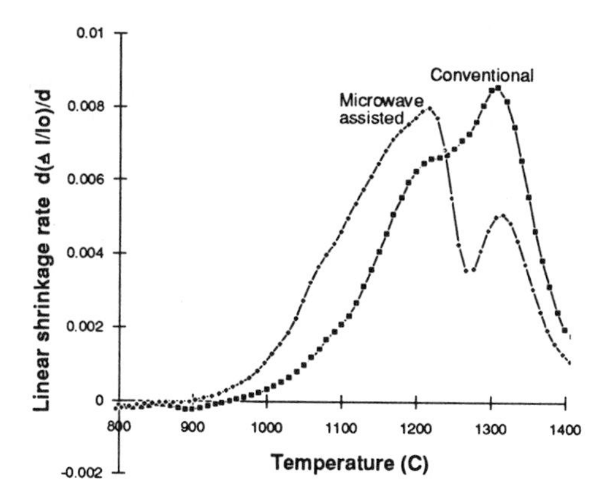

Figure 3: Normalised linear shrinkage rate (min^{-1}) of zirconia plotted as a function of temperature for conventional (radiant heat only) and microwave-assisted sintering.

function of sintering temperature for the two processing routes. In the conventional case, the peak in the sintering rate occurs at about 1300°C, with a minor peak around 1200°C. When microwaves are applied the peak sintering rate is shifted down in temperature to 1200°C.

However the second peak in the sintering rate at 1300°C is still evident (although it is much smaller) when the microwave enhancement is reduced, and the densification is mainly driven by the conventional capillary forces.

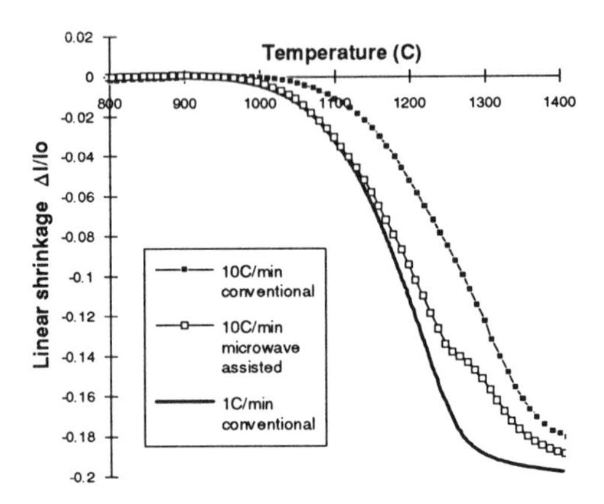

Figure 4: Normalised linear shrinkage of zirconia plotted as a function of sintering temperature for 10°C/min and 1°C/min heating rates. Microwave-assisted sintering at 10°C/min displays similar characteristics to the much slower (1°C/min) radiant only heating.

It is interesting to compare the effect of microwave power on the sintering process with the effect of conventional sintering at a slower heating rate. In both cases, the shrinkage occurs at lower temperatures, and there is an increase in the final sample density. It seems as though the microwave-assisted sintering is effectively proceeding in a similar way to conventional sintering, but at a much faster heating rate. This conclusion is emphasised in Figure 4, where the shrinkage rate for microwave-assisted sintering at 10°C per minute is compared with conventional sintering at 1°C (conventional sintering at 10°C per minute is also shown for reference). In the initial stages of sintering the microwave-assisted process almost exactly follows the 1°C per minute conventional curve. The possibility of faster (and therefore more efficient) processing is very attractive to the ceramics industry and, in many cases, should lead to microwave-assisted sintering becoming the accepted processing route. This is the main reason for the interest being shown in the technology by the ceramics industry world-wide.

Figure 5 shows the effect of switching the microwaves off at 1080°C, when the microwave enhancement of densification is already established. Quite clearly, the densification slows down substantially until the sintering temperature is high enough for conventional densification to take over. However, the final sample density is still higher than the purely conventionally sintered sample.

Figure 6 shows the results of the complimentary experiment, with the microwaves switched on at about 1080°C. At this point, densification proceeds extremely rapidly and eventually "catches up" with the continuous microwave-assisted sample. The final sample density was slightly higher in the sample in which the initial stages of sintering proceeded conventionally.

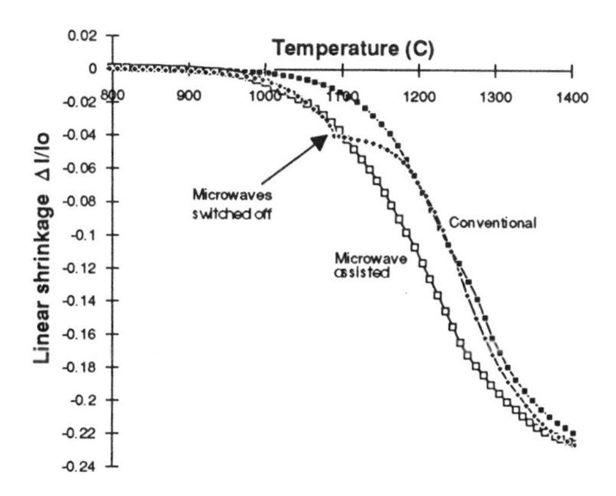

Figure 5: Normalised linear shrinkage of zirconia plotted as a function of sintering temperature for conventional and microwave-assisted sintering, showing the effect of switching off the microwaves during the process (~1080°C).

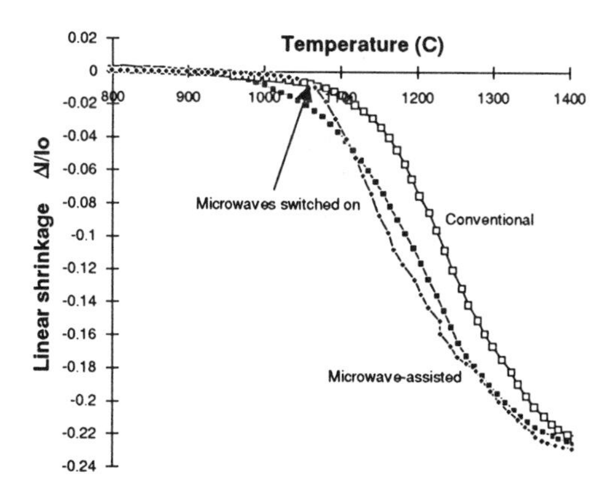

Figure 6: Normalised linear shrinkage of zirconia plotted as a function of sintering temperature for conventional and microwave-assisted sintering, showing the effect of switching on the microwaves during the process (~1080°C).

CONCLUSIONS

In all cases, the microwave-assisted sintering of partially stabilised zirconia demonstrates an enhancement relative to conventional sintering in the same furnace. Typically, the microwave-assisted processing leads to a reduction of ~80-100°C in the sintering temperature for a given value of linear shrinkage. Moreover, the final density of the samples sintered in a microwave field is slightly higher. This enhancement is non-thermal in nature; even when the microwave power is held constant throughout the whole process, there is clear evidence that the enhancement falls away towards higher temperatures. This is consistent with the enhancement being dependent on E (non-thermal) rather than P_V (thermal). More dramatically, when the electric field is switched off in the initial stages of densification, the shrinkage almost halts before resuming along the conventional curve. This supports the theory of an additional driving force induced by the microwave field. If this force is removed, the vacancy flux again is given by Fick's law, and becomes that corresponding to the conventional sintering temperature.

There is some evidence to suggest that the majority of the enhancement occurs during the densification stage of sintering, rather than the initial stage. This implies that the microwave field enhances either the volume or grain boundary diffusion mechanisms more than the surface diffusion process which dominates at lower temperatures. Since the induced driving field is likely to be due to the accumulation of space charge at boundaries, the microwaves will preferentially increase the flux of vacancies within grain boundaries in the sample. In most cases, the mobility of species in grain boundary diffusion is higher than that for volume diffusion due to the lower activation energy for vacancy motion in the boundary regions. The increased driving force, coupled with the preferential enhancement of a mechanism with a lower activation energy, could account for the large reductions in effective activation energy measured previously. However, the actual activation energies for any given diffusion process are not affected by the microwave field.

ACKNOWLEDGEMENTS

This work has been undertaken by EA Technology through its Core Research programme on behalf of its members. The authors would like to thank the Directors of EA Technology for permission to publish it here.

REFERENCES

1. C. Kittel, Introduction to Solid State Physics, Wiley, (1976).
2. J. Wilson and S.M. Kunz, J. Am. Ceram. Soc. **71**(1), c40-41, (1988).
3. D. Kim and C.H. Kim, J. Am. Ceram. Soc. **75**(3), 716-718, (1992).
4. M.A. Janney and H.D. Kimrey, Mat. Res. Symp. Proc. Vol. 189, ©1991 Materials Research Society.
5. M.A. Janney, H.D. Kimrey, M.A. Schmidt and J.O. Kiggans, J. Am. Ceram. Soc., **74**(7), 1675-81 (1991).
6. S.J. Rothman, Mat. Res. Symp. Proc. Vol. 347, 9-18, 1994.
7. M.A, Janney and H.D. Kimrey, *Advances in Sintering*, ed. Bleninger and Handwerker, Am.Cer. Soc., Westerville OH, (1990).
8. J.H. Bookse, R.F. Cooper, I. Dobson and L. McCaughn, Ceramics Transactions, (21), "Microwaves: Theory and Application in Material Processing", Am. Ceram. Soc., Westerville OH, 185-192, (1991)
9. S. Freeman, J. Bookse, R. Cooper and B. Meng, Mat. Res. Symp. Proc. Vol. 347, 479-485, 1994.
10. K.I. Rybakov and V.E. Semenov, Phys. Rev. B. **49**(1), 64-68, 1994.
11. EA Technology Ltd., International patent application, No. PCT/GB94/01730

Waste Remediation

MICROWAVE TECHNOLOGY FOR WASTE MANAGEMENT APPLICATIONS INCLUDING DISPOSITION OF ELECTRONIC CIRCUITRY

G.G. Wicks
Westinghouse Savannah River Technology Center
Aiken, SC

D.E. Clark, R.L. Schulz and D.C. Folz
University of Florida
Gainesville, FL

ABSTRACT

Microwave technology is being developed nationally and internationally for a variety of environmental remediation purposes. These efforts include treatment and destruction of a vast array of gaseous, liquid and solid hazardous wastes as well as subsequent immobilization of selected components. Microwave technology provides an important contribution to an arsenal of existing remediation methods that are designed to protect the public and environment from undesirable consequences of hazardous materials. Applications of microwave energy for environmental remediation will be discussed. Emphasized will be a newly developed microwave process designed to treat discarded electronic circuitry and reclaim the precious metals within for reuse.

INTRODUCTION

Significant quantities of hazardous wastes are generated from a multitude of processes and products in today's society. This waste inventory is not only very large and diverse, but is also growing at a disturbing rate and shows no signs of subsiding. In order to minimize dangers presented by constituents in these wastes, technologies are being investigated to minimize the waste generated and to provide safe handling, transportation, storage, disposal and destruction of hazardous components.

Microwave technology is being explored as one method to assist in remediation of these wastes. This includes development of a variety of microwave systems designed and tailored for a wide array of wastes and applications [1-5]. In the United States, waste remediation studies are being conducted primarily at Department of Energy installations. This includes studies are being conducted primarily at Department of Energy installations. This includes Oak Ridge National Laboratory, Argonne National Laboratory, the Rocky Flats Plant, Los Alamos National Laboratory, and the Westinghouse Savannah River Site. Some of these programs represent joint undertakings with academia, such as the program at Savannah River being conducted with the University of Florida. International participants involved in microwave waste remediation efforts include the Geopolymer Institute in France and the Batelle Institut in Germany, Ontario Hydro's Research Division in Canada, Kobe Steel in Japan, and others.

There are many reasons for using microwave processing for waste remediation compared to other methods. In general, advantages of using microwave technology for waste remediation may include some or all of the following features listed in Table I.

Table I. Potential Advantages of Microwave Processing for Waste Remediation

Potential Advantages
Waste volume reduction
Rapid heating
High temperature capabilities
Selective heating
Enhanced chemical reactivity
Ability to treat wastes in-situ
Treatment or immobilization of hazardous components to meet regulatory requirements for storage, transportation or disposal
Rapid and flexible process that can also be made remote
Ease of control
Process equipment availability, compactness, cost, maintainability
Portability of equipment and process
Reduction in personnel radiation exposure for rad wastes (ALARA)
Energy savings
Cleaner energy source compared to some conventional systems
Overall cost effectiveness/ savings

The advantages to be realized depend on many factors, especially the type and characteristics of the wastes or conditions to be treated.

MICROWAVE REMEDIATION APPLICATIONS

There are many types of hazardous wastes that are potential candidates for microwave treatment. Wastes currently under study and of special interest include radioactive wastes and sludges (high, low, and intermediate level wastes, transuranic and mixed wastes), contaminated soils and sediments, incinerator ashes, industrial wastes and sludges, medical and infectious wastes, asbestos, groundwaters, volatile organic compounds (VOC's), and discarded electronic circuitry. A variety of applications using microwave energy for treatment

treatment of these wastes are discussed in fine overview papers [1-5] and the reader is recommended to review these references for more detail. The following will be a brief discussion of some of the more interesting uses of microwave energy for waste remediation along with a more detailed discussion of using microwave energy for remediation of discarded electronic circuitry.

Gaseous & Liquid Treatments

Microwave Induced Plasma Technology is being developed by Argonne National Laboratory and a Microwave Fluidized Bed Reactor by Los Alamos National Laboratory for destruction of volatile organic compounds or VOC's. Microwave energy is also being investigated for treatment of nitrogen oxide as well as a variety of other gaseous components.

Destruction of Volatile Organic Compounds: Microwave induced plasma technology has the advantage of producing extremely high temperatures for selected waste management applications. Temperatures in excess of 10,000°C can be obtained if desired, although lower temperatures are used most often. Argonne National Laboratory has been involved with a microwave induced plasma reactor and environmental remediation efforts in a low temperature plasma range of less than 5,000°C [6]. In this application, trichlorethylene (TCE) or trichloroethane (TCA) in concentrations of 100-10,000 ppm have been fed into the plasma along with other constituents. Free electrons and radicals are then generated in the plasma and reactions are started which destroy the TCE or TCA and convert it into less noxious components. This conversion can be better than 99% effective. This technology is also believed to be applicable to many other multicomponent mixtures, including aromatic and aliphatic chlorinated compounds.

Los Alamos National Laboratory has investigated use of a microwave fluidized bed reactor to produce chemical oxidation reactions used for waste remediation [7,8]. In this process, microwave energy is used to selectively heat the surface of silicon carbide particles to about 500°C without heating reactor walls. This radiation also activates the silicon carbide surface to promote surface oxidation reactions. Work performed on trichloroethane-air mixtures has demonstrated as much as a 98% conversion of TCA.

NO_X Reduction: Dow Chemical and the CHA Corporation have been studying NO_X reduction on Char using microwave energy [9]. There are millions of tons of nitrogen oxide (NO and NO_2) that are discharged to the air each year with the potential of contributing to acid rain and street level ozone. These are especially important considerations in large industrial areas of our country. Existing NO_X abatement technologies are believed to be inadequate for long term needs. A form of coal, called Char, can absorb NO_X from waste gas and when microwave energy is used, can convert the NO_X and carbon to nitrogen and carbon dioxide via reduction.

Toluene and Xylene: Toluene and xylene are used as organic solvents in many industrial synthesis processes. These aromatic hydrocarbons appear as hazardous contaminants in volatile organic compounds found in soil and groundwaters. Microwave energy has been used by Mississippi State University for decontaminating soils containing these contaminants [10]. The removal rate of toluene and xylene was seen to improve by simply adding water to the system, which generates additional heat when exposed to microwave radiation. With removal of these organic soil contaminants, contaminated soil could be disposed of safer, easier and more cost effectively in landfill disposal sites.

Microwave energy has also been used for removing liquids from waste mixtures by Ontario Hydro's Research Division of Canada. These efforts include solvent waste treatment for the automotive industry and dewatering of sludges existing in the nuclear field.

Automotive Plastic Solvents: Instrument panels used in the automotive industry are made from ABS plastic, vinyl and polyurethane foam [1,11]. The machinery which produces these materials must be periodically cleaned, which produces a hazardous byproduct consisting of foam and the cleaning solvent, methylene chloride. The waste generated due to the cleaning operation is currently stored in drums.

Because the foam in the waste mixture is a good insulator, conventional electric resistance heating methods used to condense and recover the solvent in this mixture, and reduce the waste volume, were not effective. Because of the properties of the foam/ solvent waste mixture, microwave heating was examined. It was shown that the microwaves could selectively heat the solvent within the drum which quickly distilled the methylene chloride without overheating the foam. The treatment was successful and waste volume significantly reduced, which contributed to important cost savings.

Dewatering of Low Level Nuclear Waste: Using a similar concept to the successful treatment of solvents used in the automotive industry, dewatering of low level nuclear waste in-situ was also examined [1,11]. Microwave energy was used to dewater the sludge in the same waste storage vessel that the waste was contained, which resulted in minimizing handling of the radioactive waste and reducing the waste volume by about 5%. The microwave treatment designed to dewater the low level radioactive sludge resulted in a product which met Canadian Atomic Energy Control Board acceptance criteria.

Solid Treatments

Microwave energy is being used for immobilization of hazardous inorganic and organic wastes. One of the most desired waste forms produced is glass, which has exhibited an outstanding ability to retain hazardous components under adverse conditions and for long periods of time [12, 13]. There have been a variety of waste glass forms produced using a vast array of waste compositions. These glass products meet regulatory requirements and possess excellent chemical durability, mechanical integrity as well as good thermal and radiation stability. In addition to glass, other waste forms such as cementious products have been produced using microwave energy.

Vitrification of Inorganic Wastes: Since 1985, microwave technology has been under investigation and development at the Rocky Flats Plant in Golden, CO. Initial efforts involved using microwave energy to simply dry sludge. The technology then progressed into a larger effort involving using microwave energy to treat and solidify a variety of hazardous and radioactive wastes.

In the Rocky Flats process, microwave energy is used to melt waste contained inside 30-gallon drums [14,15]. A waveguide directs the energy from a 60 kilowatt, 915 megahertz microwave generator to the drum, which heats its contents to about 1000°C, thus melting the inside material. After melting is initiated on a small waste batch, the remaining waste is then continuously fed into the drum which will ultimately contain about 300 kilograms of melted waste. The most significant advantage of this technology compared to other means is that very significant volume reductions can be obtained ($\sim$80%) which translates into significant storage and disposal cost savings. In addition, the waste glass produced from the microwave vitrification process meets current transportation and disposal regulations. Wastes of interest to the Rocky Flats site include hydroxide precipitation sludge, nuclear storage tank materials, ash and soil. This technology has already been demonstrated on a laboratory scale, pilot plant scale and most recently, a full-scale demonstration unit which is now operable. The technology is believed to be applicable to most solid, inorganic waste compositions.

Incinerator/ Microwave Treatment of Plutonium: Plutonium-contaminated solid wastes are generated during mixed oxide (MOX) fuel fabrication. A process has been developed by Kobe Steel, Ltd. for treatment of this waste at the Tokai Works of the Power Reactor and Nuclear Fuel Development Corporation (PNC) of Japan [16, 17]. The process uses an incinerator coupled with a microwave unit. The waste is initially incinerated and resulting ash fed into an adjacent microwave melter where it is vitrified in a batch process. The system is relatively compact so that it may be contained within a glovebox which is necessary for safe processing, due to the radioactivity of the waste. A generator external to the glovebox, produces microwaves which are transmitted into the glovebox by a waveguide to conduct the melting operation. The waste is melted and vitrified in crucibles which serve as both the melting vessels as well as the ultimate storage vessels. This technology is being expanded to include treatment of liquids, sludges, asbestos, residues of acid digestion, concretes, contaminated soils and other radioactive waste compositions.

Non-Radioactive Ash: The University of California and Los Alamos National Laboratory have been studying the melting and immobilization of non-radioactive ash using microwave energy [18]. The ash was mixed with additives and different waste loadings were studied. The composition of the ash was primarily silica, titania, calcia, alumina and carbon and the additives consisted of magnetite, lithium and sodium carbonate and boron oxide. Among the results noted was that melted samples with more than 30 wt. % additives had an undetectable leaching rate ($<$ 0.1 ppm of lead).

Cement Waste Forms: Geopolymers are also being examined for containment of radioactive and hazardous wastes by institutions including the Geopolymer Institute in France and the Batelle Institut in Germany [19]. Mineral Geopolymer materials are new cementitious materials with zeolitic properties developed by the Group of Companies "GEOPOLYMERE" in Europe. They are alkali-silico-aluminates. Microwave processing using commercial units has been used to produce rapidly a stable waterfree ceramic waste form to contain hazardous components.

Concrete Decontamination by Microwave Spalling: Oak Ridge National Laboratory is developing a microwave spaller for decontaminating concrete currently present throughout the DOE complex [20]. The spaller uses microwaves to heat water absorbed in the outer layer of the concrete. This creates induced stresses which cause the surface to burst or spall off. A vacuum then collects the particles produced. This operation has the potential of producing less airborne dust than other techniques and less secondary waste.

DESTRUCTION OF ELECTRONIC CIRCUITRY AND RECLAMATION OF PRECIOUS METALS

Microwave technology has been examined by the University of Florida and the Westinghouse Savannah River Technology Center to treat a variety of potentially hazardous materials and components [21,22]. One application of special interest is the destruction and vitrification of electronic components and recovery of important metals for reuse [23]. Advantages of this technology include simplicity of operation, significant waste volume reduction, and production and separation of metal and waste glass forms. The waste glass product produced ties up hazardous components and meets environmental leaching standards and the metal product formed allows precious and other metals to be conveniently reclaimed for recycling. The waste glass forms have also been successfully produced without the need of any additives.

Electronic components and circuits are indispensable parts of our society today and are vital components in a multitude of consumer products and services. Each year, many of these products containing electronic circuitry are retired. Hence, millions and millions of electronic components and circuit boards must be disposed of in a cost effective and environmentally safe manner. At present, many consumer products containing electronic components are discarded in landfills throughout our nation. This results in a number of concerns. First, because of the very large volume of these materials, many landfills are now filling up, resulting in the need for new "dumps" or disposal sites and increased costs to all. Next, there are a variety of hazardous materials that can be contained within electronic components. In landfills, these elements can leach from the waste and make their way into groundwaters, which can result in undesirable public and environmental consequences. Finally, because the products are simply thrown away, there is no attempt to reclaim useful materials within the circuits and hence, natural resources in these wastes, including precious metals, are discarded and cannot be reused or recycled.

The joint program involving laboratory-scale experiments have produced the following important results:

- A wide array of electronic components were able to be treated by a relatively simple, 1-step, hybrid-heated microwave process
- Actual waste volume reductions of >50% were achieved with geometric volume reductions significantly greater
- As a result of the controlled microwave process, important metal components were readily separated from waste glass
- A waste glass product was able to be fabricated without the use of any additives and the glass produced retained hazardous components and met important environmental leach standards
- Precious metals, including gold and silver, were separated effectively and reclaimed for reuse
- The entire process can be mocked-up to a larger scale and also made mobile.

This work, along with a unique tandem microwave waste remediation system, designed to not only treat electronic circuitry but also the off-gases produced, will be described in more detail at this symposium [24]. This system is depicted in Figure 1 and in Figure 2, key operations involved in the destruction of the electronic components and reclamation of important metals are summarized.

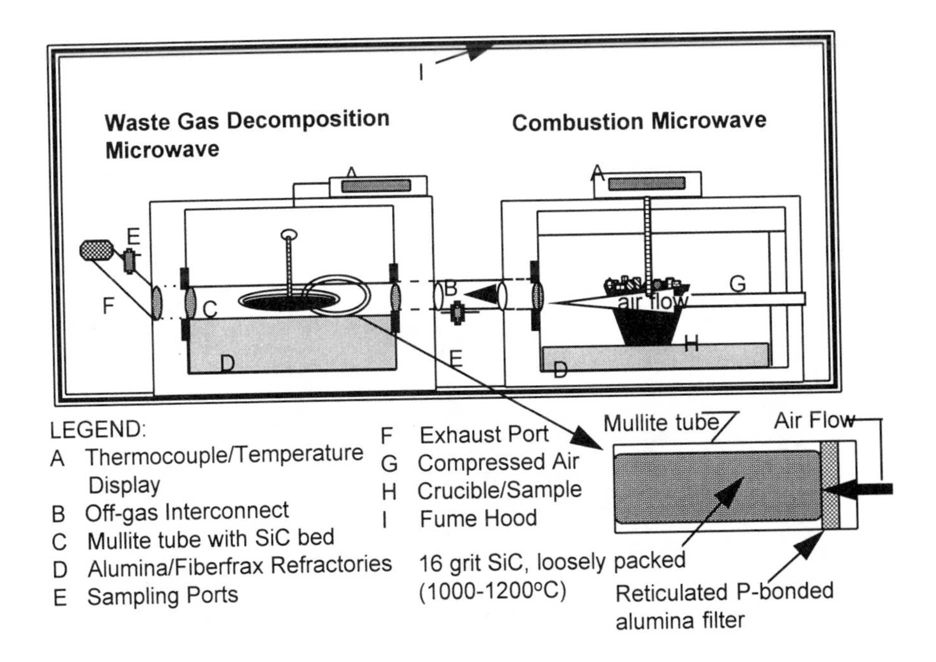

Figure 1. Schematic representation of the University of Florida/Westinghouse Savannah River Technology Center tandem microwave processing system (Ref. 24).

SYSTEMS APPROACH/ COMMERCIALIZATION CONSIDERATIONS

Considering the many potential advantages of using microwave energy for remediation of both radioactive and non-radioactive wastes, it may seem surprising to some that there are not more commercial systems in operation. There are two main reasons for the very slow adoption of microwave technology for waste remediation. First, the concept of "if it ain't broke, don't fix it" illustrates the difficulty of trying to replace existing and more conventional technology with new and mainly unproven methods. Second, most of the efforts of using microwave energy for treating hazardous wastes have been scattered and directed to

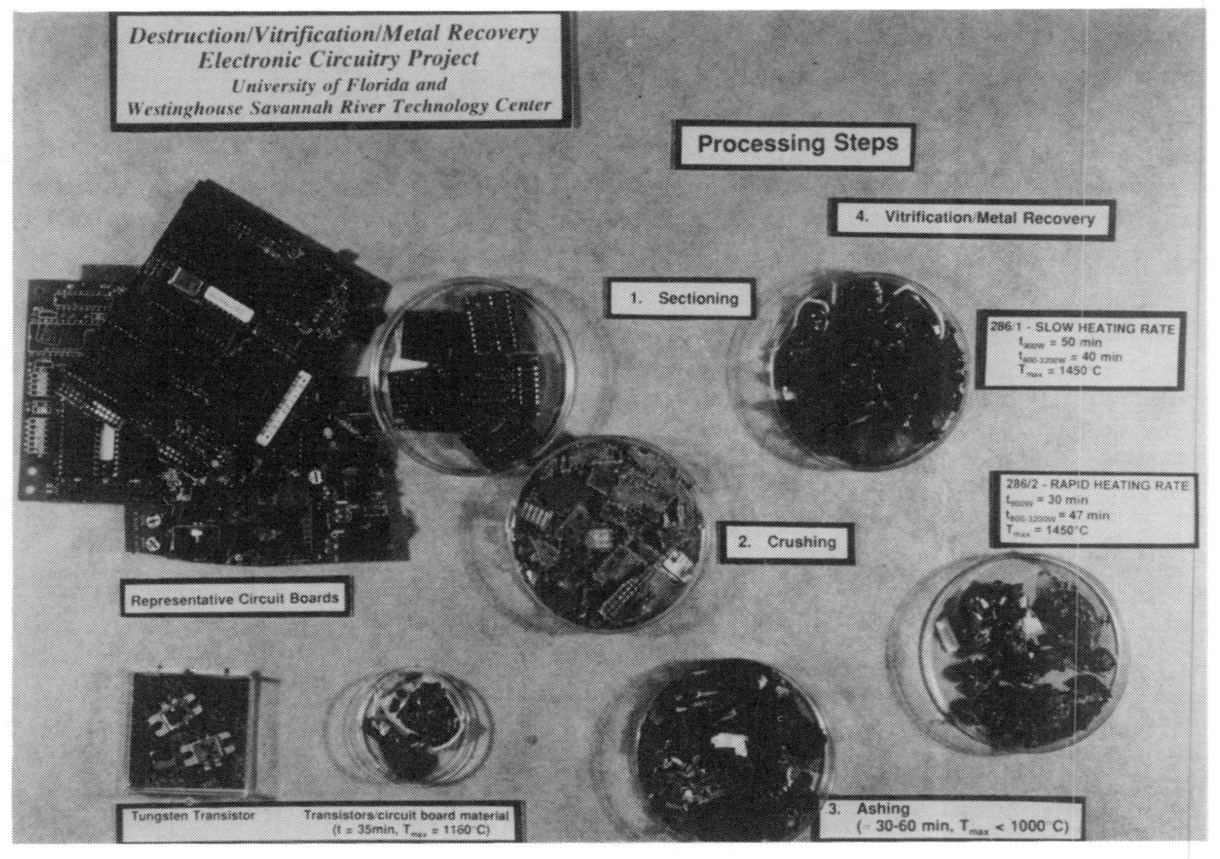

Figure 2. Steps in the processing of electronic components using hybrid microwave heating (Ref. 24).

very specific or niche markets. What is often lacking is an integrated systems approach, using interdisciplinary teams, to address the entire problem. An example of a "dream team" might include representatives from academia (technical leadership and support), a federal or national laboratory (engineering, systems and regulatory support, potential duel use definition of the technology, and financial support), microwave/ equipment manufacturer (mocking up/ tailoring of existing equipment for a potential new market), and the customer for the service (knowledge of the market place including production rates and product specifications, potential financial support, and profit motive driver).

In the case of waste remediation, expertise in microwave technology alone is insufficient to move the technology into the commercial section. For example, for vitrification of hazardous wastes using microwave energy, knowledge of glass melting processes and chemistry is very important, analytical support and an understanding of regulatory requirements are needed, off-gas treatment and the ability to manage and show compliance of both primary and secondary wastes essential, and a market-type analysis is necessary to provide a cost incentive for the service. Until teams of this type and a systems approach is fully utilized, commercialization on a large scale will continue to be very slow.

SUMMARY

Microwave energy provides important potential advantages for remediation of a wide range of radioactive and non-radioactive hazardous wastes. It provides an important contribution to an already existing and growing arsenal of remediation technologies. However, before its potential can be fully recognized and ultimately commercialized, an interdisciplinary, integrated systems approach, would be beneficial.

REFERENCES

1. Oda, S.J., "Dielectric Processing of Hazardous Materials- Present and Future Opportunities," Microwave Processing of Materials IV, (M.F. Iskander, R.J. Lauf and R.L. Beatty, eds.), Materials Research Society, Vol. 347, pp. 371-382 (1994).
2. Oda, S.J., "Microwave Remediation of Hazardous Waste: A Review," Microwave Processing of Materials III, (R.L. Beatty, W.H. Sutton and M.F. Iskander, eds.), Materials Research Society, Vol. 269, pp. 453-464 (1992).
3. R.L. Schulz, D.C. Folz, D.E. Clark, R.M. Hutcheon and G.G. Wicks, "Applications of Microwave Energy for Waste Remediation," proceedings of the 28th Microwave Power Symposium, International Microwave Power Institute, Montreal, Canada (1993).
4. Krause, R.T. and Helt, J.E., "Applications of Microwave Radiation in Environmental Remediation Technologies," Microwaves: Theory and Application in Materials Processing II, Ceramic Transactions (D.E. Clark, W.R. Tinga and J.R. Laia, eds.), American Ceramic Society, Vol. 36, pp. 53-59 (1993).
5. Dauerman, L., Windgasse, G., Zhu, N. and He, Y., "Microwave Treatment of

Hazardous Wastes: Physical Chemical Mechanisms," Microwave Processing of Materials III, (R.L. Beatty, W.H. Sutton and M.F. Iskander, eds.), Materials Research Society, Vol. 269, pp. 465-469 (1992).

6. Krause, R.T. and Helt, J.E., "Chemical Detoxification of Trichlorethylene and 1,1,1 Trichloroethane in a Microwave Discharge Plasma Reactor at Atmospheric Pressure," Emerging Technologies in Hazardous Waste Management III, American Chemical Society (D.W. Tedder and F.G. Pohland, eds.), No. 518, pp. 393-410 (1993).
7. Varma, R., Nandi, S.P. and Katz, J.D., "Detoxification of Hazardous Waste Streams Using Microwave-Assisted Fluid-Bed Oxidation," Microwave Processing of Materials II, (W.B. Snyder Jr., W.H. Sutton, M.J. Iskander and D.L. Johnson, eds.), Materials Research Society, Vol. 189, pp. 67-68 (1991).
8. Varma, R. and Nandi, "Oxidation Degradation of Trichloroethylene Absorbed on Active Carbons: Use of Microwave Energy," Microwaves: Theory and Application in Materials Processing, Ceramic Transactions (D.E. Clark, F.D. Gac and W.H. Sutton, eds.), American Ceramic Society, Vol. 21, pp. 467-73 (1991).
9. Buenger, C.W., Pererson, E.R. and Cha, C.Y., "NO_X Reduction on Char Using Microwaves," Microwave Processing of Materials IV, (M.F. Iskander, R.J. Lauf and R.L. Beatty, eds.), Materials Research Society, Vol. 347, pp. 395-399 (1994).
10. George, C.E., Jun, I. and Fan J., "Application of Microwave Heating Techniques to the Detoxification of Contaminated Soils," Nuclear Waste Management IV, Ceramic Transactions (G.G. Wicks, D.F. Bickford and L.R. Bunnell, eds.), American Ceramic Society, Vol. 23, pp. 761-768 (1991).
11. Oda, S.J., "Microwave Applications to Process Nuclear and Hazardous Waste," Microwaves: Theory and Application in Materials Processing II, Ceramic Transactions (D.E. Clark, W.R. Tinga and J.R. Laia, eds.), American Ceramic Society, Vol. 36, pp. 73-79 (1993).
12. Peterson, R.D., Johnson, A.J. and Swanson, S.D., "Application of Microwave Energy for Solidification of Transuranic Waste," Waste Management '88, Proceedings of Symposium on Waste Management (R.G. Post, ed.), pp. 325-330 (1988).
13. Peterson, R.D. and Johnson, "In-Drum Vitrification of Transuranic Waste Sludge Using Microwave Energy," Trans. Amer. Nucl. Soc., Vol. 60, pp. 132-133 (1989).
14. U.S. Environmental Protection Agency, Handbook- Vitrification Technology for Treatment of Hazardous and Radioactive Waste, EPA/625/R-92/002 (1992).
15. Wicks, G.G., "Nuclear Waste Glasses", Treatise on Materials Science and Technology, Glass IV, M. Tomozawa and R.H. Doremus, eds., Vol. 26, pp. 57-118 Academic Press, Inc. (1985).
16. Miyata, K., Ohuchi, J., Inada, E. and Tsunoda, N., "Incineration and Ash Melting for Plutonium-Contaminated Wastes at PWTF," presented at the Incineration Conference, Knoxville TN, May 1-5, (1989).
17. Ohuchi, M., Miyata, K., Ohuchi, J., Inada, E. and Tsunoda, N., "Melting of Pu-Contaminated Residues by Powering Microwave," Nuclear Waste Management III, Ceramic Transactions (G.B. Mellinger, ed.), American Ceramic Society, Vol. 9, pp. 113-121 (1990).

18. Morita, K., Nguyen, V.Q., Nakaoka, R., and Mackenzie, J.D., "Immobilization of Ash by Microwave Heating," Microwave Processing of Materials III, (R.L. Beatty, W.H. Sutton and M.F. Iskander, eds.), Materials Research Society, Vol. 269, pp. 471-476 (1992).
19. Davidovits, J., Schmitt, R.E. and Friehmelt, V., "Microwave Processing of Geopolymer- Cement Based Waste Forms," Microwaves: Theory and Application in Materials Processing II, Ceramic Transactions (D.E. Clark, W.R. Tinga and J.R. Laia, eds.), American Ceramic Society, Vol. 36, pp. 61-72 (1993).
20. White, T.L, Grubb, R.G., Pugh, L.P. Foster, D. and Box, W.D. "Removal of Contaminated Concrete Surfaces by Microwave Heating-Phase I Results," Waste Management '92, Proceedings of the Symposium on Waste Management (R.G. Post, ed.) pp.745-748 (1992).
21. Schulz, R.L., Fathi, Z. and Clark, "Microwave Processing of Simulated Nuclear Waste Glass," Microwaves: Theory and Application in Materials Processing, Ceramic Transactions (D.E. Clark, F.D. Gac and W.H. Sutton, eds.), American Ceramic Society, Vol. 21, pp. 451-458 (1991).
22. Schulz, R.L., Clark, D.E., Hutcheon, R.M. and Wicks, G.G., "Microwave Processing of Simulated Nuclear Waste Glass II," Microwaves: Theory and Application in Materials Processing II, Ceramic Transactions (D.E. Clark, W.R. Tinga and J.R. Laia, eds.), American Ceramic Society, Vol. 36, pp. 89-97 (1993).
23. Schulz, R.L., Folz, D.C., Clark, D.E. and Wicks, G.G., "Microwave Destruction/ Vitrification of Electronic Components," Microwaves: Theory and Application in Materials Processing II, Ceramic Transactions (D.E. Clark, W.R. Tinga and J.R. Laia, eds.), American Ceramic Society, Vol. 36, pp. 81-88 (1993).
24. Schulz, R.L., Folz, D.C., Clark, D.E. and Wicks, G.G., "Microwave Treatment of Emissions from the Destruction of Electronic Circuitry," to be presented at the 97th ACerS Annual Meeting in the symposium on "Microwaves: Theory and Application in Materials Processing III" and to be published in the proceedings, Cincinnati OH, April 30-May 3 (1995).

RADIO FREQUENCY ENHANCED SOIL VAPOR EXTRACTION FOR IN SITU REMEDIATION

John A. Pearce, David E. Daniel , Jana Boerigter, and Andres Zuluaga, The University of Texas at Austin, Austin TX, 78758

ABSTRACT

Traditional forms of soil vapor extraction seek to volatilize and remove organic contaminants by percolation of environmental air through the contaminated soil volume. As a practical matter this is effective for contaminants with vapor pressures in excess of about 1 torr at soil temperatures, below about 20 °C. The list of potentially susceptible contaminants can be expanded by heating the soil to increase the vapor pressure. Steam heat may be used, but its effect is limited to established vapor passageways. Radio frequency heating has many advantages because of its ability to heat volumetrically. However, the RF absorption of soil types, and thus the observed heating rate, varies widely depending on soil constituents and moisture content.

We present measurements of soil RF absorption and contaminant removal rates, as projected from measurements of soil water removal, in a measured RF electric field at 27 MHz. Heat transfer effects in the small laboratory fixture are estimated from additional experiments and applied to estimate the temperatures which might be achieved in a large volume field test at the same electric field. Though the uncertainty is high, the desired temperature range (in excess of 150 °C) is achievable in most soil types.

INTRODUCTION

Soil Vapor Extraction (SVE) has been used successfully to remove volatile organic compounds from contaminated soils. The initial applications of this method utilized ambient air percolated through quasi-porous soils driven by suction pressures applied to extraction well bore holes [1]. Results vary significantly among test sites; however, volatile organic compounds can be removed *in situ* with this technique.

The dominant mechanisms are: 1) direct vaporization of water and organic compounds, and 2) facilitation of indigenous soil aerobic microbe activity (i.e. bio-remediation). Both mechanisms may be enhanced by heating, though to very different temperature ranges. Bioremediation, using indigenous or implanted

---

microbes, is enhanced at temperatures up to about 35 to 40 °C. Volatilization of contaminants may require temperatures up to 150 to 200 °C depending on the vapor pressure of the material.

Microbic activity is enhanced by temperature elevation; however, the elevation must be gradual and sustained for long times so that the microbes can acclimate to the higher temperatures. Above the 35 to 40 °C range most biological organisms develop heat shock proteins and, at higher temperatures, biological activity is reduced as enzymatic processes degrade and cell death occurs. Even if temperatures are kept low, many months (and several generations) may be required for the organisms to acclimate to the elevated temperatures. Consequently, radio frequency (RF) heating is not an economically attractive method for raising soil temperatures to apply bioremediation.

Contaminant vaporization is another matter. Vaporization SVE works well when the vapor pressure of the target contaminant is above about 1 torr. As can be seen in Table I, there are many common compounds for which SVE may be used even in cool soils (environmental soil temperatures rarely exceed 20 °C). Substantial heating is required to remove several of the important contaminants, however.

Table I: Volatility of Typical Organic Contaminants

Compound	Vapor Pressure (torr) At 20 °C	Temperature (°C) for Vapor Pressure = 1 torr
Methylene Chloride	349	-70
Vinyl Acetate	58	-44
Benzene	76	-37
Toluene	14	-27
1,2 Dichlorobenzene	1	20
Phenol	0.2	40
Naphthalene	0.05	53
Antracene	0.0002	145

In field studies SVE has been shown to be enhanced by electromagnetic heating, most practically at radio frequencies (6.78, 13.56 or 27.12 MHz). Two applicator designs have been used in field tests: the IITRI "triplicate" array [2], and the KAI dipole antenna applicator [3]. For either electrode design the results depend critically upon the electric field strength in the soil.

Interpretation of the results obtained in the experiments can be augmented by a quick review of the governing relations — a more substantial treatment may be found in reference [4]. In the absence of phase transition and surface losses, the soil temperature is determined by the relative importance of the volumetric power deposition (Q_{gen}, W/m^3), local heat transfer (first term on the right hand side) and soil thermal properties according to the First Law of Thermodynamics:

$$\rho c \frac{\partial T}{\partial t} = \nabla \bullet \{k \nabla T\} + Q_{gen} \tag{1}$$

where: T is the soil temperature (K), ρ the density (kg/m^3), c the specific heat (J/kg/K) and k the thermal conductivity (W/m/K). Heating in an electromagnetic field may be due to resistive (Ohmic) dissipation, dielectric heating and magnetic heating. In most soils and for most RF applicators we may neglect heating due to magnetic permeability and magnetic field terms. The dominant volume power deposition then comes from a combination of Ohmic and dielectric heating:

$$Q_{gen} = \mathrm{Re}\{ -\nabla \bullet \mathbf{S}\} = \mathrm{Re}\left\{(\sigma + j\omega\varepsilon)\,|\mathbf{E}|^2\right\} \tag{2}$$

where: **S** is the electromagnetic field fluence rate (Poynting vector, W/m^2), σ is the soil electric conductivity (S/m), ω is the angular frequency (r/s), ε is the soil complex electric permittivity (F/m), **E** is the electric field (V/m) and Re{•} refers to the real part of the complex argument. We may recast the Poynting Power Theorem into a form more easily applied to the data by using the imaginary part of the complex electric permittivity ε'', where $\varepsilon = (\varepsilon_r' - j\varepsilon_r'')\,\varepsilon_0$ and ε_r'' is the so-called "loss factor":

$$Q_{gen} = \left(\sigma + \omega\varepsilon_r''\varepsilon_0\right)|\mathbf{E}|^2 \tag{3}$$

Even though conductivity only describes losses due to translational motion of charge carriers and the imaginary part of the electric permittivity only includes losses due to rotational and vibrational motion, one cannot separate the two loss mechanisms in experimental studies. Consequently, the heating term, Q_{gen}, is often expressed using "effective" conductivity, $[\sigma]_{eff}$, or loss factor, $[\varepsilon_r'']_{eff}$ which combines the separate mechanisms:

$$Q_{gen} = [\sigma]_{eff}\,|\mathbf{E}|^2 = \omega\left[\varepsilon_r''\right]_{eff}\varepsilon_0\,|\mathbf{E}|^2 \tag{4}$$

In many RF drying studies, conductivity losses dominate the heating by a wide margin and dielectric heating is usually ignored. In soil heating, the two mechanisms co-exist and dominate at different times during an experiment. Initially, when the soil is moist, translational motion of the ions through the soil matrix (Ohmic heating) is the dominant heating mechanism. As the soil dries out and the indigenous ions are not free to move, rotational motion assumes the dominant role. The distinction is important to this discussion for several reasons. First, the soil water is thermodynamically important since it prevents temperatures from exceeding 100 °C by very much until it has been completely vaporized. In fact, one may develop some insight into the binding energy of hygroscopic soils by noting the temperature at which steady state boiling occurs — high binding energy soils boil at temperatures up to the 110 to 140 °C range. Second, all soils heat much more effectively when moist since conductivity losses dominate dielectric loss by at least an order of magnitude at RF frequencies, as will be shown in the experimental studies. Consequently, very high electric field strengths may be required to achieve the desired temperatures for the low volatility organic compounds of Table I. In fact, it is entirely likely that some soil/contaminant combinations may not be amenable to RF-SVE; either because underground water migration limits the maximum achievable temperature thermodynamically, or the loss factor of dry soil is too low to permit heating to exceed thermal conduction losses once the soil moisture is exhausted.

The objectives of this study were to: 1) study the effect of soil moisture on achievable temperatures and heating rates, 2) to evaluate the relative importance of heat transfer processes in the laboratory samples, 3) to compare results obtained in laboratory scale heating experiments to field data, and 4) to compare heating histories in hygroscopic and non-hygroscopic soils.

LABORATORY SCALE EXPERIMENTS

Experiments were conducted on soil samples of: glass beads (1 mm dia), builder's sand, fine sand, roadbase, sandy loam, 15% kaolin clay-sand and landfill clay with varying amounts of water. Cylindrical soil samples of 513 cm^3 volume (about 700 g) were heated in uniform RF electric fields (at 27.12 MHz) of 1 to 6 kV/m (rms) generated by a Burdick medical diathermy device (27 MHz). Applied voltages were measured by means of a differential capacitive voltage divider. The power densities used in the tests ranged from 2 to 6 kW/m^3 in dry soils and 170 to 250 kW/m^3 in soils with 10% moisture. These values compare favorably to volume-average field test levels which range from 593 W/m^3 (Savannah River Site) to 2.8 kW/m^3 (Volk AFB site). Temperature (3 Luxtron probes), mass, air flow rates and sample permeability were recorded. Figure 1 illustrates the experiment geometry. Airflow rates were between 0.3 and 2.5 L/min., about 10 to 200 times the comparable rates achieved in soil field tests.

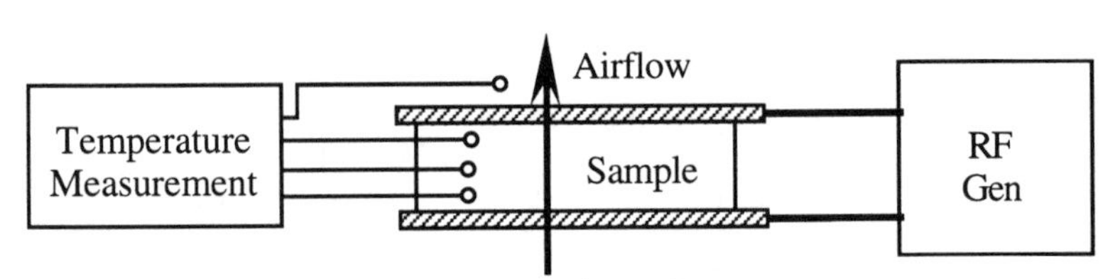

Figure 1: Laboratory experiment geometry.

Overall, significant drying usually did not occur until 100 °C was approached in the sample. Temperatures in excess of 100 °C were observed after the free water was exhausted and bound water vaporized. After all soil water was lost the temperatures decreased significantly, approaching a steady state value somewhat below 100 °C. Surface loss heat transfer in the small laboratory sample dominated power deposition, so the steady state temperatures were less than would be observed in a field test.

Typical heating curves are shown in Figure 2 for builder's sand. All three experiments in the figure were conducted at the same electric field strength, 4 kV/m (rms). Thus, the steady state temperature was the same for the three experiments, about 40 °C. The drying curves (not shown in the figure) were clearly in the falling rate period for all tests conducted owing to the low initial water concentration. A decrease in soil RF absorption coefficient manifests as a decrease in temperature in the curves for 5% and 10% soil moisture. This is directly traceable to the elimination of soil water, as previously described. For this material the dielectric loss is much less than the conductivity loss in moist soil.

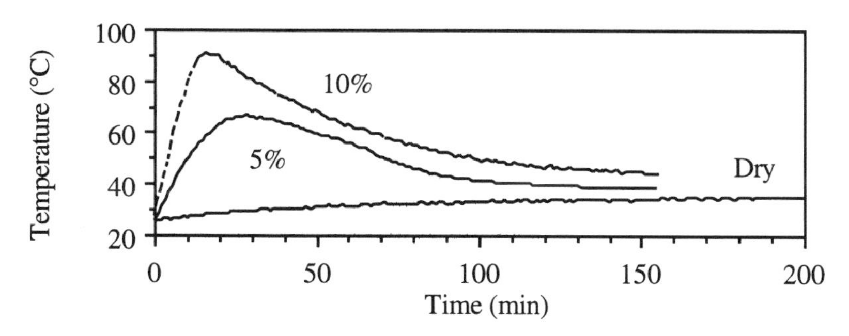

Figure 2: Sample heating data for dry builder's sand and with 5% and 10% water (dry weight basis) ($|\mathbf{E}| = 4$ kV/m).

These small samples are very sensitive to local heat transfer phenomena. Deposited RF energy is lost to the sample airflow and by heat transfer from the sample chamber surfaces. The samples were about 6 cm thick and convection heat transfer from the sample surface amounted to about 70% of the absorbed power in the dry soil, which is the origin of the decrease in temperature after elimination of the soil water. We expect heating rate decreases in soils as a natural consequence of the drying process. Because of the much larger volumes and different geometry involved higher steady state temperatures would be obtained at a field site than were measured in this experiment.

Hygroscopic soils exhibit substantial binding energies. The heating curves in Figure 3 show the excess binding energy of clay, which results in a much higher maximum temperature (130 °C) than in either roadbase or fine sand, even though the field strength was much less. The RF dielectric loss of dry clay is also much higher than the other materials as can be seen from its higher steady state temperature.

It might at first appear that the soil temperatures which can be achieved using RF heating are inadequate to treat the low volatility materials in Table I. Since dielectric loss is very weak in dry soils, the small scale of the laboratory measurements has somewhat distorted the heating picture. The steady state temperatures which might be expected in a field test were estimated from the small sample measurements from tests on dry soils in the same chamber. In these calculations the power generation, Q_{gen}, was determined from the heating rate during the initial heating phase.

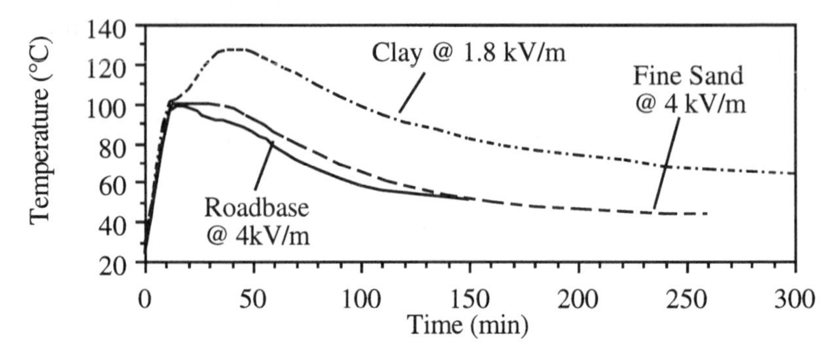

Figure 3: Heating data for fine sand, clay and roadbase at 10% water content.

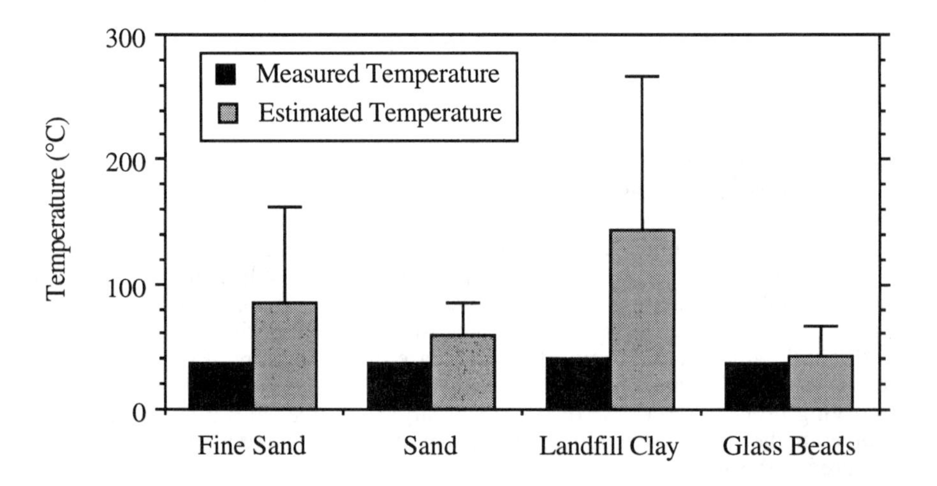

Figure 4: Estimated steady state soil temperatures in the absence of heat transfer losses. Measured steady state temperatures were determined at 2 ± 0.4 kW/m^3.

At the beginning of heating the Fourier conduction term in Equation 1 is negligible since the local thermal gradient, ∇T, is small. The temperature thus increases linearly with time, initially, and Q_{gen} may be calculated if ρ and c are known.Likewise, the combined surface and airflow loss terms can be determined from the steady state temperature since the left hand side of Equation 1 is zero. We have measured inlet and outlet air temperatures so that the RF power used to heat the air may be subtracted from the total loss to estimate the sample chamber heat transfer. The calculations of negligible loss steady state temperatures were made assuming losses to local airflow

These calculations may be used to establish an expected upper bound steady state temperature for the case of a field test (see Figure 4). The calculations represent a maximum upper bound because Fourier conduction and surface losses are less important but not zero in a large volume field test. The uncertainty in the calculation is large (bars), but much higher steady state temperatures are expected if the heat transfer losses are reduced. The uncertainty was estimated from analysis of variance by applying the appropriate functional relationships. The majority of the uncertainty arises from uncertainty in soil thermal properties.

CONCLUSIONS

The RF absorption coefficient, and thus heating rate, of soil is very sensitive to local water content. Heat transfer and water vaporization processes are very complex and heterogeneous in a typical soil and represent limiting factors in both laboratory scale and field scale studies. Field tests indicate that RF significantly enhances soil vapor extraction processes. In this application, the technical factors favor RF heating over conventional heating and the economic factors are strongly in favor of *in situ* treatment vs. removal of the soil. So, providing that the soil water mass transfer, electrical, and thermal properties are appropriate, RF enhancement of soil vapor extraction represents a viable option in the remediation of contaminated soils.

ACKNOWLEDGMENT

This project has been funded in part by Federal Funds as a part of the program of the Gulf Coast Hazardous Substance Research Center which is supported under cooperative agreement R815197 with the United States Environmental Protection Agency. The contents do not necessarily reflect the views and policies of the US. EPA, nor does mention of trade names or commercial products constitute endorsement or recommendation for use.

REFERENCES

1. Hutzler N.J., Murphy B.E. and Gierke J.S.; "State of technology review: Soil vapor extraction systems" *US. EPA Risk Reduction Engr. Laboratory*, Cincinnati OH, Publ. No. EPA/600/2-89/024 (87 pages), 1989.
2. Dev, H. "Radio frequency enhanced in situ decontamination of soils contaminated with halogenated hydrocarbons" *Proc. US. EPA 12th Ann. Res. Symposium*, Publ. No. EPA/600/9-86/022, pp402-412, 1986.

3. "VOCs in non-arid soils, integrated demonstration" *US. DOE Office of Environmental Management, Office of Technology Development* Report No. DOE/EM-0135P, February, 1994.
4. Roussy G. and Pearce J.A.; Foundations and Industrial Applications of Microwaves and Radio Frequency Energy: Physical and Chemical Processes. Wiley-Interscience-Europe, Chichester UK, 1995.

MICROWAVE REGENERATION OF ACTIVATED CARBON FOR GOLD PROCESSING APPLICATIONS

J.T. Strack, I.S. Balbaa, B.T. Barber
Ontario Hydro Technologies
800 Kipling Avenue, Toronto, Ontario, Canada M8Z 5S4

ABSTRACT

An advanced microwave system for continuously reactivating carbon, used in gold processing, has been demonstrated at a gold mine near Kirkland Lake, Ontario, Canada. The pilot unit produced reactivated carbon at a rate of 12 kg/hr. Test results indicate the microwave regeneration technology has several clear advantages over conventional regeneration kilns.

The microwave regeneration process was developed by Ontario Hydro Technologies and CANMET, in conjunction with two Canadian companies: Lochhead Haggerty Engineering and Manufacturing Co. Ltd, and Barrick Gold Corporation. This work was conducted with financial support from the Canada/Northern Ontario Development Agreement (NODA) and Ontario Hydro Technologies. It has application in gold processing and other industries such as water treatment.

BACKGROUND

In gold mining, granular activated carbon is a key ingredient used in the gold concentration process. Gold is leached from the ore and adsorbed by the activated carbon, as a gold-cyano complex, using a technique known as the "carbon-in-pulp" (CIP) process. The "loaded" carbon is removed from the adsorption circuit and the gold-cyano complex is eluted from the carbon. Before being returned to the gold recovery circuit, the carbon passes through a regeneration step which uses elevated temperatures (~700°C) to condition it for the next adsorption cycle (1). The high temperature drives out organic compounds, and expands the carbon pores, which facilitates gold adsorption. Regeneration is carried out in kilns using electricity, natural gas, or other fuel. Regeneration kilns use significant amounts of energy, and contribute to carbon losses, mainly through breakage of the granular material. Replacement carbon is valued at over $2000[1] per ton, and represents a significant operating cost.

With financial support from CANMET, the Province of Ontario, through the Northern Ontario Development Agreement (NODA), and Ontario Hydro Technologies, the Canadian development team designed, assembled and operated a new microwave-based regeneration kiln. The final evaluation took place at Barrick's Holt-McDermott gold mine, near Kirkland Lake, Ontario. The pilot plant was operated on a continuous basis and handled about 10% of the total carbon requiring reactivation at the gold mill. The balance of the carbon was reactivated through the existing electrically-heated conventional rotary kiln. It was installed in the mill building next to the conventional kiln. Operating the two regeneration kilns side-by-side, allowed the direct comparison of the two technologies, and eliminated many of the variables which normally makes performance comparisons difficult.

PILOT-PLANT DESIGN AND OPERATION

The pilot-scale carbon regeneration circuit is designed to deliver 12 kg/h of regenerated carbon. This process is based on pre-drying the spent-carbon using hot air, followed by heating to reactivation temperatures using microwaves. The pilot-plant skid is illustrated in Figure 1. The skid is fabricated from structural steel sections to form two levels. The upper level holds the feed-hopper, electric furnace, pre-dryer, discharge hood, fan, and associated controls. The lower level holds the microwave power supply and waveguide, the microwave reactivation vessel, carbon control valve, and a quench tank.

The pre-dryer is based on a hot-air rotary kiln and electric furnace design. The feed hopper is loaded with batches of moist carbon (~40%, wet basis) typically weighing 530 kg. Any excess water in the carbon drains from the hopper through a pipe located at its bottom. A feed screw delivers carbon from the bottom of the feed hopper into the rotary drum where it mixes with hot air from the 15 kW electric furnace. Carbon leaves the dryer drum (~8% moisture content) and drops into a discharge hood, which also functions as an intermediate storage vessel between the pre-dryer and the microwave reactivation chamber. In the microwave applicator, the carbon is exposed to about 10 kW of 915 MHz microwave power while flowing from the top to the bottom of the applicator. The moist carbon is completely dried and then reactivated at temperatures near 700°C. The rate of carbon flow through the microwave applicator is controlled by a special high temperature rotary valve. Hot reactivated carbon is then discharged into water in the quench tank.

---

[1] All dollar values in this paper are in 1994 Canadian dollars.
Currently, $1.00 Canadian = $0.73 U.S.

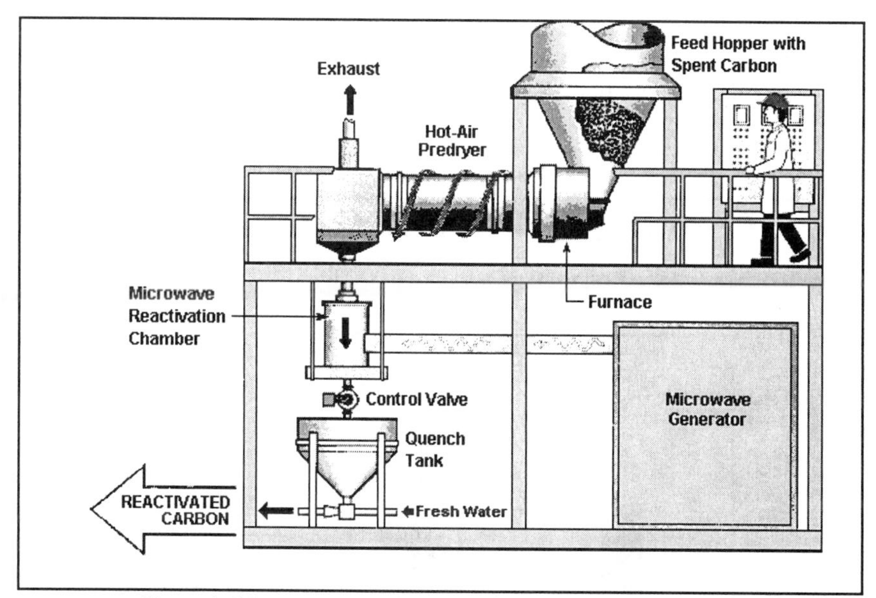

Figure 1: Microwave Reactivation Kiln with Hot-Air Pre-drying

CARBON PROPERTIES

Tests on the reactivated carbon from the two kilns were performed at the mine. The tests included carbon activity, sizing and hardness. A summary of the test results is shown in Figure 2.

Both methods proved to be equally effective in regenerating the carbon for use in the gold recovery circuit (Note: values of greater than 95% are considered good; values of greater than 100%-virgin indicate regenerated carbon is capable of performing better than new material). However, in the area of carbon losses, the kilns behaved differently. Barrick's laboratory sizing measurements on carbon samples from the two kilns indicate that the microwave process is gentler on the carbon during reactivation than the conventional kiln. As can be seen in Figure 2, carbon from the conventional kiln lost an average of 11% to undersize (14 mesh), while carbon from the microwave system lost an average of 9% to undersize; a reduction of 2%. Similarly, the microwave kiln produces a tougher, or harder carbon product than the conventional kiln, which is expected to resist attrition or breakage as it passes through the CIP circuit. Laboratory measurements of attrition losses (i.e., 100% - %hardness) after reactivation gave average values of 4.5% losses for conventionally reactivated carbon, and 2.7% losses for microwave reactivated carbon; a reduction of 1.8%. In actual operation, with the conventional kiln, Barrick loses about 7.5% of the granular carbon each regeneration cycle. Overall, the laboratory measurements indicate that the microwave kiln will reduce these losses by 3.8% absolute (i.e., 2% to undersize plus 1.8% to attrition). This approximate halving of the carbon losses represents a savings of 40 kg of carbon per ton processed.

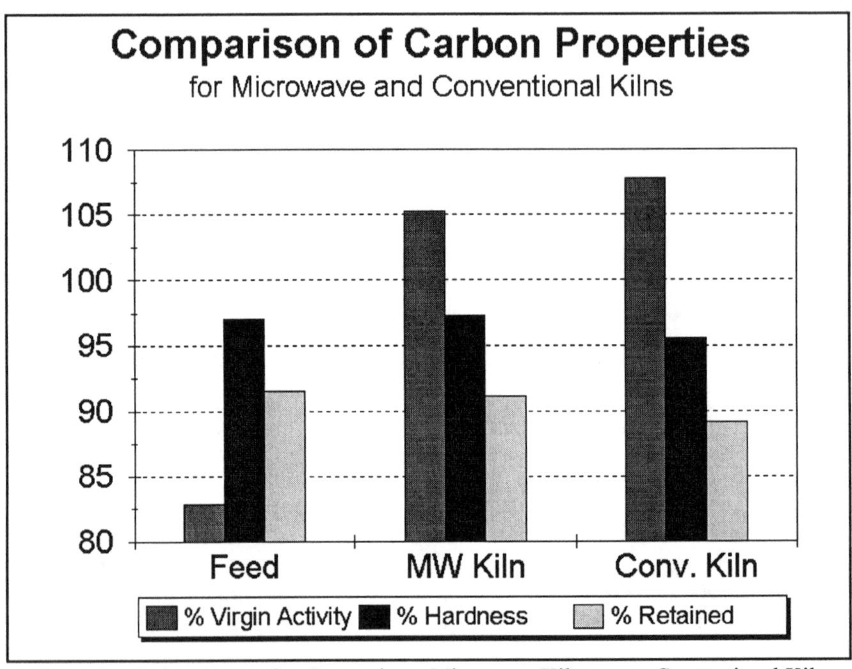

Figure 2: Carbon Properties Comparison: Microwave Kiln versus Conventional Kiln

ENERGY CONSUMPTION

An important aspect of evaluating the microwave technology is the comparison of the energy consumption for the two side-by-side systems. The microwave pilot plant was equipped with two kilowatt-hour meters which measured the electricity consumption for the microwave generator and the total pilot-plant. Kilowatt-hour meter readings were taken at the start and the end of four test runs using carbon from four different batches. Also, the main power feed to the conventional kiln was fitted with a kilowatt-hour meter. Meter readings were also taken, at the same time the meters of the pilot plant were read, to determine the kiln's energy-usage characteristic. The kiln is used to regenerate batches of carbon, each with a mass of 2700 kg (±100 kg). Each batch is processed continuously through the kiln at a flow rate of about 120 kg/h over a period of about 22½ hours. However, because the pilot scale microwave kiln produced only 10% of the carbon produced by the conventional kiln, this resulted in somewhat of an *"apples-and-oranges"* comparison. To compensate for this size mismatch, an estimate was made for the performance characteristics of a 120 kg/h microwave reactivation system based on the measurements taken and accounting for efficiency improvements resulting from scale-up.

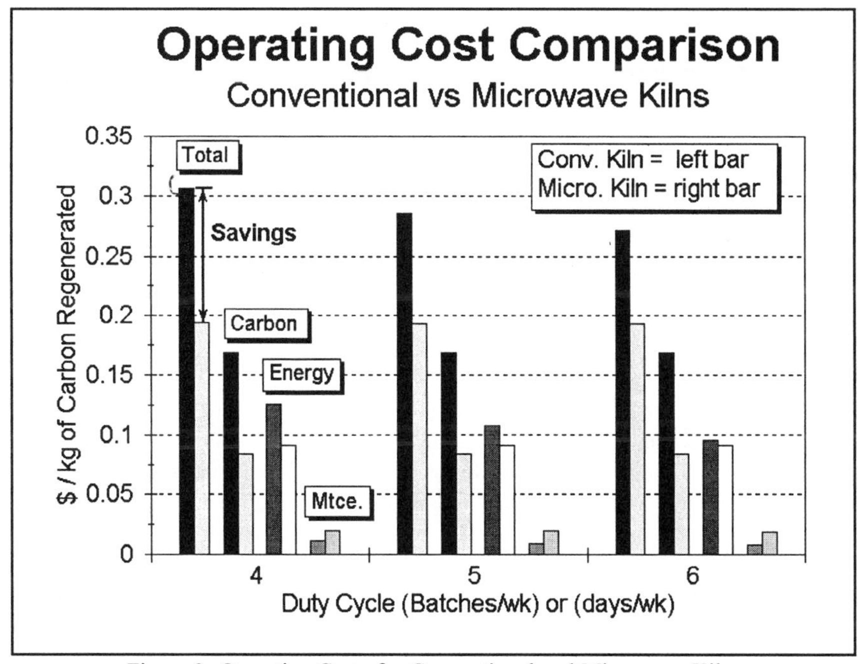

Figure 3: Operating Costs for Conventional and Microwave Kilns

The measured specific-energy consumption (i.e., energy used to reactivate one unit of carbon) of the microwave pilot-plant system was 2 kWh/kg of carbon processed, at a carbon flow rate of 12 kg/h. The measured energy split between the microwave generator and the pre-dryer was roughly equal. The microwave generator used 1.1 kWh/kg and the pre-dryer used 0.9 to 1.1 kWh/kg. When scaled-up to 120 kg/h, the microwave kiln is estimated to have an average specific-energy consumption of 1.65 kWh/kg of carbon, at annual production rates of 500 to 800 tonnes. The energy split in this case is estimated to be 0.99 kWh/kg for the microwave unit and 0.66 kWh/kg for the pre-dryer unit. In comparison, the calculated average specific-energy consumption for the conventional kiln is 2.3 kWh/kg at 520 tonnes per year (present operation: 4 days/week). For a production of 780 tonnes per year (6 days/week operation), the energy consumption is expected to be 1.7 kWh/kg. A significant factor affecting the conventional kiln's specific energy value is the standby loss which becomes a greater proportion of total consumption at lower duty cycles. The microwave kiln can be turned off when not in use, and requires only a short heat up time at the start of each regeneration cycle.

During the test period, the conventional kiln was regenerating four batches of carbon per week. When the kiln was not regenerating carbon, it was kept at temperature, or "idled." Power consumption dropped during the idling period, but was still significant. Kilowatt-hour meter readings were taken on both a daily basis and a batch-to-batch basis to determine the specific energy consumption, and the idling power for the kiln. The specific energy usage for the conventional kiln during the regeneration cycle of carbon (~22.5 h) was about 1.5 kWh/kg. When the kiln was kept at temperature but not used for carbon reactivation, the actual energy consumption was 110 kWh per hour of idle time.

On the other hand, the microwave kiln does not need to be kept at operating temperature, or idled, during the time between regeneration cycles, but rather can be shut down to lower overall energy consumption. Before each regeneration cycle, the unit's pre-dryer was preheated for about two hours. The microwave reactivation chamber did not need to be preheated, and the microwave system was started once carbon flow was established.

OPERATING COST COMPARISON

Operating costs for both the microwave and conventional rotary kilns were calculated for three different operating scenarios: 4, 5, and 6 day/week operation. Each kiln was assumed to have a 120 kg/h rating and the performance characteristics described earlier. Operating costs include price of replaced carbon, energy consumed and other consumables such as kiln parts and maintenance. The results are summarized in Figure 3. The microwave kiln has lower operating costs for all three scenarios as compared to the conventional technology. It has lower energy consumption and reduces carbon losses during reactivation. The cost of consumables is projected to be higher for the microwave system (magnetron tubes, and pre-dryer elements & seals, etc.) than the conventional kiln (electric elements, and kiln seals, etc.). This increased cost for microwave maintenance is more than compensated for by savings in energy and carbon consumption. At the current four-batch per week operation, the average cost of carbon regeneration for the gold mine would drop by a 1/3, from about $0.31/kg to $0.20/kg of carbon. This represents annual savings of $58,500 at processing levels of 520 tonnes per year.

EQUIPMENT COST

Equipment cost for the full scale microwave system was also estimated based on information from equipment suppliers[2]. The installed cost of the new technology is estimated at about $650,000 for a unit with a carbon processing capacity of 120 kg/h. This cost includes the pre-dryer price ($140,000); a 90 kW microwave system and

---

[2] Capital cost estimates have a confidence factor of ±30%.

associated components ($130,000); reactivation chamber ($40,000); installation, connections, utilities, and site preparation ($165,000); project design, and supervision ($100,000); and a 15% contingency budget ($75,000). On the other hand, the installed cost of the conventional kiln is estimated, in a similar way, to be $575,000. The additional cost for the microwave equipment would be recovered from savings in the first 15 months of operation.

ACKNOWLEDGEMENTS

The authors would like to acknowledge the contributions of K. Haque (CANMET) for his overall guidance in the project, and his assistance in assessing carbon properties; J. von Beckmann (Lochhead Haggerty) for sharing his experience with carbon processing, and his assistance throughout the project; L. Buckingham and W. Lupton (Barrick) for their assistance at the mine site; M. Rosen, R. Solczyk and A. Goldenberg (OHT) for their assistance in equipment preparation.

REFERENCES

1. Haque, K.E. & MacDonald, R.J.C. (CANMET); Oda, S.J. & Balbaa, I.S. (Ontario Hydro Research Division), **The Microwave Regeneration of CIP Spent Carbon**, paper presented at the annual Canadian Mineral Processors (CMP) Conference, Ottawa, (1993).

MICROWAVE TREATMENT OF EMISSIONS FROM THE DESTRUCTION OF ELECTRONIC CIRCUITRY

R.L. Schulz, D.C. Folz and D.E. Clark
University of Florida
Department of Materials Science & Engineering
Gainesville, FL 32611

C.J. Schmidt
University of Florida
Department of Environmental Engineering
Gainesville, FL 32611

G.G. Wicks
Westinghouse Savannah River Co.
Aiken, SC 29802

ABSTRACT

Microwave energy has been used successfully to decompose organic constituents, vitrify remaining materials and recover metal resources from a wide variety of waste electronic circuitry components. During the combustion phase, large amounts of volatiles are emitted. A microwave off-gas system has been fabricated in an attempt to destroy or breakdown (to less harmful forms) these emissions. Gas chromatography analysis results on emissions prior to entering and after leaving the microwave off-gas treatment chamber are presented.

INTRODUCTION

The objective of this study was to determine the effects of microwave heating on off-gas emissions from the combustion of printed circuit boards

(PCBs). During previous experiments conducted to recover metals from the PCBs, extensive fumes and smoke were observed during initial combustions stages. To offset this potentially hazardous situation, a liquid nitrogen cold-finger apparatus was attached to a port on the microwave and fumes were pulled through the trap using a vacuum pump [1]. While this set-up was suitable for laboratory experiments, it did not appear to be practical for industrial-level operations. Therefore, a decision was made to construct a microwave off-gas treatment system [2-3]. Several other researchers have investigated the use of microwave energy to treat hazardous/toxic gases [4-11]. Some of the hazardous gases successfully treated using microwave energy include: dioxins [4], hydrogen sulfide waste gases produced as a result of oil refinement and purification of natural gas [7], sulfur dioxide and NO_x emissions from coal-fired utility stacks [8] and trichlorethane [11].

EXPERIMENTAL

The procedures used to prepare the PCBs for combustion/vitrification and metal recovery are described in the literature [1-3]. Prior to combustion, the waste boards are sectioned with a ban saw, crushed using a Braun Chipmunk pulverizer and weighed. The crushed mixture is placed in a fused silica crucible for processing. Approximately 80-100g of mixture can be processed in 30 minutes with the current set-up. This is limited primarily by the size of the available crucibles. Figure 1 shows the microwave combustion and off-gas treatment system. The system consists of two 900W, 2.45GHz microwave units connected in tandem. Organic components of the PCBs are volatized in the combustion microwave. During the burning operations, compressed air is forced into the cavity to facilitate removal of the gases from the combustion chamber. Air is directed at the sample and the opening to the off-gas interconnect using a stainless steel braided wire sheathed tube. As combustion occurs, the emissions are forced out of the combustion chamber, through the off-gas interconnect and into the microwave off-gas treatment system. The off-gas treatment system consists of a mullite tube partially filled with 16 grit β-SiC particles. Phosphate bonded reticulated alumina filters were placed at either end of the mullite tube to increase emission dwell time. The SiC was pre-heated to temperatures between 1000-1200°C prior to beginnig the combustion run. Two sampling ports allow collection of gases prior to and after microwave treatment.

Several different methods were used to collect the gas samples in order to determine which method was best suited for this study. Tedlar bags, gas-tight syringes and Tenax air trap tubes were tested. Gas bags are constructed of 2 mil

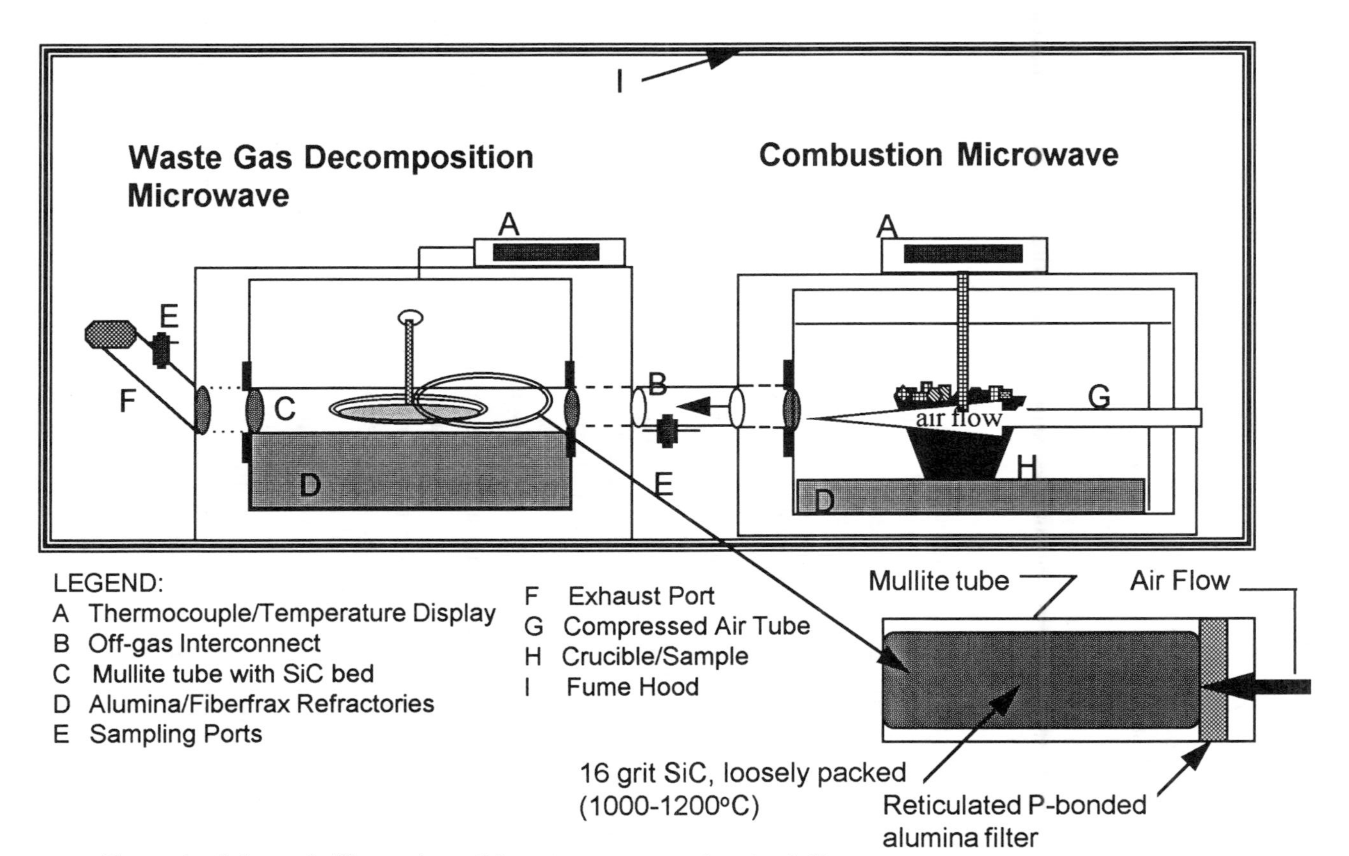

Figure 1. Schematic illustration of the microwave combustion/off gas treatment system.

Tedlar film that is unaffected by most compounds and impermeable to gas exchange. The bags are equipped with a septum for sample extraction using a syringe and an on/off valve to permit sample collection through a 6.0 mm I.D. Nalgene tube. Luer-lock gas-tight syringes (50ml) fitted with an off/off valve were used to collect and store gases prior to analysis. The air traps consist of a borosilicate glass tube packed with a polymeric material referred to as Tenax. Prior to use, the air tubes are conditioned in a nitrogen atmosphere at 210° for an hour to remove any binders that may remain after initial processing.

Although successful in obtaining a sample, gas syringes were eliminated because it was decided that a random specimen would probably not be representative of the total emissions evolved during the burning operation. Tedlar bags permitted collection throughout the run, however, the air flow through the system was slow, and only small sample volumes could be collected. Tenax air traps may be the most suitable collection method for this operation. Following the conditioning treatment described earlier, the air traps were connected to the sampling ports where the gases passed through and were absorbed in the polymer material throughout the entire run (approximately 30 minutes). Once the run was complete, the samples were placed on ice and transported to the gas chromatography lab for analysis. During analysis, the tubes are re-heated to desorb the gases.

A Hewlett-Packard 5840 gas chromatograph (GC) interfaced capillary direct to a Hewlett-Packard 5985B mass spectrometer was used for sample analyses. The GC column was a J&W DB624 (30m x 32mm (i.d) x 1.8μm film thickness) cryofocused to -80°C using a Ruska 9730 cryogenic focusing system. The basic operating parameters for the GC were as follows: initial temperature = 30°C, injector temperature = 220°C, interface temperature = 200°C, ramp rate = 5°C/min and a final temperature of 165°C. The mass spectrometer was operated in the electron impact positive mode and neutral molecules were ionized with 70eV of energy. Additional instruments were required to analyze the emissions contained in the Tenax air tubes. The air tubes were attached to a OI Analytical Discrete Purging Module (DPM-16) and were heated using an OI Analytical Multiple Heater Controller (MHC-16). The purge and trap cycle was controlled using an OI Analytical #4560 sample concentrator. The trap within the concentrator was the VOCARB 3000. During the purge cycle, the air tube is purged and heated to remove any oxygen or water vapor accumulated in the tube during the combustion process.

RESULTS

Results of the run where the emissions were collected using two techniques (gas syringe (pre-treatment) and Tedlar bag (post-treatment)) are summarized in Table I. As can be seen from the data, there was a significant reduction in the concentration (at least two orders of magnitude) and number or organic components found in the post-treatment samples. This would indicate that the process was fairly successful. Results of the gas chromatography data where both samples were collected using air traps are summarized in Table II. It was observed that concentrations for the air trap samples both pre- and post-microwave treatment are much higher than those observed for the previous two techniques used. This is most likely due to the fact that characterization was not limited by a random 25ml sample, as was the case for both the Tedlar bag and gas syringe, because gases were absorbed in the Tenax throughout the entire combustion process. The trend toward a reduction in concentration of 2 orders of magnitude is consistent with the data in Table I. However, this was the initial test using this collection method (for the microwave researchers and GC personnel) and additional experiments are necessary to verify these preliminary results.

It is unknown at this time if a portion of the organic constituents are condensing in the off-gas interconnect tube prior to treatment. It has been suggested that heating this tube would assure that condensation of the gases does not occur. Inspection of the inter-connect did not reveal a significant build-up of combustion by-products. However, the inner surface of the tube was rusted, indicating its exposure to corrosive compounds, water vapor and heat.

SUMMARY

In summary, this following points can be made:

- ➢From preliminary GC mass spectroscopy data, it appears that microwave treatment of emissions from the combustion of electronic circuitry reduced both the concentration (by approximately two orders of magnitude) and number of organic constituents found in the gas wastestream.
- ➢ Additional experiments are necessary to verify as well as quantify these results.
- ➢Continued design improvements are necessary to further increase the efficiency of the process.

Table I. GC Spectroscopy Analysis of Emissions Prior To and After Microwave Treatment.

Compound	Concentration ($\mu g/m^3$)	
	BEFORE *(gas syringe)*	AFTER *(Tedlar bag)*
1,3 Butadiyne	30.30	nd**
Chloromethane	18,73	nd
Bromomethane	36.89	nd
Ethylbenzene	6.25	nd
Styrene	5.71	nd
2-Butene, (E)	9.66	nd
1,3 Butadiene, 2-methyl	2.91	nd
1,2 Pentadiene	2.73	nd
Ethane, bromo	3.93	nd
1-Pentene, 2-methyl	1.99	nd
1-Propene, 1-bromo	2.53	nd
1-Propene, 2-bromo	1.58	nd
Furan, tetrahydro	6.12	0.09
Furan, 2-methyl	33.94	nd
Butane, 1-bromo	18.49	nd
3-Octene, (E)	2.15	nd
p-xylene = dimethyl benzene	11.40	0.10
Benzene, 1,3,5 trimethyl (or)*	**1.69**	**0.09**
Benzene, 1-ethyl-3-methyl (or)	**1.69**	**0.08**
Benzene, 1,2,3-trimethyl	**1.69**	**0.09**
alpha-methylstyrene	1.70	nd
Benzene 1-chloro-3-methyl (or)	--	**0.13**
Benzene, 1chloro-2-methyl	--	**0.13**
Methylene Chloride	nd	0.07
Benzene (various forms)	--	0.04-0.11
Toluene	17.80	0.08

* Computer could not differentiate between compounds.

** Not detected

Table II. Preliminary GC Spectroscopy Data from Air Tube Collection Method Before and After Microwave Treatment.

Target Compound	Concentration ($\mu g/m^3$) Before (Tenax air tube)	Concentration ($\mu g/m^3$) After (Tenax air tube)
Chloromethane	3440	nd
Benzene	3573	nd
1,2,4-Trimethylbenzene	117	37
1,3-Dichlorobenzene	615	nd
1,2-Dichlorobenzene	nd	42
Chlorobenzene	nd	21
Napthalene	nd	3
1,1-Dichlorethane	nd	3

ACKNOWLEDGMENTS

The authors thank Westinghouse Savannah River Co. for partial financial support under subcontract AB-46395-0 and OI Analytical of College Station, Texas for technical assistance.

REFERENCES

1. R.L. Schulz, D.C. Folz, D.E. Clark and G.G. Wicks in, Microwaves: Theory and Application in Materials Processing II (D.E. Clark, W.R. Tinga and J.R. Laia, eds.), Ceramic Transactions, Vol. 36, The American Ceramic Society, Westerville, OH, pp. 81-88 (1993).
2. R.L. Schulz, D.C. Folz, D.E. Clark, G.G. Wicks and R.M. Hutcheon in, Proceedings of the 28th Microwave Power Symposium of the International Microwave Power Institute Symposium, Montreal, Quebec, pp. 9-18 (1993).
3. R.L. Schulz, D.C. Folz, D.E. Clark and G.G. Wicks in, Microwave Processing of Materials IV, Vol. 347 (M.F. Iskander, ed.), Materials Research Society, Pittsburgh, PA, pp. 401-406 (1994).
4. Chemistry & Industry (anonymous article), p. 440, June 21 (1993).
5. L. Dauerman, G. Windgasse, N. Zhu and Y. He in, **Microwave Processing of Materials III**, (R.L. Beatty, W.H. Sutton and M.F. Iskander, eds.),

Proceedings of The Materials Research Society, Vol. 269, The Materials Research Society, Pittsburgh, PA pp. 465-469 (1992).
6. L. Dauerman, G. Windgasse, N. Zhu and Y. He in, **Microwave Processing of Materials III** (R.L. Beatty, W.H. Sutton and M.F. Iskander, eds.), Proceedings of The Materials Research Society, Vol. 269, The Materials Research Society, Pittsburgh, PA pp. 465-469 (1992).
7. Mechanical Engineering (anonymous article), 113[11], p. 18, November (1991).
8. Design News (anonymous article), p. 30, July 4 (1988) .
9. Science (anonymous article) Vol. 257, No. 11, p.1479 (1992).
10. Process Engineering, anonymous article, p. 27, February (1991).
11. R. Varma, S.P. Nandi and J.D. Katz in, **Microwave Processing of Materials II**, Materials Research Society Proceedings, Vol. 189 (W.B Snyder, Jr., W.H. Sutton, M.F. Iskander and D. L. Johnson, eds.), Materials Research Society, Pittsburgh, PA, pp. 67-68 (1991).

Processing Equipment

MATERIALS PROCESSING VIA VARIABLE FREQUENCY MICROWAVE IRRADIATION

Richard S. Garard, Zak Fathi and J. Billy Wei, Lambda Technologies, 8600 Jersey Court, Suite C, Raleigh NC 27612. (919) 420-0275

ABSTRACT

Variable Frequency Microwave (VFM) irradiation is a new technique applied to advanced materials to overcome the reported problems with uniform scale-up and repeatable control of microwave energy inside a material. A broadband, variable frequency microwave source is used to provide tuning to an optimum incident frequency and then sweeping around that centeral frequency, as required, for uniform distribution of the microwave energy. This approach has been employed in evaluating microwave processing benefits for several advanced material systems. Results of these trials plus the comparison of variable to fixed frequency processing is reviewed.

INTRODUCTION

Alternative processes are being sought to improve on the cost effectiveness and the processing cycles of various advanced materials. Microwave irradiation has been investigated as an alternative processing means by which short process cycles are often obtained. The body of work carried out up to date demonstrates several advantages brought about by the presence of a microwave electric field. High frequency heating is an efficient means by which short process cycles, fast heating rates and enhanced reaction kinetics are obtainable. Microwave heating is more efficient than conventional thermal processing because microwave (electromagnetic) energy is directly coupled into the material at the molecular level, bypassing the air medium in the cavity chamber.

Research teams worldwide are investigating the use of microwave energy in the area of materials processing -- all of which use either single-mode or multi-mode, single-frequency microwave processing techniques. The demonstrated "global advantages" of microwave heating over conventional heating include direct and volumetric heating, high selectivity, fast heating rates, and reaction kinetics enhancement. This has been demonstrated in numerous systems on a laboratory scale. Unfortunately for advanced materials, these potentials have not yet been attained on an industrial scale.

Using variable frequency irradiation, the processing of high-performance materials has been targeted such as polymers, ceramics and monolithic composites. Variable frequency irradiation provides solutions to the inherent scale up limitations of fixed-frequency microwave irradiation while simultaneously producing high energy efficient coupling to the material to be processed.

THE VARIABLE FREQUENCY MICROWAVE CONCEPT

The concept behind variable frequency processing involves the control and distribution of microwave energy throughout the material being processed. There are four controllable processing parameters that distinguish Variable Frequency Microwave (VFM) from conventional, single frequency microwave based technologies. These controllable processing parameters include; central frequency, frequency bandwidth, frequency sweep-rate and power. Variable frequency processing enables precise frequency tuning to optimize the energy coupling efficiency during initial heating. During heating, continuous and/or selective sweep through several cavity modes within a period of a few milliseconds results in a time-averaged uniformity of heating throughout the load. Frequency sweeping improves heating uniformity through the superposition of many modes within the microwave cavity. This uniformity facilitates process scale up either to large components or to large batches of small components. The bandwidth required to achieve microwave energy uniformity inside the cavity depends on the operating frequency, the material chemistry (dielectric properties) and the size of the cavity. At high central frequencies, only a narrow bandwidth may be required to provide uniform energy distribution.

During processing, the central frequency irradiated inside the microwave cavity can be tuned to increase the coupling efficiency with the material to be processed. This frequency tuning capability is especially important for materials with frequency sensitive dielectric behavior. Thus, the heating rate can be increased without changing the power. The combination of bandwidth and sweep rate around the selected central frequency, utilized during processing, provides the necessary distribution of microwave energy to carry out uniform heating throughout the workload. The sweep rate is electronically controlled and can be varied from minutes to milliseconds. The microwave incident power can be pulsed or continuously varied to provide additional control over the heating profile of the workload.

The dielectric loss measurements, of a given material, over the range of temperatures and frequencies of interest provides the necessary feedback for frequency selection during heating in order to optimize the processing of different materials. In general, the orientation polarization loss peak (maximum losses) shifts to higher frequencies as the temperature of the material is elevated. Thus, there is a definite advantage in having the ability to initially tune and then vary the frequency of the irradiated microwave signal as the material being processed is heated.

The shift in frequency of the loss peak maxima associated with materials is temperature dependent. While many polymers and ceramics do not exhibit a strong dependence on the incident frequency, there are definite advantages in frequency tuning capability when processing strong microwave absorbers as is the case of drying applications and/or in a composite form such as processing of graphite fiber reinforced polymeric materials.

MATERIALS PROCESSING & RESULTS

An array of materials have been processed using variable frequency irradiation with emphasis on epoxies and polymer matrix composites. Drying, binder burnout as well as ceramic and ceramic composite, high temperature sintering are also being investigated. Of special interest in the ceramic field is the co-firing capabilities offered by controlling variable frequency irradiation. Microwave assisted CVD and other plasma based processes are also of interest for applying variable frequency microwaves. The ability to control plasma location and/or generate multiple plasma balls by launching multiple incident frequencies has generated interest in various industrial applications.

Experimental: The current microwave source used for variable frequency irradiation is a traveling wave tube (TWT). Variable frequency microwave systems ranging from 100 to 1,500 watts and over the frequencies of 0.9 to 18 Ghz, typically in one to two octave bandwidths are available. Model LT-0200 VFMF system has been utilized to process polymers and polymer composites in the 2.5 - 7.5 GHz frequency range. This furnace can be operated at a fixed frequency, or over the entire bandwidth, or any portion thereof may be swept at interval rates of 100 msec to 2 minutes. The furnace cavity is nominally a 12"x12"x10" stainless steel enclosure. A thin sheet of polyimide film was placed over the microwave launcher to prevent any volatiles from condensing in the waveguide, and potentially contaminating the power source and/or sensors.

Epoxies: A stoichiometric mixture of an epoxy resin ERL-2258 (Union Carbide) with M-phenylenediamine hardener (Dupont) (25 g MPDA per 100g resin) was cured using variable frequency irradiation. The MPDA was dissolved into the resin by mixing at 65 °C. The properties of the processed materials are reported elsewhere [1]. The recommended conventional curing for these epoxy samples is illustrated in Figure 1. The heating in this particular system is hard to control and slow heating rates are used because of the reaction exotherm, avoiding materials degradation and temperature excursions. In general, microwave energy can be pulsed as to eliminate exothermic reactions without slowing the down the heating rate, and has been shown to substantially shorten the process cycle [2]. The variable frequency approach was found to provide a means by which uniform and controllable heating can be achieved. The curing cycles utilized during variable frequency processing were 8 times faster than the conventional cycles. To illustrate uniformity, eight samples were configured as shown in Figure 2. The temperature was monitored in one representative sample. The temperature versus time profile is illustrated in Figure 3. The curing was uniform for all eight samples and within each and every sample.

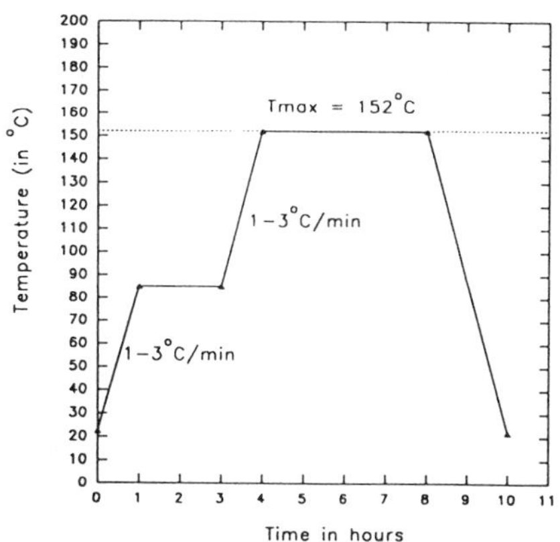

Figure 1. The recommended conventional curing cycle for the Union Carbide Epoxy Resin -- ERL 2258.

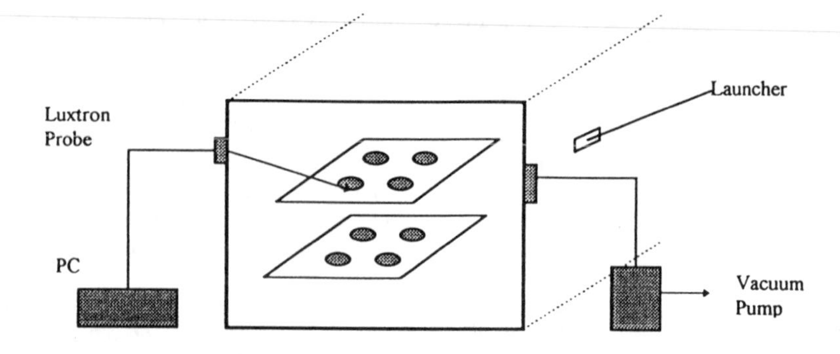

Figure 2. The variable Frequency Microwave Setup. Eight samples have configured as illustrated, the temperature monitoring was performed in one representative samples.

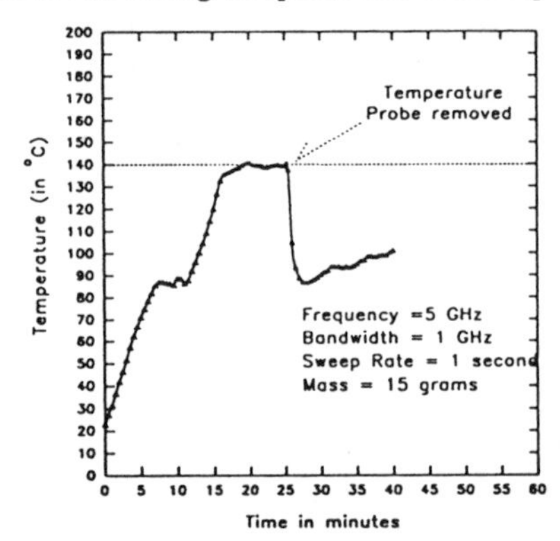

Figure 3. The thermal profile during processing at 5.0±0.5 Ghz variable frequency.

As a scale up illustration, an epoxy sample weighting 530 grams has been controllably cured using variable frequency irradiation. Figure 4 illustrates the microwave cured 10.5" x 6.5" x 0.6" , 530 g epoxy sample.

Polymer Matrix Composites: Glass and graphite fiber reinforced polymer matrix composites were successfully processed using variable frequency microwave energy.

The chemistry of an isocyanate/epoxy composite was evaluated. Specific physical and chemical characteristics (glass transition temperature, flex modulus, flex strength and dielectric behavior) of the conventionally cured PMCs have been reported [3,4]. The conventional post-curing cycle requires 8 hours at temperatures of up to 240°C. Post-curing cycles of 50 minutes have been demonstrated with variable frequency microwave energy. Variable frequency heating of a series of plate configurations were performed, including 1,2,3,4, and 8 plates stacking. The mechanical properties of the samples processed using variable frequency microwave energy are comparable to those of the samples processed using conventional heating (refer to Table 1).

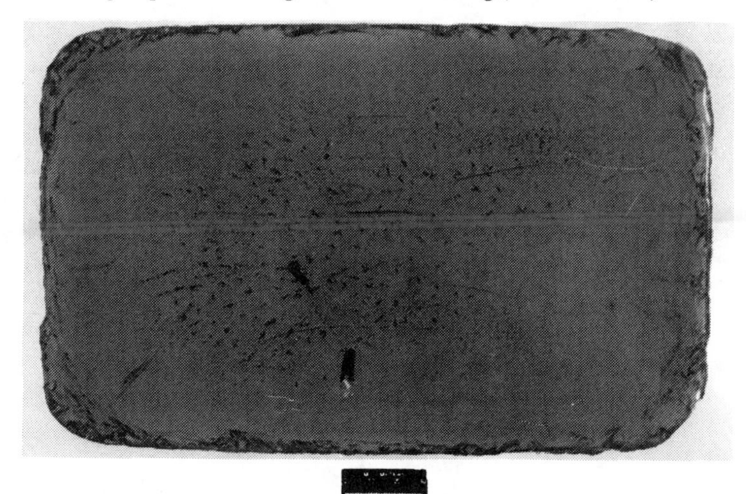

Figure 4. Illustration of a 530 grams ERL 256 epoxy processed using variable frequency.

Tbale 1. Mechanical Behavior

	Flex Modulus(GPa)	Flex Strength(MPa)	Impact Strength (KJ/m ^{2})
Microwave	5.77 ± 0.15	197 ± 15.5	5.7
Conventional	5.75 ± 0.14	197 ± 11.5	5.7

Scale-up benefits of variable frequency was demonstrated by processing 8" x 8" PMC plates. The variable frequency heating was directly contrasted with fixed frequency heating. The fixed frequency irradiation at 4.5 GHz (no mode stirring) of a 4"x 4" sample leads to non-uniform curing, as is illustrated in Figure 5a. The non-uniform curing is the results of a non-uniform distribution of the electric field within the microwave cavity. The sample processed using variable frequency (4.5±0.5 GHz) is fully uniform as shown in 5b.

Unidirectional and crossply graphite-reinforced PMCs were also investigated. Carbon Fiber Composite (CFC) of up to 8" x 8" and up to 32 ply have been successfully processed with variable frequency. Figure 6 illustrates the heating profile of 4" x 4" unidirectional and crossply samples. As illustrated, the heating rates generated within unidirectional samples are generally higher than those achieved within crossply samples. To investigate the uniformity of cure after heating with variable frequency microwave energy, glass transition measurements were performed throughout the samples dimensions. The glass transition values summarized in Figure 6 reveal uniform curing throughout the sample.

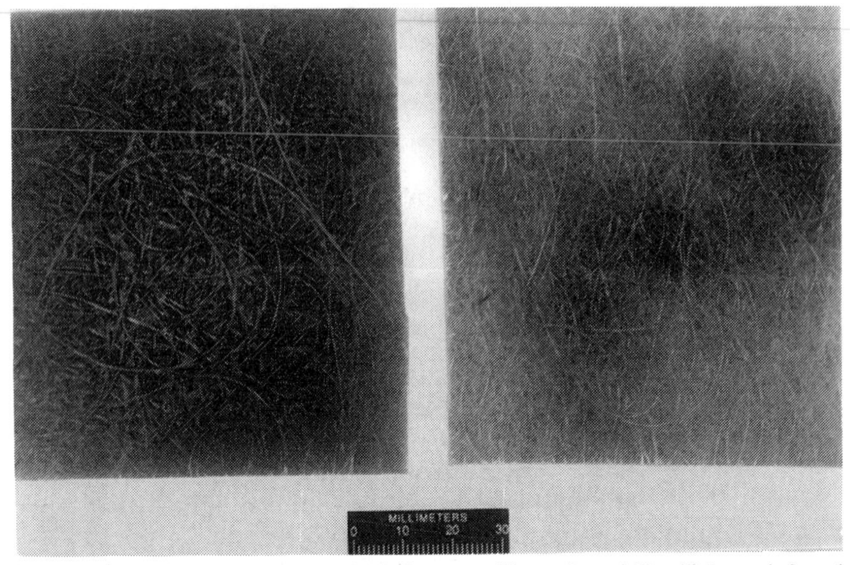

Figure 5. Fixed frequency versus variable frequency illustration of 4" x 4"glass reinforced isocyanate/epoxy composites: a- variable frequency processing and b- fixed frequency processing.

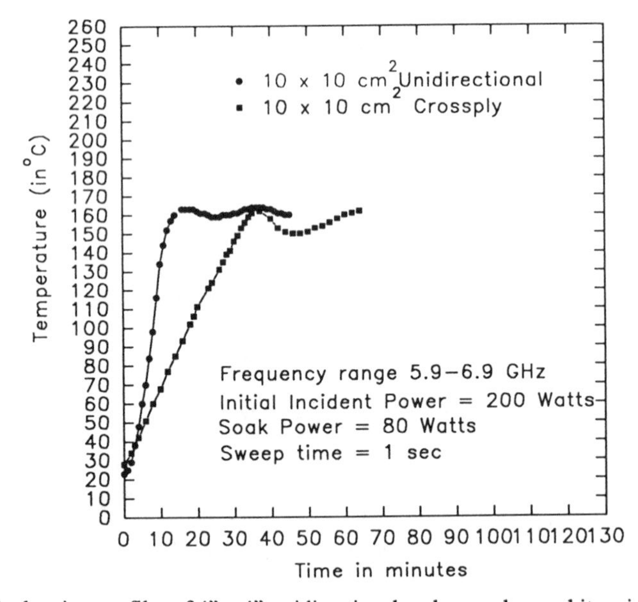

Figure 6. The heating profiles of 4" x 4" unidirectional and crossply graphite reinforced PMC along with the glass transition measurements as a function of location within the variable frequency microwave processed samples. Uniform curing is achieved.

Plasma Processing: Plasma generation and plasma position control have been investigated using variable frequency microwave energy. By changing the excitation frequency of a plasma in a multi-mode chamber, the locations of the maxima and minima of the power density within the chamber are also changed. This allows localization of the plasma discharge in the precise processing area desired in order to increase the process' efficiency. In a similar fashion, sweeping of the processing frequency during the process cycle can theoretically "scan" the plasma across an arbitrarily-shaped target surface. A series of experiments was conducted to evaluate the flexibility of variable frequency microwave energy as a plasma process tool. The same basic LT 0200 variable frequency source was used with a bell jar adapter and vacuum system. The gas used for this study was a solution of 50% ethanol and 50% water vapor. Pressure control was maintained between 15 and 25 Torr. The results of these tests are presented elsewhere [5].

Clay Drying: Two series of clays with different water contents (57.18 % and 46.42%) were dried under different experimental conditions over the 2.5 - 7.5 GHz range. The benefits of varying the frequency of the incident signal were clearly illustrated. The weight loss (dehydration) and heating rates were derived for several samples and were found to be directly related to the incident frequency. The microwave absorption characteristics of water depend on both temperature and frequency [6]. The experimental results suggest that the moisture removal is more efficient at higher frequencies and powers. The various clay samples were weighed and placed inside microwave transparent plastic jars, the samples were then inserted inside the microwave furnace for drying. A Luxtron temperature probe was used to monitor temperature during irradiation. The temperature and time were then recorded for each of the samples, the thermal profiles generated were then plotted as a function of time. The samples were weighed at the end of the drying process, and the water loss was then calculated for each of the samples. Figure 7a illustrates the temperature profile generated within clay samples under different microwave incident frequencies. The heating rates as well as the weight loss increase with increasing frequency. Figure 7b illustrates the weight loss as a function of frequency for clay samples processed under otherwise similar experimental conditions.

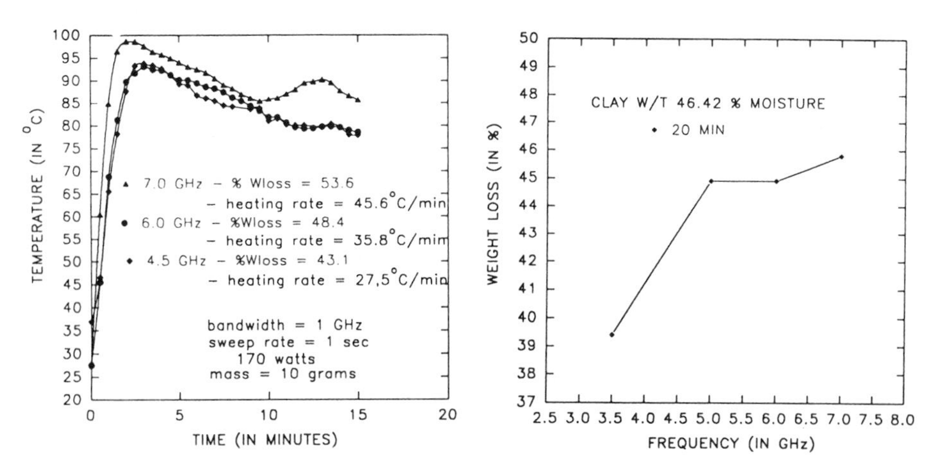

Figures 7. a- Illustration of the thermal profiles generated during drying clays as a function of incident frequencies and b- the accompanying weight loss.

Pending & Future Work: Preliminary work in the areas of binder-burnout reveal a good potential for effective microwave/materials coupling via variable frequency microwave energy. The details of this work will be revealed in future publications. The heating and sintering of Alumina, SiC and Si3N4 are being investigated under controlled and uncontrolled atmospheres. The high temperature processing of B4C (1" diam x 1" long), porous tantalum capacitors, and several samples of tungsten alloy (W-Ni-Fe) have been achieved.

CONCLUSIONS

The results discussed in this paper illustrate the use of variable frequency microwave energy as a controllable and flexible processing tool. Variation of the processing frequency has shown to generate uniform heating and curing profiles within various materials. These results indicate that it is possible to rapidly and uniformly process a variety of polymer matrix composites using variable frequency microwave energy. The benefits of frequency sweep to achieve uniform heating has been demonstrated. The benefits of frequency tuning has been demonstrated to be limited to specific materials, but of particular advantage for addressing materials of complex composition. The benefits of frequency agile processing for clay drying was clearly demonstrated. Variation of the processing frequency was shown to generate intense, controllable, localized water-ethanol vapor plasmas at pressures of 5 to 10 Torr in virtually every location within a quartz bell jar. By increasing the swept-frequency bandwidth about a center processing frequency, the ability of variable frequency microwave energy to generate uniform, large-volume plasmas was shown. Future work will expand into processing high temperature advanced materials.

REFERENCES:

1. A.D. Surrett, R. Lauf, F. Paulauskas and A.C. Johnson,"Polymer Curing Using Variable Frequency Microwave Processing", **Microwave Processing of Materials IV**, eds. M.F. Iskander, R.J. Lauf, and W.H. Sutton, MRS 347 , 691 (1994).
2. Jianghua Wei, Ph.D. dissertation, "Microwave Processing of Epoxy Resins and Graphite fiber/Epoxy Composites in a Cylindrical Tunable Resonant Cavity", Department of Chemical Engineering, Michigan State University, (1992).
3. DeMeuse, M.T., and Johnson, A. C., "Variable Frequency Microwave Processing of PMCs", **Microwave Processing of Materials IV**, eds. M.F. Iskander, R.J. Lauf, and W.H. Sutton, MRS 347, 723 (1994).
4. Zak Fathi, Mark. T. DeMeuse,John Clemens and Craig Saltiel, "Characterization & Numerical Modeling of Variable Frequency Microwave Processed Materials", Spring 95 ACerS Conference, (1995).
5. A.C. Johnson, R.A. Rudder, W.A. Lewis, and R.C. Hendry, "Use of Variable Frequency Microwave Energy as a Flexible Plasma Tool", **Microwave Processing of Materials IV**, eds. M.F. Iskander, R.J. Lauf, and W.H. Sutton, MRS 347, 617, (1994).
6. W.H. Sutton,"Key Issues in Microwave Processing Technology", **Microwaves: Theory and Application in Materials Processing II**, eds. D.E. Clark, J.R. Laia, and W.R. Tinga , Ceramics Transactions, 36, 3, (1993).

USE OF A VARIABLE FREQUENCY SOURCE WITH A SINGLE-MODE CAVITY TO PROCESS CERAMIC FILAMENTS

G.J. Vogt, A.H. Regan, A.S. Rohlev, and M.T Curtin
Los Alamos National Laboratory, P.O. Box 1663
Los Alamos, New Mexico 87545

ABSTRACT

Rapid feedback control is needed for practical microwave processing of continuous ceramic oxide filaments to regulate the process temperature where the dielectric properties of the filaments change rapidly with temperature. A broadband traveling wave tube (TWT) amplifier provides a highly versatile process control platform for filament processing. By comparing a RF signal from the cavity to a reference signal from the TWT, phase information can be used in a negative feedback loop to allow the oscillator to track the cavity frequency as it shifts due to the changing dielectric constant in the filaments being heated. By sampling the electric field level in the cavity with a detector, amplitude control can be done to maintain a constant absorbed power in a fiber tow, which is important for controlling the tow heating and temperature. The system design will be discussed along with application data for commercial ceramic samples.

INTRODUCTION

Microwave heating of ceramic filaments, cylinders, and pellets in a single-mode resonator can be difficult to control due to large dielectric changes in the ceramic load during heating. Large dielectric changes can result from the rapid increase in the complex permittivity with increasing temperature and from the production of new ceramic phases with a complex permittivity significantly different than the starting phases. These dielectric changes can produce large rapid changes in the resonant frequency, the reflectivity, and the power density of the cavity. To control heating of the ceramic load requires adjustment in input microwave frequency and power on a millisecond time scale.

An electronic control scheme is necessary to effect the rapid response needed to control the temperature of a rapidly cooling ceramic load like filament tows. The traveling wave tube (TWT) is an excellent variable microwave source suitable for desired electronic control [1-3]. Unlike the magnetron, the output TWT frequency and power are functionally independent and, therefore, can be controlled independently. Although the 2.45 GHz magnetron is typically limited to a ±20 MHz bandwidth, a TWT is a broadband microwave source with a typical 2 to 8 GHz bandwidth. This paper will describe the design and testing of feedback controller with a TE_{10n} resonator driven by a commercial TWT.

FEEDBACK CONTROL DESIGN

The purpose of the electronic control system is to regulate and measure the RF power absorbed by a material placed at an electric-field maximum in a resonant cavity. The overall goal of this system is to control the temperature of the fiber. However, a control system that attempts to keep the measured temperature constant can produce erratic results, because frequent changes in the position of the heated zone on a moving fiber tow can greatly influence the temperature measurement by optical thermometry. Absorbed power was chosen to be the controlled variable since, in the steady state, constant absorbed power will produce a constant temperature in the fiber. Also, RF power rather than RF field was controlled since fiber heating in the cavity can change with fiber temperature without a corresponding change in the RF electric field.

A single mode resonant cavity is used to heat filament tows. The fiber tow presents a load to the cavity and, as it is heated, both its reactive and resistive component changes. Since a change in the reactive component of the load will shift the resonant frequency of the cavity, a frequency control loop was added to guarantee that the cavity remains on resonance as the fiber is heated. Keeping the cavity on resonance improves the RF power coupling into the fiber and, consequently, the efficiency of the system.

Figure 1 shows a block diagram representation of the frequency control loop. A well established method [4] is used to maintain the resonance condition in the cavity which occurs when the phase of the forward and cavity RF signals differ by $\pi/2$. A mixer is used as a phase detector which produces a DC voltage at its IF port proportional to the phase difference between its RF and LO ports. This voltage is nominally zero when the phase difference $\Delta\phi$ is

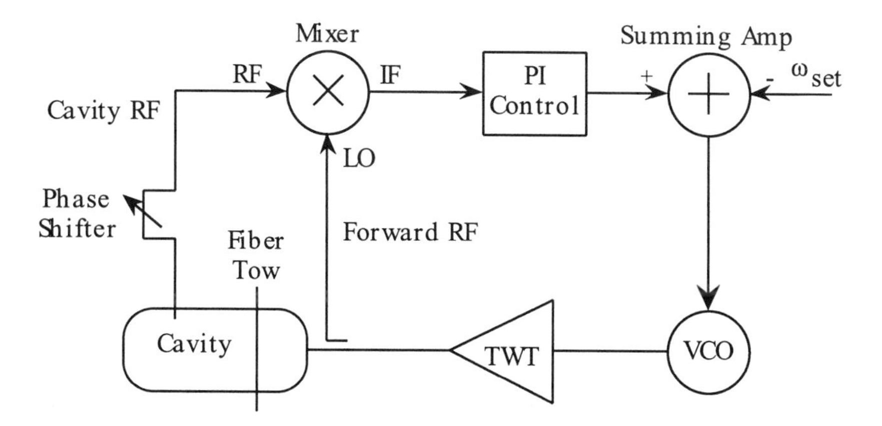

Figure 1. Control structure used to maintain cavity resonance as fiber tow is heated.

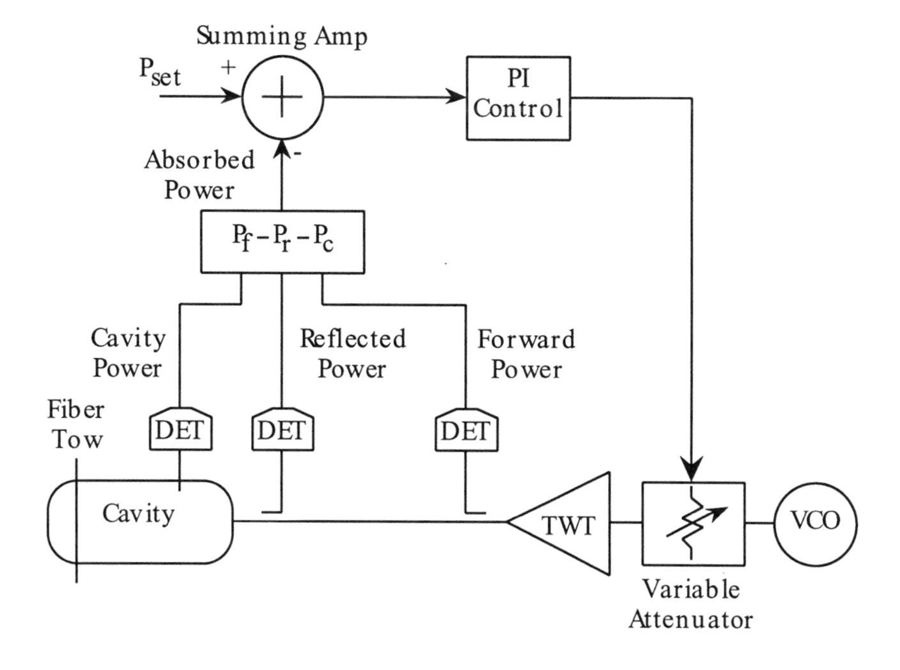

Figure 2. Control structure used to maintain constant absorbed power in the fiber tow.

$$\Delta\phi = \phi_{RF_c} - \phi_{RF_f} = \frac{\pi}{2} \qquad (1)$$

where ϕ_{RFc} and ϕ_{RFf} are the RF phase of the processing cavity and TWT forward power, respectively. A proportional, integral (PI) controller drives a voltage controlled oscillator (VCO) which shifts the operating frequency so that $\Delta\phi$ remains at $\pi/2$.

Figure 2 shows a block diagram of the power control structure. Three crystal detectors (DET) are used to convert the forward (P_f), reflected (P_r), and cavity (P_c) powers to DC voltages. The signals proportional to the reflected and cavity power are then subtracted from the forward power signal. The resulting signal which is proportional to the absorbed power in the fiber tow is subtracted from a set point producing an error signal. A PI controller drives a variable attenuator which modulates the forward power from the TWT so as to reduce the error signal.

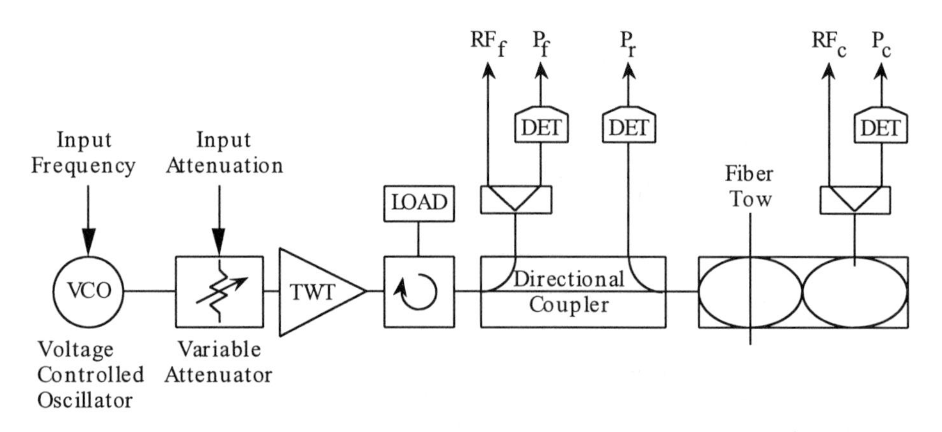

Figure 3. Simplified schematic of fiber tow heating system.

Figure 3 shows a simplified schematic of the fiber heating system. The forward power (RF_f) and cavity (RF_c) signals are RF signals used for the frequency control loop. Measured by the crystal detectors (DET), the P_f, P_r, and P_c signals are DC voltages proportional to the forward power from the TWT, reflected power from the cavity, and power in the cavity, respectively. The P_f, P_r, and P_c signals are input voltages used in the power control loop. The two inputs to the system are DC voltages used to set the VCO frequency and the forward RF power.

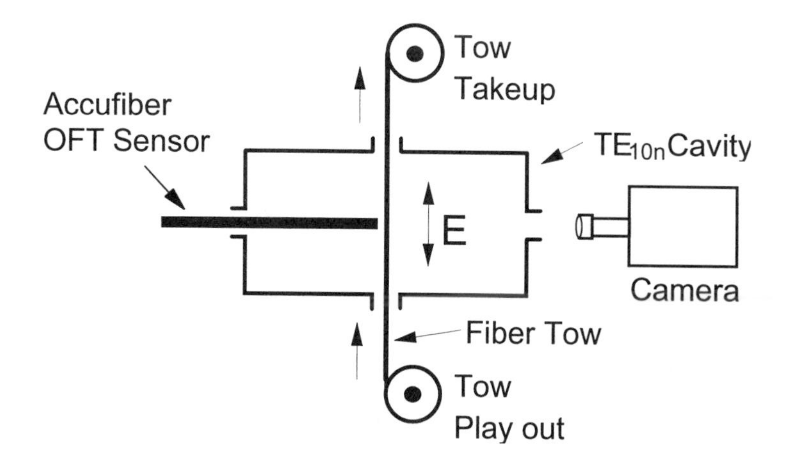

Figure 4. Cross-sectional view of the TE_{10n} cavity, showing the tow position with OFT temperature sensor and video camera.

EXPERIMENTAL TESTING

The feedback controller was tested by microwave heating stationary and moving samples to examine the ability of the controller to regulate the sample temperature. Heating tests were conducted with a TE_{10n} single-mode cavity with a fixed iris and "beyond-cutoff" coupler in the system configuration shown in Figure 3. The TE resonant cavity consisted of a water-cooled section of copper WR284 waveguide with a coupling iris at the inlet and an adjustable short. The TWT* had an output frequency range of 2.5 to 8.0 GHz and an output power of 0-300 Watts. The test cavity was tuned to operate near 3 GHz.

The cavity was fabricated with two pairs of opposing circular ports, as shown in Figure 4, positioned at an electric field maximum. Test samples were loaded through the paired ports aligned with the transverse electric field in the cavity. The pair of ports, perpendicular to the transverse electric field in Figure 4, provided for insertion of an optical fiber thermometer (OFT) and for video recording of the heated sample. The OFT unit** had a dual-wavelength module and a lightpipe sensor to measure the mean sample temperature [5]. The ratio temperatures reported in this paper are two-color temperatures computed from the ratio of light intensities measured at 800 nm and 950 nm.

---

* Logimetric Model A600/EH-885
** Luxtron/Accufiber Optical Fiber Thermometer Model 100C

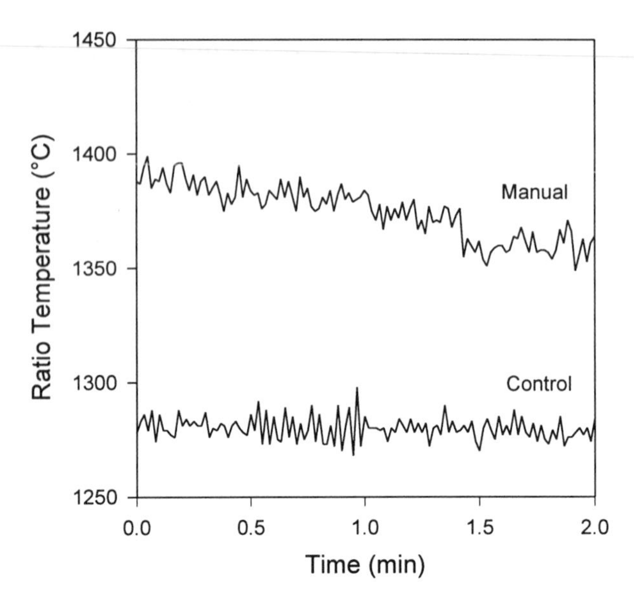

Figure 5. Measured ratio temperature versus time for stationary NICALON tows under feedback control and manual control at 2.92 GHz.

Test samples included NICALON silicon carbide filament tows. The NICALON tows were a continuous carbon-coated fiber bundle with PVA sizing.

TESTING RESULTS

Figure 5 compares the time-temperature response of stationary NICALON tow for feedback controlled and manually controlled heating at 2.92 GHz. Feedback controlled heating was performed with the feedback controllers outlined in Figures 1 and 2 with rapid adjustment in the resonant frequency and TWT power. Manual heating was carried out by operator control of the TWT frequency and power through a sweep generator (VCO). As seen in Figure 5, NICALON temperature under manual control drifted significantly to lower, or higher, values after a minute of heating. The NICALON temperature under feedback control was very steady. Temperature drift while heating at 2.45 GHz with a magnetron source could be substantially greater, since the combined sweep generator and TWT can be significantly more stable than the magnetron.

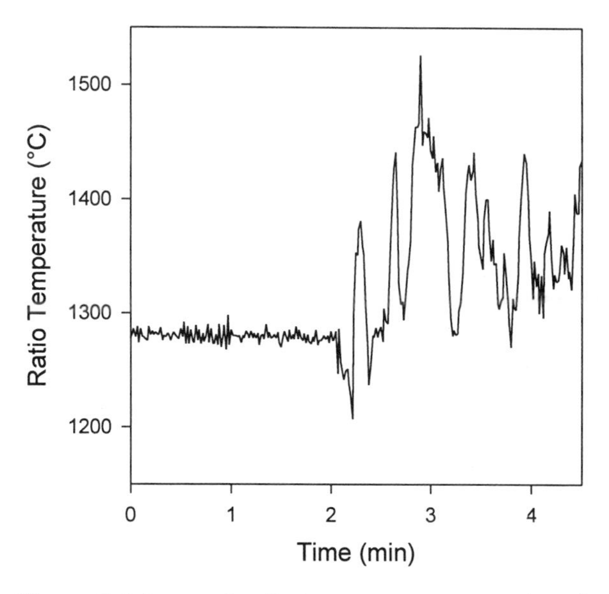

Figure 6. Measured ratio temperature versus time for a moving NICALON tow under feedback control at 2.92 GHz. Tow was pulled through cavity at 2.0 cm/min, starting at 2 minutes.

Figure 6 shows the temperature-time response under feedback control for a heated NICALON tow that was pulled through the cavity at 2.0 cm/min after stationary heating for the first two minutes. The measured temperature of the moving tow varied dramatically with large swings over a 300°C range. Surprisingly, manually controlled heating produced the same large variations in measured temperature. No significant difference in the temperature variations of the moving NICALON tows were found between feedback control and manual control. Video recordings of the heated tows suggest that the large temperature variations may, in part, be an artifact of the temperature measurement.

Video recordings showed that the NICALON tow in the resonant cavity was not heating uniformly along its length as it was pulled through the cavity. In fact, only 60 to 70% of the tow in the cavity was heated to radiant temperatures with no visible heating of the tow as it entered the cavity. As the tow was pulled through the cavity and past the OFT sensor, the leading edge of the heated zone moved in and out of view of the OFT sensor. The leading edge of the heated zone on the tow therefore changed position relative to the OFT sensor at its fixed position in the cavity. This fluctuation in the length of heated tow viewed by the OFT sensor could potentially induce a fluctuation in the measured ratio temperature. This hypothesis will be verified experimentally before further testing of the feedback control system can continue.

CONCLUSION

A feedback control system was designed to work with a TWT to regulate the temperature of a stationary or moving filament tow as heated in a single-mode resonant cavity. The control system simultaneously maintains the TWT frequency at resonance and the microwave power absorbed in the tow at a desired level. The feedback control was tested with a TE_{10n} single-mode cavity for microwave heating of NICALON filament tows. Temperature control was observed for stationary test samples, but not for moving NICALON tows. Tests with moving NICALON tows may be invalid due to measurement fluctuations in the mean sample temperature.

ACKNOWLEDGMENTS

This work at Los Alamos was supported by the Department of Energy, Office of Industrial Technologies, Advanced Industrial Materials Program.

REFERENCES

1. J. Thuery, **Microwaves: Industrial, Scientific and Medical Applications**, Artech House, Boston, (English version edited by E.H. Grant), pp. 125-127 (1992).
2. C.A. Everleigh, A.C. Johnson, R.J. Espinosa, and R.S. Garard, "Use of High-Power Traveling Wave Tubes as a Microwave Heating Source," **Microwave Processing of Materials IV**, Materials Research Society Symposium Proceedings, Vol. 347, p. 79-89 (1994).
3. V.F. Veley, **Modern Microwave Technology**, Prentice-Hall, Inc, Englewood Cliffs, NJ, Chapter 5, pp. 200-250 (1987).
4. F.M. Gardner, **Phaselock Techniques**, 2nd Edition ,Wiley, New York, p. 8 (1979).
5. G.J. Vogt, W.P. Unruh, and J.R. Thomas, Jr., "Experimental Observations of Thermal Spikes in Microwave Processing of Ceramic Oxide Fibers," **Microwave Processing of Materials IV**, Materials Research Society Symposium Proceedings, Vol. 347, p. 539-544 (1994).

THE GYROTRON SYSTEM FOR CERAMICS SINTERING.

Yu.Bykov, A.Eremeev, V.Flyagin, V.Kaurov, A.Kuftin,
A.Luchinin, O.Malygin, I.Plotnikov, V.Zapevalov
Institute of Applied Physics, Russian Academy of Sciences,
46 Ulianov St., Nizhny Novgorod 603600 Russia

ABSTRACT

The prospects of using millimeter-wave radiation for ceramics sintering stem primarily from the analysis of the production scale-up problem. A gyrotron is a unique source of microwave radiation in that its output power does not fall drastically in the millimeter-wave range. The description of the gyrotron-based system designed and manufactured purposely for high-temperature processing of materials is presented. Attention is paid to general concepts of the system which includes: the gyrotron with power up to 10 kW at 30 GHz, the oil-cooled magnet, the quasioptic transmission line, vacuum-proof microwave furnace, and the PC-based feedback control system. Main directions in the Institute of Applied Physics today's research in microwave-matter interaction are described.

INTRODUCTION

The activity of the Institute of Applied Physics (IAP, Nizhny Novgorod, Russia) in microwave processing of materials leans upon the accumulated more than thirty-year experience in the design and development of the gyrotrons - the sources of high-power radiation in the centimeter and millimeter-wave range. Since the invention of first gyrotrons, IAP has been keeping the leading position in theoretical and experimental study of the principles of gyrotron operation, which has resulted in a steady growth of the output power and broadening of the frequency range. Owing to the recent efforts, mainly of USA and Russia researchers, gyrotrons with power about 1 MW in the long-pulse regime are available now with frequencies as high as 170 GHz [1]. Their reliability and efficiency have been proved in their operation for many years in nuclear fusion installations throughout the world. Recently, gyrotrons operating in continuous-wave (CW) or pulse regimes with average output power of several tens of kilowatts have been designed and produced. Commercial availability of these sources of radiation in the frequency range that is new for industrial applications opens an opportunity for the development of innovative microwave technologies.

Mastering the new frequency range means, at the first turn, a significant extension of the number of materials which can be processed effectively with the use of electromagnetic radiation. It is well known that microwave absorption for many types of materials increases rapidly with frequency [2]. The opportunity for better utilization of the microwave energy for heating induces a reappraisal of many potential microwave applications.

Two aspects pertaining to the electromagnetic energy distribution in the applicators must be taken into account to gain a better understanding of the advantages promised by using the millimeter-wave radiation in practice. In fact, the use of multimode applicators becomes unavoidable for most industrial processes due to productivity requirements. The uniformity of microwave energy distribution within the multimode systems, which is characterized by the magnitude of ratio L/λ (L is the dimension of applicator, λ is the wavelength of radiation), is a crucial problem for most applications. Diminishing the wavelength of radiation makes it possible to exploit untuned multimode applicators with $L/\lambda \geq 100$. High uniformity of the microwave energy distribution in such applicators is achieved as the result of superposition of the electromagnetic fields of hundreds simultaneously exited modes.

On the other hand, a new approach to the microwave surface treatment of materials can be developed with diminishing the radiation wavelength. In addition to a potential for uniform processing of a large surface area within untuned multimode applicators, local treatment becomes feasible by using wave beams of the millimeter-wave radiation. The intensity of up to $\sim 10^4$ W/cm^2 can be easily achieved with contemporary CW gyrotrons by focusing their output radiation into the spot of size λ^2. Rather simple and rough quasioptical systems for transportation and concentration of the microwave energy can successfully be applied in the technologies of local processing of materials. Such technologies show up as the alternative methods to surface treatment with the beams of radiation produced by infrared lasers.

30 GHZ GYROTRON HEATING SYSTEM

By now, a unique set of facilities has been built up at IAP for study of processing of materials with millimeter-wave radiation. The available equipment is specific in a variety of the main parameters of radiation: frequencies, powers, modes of operation. Diversity of these parameters is of paramount importance when the research pursues each of the two goals: to clarify a specific effect inherent to microwave-based processes and to determine the optimal conditions which yield a high-quality final product.

Primarily, the research into processing of materials at IAP was based on the use of pulse gyrotrons designed for nuclear fusion applications. Along with producing single pulses of a megawatt power at frequencies 50 - 160 GHz, these gyrotrons are also capable of operation in a mode more suitable for processing of materials, that is in the regime of repeated short pulses with the average power of several tens of kilowatts.

Recently, a specialized 10 kW, CW, 30 GHz gyrotron for processing applications has been built by the Department of High-Power Electronics at IAP. This tube is a key component of the Gyrotron Heating System designed purposely for high temperature processing of materials with millimeter-wave radiation (Figs.1 and 2). The system consists of the following main parts:

- gyrotron with continuously controllable output power of up to 10 kW
- warm magnetic system
- 3-mirror quasioptical transmission/focusing line
- microwave furnace
- temperature diagnostic units
- PC-based processing control system
- vacuum equipment
- cooling equipment, and
- system for control, interconnection and safety.

Two engineering approaches allowed achievement of the compactness and reasonable volume of the system. One of them is the gyrotron operation at the second harmonic of the electron cyclotron resonance. This mode of operation signifies a twofold reduction of the required stationary magnetic field. As a consequence, energy consumption by the compact magnet is reduced almost 4 times. In addition, the power supply unit and cooling system for such a magnet becomes appreciably more simple. Further decreases in the overall electrical consumption of the system will require additional efforts aimed to increase the efficiency of the gyrotron itself. The possibility for such an increase follows

Fig.1 Gyrotron System for processing of materials (10 kW, CW, 30 GHz).

from the theoretical calculations and previous experiments.

The mismatch of the generator and the load which results from the variation of dielectric properties of heated materials is the acute problem facing the designers of microwave heating facility. The microwave isolation of the gyrotron and its protection from the radiation backscattered from the furnace is achieved in the system by using a quasioptical mirror transmission/transforming line. Scattered microwave radiation is dissipated within the water-cooled absorber located in the metal cabinet enclosing the line.

Formed as the wave-beam, radiation enters through the microwave-transparent window into the microwave furnace of ~50 cm in diameter and ~60 cm in height, made of stainless steel. The furnace lid is furnished with the mode stirrer. Each of the furnace flanges has O-ring seals providing protection against

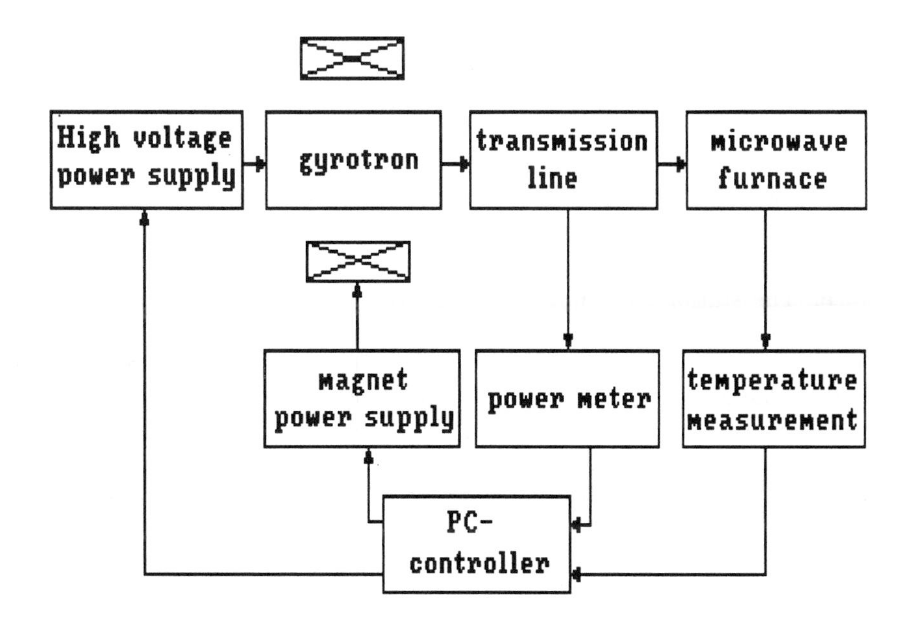

Fig.2 Diagram of the Gyrotron System.

both radiation and environment gas leakage. The vacuum facility of the system allows processes in any gas environment in the pressure range 1 - $2 \cdot 10^5$ Pa or with the gas flow through the furnace. The temperature of the workpieces can be monitored using thermocouples, an optical pyrometer and fiber optics units, for which the appropriate ports are mounted on the special flange. The flange has also 6 insulated inlets for supplying the optional resistive heaters with power of about 15 kW (when hybrid heating is desirable).

The system can be operated in both manual- and automatic-control mode. To accomplish the automatic control, the feedback loop embraces the whole system together with its main power supplies. This loop includes also the interface units suited to the exploited temperature diagnostic means and the PC-controller.

It should be noted that without any alteration in the structure, but only through modification of its main components, the operation frequency of the system can be optionally varied within the range 20 - 40 GHz. The Gyrotron Heating System is commercially available. At present, one such system is being operated at the Forschungszentrum Karlsruhe (FZK, Germany), where processing of materials with millimeter-wave radiation is studied jointly by IAP and FZK specialists.

The same structure of the equipment is the basis of installations using the gyrotrons with output power about 30 kW CW at frequencies 35 and 83 GHz. These gyrotrons (produced by the company "GYCOM" (Nizhny Novgorod, Russia)), operate at the fundamental frequency of the electron cyclotron resonance. The stationary magnetic field needed for their operation is provided by the superconducting coils cooled with liquid helium. The coil geometry and cryostat design are optimized for use gyrotrons operating at both 35 and 83 GHz. The gyrotrons contain the built-in mode converters for shaping the output radiation as a Gaussian wave-beam. The so formed radiation enters the furnace of 70 cm in diameter and 120 cm in length through the oversized waveguide. Here, in the furnace, the microwave energy can be used in each of the two ways: for ordinary volumetric heating of materials in the multimode cavity, or as a wave-beam with the field pattern suitable for the peculiar surface treatment of materials.

In summary, the complex of the equipment available now at IAP allows users to carry out the broad-scale study of both volumetric and surface processes using CW radiation with power 10 - 30 kW and pulse radiation with power up to 1 MW within the frequency range 30 - 160 GHz.

EXPERIMENTAL RESULTS

There are several directions at IAP in the field of the interaction of the intensive millimeter-wave radiation with matter. The scope of this research falls into two groups.

One of them involves processing of the solid materials heated with the millimeter-wave radiation and, primarily in ceramics sintering. The main aspects of the millimeter-wave sintering are being investigated at IAP for a broad variety of constructional ceramic materials: Al_2O_3, Al_2O_3 - $ZrO_2(Y_2O_3)$, $ZrO_2(Y_2O_3)$, $Si_3N_4(Al_2O_3$, Y_2O_3, ZrO_2, $AlN)$, and functional ceramics: $Pb(Zr,Sr,Nd)O_3$, $BaTiO_3$. In this research, the Laboratory of Microwave Processing of Materials of IAP collaborates closely with specialists in materials science from Nizhny Novgorod State University and the Institute for Microstructures, Russian Federal Nuclear Center, Research & Production Center "Tekhnologia" (Russia), Institut for Materialforschung III (Karlsruhe, Germany).

The research is mainly aimed at answering the conceptual question: does microwave processing - due to the volumetric, inertialess, controllable character of heating - have the potential for adding a new value to ceramic and composite materials? In spite of many encouraging results achieved to date in the temperature-time procedure of sintering, mechanical and functional properties of materials, disclosure of actual relations in the sequence - "processing conditions - microstructure - physical and mechanical properties" - requires further efforts.

In addition to the generally accepted approach based on comparison between the results of microwave and conventional sintering, the method of comparative sintering at various frequencies of radiation becomes possible with the diversity of facilities at IAP. This method provides a new approach to the problem of the existence of a specific non-thermal "microwave effect" in ceramics sintering, which is not solved so far. The crucial factors in this problem are the reverse temperature distributions within conventionally and microwave heated bodies, and impossibility to determine the actual temperature distribution and even the magnitude of temperature during microwave heating. Comparative study of the microwave processes performed at different frequencies provides the basis not only for detecting specific "microwave effects", but also for correlating this specificity, if any, with parameters of radiation [3].

In recent years, special attention has been paid to microwave sintering of nanophase materials. It seems that fast and controllable microwave processing has a great potential for producing the nanostructural ceramics but first results on sintering nano - TiO_2 and Al_2O_3 - ZrO_2 composite showed the difficulties which originated, primarily, from the quality of the starting raw materials [4].

Great diversity of the available microwave equipment makes possible the broad-scale verification of the theoretical predictions relating to probable mechanisms of a specific effect inherent to microwave heating. Recently, Semenov and Rybakov suggested the theoretical explanation for the "microwave effect" by considering the mean ponderomotive force affecting the transport of the charged vacancies in the near-surface layer of ionic crystalline grains [5]. This theory has begun to receive experimental support [6], and development of the theory requires data from experiments on other materials.

The study of various processes based on the local surface heating with intensive millimeter-wave beams has verified the feasibility of microwave surface treatment of materials. Among these processes, there are surface hardening of metal alloys, joining of ceramic workpieces with refractory coating, joining of high temperature superconducting layers with metal tapes. In this research, the upper bound of the wave-beam intensity has been found [7]. This limitation is imposed by self-ignition of the near-surface gas discharge which cuts off the penetration of the microwave energy to the surface. It can be said with confidence that the use of millimeter-wave beam techniques for surface processing of materials possesses a considerable potential, and only a few of possible applications have been viewed up to date.

Other research activities are associated with applications based on the processes in a gas plasma sustained by the millimeter-wave radiation. Among the most advanced research areas are plasmachemical processes in a non-equilibrium free localized discharge at moderate gas pressure [8], and the electron-cyclotron discharge as a source of multicharged ions and soft X-ray radiation [9]. The use of beams of intensive millimeter-wave radiation is studied also in application to the monitoring of atmospherical nonuniformities [10].

REFERENCES

1. V.A.Flyagin, A.L.Goldenberg, and V.E.Zapevalov, in *Proc. Int. Workshop "Strong Microwaves in Plasmas"* Ed. by A.Litvak, v.2, pp.597-600 (1994).
2. A.R.Von Hippel, *"Dielectric Materials and Application"*, John Willey and Sons, New York (1954).
3. Yu.Bykov, A.Eremeev, and V.Holoptsev, in *Mat. Res. Soc. Symp. Proc.,* v.347, pp.585-590 (1994).
4. Yu.Bykov, S.Gusev, A.Eremeev, V.Holoptsev, N.Malygin, S.Pivarunas, A.Sorokin and A.Shurov, in *Proc. Second Int. Conf. on Nanostructure Materials, Progr. and Abstr., Stuttgart,* p.130 (1994).
5. K.I.Rybakov, and V.E.Semenov, *Phys.Rev.B,* v.49, n.1, pp.64-68 (1993).
6. S.A.Freeman, J.H.Booske, and R.F.Copper, *Phys.Rev.Lett.*, v.74, n.11, pp.2042-2045 (1995).
7. Yu.Bykov, A.Eremeev, and A.Sorokin, *Fizika i Khimia Obrabotki Materialov*, n.6, pp.107-112 (in Russian) (1991).
8. Yu.Bykov, *Khimia Vysokikh Energii*, v.18, n.4, pp.347-353 (in Russian) (1984).
9. Yu.V.Bykov, S.V.Golubev, A.G.Eremeev, and V.G.Zorin, *Sov. J. Plasma Phys.*, v.16, n.4, p.279 (1990).
10. Yu.Bykov, Yu.Dryagin, A.Eremeev, L.Kukin, Yu.Lebsky, O.Mocheneva, and M.Tokman, *Fizika Atmosfery i Okeana*, (in print), (in Russian) (1995).

QUASI-OPTICAL GYROTRON MATERIALS PROCESSING AT LOS ALAMOS

J.D. Katz and D.E. Rees
Los Alamos National Laboratory
Los Alamos, NM 87545

ABSTRACT

Los Alamos has recently obtained and installed quasi-optical gyrotrons of 37 and 84 GHz with power outputs up to 35 kW. A quasi-optical gyrotron is unique in that the output is a Gaussian beam which can be focused and manipulated using mirrors. The Gaussian beam output is ideally suited for one and two dimensional materials processing applications such as joining and surface treatment.

A consortium has been formed with several National Center for Manufacturing Sciences (NCMS) member companies to investigate several materials processing applications.

INTRODUCTION

Materials processing using millimeter waves has had very limited application in the United States because millimeter-wave sources have been prohibitively expensive[1]. This technology, however, has been demonstrated on an industrial scale by the Paton Welding Institute in the Ukraine for several applications including the bonding and coating of ceramics, the bonding of plastic insulation and protective coatings to pipe, and the bonding of composite materials[2]. At the heart of this technology are quasi-optical gyrotrons that produce up to 35 kW of microwave power at 37 and 84 GHz and which are sufficiently robust for industrial use. Working with The National Center for Manufacturing Sciences (NCMS), a non-profit corporation located in Ann Arbor, MI, a consortium of industrial partners interested in applying this technology has been organized. The members of the consortium are listed in Table 1. Each of the consortium members will be allotted experimental time to test the feasibility of one of more proprietary applications.

Table 1. Consortium Members

American Telephone and Telegraph
Baxter Healthcare
Continental Electronics
Ford Motor
General Atomics
General Motors
Physical Sciences
Rockwell International
Tycom
United Technologies
Valenite

The output from a quasi-optical gyrotron is a coherent microwave beam. This is very different from the cavity devices usually used in microwave processing. The millimeter wave output can be reflected, focused, and rastered over surfaces using a metal mirror and angled dielectric lens, making it possible to direct energy precisely. The quasi-optical gyrotron allows for one and two dimensional heating of an object in contrast to the volumetric heating obtained by cavity processing. This unique ability to heat only a line or surface enables many new applications for microwaves in the areas of joining and coating.

It would be desirable to re-configure the current free space set-up into a resonant cavity for ceramic sintering applications. A resonant cavity is advantageous since it will result in a more uniform and higher electric field. A uniform electric field is beneficial since it helps minimize temperature gradients which cause cracking. This subject will be discussed further in the next section. Why a high electric field is desirable is easily shown by reference to the formula for the power dissipated per unit volume by a dielectric undergoing microwave heating:

$$P = 2\pi f E_2 \varepsilon' \tan\delta \tag{1}$$

where

P = power dissipated per unit volume,
E = electric field,
f = frequency,
$\tan\delta$ = loss tangent, and
ε' = dielectric constant.

The rate of power dissipation, and hence heating, are linearly dependent on frequency, so the very high frequencies of 37 and 84 GHz certainly help to achieve heating. The rate of power dissipation, however, increases with the square of the electric field. Consequently, the electric field has a much greater influence on heating.

DESCRIPTION OF EQUIPMENT

The experimental equipment is pictured in Figure 1. The quasi-optical gyrotrons in use at Los Alamos were manufactured by Salute in Nizhny Novgorod, Russia. The tubes are 37 and 84 GHz with power ratings of up to 35 kW. These gyrotrons have an internal mode launcher to convert to the quasi-optical mode. A cryogenic magnet is necessary to operate these tubes. The gyrotrons are currently configured with separate anode and cathode power supplies, but can also be operated with one power supply. The independent power supplies allow for more flexibility in the operation of the gyrotron and for a higher power output. Experiments are conducted in an experimental chamber which is coupled to the gyrotron output window.

The beam enters the experimental chamber with a horizontal trajectory and reflects from a Gaussian reflector down onto the experimental plafform where the material sample is located. The experimental chamber contains a rotating dielectric lens which can be inserted into the vertical beam path between the gyrotron output window and reflector which present the normally incident beam with a variable angle of incidence which translates to a varying angle of transmission by Snell's law [3]. This rotating lens is used to raster the beam. Two samples of incident energy are taken within the experimental chamber. One sample is taken through a small hole in the reflector. This hole is less than a millimeter in diameter so it acts as a cutoff section of circular waveguide to the gyrotron frequencies. The incident energy sample is transported in the circular waveguide and converted to a voltage proportional to the millimeter wave power by use of a diode. A portion of the beam is also sampled through interception with a circular waveguide at the entrance to the experimental chamber. Calorimetry of the energy dissipated in the output window can be used to measure the absolute power emitted by the gyrotron. There are two observation windows on the experimental chamber. These windows are filled with water and shielded with screens to protect the experimenter from microwave leakage. The chamber can also be flushed with an inert gas to prevent oxidation.

One of the first tasks upon bringing this system to the United States was to modify it along with its accessory equipment to conform to U.S. safety standards. The next major task was to develop the interface between the high voltage power supply and the gyrotrons. A control system was also designed and constructed which incorporates proper safeguards for both personnel and equipment and allows experimenters to establish reproducible experimental conditions.

OPERATING CHARACTERISTICS

Upon bringing the microwave beam up for the first time, several experiments were performed to develop a feel for the operational characteristics of the gyrotrons. Because of extensive experience with ceramic sintering, some of the first experiments were attempts to heat and sinter several green ceramic bodies including alumina, alumina-5vol % silicon carbide particulates and alumina stabilized zirconia. Direct heating by the 84 GHz beam is simply too intense and

Figure 1. Frontal view of the experimental apparatus. Gyrotron in cryostat at right, experimental chamber center and control system left.

localized for the typical green body. Invariably, the green body would crack violently and fall apart.

Because of their high power density, the gyrotrons are capable of "vaporizing" certain materials. Figure 2 shows what remains of an aluminasilica refractory brick* after several hits from the microwave beam. This material couples well, melting instantaneously under the beam and vaporizing to form copious amounts of yellow-green plasma. Puddles of greenish glass are visible in the bottom of the brick.

The beam nature of the quasi-optical gyrotron was graphically illustrated shortly after turning on the equipment. An interesting interference fringe pattern which is shown in Figs. 3 and 4 was observed. Figure 3 is a section from a 99% alumina refractory brick. The tan markings show a regular pattern of ripples with an approximately 2 mm spacing. (The wavelength is 3.6 mm at 84 GHz.) Figure 4 is a section of alumina 20% silica insulation board** which also has regular ripples at an approximately 2 mm spacing. It is believed that the interference fringe is due to imperfections in the reflector.

---

*Grade K20, Babcox and Wilcox Co., Pittsburgh, PA
**Zircar Products Inc., Florida, NY.

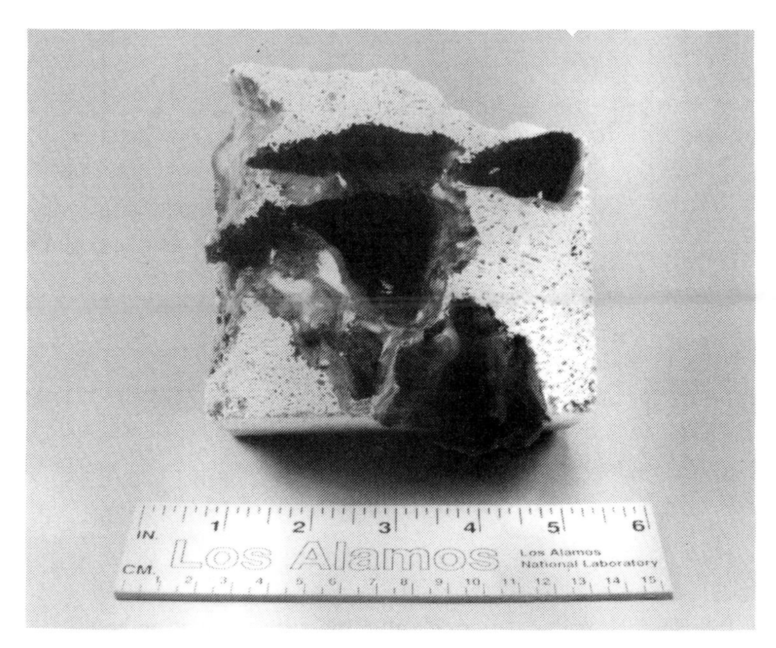

Figure 2. "Vaporized" alumina-silica refractory brick.

Figure 3. High density 99% alumina refractory brick. Note dark area with interference fringe pattern.

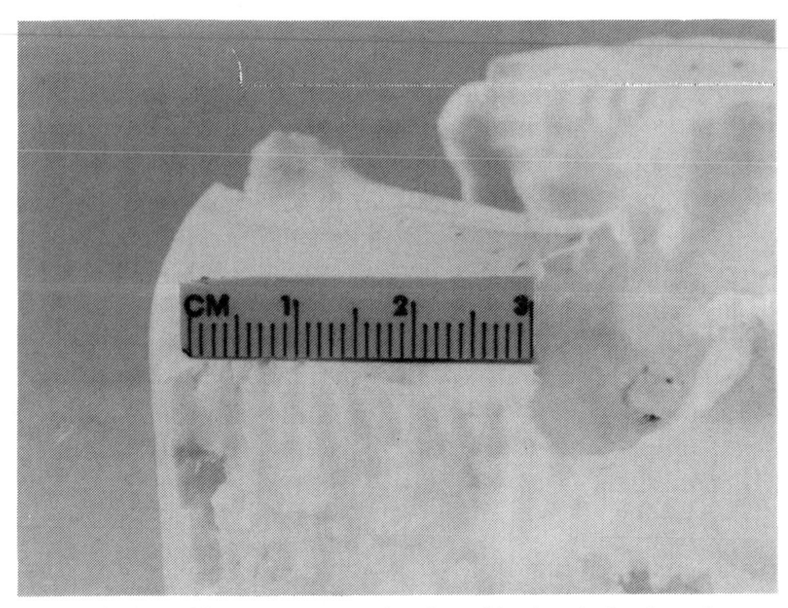

Figure 4. Interference fringe pattern on alumina-silica insulation board.

PRELIMINARY EXPERIMENTS

Some preliminary experiments were performed with pre-ceramic polymers. Pre-ceramic polymers are of interest for both coating and ceramic joining applications. The joining application involves joining sintered silicon carbide to sintered silicon carbide by the in-situ conversion of polycarbosilane to silicon carbide. After sufficient flushing of the experimental chamber with an inert gas, the 84 GHz beam was turned on and directed at a chunk of polycarbosilane. The polycarbosilane soon melted and began to emit a plasma. The polycarbosilane was converted to a black powder which x-ray analysis (see Figure 5) confirmed to be silicon carbide. Previous work at 2.45 GHz in a resonant cavity has also resulted in the conversion of polycarbosilane sandwiched between two pieces of to silicon carbide [4]. This finding, however, is proof that a direct microwave beam can be used to achieve conversion.

CONCLUSIONS

A quasi-optical gyrotron system offers unique materials processing opportunities. The operating characteristics are very different from traditional cavity processing systems. The wave nature of the microwave beam is readily evident from material interactions. A quasi-optical system holds promise for coating and joining applications but is undesirable for volumetric heating and sintering. In cooperation with the National Center for Manufacturing Sciences

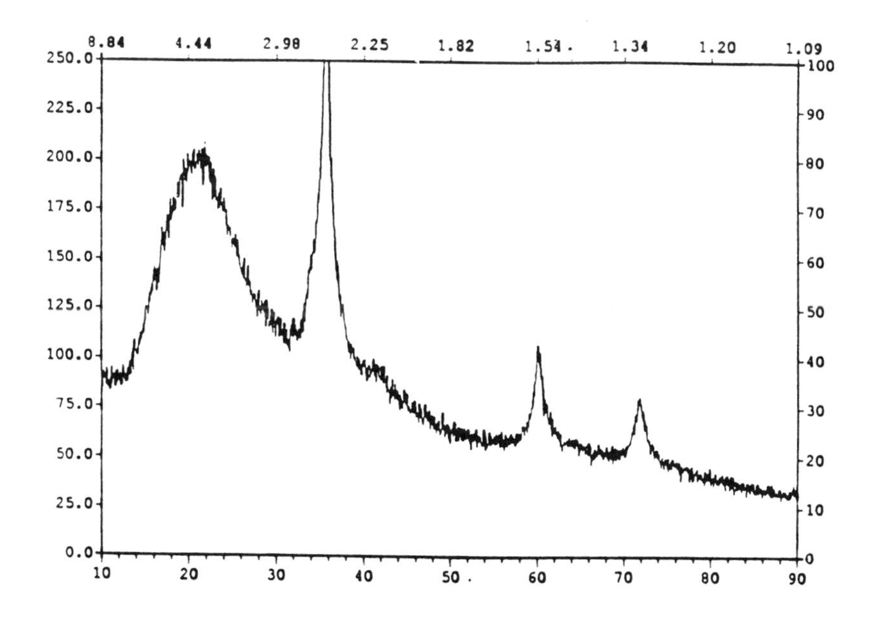

Figure 5. X-ray diffraction pattern for polycarbosilane converted to silicon carbide using 84 GHz microwave beam.

(NCMS) and several of their member companies a consortium has been formed to investigate commercial applications of this technology.

REFERENCES

1. M.A. Janney and H.D. Kimrey, Microwave Processing of Materials, Mater. Res. Soc. Symp. Proc. 189, pp. 215-28 (1991).
2. V.E. Sklyarevich and R. F. Decker, "Super High Frequency Microwave Processing of Materials - a Basis for Developing New Technologies," Industrial Heating, pp. 54-56, October (1991).
3. Page F-103, CRC Handbook of Chemistry and Physics, 68th Edition, R. C. Weast, Editor-in-Chief, CRC Press, Inc., Boca Raton, Florida (1987).
4. R. Silberglitt, Private Communication.

UNIFORM MICROWAVE HEATING OF POLYMER MATRIX COMPOSITES

V. Adegbite, L. Fellows, J. Wei, Y. Qiu and M. Hawley
Department of Chemical Engineering
Michigan State University
East Lansing, MI

ABSTRACT

Uniform heating of polymer matrix composite materials in a single-mode resonant cavity was studied using fixed frequency mode-switching, variable frequency heating, and dual-coupling heating. Fixed frequency mode-switching is based upon the alternation between two or more electromagnetic modes which have complementary heating patterns. The mode alternation is done in such a manner that microwave energy distribution across the sample is optimized and thermal gradients are minimized. Heating by dual-coupling is the simultaneous coupling of microwave energy from an axial and radial mounted coupling probes such that modes with complementary heating patterns are excited by both probes. Variable frequency heating is similar to the fixed frequency mode-switching method except for that complementary heating modes are generated by varying the frequency and not the cavity volume. These methods were demonstrated to be effective in achieving uniform heating of flat panel, V-shaped and triplanar shaped vinyl ester / glass and graphite / epoxy composites.

INTRODUCTION

Microwave processing of composite materials has been studied as an alternative to conventional composite processing techniques with demonstrated virtues in faster heating rates, enhanced polymer kinetics, selective heating, and inside-out thermal gradients [1,2]. However, one of the challenges in microwave heating of composites is uniform heating. Uniform heating by microwaves is a fundamental electromagnetic problem which is complicated even more when very lossy and anisotropic materials like composites are heated. Historically, mode stirrers and revolving tables have been used to achieve uniform heating of food in domestic microwave ovens [3]. However, the single-mode resonant cavity is designed such that it supports only one type of mode at a time, which makes this applicator different from the traditional multimode oven (domestic microwave oven) where several modes are excited at a time [3].

The characteristics and field components of these modes are derived by resolving Maxwell's equations with the appropriate boundary conditions for a specified structure[4]. The solutions provide relations of frequency as a function of cavity length and diameter for different modes. In operation, mode excitation can be achieved by fixing the frequency of the microwave source and
varying the cavity length (fixed frequency tuning) or by fixing the cavity length and varying the frequency of the microwave source(variable frequency tuning).

The microwave cavity has two adjustable parameters for the selection and control of the electromagnetic fields. Figure 1 shows a single-mode resonant microwave cavity. The sliding short can be adjusted to vary the cavity length. The coupling probe inserted through the side acts as an antenna that couples the microwave energy into the cavity. It also acts as a disturbance in the cavity which has to be adjusted to match the impedance of the cavity to achieve resonance. This impedance matching focuses the energy into the cavity and minimizes reflected power, making the single mode cavity a highly energy efficient device.

At microwave frequencies, both the magnetic and electric fields exist orthogonal to each other. Hence, in order to heat these non-magnetic materials, it is important to align the sample with the electric field component. Two types of electromagnetic fields may be present within the microwave applicator. One is a TE (transverse electric) mode, in which the electric field is transverse to the direction of propagation, and the other is a TM (transverse magnetic) mode, in which the electric field is parallel to the direction of propagation where the axial axis is taken as the direction of propagation. The TE-modes are best for sample geometry such as a panel or a disk, while the TM-modes are best suited for processing cylinders or rods.

It was demonstrated by Wei[2] that a single mode can be used to successfully heat and process composite panels with conductive fibers. This is not always the case in the heating of other materials with different electrical properties and shapes. Hence, other methods have been investigated which are fixed frequency mode-switching[5], variable frequency heating[6], and heating by simultaneous dual microwave coupling[7].

UNIFORM HEATING METHODS

Fixed Frequency Mode-Switching Heating

Mode-switching is the alternation between two or more modes with complementary heating patterns such that energy distribution across the sample is optimized and thermal gradients are minimized. [5]. This is accomplished with the intelligent selection of the modes based upon knowledge of the heating patterns generated by each individual mode for a particular material. By sequencing modes in a proper order, the microwave energy can be focused directly to where it is needed most to minimize thermal gradients. The microwave source was a magnetron that operates at 2.45 GHz with a maximum of 100 Watts of output power.

Fixed frequency mode-switching has been successfully applied to the processing of vinyl ester-vinyl toluene/glass composite panels, V-shaped samples and triplanar shaped samples[5,8]. The choice of the V-shape and the triplanar shaped parts was to demonstrate the processing of samples that do not necessarily align with either the TE or TM modes. In using the mode-switching technique, several modes were combined to achieve uniform heating with gradients of 20°C or less, where none of the single modes heated the sample with a gradient of less than 40°C. The processing temperature profiles obtained by mode-switching are presented in Figure 2 along with the various shapes.

Variable Frequency Heating

There are several variations of variable frequency heating which is defined by the method in which the frequencies are introduced. The most familiar method is the variation of the microwave frequency across a range of fixed frequencies which has been shown to heat samples more uniformly than single frequency systems [9]. In the single-mode resonant cavity the concept of variable frequency heating is different. It is based upon the heating at only selected frequencies where desirable modes are known to exist. This concept is similar to the mode-switching method discussed above except for that the cavity length is fixed and the frequency is varied as opposed to fixing the frequency and varying the cavity length.

The variable frequency system used in this work consist of an oscillator with an RF plug-in which is amplified by a TWT (Traveling Wave Tube) with a frequency ranges 1.7 GHz to 4.3 GHz (see Figure 3). The power output of the RF plug-in and the amplification ratio of TWT vary with frequency and the maximum

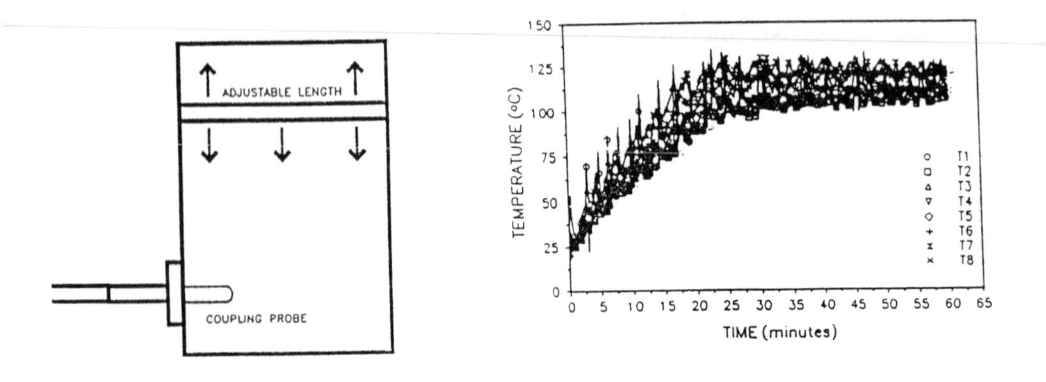

Figure 1. Single-mode resonant cavity.

Figure 2a. Temperature profile for heating vinyl ester panel by mode-switching.

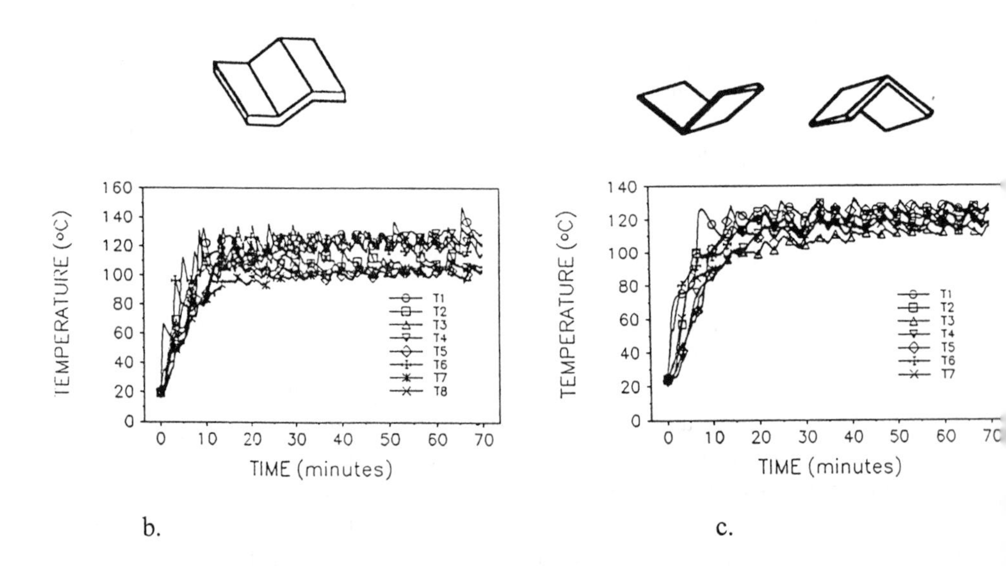

b.

c.

Figure 2b & c. Temperature profile for b) triplanar and c) v-shaped vinyl/ester glass composite.

power output ranged from 10 Watts to 40 Watts. Using the variable frequency system, a cured 48-ply unidirectional 2"x2" graphite/epoxy composite was heated at various frequencies. Thermal paper was used to map the heating patterns as well as fluoroptic thermometry for quantitative temperature profiles. Frequencies were chosen such that the input power to the cavity was maintained at 30 Watts at a fixed cavity length of 15 cm. The samples were heated until the highest temperature reached 95 °C.

Temperature profiles are shown in Figures 4-7. The highest heating rate observed was 20 degrees per minute with only 30 Watts of input power. The temperature profiles show that uniform heating can be achieved not only by a combination of modes but by a single mode. At a frequency of 2.3133 GHz, relatively uniform heating was achieved while the heating patterns at the frequencies of 2.5342 GHz and 2.2135 GHz were complementary and could be combined to achieve uniform heating.

Dual-coupling

Dual-coupling is the coupling of microwave from the radial and axial mounted coupling probes on single-mode resonant cavity (see Figure 8). In a single-mode resonant cavity, only single unique modes can be excited at a unique cavity length irrespective of the location of the coupling probe, except for certain cavity lengths where the roots of the Bessel's function are equal for the TM and TE modes [4]. At this cavity length, hybrid TE and TM modes are said to exist. Theoretically, the TM-mode should be preferentially excited from the axially mounted probe and the TE-mode from the radial mounted probe at the same cavity length cavity length.

In the dual-coupling technique, the loaded cavity lengths where hybrid modes exist were located empirically [7]. This was done by heating samples at a fixed cavity length for both coupling configurations. At the cavity length corresponding to a hybrid mode, different heating patterns were achieved for both coupling configurations. It was expected that coupling from the axial position would excite the TM component of the hybrid modes and the side would excite the TE component. Each coupling probe was connected to an independent microwave system which included a magnetron power source and power meters. While one coupling probe was used the other was retracted from the cavity while the reflected power was measured from both probes. The microwave power source was a magnetron operating at a frequency of 2.45 GHz with a maximum power of 100W. Sample temperature was measured using a fiber optic unit and thermal paper.

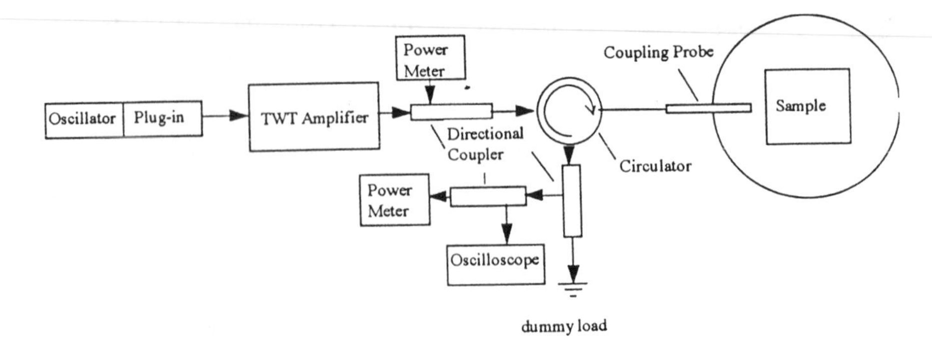

Figure 3. Variable frequency system.

Figure 4. Heating pattern at 2.2135 GHz.

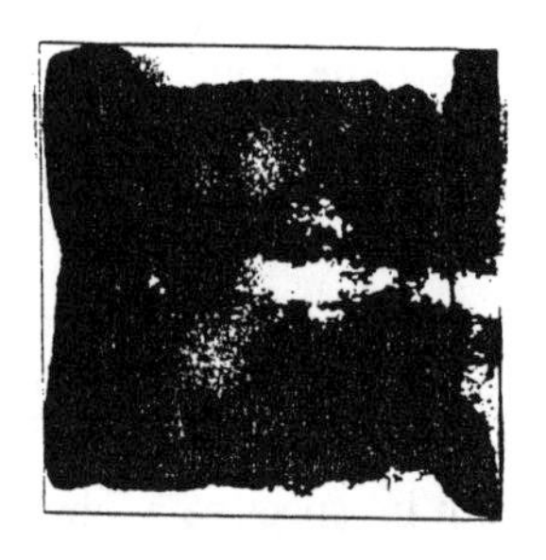

Figure 5. Heating pattern at 2.2133 GHz.

Figure 6. Heating pattern at 2.4359 GHz.

Figure 7. Heating pattern at 2.5342 GHz.

Two different heating profiles were observed for heating a 3- inch. square graphite/epoxy sample (Figures 10 and 11) at cavity lengths of 19.0 cm and 19.7 cm for the top and side coupling probe, respectively. By simultaneously coupling from the top and side at a cavity length of 19.9 cm, more uniform heating was achieved (see Figure 9) which looked like the superposition of the side and top heating modes. It was only at this cavity length where simultaneous coupling was possible where at the other cavity lengths, as one probe coupled energy into the cavity, the other probe coupled the microwave energy out of the cavity and samples did not heat.

DISCUSSION

All three methods discussed, have unique attributes that make each one a worthy technology for achieving uniform heating in different applications. The fixed frequency system could be applicable to processing systems that do not require several modes to achieve uniform heating or those that do not require rapid cavity adjustment response times. This is due to the mechanical cavity adjustments required for mode-switching, which makes this method difficult for several rapid mode-switching adjustments. The variable frequency method, on the other hand, could be applied to processes that require rapid process response times because mode-switching would be accomplished by varying the frequency electronically, providing a faster process response time. The dual-coupling method could be applied to samples of unique geometry that require both axial and radial fields to achieve uniform heating.

CONCLUSIONS

Uniform heating of polymer matrix composite materials using fixed frequency mode-switching, variable frequency heating, and dual-coupling were discussed. The fixed frequency mode-switching method was demonstrated for flat panel, V-shaped, and triplanar vinyl ester / glass composite materials. Variable frequency heating was demonstrated for 2.0-inch. square 48-ply graphite epoxy composite materials, and dual-coupling heating was demonstrated for a 3-inch square graphite epoxy composite materials. All of the methods were shown to be effective in achieving uniform heating, but each was more applicable to different processing needs. The fixed frequency method would be more applicable to systems that require few modes to achieve uniform heating by mode-switching; the

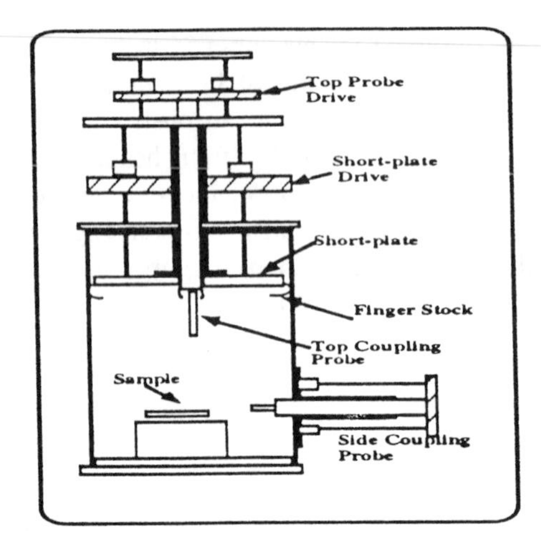

Figure 8. Single-mode resonant cavity for dual coupling.

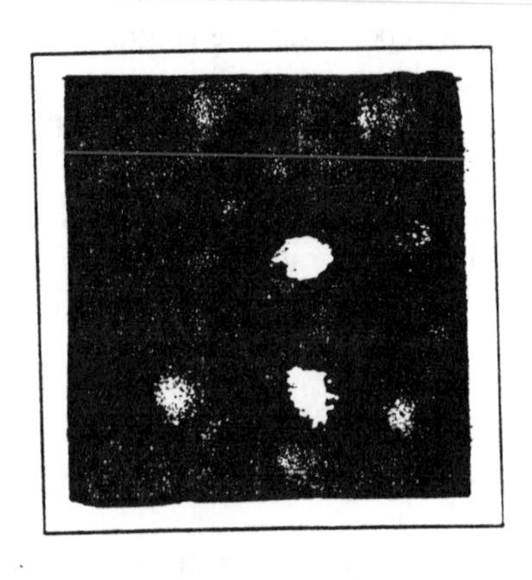

Figure 9. Thermograph from dual coupling for heating graphite epoxy composite, CL = 19.9 cm.

Figure 10. Thermograph for coupling from the side, CL = 19.7 cm.

Figure 11. Thermograph for coupling from the top, CL = 19.0 cm.

variable frequency method would be more applicable to systems that require rapid process response times; and the dual-coupling method would be applicable to processing complex shapes that require both the radial and axial electric fields to achieve uniform heating.

REFERENCES

1. J. Jow, Ph.D. Dissertation, Department of Chemical Engineering, Michigan State University (1988).
2. J. Wei, Ph.D. Dissertation, Department of Chemical Engineering, Michigan State University (1992).
3. J. Asmussen, et al., Rev. Sci. Instrum. 58[8], pp. 1477-1486 (1987).
4. R.F. Harrington, Time Harmonic Electronic Fields, McGraw-Hill, New York (1961).
5. L.A. Fellows, R. Delgado and M.C.Hawley, presented at the 26th International SAMPE Technical Conf., Altanta, GA, Oct. (1994).
6. Y. Qiu, V. Adegbite and M.C. Hawley, to be presented at the 30th Microwave Power Symposium, Denver,CO, July (1995).
7. V. Adegbite, J. Wei and M.C. Hawley, 29th Microwave Power Symposium, IMPI, pp. 156-159 (1994).
8. L.A. Fellows and M.C. Hawley, 29th Microwave Power Symposium, IMPI, pp. 916 (1994).
9. A.C. Johnson, et al., American Ceramic Society Transactions, Vol. 36, pp. 563-570 (1993).

Microwave Raytracing in Large Overmoded Industrial Resonators

L. Feher*, **G. Link**, **M. Thumm****

Forschungszentrum Karlsruhe, Technik und Umwelt
Institut für Technische Physik-ITP
P.O.Box 3640, D-76021 Karlsruhe - Germany
**** and University Karlsruhe,**
Institut für Höchstfrequenztechnik und Elektronik,
Kaiserstr. 12, D-76128 Karlsruhe, Germany

Abstract. High power millimeter waves are important for industrial processing of dielectric materials. Uniform and homogeneous electromagnetic field distributions within the resonators are desired for controllable production rates. Precise knowledge of the stationary field deposition is therefore essential for microwave oven design and application. The industrial resonators that we are interested in here, make great demands for computer simulation in memory allocation and computational speed, because of the resonator's dimensions ($\sim$ m) compared to the small microwave wavelength ($\sim$ mm). This is particulary important in three dimensions, which leads to computationally intensive numerical optimizations using standard schemes for solving Maxwell's equations, such as Finite Difference-, Finite Element methods etc.. Optical approaches promise a much more flexible engineering support for designing and optimizing resonator conditions even on smaller workstations. Investigations and results based on a generalization of Fermat's principle for wave propagation will be presented and compared to measurements performed on a 30 GHz gyrotron installation.

Microwave installation for ceramics processing

In recent years, growing interest has been attracted to high-temperature microwave processing of materials and, primarily, to such applications as ceramic sintering, joining of ceramics and metals, etc.. Volumetric heating resulting from an absorption of microwave energy within the bulk of material leads not only to savings in energy and processing time. The fast and easily controllable procedure of microwave heating offers unique capabilities for producing materials with new properties. The implementation of traditional industrial microwave techniques at frequencies of 0.915 and 2.45 GHz for high-temperature processes is restricted due to the low absorption of decimeter wavelength radiation in most advanced ceramic materials and the non-uniform distribution of the electromagnetic field within the applicator. For the processing of large size specimens a highly uniform field distribution is required. With increasing frequencies respectively, these problems are reduced, because of higher microwave absorption in dielectrics with increasing frequencies and due to the fact that the spatial non-uniformity of the field distribution is diminished with an increasing ratio of the furnace dimensions to the wavelength. Thus, millimeter wave power sources initially developed for plasma heating have been applied for processing of new materials not feasible at lower frequencies [1]. A compact Gyrotron Oscillator Technological System (GOTS) has been installed at Forschungszentrum Karlsruhe, Germany, which makes use of large experience of the Institute of Applied Physics of the Russian Academy of Sciences in Nizhny Novgorod in the development of gyrotrons. This GOTS is a reliable and versatile tool for investigations in high temperature processing of materials with millimeter-wave radiation [2]. The second harmonic gyrotron, generating millimeter-waves at 30 GHz in the TE_{02} cavity mode is operated in the continuous wave (CW) regime with a manual or PC-controllable output power up to 10 kW. The generated radiation passes through a mode converting quasi-optical transmission line, consisting of three metallic mirrors, and through a BN vacuum window into the applicator.

Figure 1: Resonator and control unit of the Gyrotron Oscillator Technological System

Theoretical aspects of stationary wave propagation

The propagation of electromagnetic waves is basically a time dependent process and is represented by Maxwells equations. Matter is described by the parameters $\epsilon_r, \mu_r, \sigma, \rho$: these are the dielectric permittivity ϵ_r, μ_r magnetic permeability, (this can be set to $\mu_r = 1$ for dielectrics), electrical conductivity σ and the charge density ρ (in general, the energies applied the applicators of interest do not cause ionizing processes; thus $\rho \to 0$). Our dielectric ceramics are expected to be isotropic, but inhomogeneous, so that we can write Maxwell equations to be

$$\nabla \cdot \vec{E}(\vec{x}, t) = 0 \qquad \nabla \cdot \vec{B}(\vec{x}, t) = 0 \tag{1}$$

$$\nabla \times \vec{B}(\vec{x},t) = \mu_0 \sigma \vec{E}(\vec{x},t) + \mu_0 \epsilon_0 \epsilon_r \frac{\partial \vec{E}(\vec{x},t)}{\partial t} \qquad \nabla \times \vec{E}(\vec{x},t) = -\frac{\partial \vec{B}(\vec{x},t)}{\partial t} \tag{2}$$

Since the optical axes are randomly oriented, it is possible in any case to approximate the permittivity tensor by an effective scalar ϵ_r in terms of its diagonalized elements $\epsilon_r = \frac{1}{3}\sum_{i=1}^{3}\epsilon_{ii}$ Reaching the stationary state for the electromagnetic field ($\approx 10^{-7}$ s) occurs reasonably fast in comparison to the thermodynamical process of sintering. On the thermodynamical time scale, the stationary electromagnetic energy distribution is determined by assigning the temperature dependent values of $\sigma(T), \epsilon_r'(T)$ for the processed material. In the domain which we consider there is no current from free electrons or ions,

Material	ϵ_r	$\tan\delta$	$\sigma[S/m]$	reference
Teflon^R [2.45GHz]	2.1	0.0019	$5.4 \cdot 10^{-4}$	[3]
Al_2O_3 [8.5 GHz]	9.31	0.00034	$1.5 \cdot 10^{-3}$	[4]
Silicon nitride [8.5 GHz]	8.63	0.0064	$2.6 \cdot 10^{-2}$	[4]
Boron nitride [24 GHz]	7.01	0.0048	$4.5 \cdot 10^{-2}$	[4]
AlN [145 GHz]	8.3	≈ 0.00075	$5.0 \cdot 10^{-2}$	[5]
BeO [145 GHz]	6.7	0.0007	$3.7 \cdot 10^{-2}$	[5]

Table 1: Examples for dielectric parameters (room temperature)

so only a current density of the dielectric matter is contributing by $\vec{j}(\vec{x},t) = \sigma\vec{E}(\vec{x},t)$. We want to take into account inhomogeneous distributions in space with $\epsilon_r' = \epsilon(\vec{x})$ or $\sigma = \sigma(\vec{x})$ for the material. We obtain immediately the equation

$$\nabla \times \ \nabla \times \vec{E}(\vec{x},t) + \mu_0 \epsilon_0 \epsilon_r' \frac{\partial^2 \vec{E}(\vec{x},t)}{\partial t^2} + \mu_0 \sigma \frac{\partial \vec{E}(\vec{x},t)}{\partial t} = 0 \tag{3}$$

Using standard identities, no free current and the ansatz for harmonic time variation of electric fields $\vec{E}(\vec{x},t) = \vec{E}(\vec{x})e^{-i\omega t}$, one obtains a vectoric Helmholtz equation as an exact second order composite of Maxwells first order equations

$$\nabla^2 \vec{E}(\vec{x}) + k^2(\vec{x})\vec{E}(\vec{x}) = 0 \tag{4}$$

The complex wave number $k(\vec{x})$ is a spatial function representing the material properties such as permittivity and absorption according the relation $k^2(\vec{x}) = \frac{\omega^2}{c^2}\epsilon_r^*(\vec{x}) = k_0^2 n^2(\vec{x})$ with $\epsilon_r^* := \epsilon_r' + i\epsilon_r''$ as the complex permittivity, while the imaginary part obeys the condition $\epsilon_r''(\vec{x}) = \frac{\sigma(\vec{x})}{\omega\epsilon_0}$. $n(\vec{x})$ describes the complex optical refractive index of the material. The loss tangent is defined by ϵ_r^* as its complex phase $\tan\delta = \frac{Im(\epsilon_r^*)}{Re(\epsilon_r^*)}$. A solution of the original problem (2) formulated in terms of the fields $\vec{E}$ and $\vec{B}$ can now be treated by solving Helmholtz's equation (4) separately for $\vec{E}$ with given stationary boundary conditions. This property can be observed as an advantage for determining the power density within the resonator or the processed material according $\rho_{power}(\vec{x}) = \frac{1}{2}\sigma(\vec{x})|\vec{E}(\vec{x})|^2$. Thus, directly solving (4) for example with finite difference techniques could save necessary computational memory and time compared to the direct time dependent formulation. But for the required sizes of very large resonators and proposed high microwave frequencies,

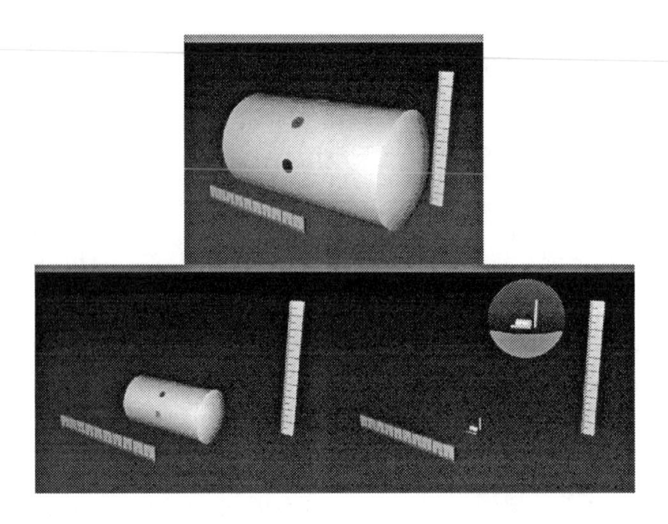

Figure 2: Resonator sizes and memory

the memory requirement still outstrips the capability of the most modern supercomputer facilities. A cylindrical resonator of length ≈ 1.6m and radius ≈ 0.8m with 30 GHz, for example, needs a calculation grid resolution of $\frac{\lambda}{10} \approx 1$mm. Unfortunately, no symmetry conditions except for very special cases can be applied for simulating the wave propagation inside the loaded resonator with realistic beam coupling. The gridsize as required in MAFIA [6] simulation codes are of the size $800 \times 800 \times 1600 \approx 1024$ Mega cells. Double precision (8 Bytes) and 3 components for the $\vec{E}$ field therefore require an allocation of 24 Giga Byte RAM for each iterative step. In addition, no precise knowledge about the stationary conditions in this field formulation is known, and iterative converging algorithms will cost a lot of computation time. Figure 3 shows a vivid example: on the top picture, the original resonator (together with 1 m rulers) with the above mentioned sizes is shown. This kind of resonator is situated at University of Stuttgart (Institut für Plasmaphysik,IPF) for investigations on industrial microwave processing. At the lower left, the size of a resonator according to the available memory on a well equiped workstation (130 MB) using Finite Difference Time Domain solvers of the MAFIA software package at 2.45 GHz is visualized. This is about a quarter of the original. Trying to simulate a 30 GHz microwave radiation input the resulting simulation geometry (Fig.2 down right, with tiny dm rulers) cannot be scaled up in a comparable realistic manner to the original hardware. We try to derive therefore a computational method from Helmholtz's equation for a consistent quantitative evaluation of the $\vec{E}$ field and obtaining the energy density distribution $\sim |\vec{E}(\vec{x})|^2$. For industrial processing it is our goal to work out a numerical technique that will be fast and flexible in memory allocation and can be used on smaller computers (workstations). We will show, that this code is well suited for use with modern parallizing facilities.

Distribution of phase inside the resonator

Our crucial determining equation (second order in space) for the stationary $\vec{E}$ field is represented by (4) and boundary conditions. At the wavelengths considered, interference is important for the energy distribution. Basic elements describing wave phenomena are phase and amplitude. At this point, we want precise knowledge about the spatial phase structure. We call $S(\vec{x})$ the according stationary phase function in space. If u is an arbitrary Cartesian component of $\vec{E}(\vec{x})$, thus getting the scalar equation

$$\nabla^2 u + k^2 u = 0 \; . \tag{5}$$

The ansatz $u(\vec{x}) = A(\vec{x}) e^{ik_0 S(\vec{x})}$, with $A(\vec{x})$ as the amplitude function of the corresponding field component leads to the coupled equations

$$k_0^2 (\nabla S)^2 = k_0^2 \epsilon_r' + (\nabla(\ln A))^2 + \nabla^2(\ln A) \tag{6}$$

$$2(\nabla S) \cdot \nabla(\ln A) = -\nabla^2 S - k_0 \epsilon_r'' \tag{7}$$

arranged for real and imaginary part. The phase function $S(\vec{x})$ was introduced by Sommerfeld [7] and is called the Eikonal (greec: picture). We have to consider this formulation of the Helmholtz equation under the aspect, that the ratio of the characteristic resonator dimension L and the wavelength of the propagating wave is large. Therefore in (6) terms with k_0 dominate, and we get the important relation

$$[\nabla S(\vec{x})]^2 = \epsilon_r'(\vec{x}) = n(\vec{x})^2 \tag{8}$$

$$\epsilon_r'' \approx 0 \Longleftrightarrow \sigma \to 0 \tag{9}$$

These are the basic equations of geometrical optics, where (8) is known as the *Eikonal equation*. The absorption (ϵ_r'') in this context does not have any influence on the phase distribution $S(\vec{x})$; this is exclusively given by the permittivity of the matter. Our limit $\lambda \to 0$ in addition leads us to the result (7), that the conductivity σ of the considered material has to be sufficiently small. This is very well fulfilled for dielectrics (see Table 1). If $S(\vec{x})$ is given as a solution of (8), wavefronts can be identified as areas of constant phase $S(\vec{x}) = const$. The curve that is orthogonal in each point to these wavefronts is called a ray with the local ray vector $\vec{t}$

$$\hat{t}(\vec{x}) = \frac{\nabla S(\vec{x})}{|\nabla S(\vec{x})|} \tag{10}$$

We can state at this point immediately, that in general such a ray propagates a curved trajectory in inhomogeneous materials. (8) can be reformulated (10) to a vector equation $n\hat{t} = \nabla S(\vec{x})$.

Boundary conditions and modelling

The ray is our basic means of transport for distributing along the wave properties phase and amplitude in space. The numerical realisation of such a scheme has been given by the MiRa-Code (**Mi**crowave **Ra**ytracing). Solving the scalar equation (5) for plane waves at plane surfaces of homogeneous media results in phase conditions for waves at boundaries known as Fresnels Formulas: obtaining refracted rays, reflected rays and rays traversing the different media. A rigorous tensor treatment for arbitrary curved surfaces and

inhomogeneous media is developed for solving the vectoric Helmholtz equation (4) determining the explicit amplitude conditions for the wave-surface interaction and propagation (as unpublished yet).
Finite boundaries, such as edges and tips yield in diffraction regions, that occur as total shadow regions tracing the reflected and transmitted rays. We call this geometrical optics wave field the primary wave field. Reflected and diffracted waves can be traced as secondary wave fields according Snell's law, Fresnels formulas and Geometrical Theory of Diffraction (GTD) [8].

An important role is played by the modelling of the radiation sources. Extended beams

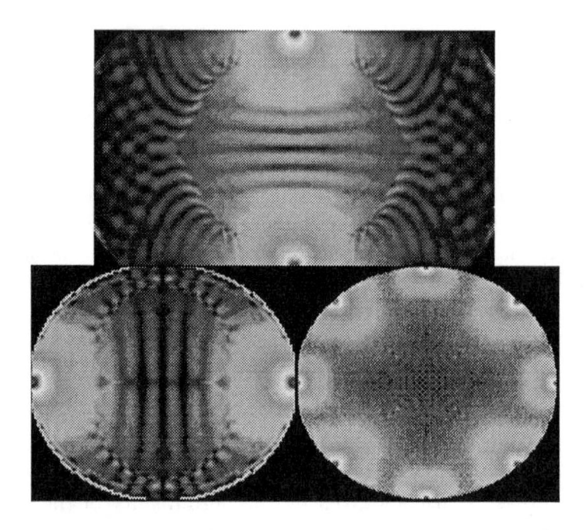

Figure 3: Results for resonator of Fig.2 (top) for 2.45 (top,lower left) and 30 GHz(lower right)

with characteristic shapes are constructed by the Huygens-Fresnel Theory [9], as a composite of different spatial point sources. A single ray out of a set of rays from an original point source is traced until the intensity decayed under a specified threshold. One must be sure, that the spatial density of the emitted rays from a point source is sufficiently high for generating enough events for each voxel.

Results on microwave raytracing

Fig.3 presents some results on the resonator design of Fig.2 with 2.45 GHz and 30 GHz. The lighter regions in this figure represent higher field intensity. In the top picture the energy density structure is shown in the mid plane along the cylinder axis with two radiation sources at 2.45 GHz facing each other. Several caustics with high intensity dividing geometrical light/shadow zones can be recognized.
The effect of interference at this frequency leads to spatially fluctuating inhomogeneous energy distributions. The picture at the lower left shows the perpendicular cross section of

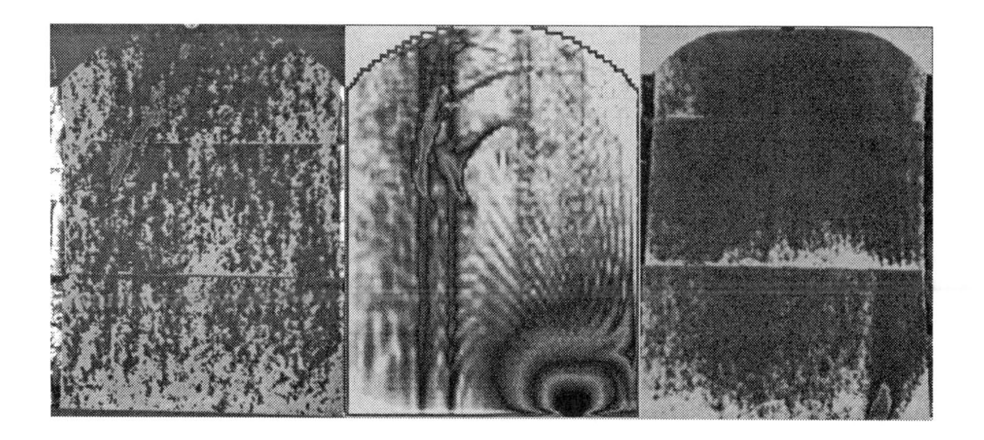

Figure 4: Comparison of measurement and simulation for the Gyrotron Oscillator Technological System (GOTS) at 30 GHz

the resonator cylinder with related cylinder caustics framing the standing wave structure between the sources. A clear advantage concerning the fluctuations can be seen at the picture lower right using 30 GHz radiation in the same plane, leading to a much smoother microwave illumination. Unfortunately, the figure does not clearly resolve the microwave fine structure of the phase.
In Fig.4, first results of a MiRa-simulation (middle) as well as measurements obtained by microwave heated thermopaper (left) inside the unloaded resonator with removed mode stirrer (left) of the GOTS and moving stirrer representing an averaged field distribution at 30 GHz are shown. The caustic due to the focusing effect of the concave shaped furnace lid is clearly visible in simulation and stirrer removed experiment (left). Several caustics caused by the cylindrical geometry of the resonator can be recognized in simulation and experiment. The intense microwave beam on the lower right of the resonator was not detected during this experiment. The thermopaper's microwave exposure is dominated by nonlinear effects, so that structures and quantitative results have to be obtained and compared by considering a set of measurements. The averaged field with stirrer (right) shows a clearly visible stationary phase structure at the right of the resonator. This remaining structure could be verified by MiRa-simulation series of the rotating stirrer.

Software engineering and conclusions

The microwave raytracing scheme is a powerful numerical technique for reconstructing stationary wave field distributions in large bounded geometries. The intention of the MiRa-Code as a tool for engineering support implies easy programmable interfaces for complex resonator geometries without any access to the code implementation for the user. The resonator is designed by a script, that itself can be controlled and programmed from outside for processing automatically e.g. geometrical studies, phase studies etc.. A set of different and easy controllable radiation sources (point source, dipole, beams,Gaussian beam) are implemented and can be handled by the user like a geometrical object. The independence of the raytracing scheme from any simulation grid in the sense of calculating a voxel in terms of its neighbour voxels as in e.g. Finite Difference schemes can be used for software implementations under PVM (Parallel Virtual Machine, available for the C and C^{++} languages) to run on UNIX clusters for parallizing the tracing of independently propagating wave fronts on different machines with joined access for the summed up complex $\vec{E}$ fields. This property, finally, helps us in a flexible way, to solve the previously discussed RAM allocation problem. It is possible with this scheme, to define detector areas of interest for evaluating the field energy without the needing to keep knowledge about the whole propagation area. With several simulation runs, the whole field inside the resonator can be reconstructed out of the single stored detector cubes.

Acknowledgements

Thanks to Prof. Edith Borie for discussions and critically reading the manuscript.
Thanks to Dr. G. Müller at University of Stuttgart, IPF for very friendly cooperation and fruitfull discussions.

References

[1] Yu.V. Bykov, V.E. Semenov, "Processing of Material Using Microwave Radiation" in "Application of High-Power Microwaves", ed. A.V.Gaponov-Grehkov and V.L. Granatstein. Artech House Inc. Norwood MA, (1994) ISBN 0-89006-699-x,319-351

[2] Yu.V. Bykov,A.G. Eremeev, V.A. Flyagin, V.V. Kaurov, A.N. Kuftin, A.G. Luchinin, O.V.Malygin, I. Plotnikov, V.E. Zapelov, L.Feher, M. Kuntze, G. Link, M. Thumm, "Gyrotron System for Millimeter-Wave Processing of Materials", Microwave and Optronics 1995, May 30-June 1, Stuttgart Sindelfingen, Germany

[3] V.E. Jenas, C.J. Malarkey, M.F. Iskander,O. Audrade, S. Bringhurst, "Dielectric Property Measurement of Ceramic Bodies", Ceramic Transaction Vol.36, Microwaves: Theory and application in materials processing II, Ed. David E. Clark, Wayne R. Tinga, Joseph R. Laia, 1993,493-502

[4] Editor William B. Westphal, "Dielectric Constant and Loss Data", Report Number AFML-TR-740-250, Part III, June' 80

[5] R. Heidinger, G. Link, The temperature dependence of the permittivity in high power and broadband mm-wave window materials, Proceedings of the 6th Russian-German Meeting on ECRH and Gyrotrons,1994

[6] T. Barts, "Maxwell's Grid Equations", Frequenz 44,1990,9

[7] A. Sommerfeld, "Optik", Verlag Harri Deutsch, ISBN 3-87144-377-8,180

[8] V.A. Borovikov, B.Ye. Kinber, "Geometrical theory of diffraction", IEEE Waves Series 37, ISBN 0-85296-830-2

[9] G. Barton, "Elements of Green's Functions and Propagation", Oxford Science Publications, ISBN 0-19-851998-2,345

VIBRATION METHOD FOR TRACKING THE RESONANT MODE AND IMPEDANCE OF A MICROWAVE CAVITY

M. Barmatz, O. Iny, T. Yiin and I. Khan
Jet Propulsion Laboratory, California Institute of Technology
Pasadena, CA 91109

ABSTRACT

A vibration technique has been developed to continuously maintain mode resonance and impedance match between a constant frequency magnetron source and resonant cavity. This method uses a vibrating metal rod to modulate the volume of the cavity in a manner equivalent to modulating an adjustable plunger. A similar vibrating metal rod attached to a stub tuner modulates the waveguide volume between the source and cavity. A phase sensitive detection scheme determines the optimum position of the adjustable plunger and stub tuner during processing. The improved power transfer during the heating of a 99.8% pure alumina rod was demonstrated using this new technique. Temperature-time and reflected power-time heating curves are presented for the cases of no tracking, impedance tracker only, mode tracker only and simultaneous impedance and mode tracking. Controlled internal melting of an alumina rod near 2000°C using both tracking units was also demonstrated.

INTRODUCTION

In recent years, there has been a concerted effort to improve the microwave heating technique for processing materials at high temperature [1]. Improved optimization of the energy transfer from a microwave source to the material being processed is required in many applications before the microwave heating technique will become commercially competitive. The most important factors that limit the transfer of energy to the material are the detuning of the microwave cavity resonance and the impedance match between the microwave source and microwave cavity as the material is heated. These detuning effects are associated with the temperature dependent variation of the complex dielectric constant of the material during processed. In the case of the impedance match, a variation in the imaginary component of the dielectric constant causes a change in the cavity load as seen by the power source driving the cavity. The variation in the real part of the dielectric constant is primarily responsible for a similar change in the cavity resonance.

For a constant frequency microwave source, such as a magnetron, correcting for variations in the cavity resonant frequency generally requires the use of a mechanical plunger to adjust one of the microwave cavity dimensions. By changing the cavity dimension during processing, one can attempt to continuously tune the cavity resonance to the fixed frequency source. The problem is that manual adjustment of the plunger is generally not an efficient method to control the microwave processing of the material.

Impedance matching between a microwave source and applicator load has been accomplished in the past by either manually adjusting the position or shape of the excitation coupler at the cavity (adjustable iris method) or manually changing the phase of the incoming microwave energy using stub tuners (phase shifter method) [2,3]. There are many cases where the rate of change of the cavity load during microwave processing is so fast that manual adjustment is too slow to maintain impedance matching. The lack of maintaining a resonance and impedance matching can lead to a significant reduction in the power transfer to the material being processed. This is particularly true for low absorbing materials where the quality factor of the excited cavity resonant mode is very large.

A new automatic vibration method for tracking the cavity resonance and impedance match was developed and tested This method optimizes microwave power transfer to a material being processed and thus allows a much larger class of low loss materials to be heated to higher temperatures starting from room temperature. This new capability should enhance the development of economical commercialization processes using microwave heating.

RESONANT FREQUENCY TRACKER

One approach for tracking a resonance is to frequency modulate the carrier signal. In this approach, the modulation in the drive signal leads to a modulation in the resonant condition of the cavity [4]. A microwave diode can be used to detect the associated electric field modulation in the cavity. By comparing the phase shift between the detected and carrier signals, one can determine the direction needed to shift the frequency of the source back to the resonant condition. This method works very satisfactorily with microwave sources that are broadband, such as a Traveling Wave Tube (TWT) source. In most processing applications, however, a fixed frequency magnetron source is used.

Our solution for a fixed frequency source was to find an alternative approach to modulate the resonant frequency of the cavity. This was accomplished by using a loudspeaker to vibrate a metal rod. The rod protruded into the cavity through a hole in the wall. The vibrating rod modulated the volume of the cavity which was equivalent to modulating the plunger and hence the resonant frequency. The most effective location for the entrance hole corresponded to a position of maximum electric field strength. This position will depend on the microwave mode being excited. For this study, a TE_{102} mode in a WR 284 waveguide cavity was used as shown in Fig. 1. The vibrating rod was placed a quarter of the way along the cavity which corresponds to a maximum electric field line for this mode. An adjustable diode crystal detector that monitored the relative power level associated with the electric field was also placed along the same maximum electric field line at the opposite wall from the rod.

An oscillator drove the loudspeaker at a low modulation frequency in the range 30 < f_1 < 100 Hz. A hollow, low mass, metal cylindrical rod was attached to the loudspeaker cone and its tip inserted ≈ 0.6 cm inside the cavity. The detected low frequency modulation of the electric field obtained by the diode was sent to an amplifier and then to a multiplier unit. The multiplier was used to measure the relative phase shift between the source and detector signals. This phase shift determined which direction the plunger should be moved.

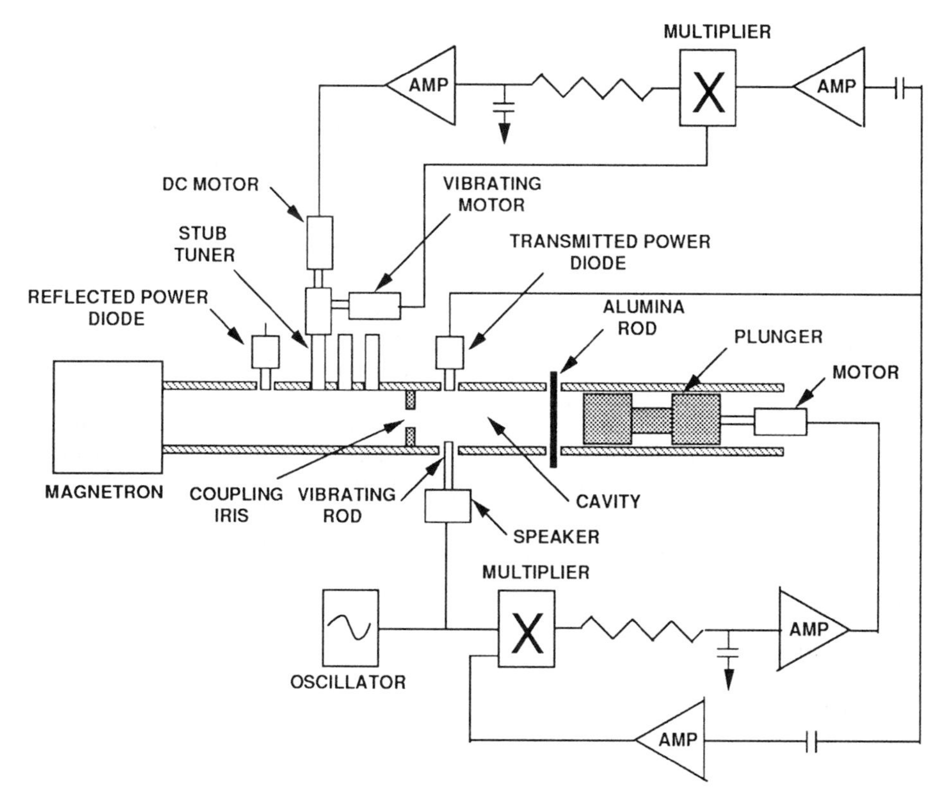

Fig. 1. Schematic of the microwave system and phase sensitive detection schemes used in the vibration tracking method.

IMPEDANCE MATCHING TRACKER

The apparatus used in the impedance matching technique is also shown schematically in Fig. 1. When the power source is critically coupled to the cavity, all the microwave energy is transmitted to the cavity and thus the transmitted power diode signal is maximum and the reflected power diode signal is zero. An ac motor driving an eccentric crank on its shaft was used to modulate one of the stub tuners. This modulation amplitude remained constant with its value being determined by the displacement of the eccentric. The modulation frequency, f_2, was controlled by the motor speed and could be varied from $0 < f_2 < 12$ Hz. The modulation frequency was measured using a light sensitive diode positioned to view a shiny surface, attached to the motor shaft, that oscillated through the diode field of view. A phase sensitive detection technique, similar to the one used to track the cavity resonance, is shown schematically in Fig. 1. By comparing the phase relationship between the modulation source and transmitted or reflected power diode signals, one could determine whether the stub tuner was above or below the optimum time averaged position for impedance matching. If the position of the stub tuner was not optimized, a positive or negative dc signal was generated to control a motor that could move the time averaged position of the vibrating stub tuner up or down. One advantage of using the reflected power diode is that it more strongly senses the stub tuner modulation than the transmitted power diode, especially at high temperatures where the field strength in the cavity is weak. The reflected power diode also more weakly senses the mode resonance modulation and thus essentially decouples the two modulation signals.

RESULTS AND DISCUSSION

The versatility of these tracking techniques was demonstrated by heating a 0.63 cm diameter, Vesuvius McDanel Company, 99.8% pure alumina rod. Before each heating test, the stub tuners and plunger position were optimized at room temperature. A constant 200 watts of forward power was used for all the tests. The first heating test was performed with both the impedance and resonant mode trackers off. An Everest IR noncontact thermometer, with a range of 0 - 1100°C, measured the surface of the rod half way down in the cavity. Figure 2 shows the sample temperature and reflected power as a function of time. As the sample temperature rose, both impedance and resonant mode mismatch occurred. Since the detuning of the impedance and mode resonance both depend on the temperature dependence of the complex dielectric constant, the sample temperature climbed asymptotically to a maximum temperature of ≈ 470°C. This maximum temperature depends on a balance between the degree of detuning and the corresponding reduction of power transmitted into the cavity. The reflected power, which is associated with both the impedance and mode resonance detuning became very large reaching an asymptotic value of ≈ 160 watts. The initial drop in reflected power at the beginning of this test was associated with a slight initial detuning of the system.

In the next test, only the resonant mode of the cavity was continuously tracked and adjusted to match the magnetron frequency. A modulation frequency of 30 Hz was used with this mode tracker. By tracking the resonant mode, one can eliminate any reduction in power transfer to the material (and increase in reflected power) primarily due to changes in the real part of the dielectric constant. Figure 3 again shows the sample temperature and reflected power under these conditions. Because of the higher temperatures measured in this run, a second measurement of the sample surface on the opposite side from the Everest (dashed curve) was made using

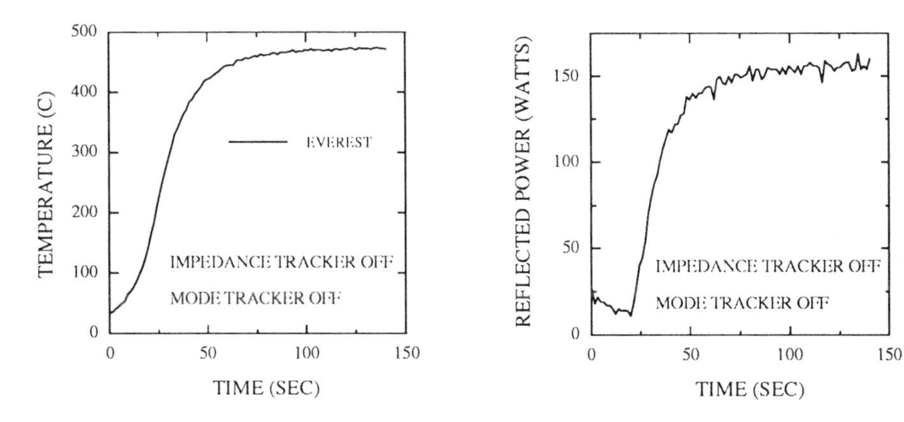

Fig. 2. Temperature and reflected power heating curves with both vibration tracking units off.

an Accufiber pyrometer thermometer (solid curve) with a range 500 - 2400°C. The maximum temperature reached ≈ 1255°C using the plunger mode tracker. Also, the reflected power reached ≈ 90 watts which is much less than with no resonant mode tracker (see Fig. 2). This 90 watts was assumed to be associated with the impedance detuning that was still present. By comparing the maximum reflected values obtained in Figs. 2 and 3, one can deduce that the detuning of the resonant mode caused ≈ 70 watts of reflected power for the present configuration.

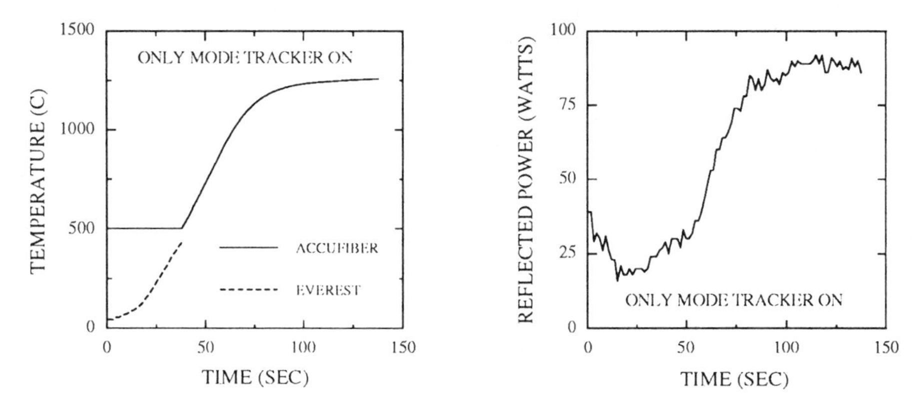

Fig. 3. Heating curves using only the cavity mode vibration tracking unit.

In the next test, the impedance mismatch between the cavity load and magnetron source was tracked and minimized by motor controlling one of the stub tuners (the most sensitive one). This tracking unit used a modulation frequency of 6.4 Hz. This frequency was chosen to be significantly different from the 30 Hz of the plunger

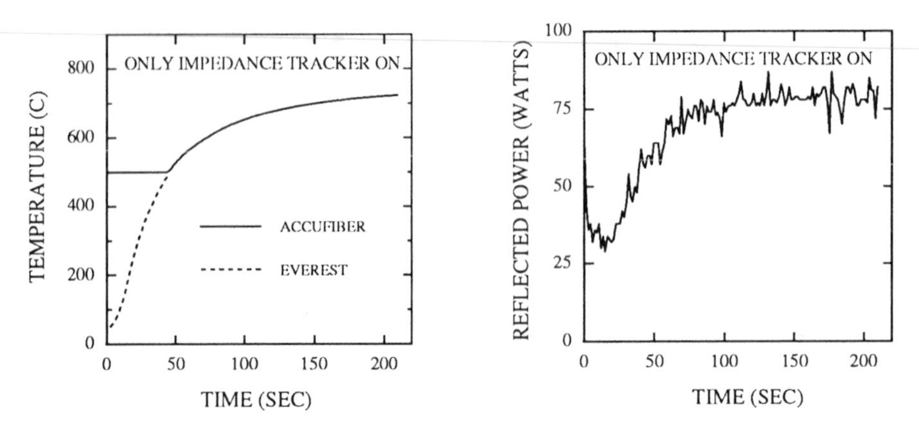

Fig. 4. Heating curves using only the impedance vibration tracking unit.

tracker and not to be a submultiple of that frequency. Again both thermometers were used. Figure 4 shows the sample temperature and waveguide reflected power versus time. The alumina rod initially heated quickly until the cavity mode frequency was sufficiently shifted away from the magnetron source frequency. As this occurred there was a gradual decrease in the transmitted power (increase in reflected power) since less microwave energy entered the cavity to heat the sample. Since this detuning of the impedance depends primarily on the temperature dependence of the complex dielectric constant, the sample temperature climbed asymptotically to a maximum temperature. This maximum temperature depends on a balance between the degree of mode detuning and the corresponding reduction of power absorbed by the sample. This time the sample surface reached a maximum temperature of ≈ 725°C which was lower than the maximum of 1255°C obtained in Fig. 3 using only the resonant mode tracker. By comparing the heating curves of Figs. 3 and 4, one can see that for the present configuration the resonant mode tracker is more important than the impedance tracker for reaching the highest sample temperature. The reflected power reached a maximum of ≈ 80 watts that is consistent with our estimate of 70 watts obtained from Figs. 2 and 3.

Figure 5 shows the sample heating curve and waveguide reflected power when both the stub tuner and the resonant mode vibrating units are in operation. In this test, the impedance tracker was turned on after ≈ 13 seconds. Thus, during this initial time interval, the reflected power began to increase as the sample temperature increased. The reflected power was returned to a minimum value within ≈ 20 seconds after turning on the impedance tracker. It took ≈ 40 seconds for the sample temperature to reach 500°C after which the sample quickly heated to ≈ 1440°C within two minutes. The heating rate of the alumina sample also increased significantly to ≈ 20°C/sec. The reason for the improvement in the peak temperature over the previous tests was that maximum power was now continuously being transmitted into the cavity. The maximum sample temperature

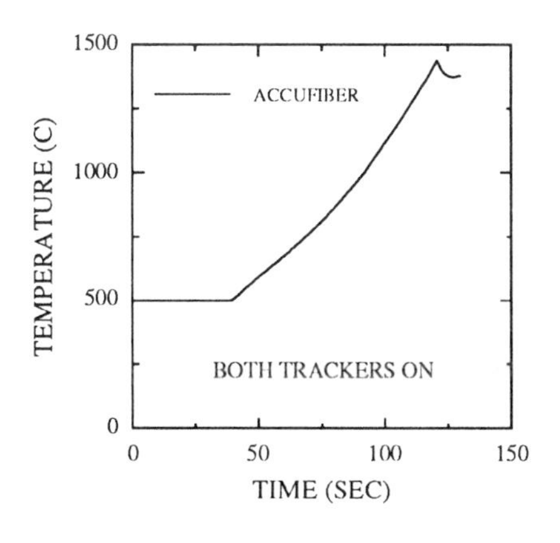

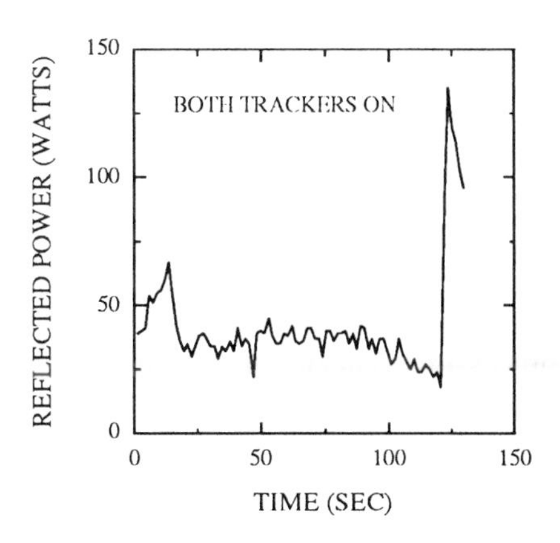

Fig. 5. Heating curves using both the cavity resonance and impedance vibration tracking units.

was also enhanced by the fact that the imaginary dielectric constant, ε", increases with temperature. This increase in ε" leads to an increase in the microwave absorption in the sample that further tends to heat the sample to even higher temperatures. The reflected power did not reach zero even with both stub tuner and resonant tracking units on. The resultant 20 watt minimum baseline, occurring at higher temperatures, was probably associated with the detuning of the impedance matching between the microwave source and the cavity due to the other fixed stub tuners.

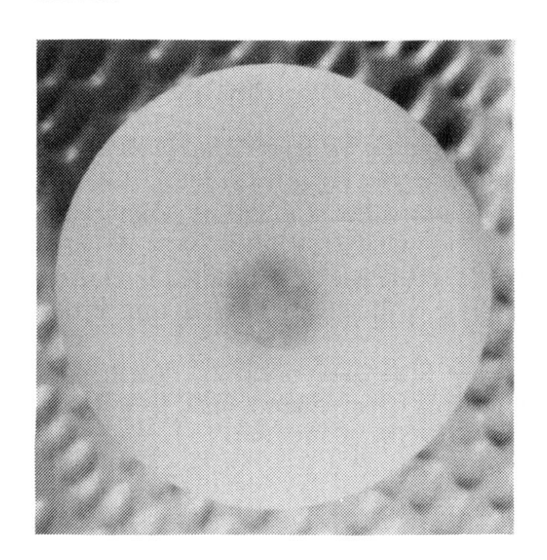

Fig. 6. Cross-section of 0.5 cm diameter rod showing internally melted region.

When the alumina rod reached approximately 1400°C there was a sudden increase in the reflected power leading to a drop in temperature. This phenomenon was associated with the onset of melting inside the sample. After this heating test, the sample was cut at the position where the temperature was measured and observed an internally melted region as shown in Fig. 6. Since the melting point of this high purity alumina is ≈ 2000°C, there was ≈ 600°C temperature gradient between the surface and center of the sample. This behavior is consistent with the predictions of microwave absorption models [5].

CONCLUSION

In conclusion, we have developed a novel vibration tracking technique. This technique was implemented to maintain a frequency and impedance match between a cavity load and magnetron source. With these tracking units in operation, a high purity alumina rod was internally melted within two minutes using only 200 watts of forward power. We successfully demonstrated the tracking of the most sensitive stub tuner for our configuration. Adjustment of up to three stub tuners may be necessary to continuously match the source to the cavity during high temperature processing. This in-situ vibration method for tracking the resonant mode and load impedance should permit enhanced process monitoring and control and lead to the high temperature processing of a large number of low microwave absorbing materials that have commercial value.

ACKNOWLEDGMENT

We thank Dr. Henry W. Jackson for many stimulating discussions regarding the vibration tracking technique and R. Zanteson for technical assistance in fabricating the vibration tracking units. The research described in this article was carried out at the Jet Propulsion Laboratory, California Institute of Technology, under contract with the National Aeronautics and Space Administration.

REFERENCES

1. W. H. Sutton, M. H. Brooks, and I. J. Chabinsky, Editors *Microwave Processing of Materials*, Mat. Res. Soc. Symp. Proc., **124**, Pittsburgh, PA (1988).
2. H. Fukushima, T. Yamanaka and M. Matsui, "Microwave Heating of Ceramics and its Application to Joining," Mat. Res. Soc. Symp. Proc., **124**, pp. 267-272 (1988).
3. A. C. Metaxas, "Rapid Feasibility Tests Using a TE_{10n} Variable Aperture Resonant Applicator," J. Microwave Power and Electromagnetic Energy, **28**, (1), pp. 16-24 (1990); and "Applicators For Industrial Microwave Processing," Ceramic Transactions, **36**, pp. 549-562 (1993).
4. O. Iny and M. Barmatz, "Decay Method for Measuring Complex Dielectric Constants During Microwave Processing," Mat. Res. Soc. Symp. Proc., **347**, pp. 241-245 (1994).
5. M. Barmatz and H.W. Jackson, "Steady State Temperature Profile in a Sphere Heated by Microwaves," Mat. Res. Soc. Symp. Proc., **269**, pp. 97-103 (1992).

Microwave/Materials Interactions

MICROWAVE ABSORPTION IN IONIC CRYSTALLINE CERAMICS WITH VARIOUS CONCENTRATIONS OF POINT DEFECTS

Binshen Meng[a], John Booske[a,c], Reid Cooper[b,c], and Benjamin Klein[a]
[a] Department of Electrical and Computer Engineering
[b] Department of Materials Science and Engineering
[c] Materials Science Program
University of Wisconsin, Madison, WI 53706

ABSTRACT

A theoretical model of microwave absorption in ionic crystalline solids has been synthesized and specifically applied to $NaCl$ single crystals by considering all relevant mechanisms, including ionic conduction, ion jump dipole relaxation and multiphonon processes. The loss factor ϵ'' has been measured by a cavity perturbation technique for nearly-pure and Ca^{++} doped $NaCl$ single crystals in the frequency range of $2 - 16\,GHz$ and for the temperatures between 300^0K and 700^0K. The experimental results are in good agreement with the theoretical model.

INTRODUCTION

An understanding of interaction mechanisms between microwave radiation and ionic crystalline solids is crucial for several materials applications of growing interest and importance, namely: (1) materials selection and fabrication for advanced radomes as well as window and insulator structures in high power coherent microwave sources, (2) microwave processing of ceramics, and (3) complex dielectric properties of ceramic substrate materials for microwave and high speed digital circuits.

In this paper, we describe a composite theory on the relative roles of fundamental mechanisms for absorption of microwave radiation in ionic crystalline solids, as well as the experimental results for $NaCl$ ionic single crystals with varied concentrations of cation impurities (Ca^{++}) in the temperature range ($300 - 700^0K$) and the frequency range ($2 - 16\,GHz$). A particular understanding of the influence of point defects (vacancies, impurities) on microwave absorption processes was an original motivation for this work.

Electromagnetic energy loss mechanisms in single crystals with various point defect concentrations have previously been studied in the low frequency range of

0.3 – 300 KHz [1, 2] and to a more limited extent in the millimeter-to-infrared frequency range[3, 4, 5]. However, an unmet need exists in the microwave region for systematic, simplified experiments that isolate and quantitatively identify the precise roles of microwave absorption mechanisms including those related to the presence of various types of crystal defects.

Microwave energy absorption in ionic crystalline solids results from ohmic loss and dielectric loss. The former is due to electrical conduction of mobile charges and the latter is due to the relaxation of polarized bound charges. In general, the time-averaged power dissipated per unit volume in a material can be expressed as $P_{loss} = \frac{1}{2}\omega\epsilon_0\epsilon'' E^2$, where ϵ'' is the imaginary part of the complex dielectric constant of the material, ω is the angular frequency of the electric field E, and ϵ_0 is the permittivity of free space. The complex and frequency-dependent dielectric constant $\epsilon = \epsilon_0(\epsilon' - i\epsilon'')$ is determined by the structural properties and thermodynamic state of the material.

In ionic crystalline solids, the dominant absorption or relaxation mechanisms can be expected to change as a function of the microscopic defect conditions. In frequency regimes outside the microwave region, three mechanisms have been experimentally verified. At frequencies above 30 GHz and in relatively pure single crystals (e.g., with impurity levels $\leq 10ppm$) microwave absorption is dominated by quasi-resonant multi-phonon interactions[3]-[5]. At much lower frequencies–i.e., between 1 and 10 KHz–and higher point defect concentrations, ionic conduction and ion jump relaxation mechanisms have been extensively studied and experimentally characterized [1, 2, 6]. However, we still do not possess a quantitative, experimentally-derived knowledge of the conditions (frequency, temperature, and type and density of point defects) under which these different mechanisms dominate the absorption of microwave radiation. Hence, our studies will establish this fundamental knowledge for single crystal ionic solids with varying levels of point defect concentrations.

MICROWAVE ABSORPTION MECHANISMS

On the basis of existing theories, the total absorption of microwave energy in single-crystal wide band-gap ionic solids is hypothesized to result from the simultaneous contributions of ionic conduction, ion jump relaxation, and multi-phonon processes. These mechanisms should exchange the dominating role as conditions such as defect concentration, frequency, and temperature vary. Thus, the imaginary part of the total complex dielectric constant ϵ'' can be expressed as

$$\epsilon'' = \epsilon''_c + \epsilon''_j + \epsilon''_{mp}, \tag{1}$$

where ϵ''_c, ϵ''_j and ϵ''_m represent contributions from ionic conduction, ion jump dipole relaxation and multi-phonon processes, respectively. Theoretical formulas for quantitatively predicting the contributions to microwave energy absorption from each of the mechanisms are presented below. Specific illustrative examples based on $NaCl$ are discussed.

Ionic Conduction

In ionic crystals, the most important mechanism for electrical conduction results from ion migration in the presence of an external electric field due to the existence of both intrinsic and extrinsic defects allowing both cations and anions to jump among their equivalent lattice sites. However, since the charge carriers contributing to electrical conductivity are those with neighboring vacancies, we can equivalently treat the vacancies as the charged carriers. Calculations based on this model lead to a thermally activated expression for conductivity , which yields[7]

$$\epsilon_c'' = \frac{ne^2a^2\nu}{\omega\epsilon_0 kT}e^{-\frac{\Delta g_m}{kT}}, \tag{2}$$

where, n is the mobile charge carrier number density, and for $NaCl$, $a = 3.99\mathring{A}$ is the ion jump distance, $\nu \approx 2 \times 10^{13} s^{-1}$ is the lattice vibrational frequency, and $\Delta g_m \approx 0.69eV$ is the Gibbs free energy for ion motion.

To obtain the equilibrium number density of free cation vacancies n, one simultaneously solves the reaction equations for Schottky vacancy formation, vacancy-vacancy $v - v$ pair formation, and vacancy-impurity $v - i$ pair association along with the mass action law and charge neutrality[2, 7].

From Eq. (2) one can see that ϵ_c'' has the exponentially-sensitive temperature dependence expected of electrical conductivity in a thermal equilibrium system characterized by Maxwell-Boltzmann energy statistics.

Ion Jump (or Dielectric) Relaxation

As discussed previously, at a given temperature a certain fraction of vacancies and divalent impurities will be associated to form pairs (both $v - v$ pairs and $v - i$ pairs) because of attractive Coulomb forces. Each pair consists of two charge carriers with opposite polarities so that the pair can be treated as a dipole. When an external electric field is applied, the dipoles will be aligned in the direction of the electric field. If the field is alternating, once the field changes direction the dipoles will be reoriented. This is referred to as ion jump relaxation. During this process, a certain amount of electromagnetic energy is irreversibly transferred to the dipole system (dielectric loss). Hence, the microwave absorption results from the interaction of the dipoles and the microwave radiation, which can be modeled by classical dynamics and statistics [1, 6]. Following Ref. [1, 6], we can obtain the imaginary part of the dielectric constant due to ion jump relaxation contributed by both $v - v$ pairs and $v - i$ pairs in a ionic crystal having the rock salt structure as

$$\epsilon_j'' = \frac{Nx_{vp}e^2a^2(\epsilon_\infty + 2)^2}{54\epsilon_0 kT}\frac{\omega\tau_{vp}}{1+\omega^2\tau_{vp}^2} + \frac{Nx_a e^2a^2(\epsilon_\infty + 2)^2}{28\epsilon_0 kT}\frac{\omega\tau_a}{1+\omega^2\tau_a^2}, \tag{3}$$

where N is the number density of normal cation ion sites, a is the lattice parameter ($a = 5.639\mathring{A}$ for $NaCl$), $\epsilon_\infty = \epsilon'(\omega \to \infty)$, x_{vp} is the mole fraction of associated $v - v$

pairs, x_a is the mole fraction of associated $v-i$ pairs, and τ_{vp} and τ_a are relaxation times for $v-v$ pairs and $v-i$ pairs, respectively, and given by $\tau_{vp} = \frac{1}{4\nu_0} e^{\frac{\Delta g_m}{kT}}$ and $\tau_a = \frac{1}{2\nu_0} e^{\frac{\Delta g_m}{kT}}$ where ν_0 is the characteristic (reststrahlen) phonon frequency ($\nu_0 \approx 2 \times 10^{13}$ for $NaCl$) and Δg_m is Gibbs free activation energy for motion. As above for ionic conduction, the quantities x_{vp} and x_a are obtained by simultaneously solving the reaction equations relating to the formation and annihilation of $v-v$ pairs and $v-i$ pairs along with the mass action law and the requirement of charge neutrality[2, 7]. These calculations properly include the activation energy parameters and the temperature dependency. Similar to the contribution from ionic conduction discussed earlier, the contribution from the ion jump relaxation process is very temperature sensitive. However, the magnitude of ϵ_j'' is typically much greater than ϵ_c'' by several orders of magnitude.

Multi-Phonon Quasi-Resonant Process

In addition to ion jump processes, photon-phonon interactions are another mechanism for absorption of microwave energy, especially in the high frequency range (millimeter-wave and far-infrared). However, its contribution to the frequency range of $2-20\,GHz$ cannot be ignored, particularly in relatively pure crystals where it could even be the dominant process. The interaction between phonons and photons resulting from quantization of the electromagnetic fields and lattice vibrational dynamics leads to microwave absorption in crystals[5]. For the specific case of crystalline materials with the rock-salt structure, multi-phonon absorption of microwave energy is dominated by a lifetime-broadened two-phonon difference process [3, 4]. Based on the work in [3, 4], ϵ_{mp}'' can be calculated as:

$$\epsilon_{mp}'' = \frac{(\epsilon_0 - \epsilon_\infty)\omega_f^3\Gamma}{(\omega^2 - \omega_f^2)^2 + (\omega_f\Gamma)^2}, \tag{4}$$

where ω_f is the resonant frequency of the fundamental reststrahlen transverse-phonon mode. The relaxation rate constant Γ corresponding to this reststrahlen mode is given by $\Gamma = \int_0^{K_{BZ}} dk\, g(k)\, L(\mathcal{E})$, where K_{BZ} is the Brillouin zone edge in the [111] direction, $g(k)$ is determined by the geometry and structure of the crystal, and $L(\mathcal{E}) = \frac{\gamma/\pi\hbar}{(\omega_{ji}-\omega)^2+\gamma^2}$ is a normalized Lorentzian function with $\mathcal{E} = \hbar\omega - \hbar\omega_{ji}$, $\omega_{ji} = \omega_j - \omega_i$. Calculations with the above expressions predict absorption as a function of temperature in good agreement with experimental results in the millimeter-wave frequency range, $f > 30\,GHz$. Our own extrapolations down into the lower microwave region, $2-20\,GHz$ show that ϵ_{mp}'' varies much less sensitively with temperature than contributions from ionic conductivity or ion jump relaxation. In particular, ϵ_{mp}'' varies approximately linearly with both temperature and frequency in this frequency regime.

A Composite Theory

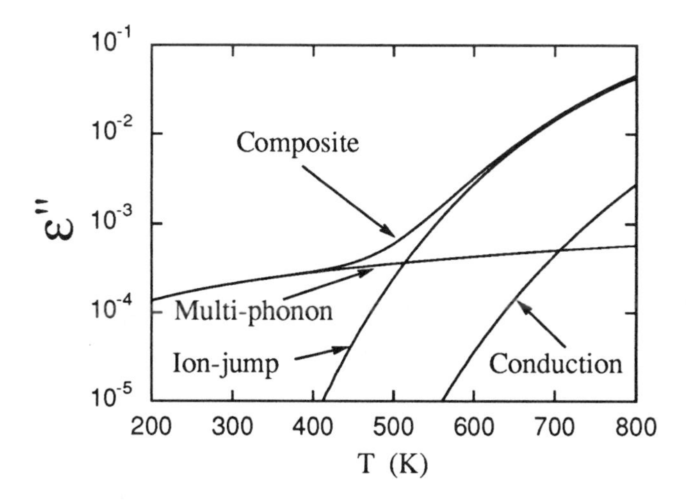

Figure 1: The imaginary part of the dielectric constant as a function of temperature at a frequency of $2\,GHz$ and $100\,ppm\ Ca^{++}$ doping level.

Combining Eqs.(2)-(4), we obtain an analytical expression for the overall imaginary part of the dielectric constant (i.e., Eq. (1)). Experimental evaluation of the accuracy of this composite theoretical model is the point of initial inquiry for the establishment of a quantitative predictive capability for microwave absorption in ionic crystalline solids including the role of crystal defects. To illustrate,in Fig. 1 we have plotted the theoretically predicted microwave absorption properties (specifically, the imaginary part of the complex dielectric constant) versus temperature and frequency for a single crystal $NaCl$ sample with a $100\,ppm$ concentration of Ca^{++} impurities. Generally, the multi-phonon mechanism is expected to dominate microwave absorption at higher frequencies and lower temperatures. Doping with modest levels of Ca^{++} ($100\,ppm$) should have a significant effect on the net microwave absorption at low frequencies (e.g., $2\,GHz$) and high temperatures (above 500^0K) by increasing the contributions from ion-jump relaxation of defect complexes (both $v-v$ and $v-i$ dipole pairs).

EXPERIMENTAL METHODS AND RESULTS

To experimentally evaluate microwave absorption theories for $NaCl$ in the $2-20\,GHz$ range we employed resonant cavity spectroscopy with an improved cavity perturbation method[8]. The measurement system has been described in Ref.[9]. A Hewlett-Packard 8637A synthesized signal generator is used as the microwave source ($2-26\,GHz$). Two resonant cavities are used to provide measurements at five frequencies for spanning the range of $2-20\,GHz$. To ensure that the $NaCl$ specimen is isothermal, the entire cavity is resistively heated, surrounded by thermal insulation, and placed inside a controlled gas chamber to prevent oxidation. The imaginary part of the dielectric constant ϵ'' is determined from the change of Q factor due the presence of the $NaCl$ sample, as measured from the 2-port transmission coefficient of the cavity.

We have experimentally measured ϵ'' in nearly-pure commercial $NaCl$ single crystal samples (Harshaw) and single crystal $NaCl$ samples doped with approximately $200 - 400\, ppm$ Ca^{++} ions for each cavity. The doped specimens started as nearly-pure samples obtained from the same melt as the undoped control specimens. The supplier specifications quote initial impurity levels of approximately $10\, ppm$ (or better). The measurements were performed over the temperature range of $300 - 700^0 K$ at the frequencies of 2.2, 4.8, and $7.4\, GHz$ for the first cavity, and 4.8, 10.2, and $15.5\, GHz$ for the second cavity. The redundancy at $4.8\, GHz$ was intentional to cross-check that both cavities yield identical data with undoped samples at $4.8\, GHz$. The doped samples for the two cavities were prepared seperately so that slightly different concentrations of Ca^{++} in them was expected. The results are plotted in Fig. 2 along with theoretical predictions based on Eqs. (2)-(4). The only theoretical parameter that was adjusted to fit the prediction to the experimental data was the impurity ion concentration. In particular, it was found that best agreement resulted with values of 40 and $370\, ppm$ for the nearly-pure and doped samples, respectively, for the first cavity, and 40 and $160\, ppm$ for the second cavity. The values of $370\, ppm$ and $160\, ppm$ for the doped samples are in excellent agreement with separately measured concentrations obtained using inductively coupled plasma spectroscopy ($375 \pm 30\, ppm$ and $152 \pm 20\, ppm$, respectively). Although slightly higher than the manufacturer's specifications, the $40\, ppm$ value for the nearly-pure sample is considered a reasonable level of residual impurity concentration acquired during the process of growing the pure crystal from the melt. The one caveat is that our model currently assumes a single divalent impurity ion with activation energies for dissociation equivalent to those of Ca^{++} in $NaCl$. Hence, our value of $40\, ppm$ would represent an *effective* Ca^{++} impurity concentration equivalent to the cumulative effect of the probably multiple species of residual impurities in the undoped sample. Measurements to confirm this effective concentration using inductively coupled plasma spectroscopy are currently in progress.

An important distinguishing feature in all of the theoretical curves is the existence of a transition temperature T_t at which there is a transition from the multiphonon process (below T_t) to ion-jump relaxation of defect complexes (above T_t). Generally, T_t varies with both the dopant concentration and the microwave frequency. However, the experimental data and the theoretical predictions are in excellent agreement regarding the value of T_t in all cases. Another noticable feature is that the curves are flatter at the higher frequency ($15.5\, GHz$) as theoretically predicted.

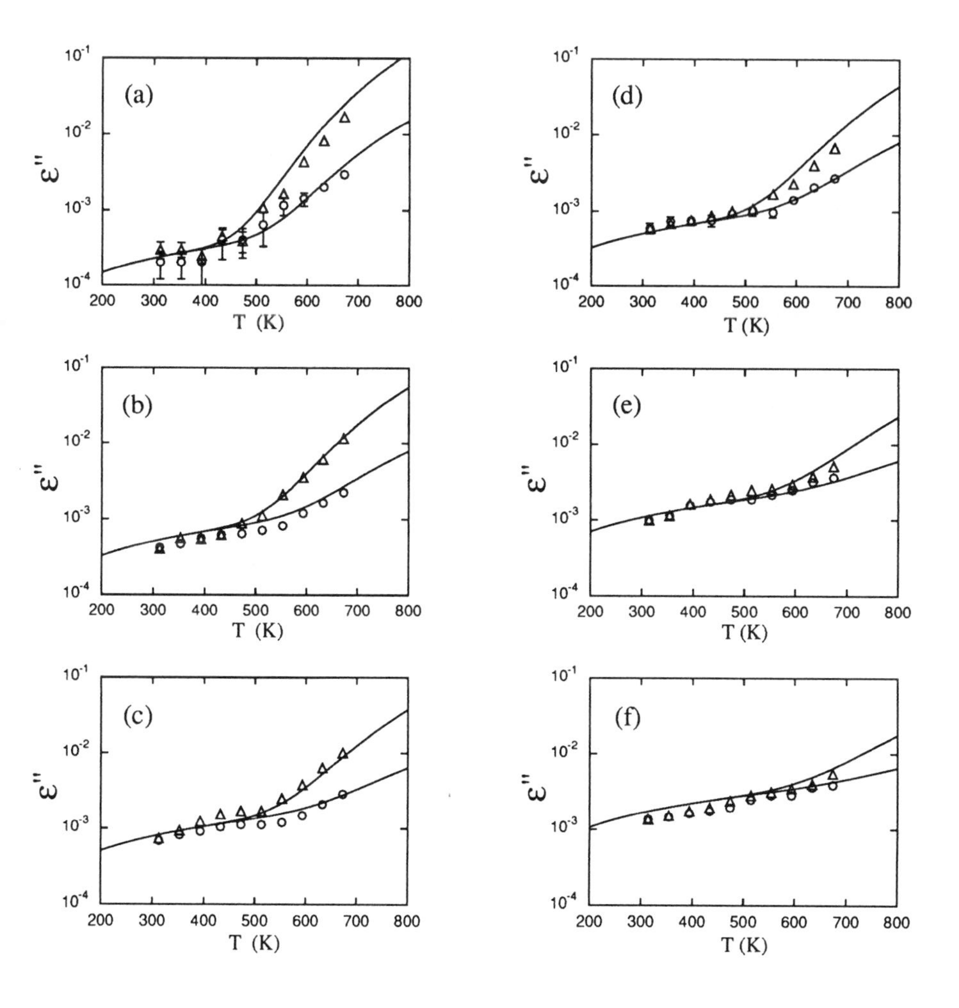

Figure 2: The experimentally measured ϵ'' and theoretically predicted ϵ'' in $NaCl$ single crystals at the frequencies of (a) $2.2\,GHz$, (b) $4.8\,GHz$, and (c) $7.4\,GHz$ versus temperature for the first cavity, and (d) $4.8\,GHz$, (e) $10.2\,GHz$, and (f) $15.5\,GHz$ versus temperature for the second cavity. In all the plots, o: undoped crystal (nearly-pure crystals), $\triangle$: doped crystals. The solid curves represent the theoretical predictions. A $40\,ppm$ effective Ca^{++} concentration was used for all undoped samples. For doped samples, a theoretical fit was obtained with $370\,ppm$ Ca^{++} in (a), (b), (c) and $160\,ppm$ in (d), (e), (f).

ACKNOWLEDGEMENTS

The authors gratefully acknowledge the support of the Electric Power Research Institute, the Wisconsin Alumni Research Foundation, the National Science Foundation through a Presidential Young Investigator Award, and Nicolet Corp. for donation of the digital oscilloscope used in the experiments. Valuable discussions with Sam Freeman and other colleagues contributed to this work. The inductively coupled plasma spectroscopy measurements of Ca^{++} impurity concentrations were performed by the Wisconsin State Laboratory of Hygiene.

REFERENCES

[1] R.G. Breckenridge, *Imperfections in Nearly Perfect Crystals*, Ed. W. Shockley, J.H. Hollomon, R. Maurer, and F. Seitz (J. Wiley and Sons, N.Y., 1952), Chap. 8.

[2] I.E. Hooton and P.W.M. Jacobs, *J. Phys. Chem. Solids* **51**(10), 1207(1990); Hooton and Jacobs, *Can. J. Chem.* **66**, 830 (1988).

[3] M. Sparks, D.F. King, and D.L. Mills, *Phys. Rev.* **B26**, 6987(1982), and references contained therein.

[4] K.R. Subbaswamy and D.L. Mills, *Phys. Rev.* **B33**, 4213(1986).

[5] V.L. Gurevich and A.K. Tagantsev, *Adv. in Phys.* **40**, 719-767(1991).

[6] I. Bunget and M. Popescu, *Physics of Solid Dielectrics*, (Elsevier, Amsterdam, 1984).

[7] W.D. Kingery, H.K. Bowen, and D.R. Uhlmann, *Introduction to Ceramics* (Wiley and Sons, NY, 1976).

[8] B. Meng, J.H. Booske, and R.F. Cooper, submitted for publication to *IEEE Trans. Microwave Theory Tech.*, (1994).

[9] B. Meng, J.H. Booske, and R.F. Cooper, *Rev. Sci. Instrum.* **66**(2), 1068-1071(1995).

IONIC TRANSPORT IN SODIUM CHLORIDE AND SILVER CHLORIDE DURING MICROWAVE IRRADIATION

S.A. Freeman, Mat. Sci. Program, Univ. of Wisconsin, Madison, WI 53706
J.H. Booske, ECE Dept., Univ. of Wisconsin, Madison, WI 53706
R.F. Cooper, MS&E Dept., Univ. of Wisconsin, Madison, WI 53706

ABSTRACT

Our experimental results have shown that activation energies for ionic diffusion are not affected by microwave fields. However, results show that intense microwave fields can *drive* ionic currents in NaCl and AgCl crystals. The results are compared to a theoretical model for radiation pressure that causes a near-surface flux of ionic defects. The results for AgCl are also compared to similar experiments performed some twenty years ago on AgCl. The potential role of the radiation pressure on diffusion, surface reactions, and sintering is discussed.

INTRODUCTION

To date, there has been no thorough and satisfying explanation for the enhanced diffusion and reaction rates reported in microwave processed ceramics [1]. The initial focus was on possible ionic/defect mobility enhancements, but preliminary experiments of ionic currents in sodium chloride (NaCl) proved that defect mobilities in ionic crystalline ceramics are unaffected by intense microwave radiation [2].

However, a recent theory about radiation pressure [3] did introduce a microwave-material interaction mechanism that had not been previously considered. Specifically, the mechanism does not affect the mobilities of ionic species or defects. Instead radiation pressure acts as a driving force for defect fluxes near crystal surfaces, much like a mechanical pressure will cause defect fluxes in the cases of Nabarro-Herring creep and Coble creep. On a microscopic level, the oscillating electromagnetic field causes an oscillation of charged defects that is rectified near surfaces and boundaries, resulting in a net current. Again, our preliminary studies with NaCl showed that such microwave-induced currents do exist, and their behavior is consistent with the radiation pressure model [4,5].

We have since discovered a very similar research effort undertaken some 20 years previously in which the experimenters studied the effects of microwave radiation on silver chloride (AgCl) [6]. The authors observed small enhancements of ionic currents through AgCl immediately after they irradiated the crystal. They attributed the enhancement to an increase in the defect concentration (Ag interstitial, Frenkel defects) caused by the microwave fields. Since it is also possible that their results could be attributed to the radiation pressure theory, we sought to reproduce these experiments with a slightly more sophisticated experimental configuration and with a broader interrogation of the experimental variables.

EXPERIMENTAL APPROACH

Three fundamental concepts went into the design of our ionic current experiments. First, we eliminated the possibility of temperature gradients *within* the sample. This was accomplished by using conventional heating to heat and control the temperature of the sample, and by irradiating the crystal in pulses that are too short to provide microwave heating. Thus, our experimental results are not complicated by the possibility of temperature gradients that can provide an additional driving force for ionic transport, or by the possibility of inaccurate temperature measurement affecting the highly-temperature-sensitive ionic mobilities.

Second, we maintained a constant driving force throughout the experiments. For tests involving sintering, grain growth, diffusion, etc., the driving force is changing during the course of the experiment, thus complicating the analysis. In our case, the driving force for defect diffusion is a non-varying electrical bias applied across the ends of the salt crystal.

Third, we chose ionic materials with simple and well-characterized defect chemistries. Specifically, we chose NaCl, dominated by vacancy defects (Schottky), and AgCl, dominated by interstitial defects (Frenkel). Since these materials have lower melting temperatures than oxide ceramics, the relevant diffusion processes can be achieved at much lower temperatures.

Further details of the experimental configuration can be found in previous work [4,7].

PRELIMINARY RESULTS WITH AgCl

Figure 1 shows a typical response of a AgCl crystal to a microwave pulse. In this case no external bias is applied to the sample. As we found with NaCl, the ionic current resulting from the external bias and the ionic current resulting from the microwave fields are simply superimposed on one another and do not affect one another. We have therefore left the external bias off for the experimental data depicted here.

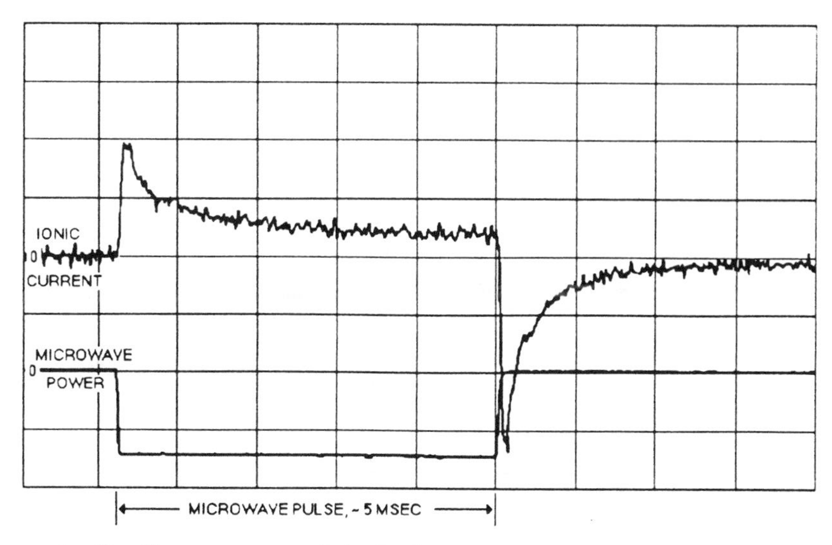

Figure 1: Oscilloscope trace of the ionic current response (top trace) of a AgCl single crystal to a 5 millisecond microwave pulse (bottom trace). The horizontal scale is 1 millisecond/division. The vertical divisions correspond to 200 picoamps.

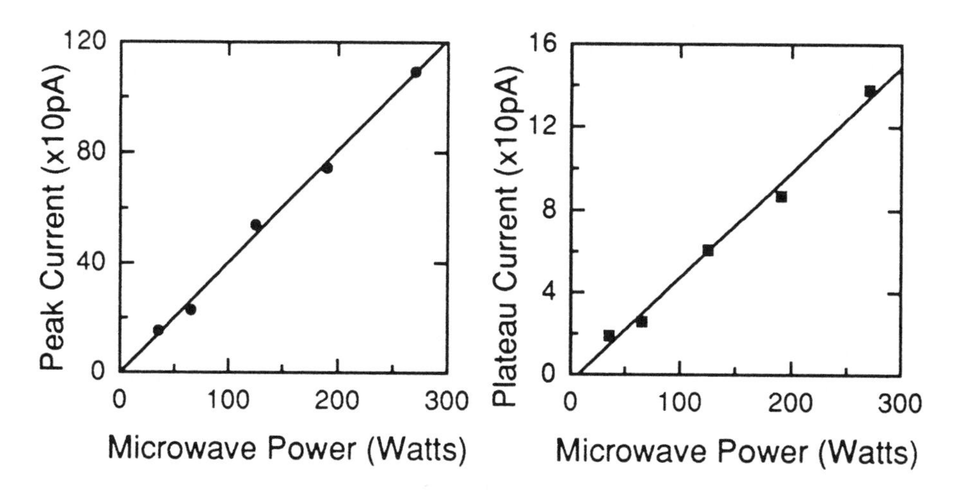

Figure 2: A linear relationship is observed between the microwave power and both the peak current and the plateau current. The peak current is determined from the height of the initial response seen in Figure 1. The plateau current is the current immediately before the end of the pulse.

Several important features of the microwave-induced current are seen in Figure 1. There is a peak current followed by a decay to a pseudo-plateau value. The magnitude of both the peak and plateau currents increase linearly with microwave power, which is consistent with the radiation pressure theory. Figure 2 shows this relationship. Following the microwave pulse, the current swings back in the opposite direction and then decays back to zero. The magnitude of the swing-back depends on the length and power of the microwave pulse. These observations are consistent with a build up (during the microwave pulse) and then relaxation (after the pulse) of a near-surface space-charge layer. The response times (decay times) indicate that this is an ionic charge layer, not an electronic charge layer.

The results seen in AgCl echo the results we had previously seen and continue to see with NaCl. There are some minor differences between the responses of each material which can be attributed to several factors. First, the two salts have different defect chemistries, and interstitial defects are much more mobile than vacancies. Second, AgCl has a greater defect concentration than NaCl. Finally, the AgCl crystals have "reacting", Ag-paint electrodes, while the NaCl crystals have "non-reacting", Ag-paint or Pt-sputtered electrodes.

CONTINUING EXPERIMENTS WITH NaCl

Continued experiments with NaCl crystals have led to further characterization of the microwave effect and enhanced understanding of its nature. By reconfiguring our system, we have been able to explore longer pulse lengths and the dependence of the effect on the pulse length. Figure 3 shows the response of a NaCl crystal to a microwave pulse that is approximately 30 times longer than in our preliminary experimentation.

Figure 3 shows another important feature consistent with the radiation pressure theory. The response has a transient nature to it, and it in fact drops to zero if the pulse length is long enough (as depicted in Figure 3). The radiation pressure model predicts a flux of charged defects to or from the surface (depending on charge polarity) that would build up a electro-chemical gradient to oppose further flux. This is exactly what is seen in Figure 3. An integration of the current versus time yields a total charge of approximately 10^{-11} C. This number agrees well with what we have calculated based on the defect concentrations, concentration gradients, and space-charge-layer thicknesses one would expect to find in NaCl.

There are many aspects of this microwave-material interaction that we are still investigating. Some evidence suggests that the end conditions of the crystals are important, consistent with a phenomena that is sensitive to the levels, types, and distributions of defect concentrations at the ends. The effect is more pronounced in crystals that have had platinum plasma-sputtered electrodes, for example, than in crystals that have silver-painted electrodes. The stress state and stress history of the crystals may also play an important role.

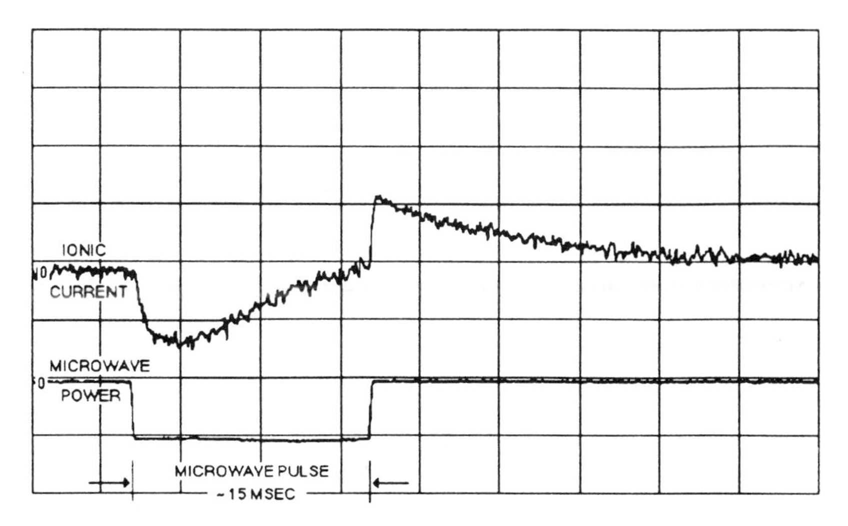

Figure 3: Oscilloscope trace of the ionic current response (top trace) of a NaCl single crystal to a 15 millisecond microwave pulse (bottom trace). The horizontal scale is 5 millisecond/division. The vertical divisions correspond to 100 picoamps.

DISCUSSION OF RESULTS

In the (recently discovered) work of Rind and Martienssen [6], the authors performed very similar experiments to ours. They established an ionic current through a AgCl crystal, irradiated it for 1 microsecond, and saw an increase in the ionic current for microseconds to milliseconds after the pulse. They assumed that their applied bias was the sole driving force, and they therefore interpreted the enhancement of current to be an increase in the Frenkel defect concentration caused by the irradiation. As we understand their experimental techniques, they never changed the applied bias to see if the enhancement scaled linearly with it (as one would expect for a mobility enhancement or a defect concentration enhancement). This is understandable since their research goal was to investigate defect relaxation kinetics in photographic materials. At that time nothing was known about microwave effects on ceramic processing.

We have repeated their experiments (albeit with longer pulses, 500 microseconds to several milliseconds), but we further investigated the effect at different bias voltages. The most telling example is Figure 1 where no bias at all is applied. The effect is still seen even with no externally applied driving force. This offers evidence that the effect is related to radiation-induced driving forces and not to increased defect concentrations.

Since strong experimental evidence for the radiation pressure theory exists, some thought has gone into how this pressure would affect ceramic processing and whether it could explain the enhancements seen in diffusion, sintering, etc. Since the radiation pressure can draw vacancies towards a surface, it is possible that it will assist the transport of ions via vacancies into the material. We cannot speculate, though, whether this assistance would extend beyond the near-surface layer of several to hundreds of atomic layers. On the other hand, it seems likely that the change in defect concentration near the surface could change the chemical activity at the surface, and some effect on surface reactions may be seen. Sintering can involve transport of ions and defects along a grain boundary or through the bulk, so it is not clear how a near-surface effect like radiation pressure would affect ceramic sintering rates.

CONCLUSIONS

Preliminary tests with AgCl and further tests with NaCl continue to show that a microwave effect exists in these materials. The effect is not an increase in the defect concentration or mobility; instead it is a driving force for ionic transport caused by a radiation pressure on the surface of the crystal. In effect, the theory predicts that the high-frequency electromagnetic field causes a high-frequency oscillation of charged defects that is rectified near surfaces, resulting in a net current. Many aspects of the effect are consistent with near-surface space-charge layers built up during the microwave irradiation. The results are in quantitative and qualitative agreement with the theory. It is possible that the radiation pressure could affect diffusion penetration and chemical reactions at the surface, but it is not clear whether this effect would enhance ceramic sintering.

ACKNOWLEDGMENTS

We would like to acknowledge EPRI, NSF (PYI Program), NASA, and the useful discussions we have had with Vladimir Semenov and Kirill Rybakov.

REFERENCES

1. S.J. Rothman, "Critical Assessment of Microwave-Enhanced Diffusion," *Mat. Res. Soc. Symp. Proc.Vol. 347: Microwave Processing of Materials IV*, edited by M.F. Iskander, R.J. Lauf, and W.H. Sutton, pp. 9-18 (1994).
2. S.A. Freeman, J.H. Booske, R.F. Cooper, B. Meng, J. Kieffer and B.J. Reardon, "Studies of Microwave Field Effects on Ionic Transport in Ionic Crystalline Solids", *Microwaves: Theory and Application in Materials Processing II, Ceramic Trans., Vol. 36*, Amer. Cer. Soc., edited by D.E. Clark, W.R. Tinga, and J.R. Laia, pp. 213-220 (1993).
3. K.I. Rybakov and V.E. Semenov, "Possibility of plastic deformation of an ionic crystal due to the nonthermal influence of a high-frequency electric field," *Phys. Rev. B*, v49, n1, pp. 64-68 (1994).
4. S.A. Freeman, J.H. Booske, R.F. Cooper, B. Meng, "Microwave Radiation Effects on Ionic Current in Ionic Crystalline Solids," *Mat. Res. Soc. Symp.*

Proc.Vol. 347: Microwave Processing of Materials IV, edited by M.F. Iskander, R.J. Lauf, and W.H. Sutton, pp. 479-85 (1994).

5. S.A. Freeman, J.H. Booske, R.F. Cooper, "Microwave Field Enhancement of Charge Transport in Sodium Chloride," *Phys. Rev. Lett.*, v74, n11, pp. 2042-5 (1995).
6. W. Rind and W. Martienssen, "Relaxation measurements on Frenkel defects in AgCl and CuCl single crystals," Photographic Science and Engineering, v17, n1, pp. 2-7 (1973).
7. S.A. Freeman, J.H. Booske, and R.F. Cooper, "A Novel Method for Measuring Intense Microwave Radiation Effects on Ionic Transport in Ceramic Materials," *Rev. Sci. Int.*, v66, n6 (1995).

THE ROLE OF SPACE CHARGE FOR MICROWAVE SINTERING OF OXIDES

M. Willert-Porada, University of Dortmund, Dept. Chemical Engineering, D-44221 Dortmund, F.R. Germany

ABSTRACT

For compacted t-ZrO_2 powders with different specific surface area, the influence of space charge for the low temperature microwave heating behavior is discussed. In isothermal sintering experiments of >99.6% Al_2O_3 ceramics doped with Mg^{2+}, indirect evidence is found for a space charge dependent transport mechanism in the early and middle stage of microwave sintering, which may be responsible for the enhanced densification observed in microwave sintering experiments.

INTRODUCTION

Microwave sintering of ceramics is expected to provide new materials, with properties not achievable by other processing methods. Since the beginning of research in this area the question of microwave specific transport phenomena, as well as the introduction of a general mechanism by which microwave radiation could influence the sintering process of ceramics, were subject of controversial discussion. Because of enhanced densification and decreased sintering temperatures observed upon microwave sintering, diffusion coefficients were among the first parameters under investigation in the search for a "microwave effect". However, experimental evidence for a direct effect of microwaves on diffusion is scarce [1]. Further mechanisms considered are, e.g., thermal gradient driven diffusion [2,3], a transient liquid phase increasing the capillary pressure under the action of a microwave field [4], electroponderomotic forces within an amorphous layer around each particle [5], and ionic currents induced by a microwave field [6].

In a real sintering process material transport occurs by numerous mechanisms, which could be more or less susceptible to electrodynamic effects caused by the ac-field present inside a low to medium loss material heated in or by a microwave field. Therefore, at the University of Dortmund, microwave sintering of alumina ceramics was studied in the initial, middle and late stage of microwave sintering, separately [7-12]. Indirect evidence for an influence of the microwave field on the sintering mechanism in the initial and middle stages of sintering was found, based on an accelerated surface area reduction, increased densification and delayed pore closure [10, 11], as well as a different pore size distribution [12] upon microwave sintering as compared to conventional sintering. For the late stage of microwave sintering, a decreased sintering temperature and grain growth was found for aliovalent dopants only [8]. For nanocrystalline powders, increased susceptibility to thermal runaway [13] prevented an analysis of the sintering mechanism.

In order to explain these experimental observations, an electric field effect on material transport comparable to the action of a dc-field on ion transport has been assumed [9]. The dc-like mechanism was attributed to the difference between the time scale of thermally activated atomic jump processes (10^{-13} s) and the applied microwave frequency (10^{-9} s). Because sintering is an irreversible process, flow of material supported by the microwave field during one ac-cycle can not be reversed by the changing field direction in the following cycle [9]. From there it has been concluded, that the enhanced densification upon microwave sintering would result from a superposition of thermally activated transport phenomena with an electrical potential [9]. In binary or ternary compounds with a high degree of ionic bonding, the difference in formation energy for cation and anion vacancies causes an "intrinsic" electrical potential, known as space charge [14, 15], at the grain boundary. In oxide ceramics, space charge of 0.5 - 2.0 V is calculated [15] or measured due to segregation of dopants [16]. In conventional sintering, grain boundary mobility and grain growth [17], as well as impurity segregation [16], are controlled by this space charge. In the present paper, the interaction of this intrinsic space charge [14] with the ac-field existing in the volume of an oxide ceramic processed in a microwave field will be discussed in terms of heating behavior and material transport upon sintering.

To distinguish between surface or grain boundary defects and volume defects, heating experiments on t-ZrO_2 powder compacts of different specific surface area and dopant level were performed within the study presented here. In nanocrystalline powders, the amount of grain boundary phase is significantly increased; therefore, surface defects or space charge contributions to ac conductivity could become important [14]. By comparison of the results of dielectric and mechanical loss measurements, the nature of defects responsible for the dielectric loss behavior at low and high temperatures is analyzed. Based on this investigation, a mechanism which is not directly responsible for the heating behavior but which could facilitate densification due to the electrochemical potential active in microwave sintering experiments will be discussed.

EXPERIMENTAL

Y-free, t-ZrO_2 powders with specific surface area >150 m^2g^{-1} were prepared by microwave pyrolysis of emulsified $Zr(O\text{-}iPr)_4$, as described in [18]. 8Y-ZrO_2-powders with specific surface area >80 m^2g^{-1} were prepared by the same method, using $Y(NO)_3 \bullet 2\ H_2O$ in EtOH as an additional precursor [19]. Medium to low specific surface area 8Y-ZrO_2 powders were obtained by a conventional thermal method, using $Y(NO)_3$ as the Y-source and m-ZrO_2 (Hereaus, 10 m^2g^{-1}). After solution of Y_2O_3 in the ZrO_2-powder at 1100°C the product was crushed and milled to the desired quality using Ca-ZrO_2 milling media. Contamination due to abrasion was estimated by gravimetric analysis. For <3h milling time, 1% total weight increase is found. To establish comparable dopant and impurity levels, samples for mechanical loss measurements were prepared from milled 8Y-ZrO_2 nanocrystalline and conventional powders, and conventionally sintered at 1550 °C for 2h to 97% of theoretical density (TD).

For dielectric measurements, powders with different specific surface areas were compacted at 300 MPa isostatic pressure into flat shapes or at 1 GPa axial pressure into cylinders. Measurements at 10 kHz-20 MHz were performed on ϕ = 30 mm, h = 2 mm plates polished with diamond paste to enable sufficient contact to the electrodes, using an HP 4275A impedance meter and an apparatus designed for measurements on low loss materials (Dr. Heidinger, KFA Karlsruhe). The samples were heated to 430°C and kept at temperature for several hours, before the measurements were taken during the cooling cycle. Measurements at 2.45 GHz were taken at ambient temperature and pressure in a cavity, using quartz tubes. Corrections for the tube losses were made.

Isothermal microwave sintering experiments on > 99.6 % alumina ceramics doped with MgO (400 ppm) were performed using preheated alumina supports (> 99.6 Al_2O_3) and 0.3-0.4 gcm^{-3} alumina fiber board (Siemens) caskets, as described in [7]. Microwave cold testing (T_{max} 125°C) experiments were performed in a multi mode laboratory oven at ambient pressure and atmosphere. To remove solvents or moisture adsorbed to the surface of the powder, at least three subsequent measurements were taken. Silicone was used as a dummy load to establish a reproducible field pattern. The samples, each weighing 8g and having a cube-geometry (comparable surface/volume ratio), were embedded in fibrous alumina. Temperature readings were taken by a low temperature pyrometer (Keller).

RESULTS AND DISCUSSION

Defects Responsible for Dielectric and Mechanical Loss Behavior of t-ZrO_2-Greenbodies and Ceramics

In powdery or polycrystalline materials with a high degree of ionic bonding, a potential difference exists between the interior of the grain (charge balance is maintained) and the surface (thermal equilibrium is maintained), known as space charge [14,17]. In this region, increased dc-ionic conductivity (Lehovec [14] created the term "space charge conduction") will exist, which decreases at higher temperatures due to the decreasing potential difference between the bulk and the grain boundary region of a crystallite. When an alternating field is applied, the displacement of charges within the space charge layer could cause conduction losses, decreasing with increased frequency. In addition, depending upon the grain size, agglomeration and adsorbed impurities, displacement of the space charge layer itself under the action of the ac-field would induce dipoles, which could cause losses due to dipole relaxation, as shown in Figure 1a. The displacement of the space charge layer by an ac-field could furthermore enhance the conduction and ion jump relaxation loss. In the bulk of a crystallite, oxygen vacancies attracted to the dopant cation form dipoles, as shown in Figure 1b. In an ac-field (E) as well as upon alternating mechanical (F) stress [20], dipole relaxation losses are caused by this type of defect. At 2.45 GHz, the Debye temperature for the latter loss mechanism is ≈1600°C [19]. Therefore, at high temperature dipole relaxation of volume defects will govern the heating behavior.

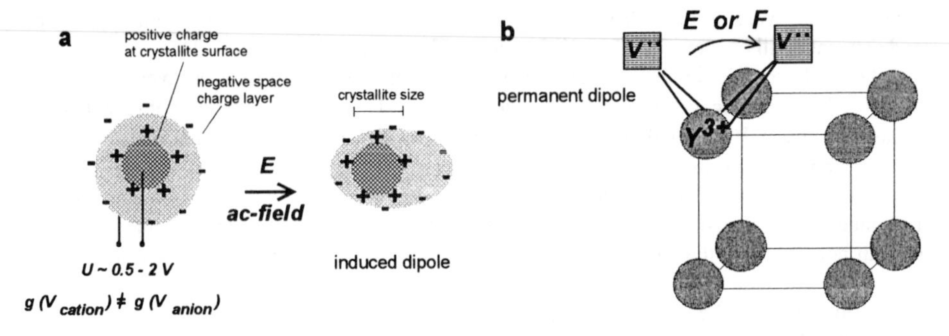

Figure 1. Dipole relaxation in :a) crystallites of oxide ceramics due to the space charge layer surrounding each grain; *for clarity the size of the space charge layer is exaggerated*, b) zirconia ceramics due to O^{2-}-vacancy -Y^{3+} -pairs aligning in an applied mechanical or electrical field

In Figure 2 the microwave heating behavior of different t-ZrO_2 powders at low temperature is shown. The compacted density, ρ, of the samples was used to calculate the specific power absorbed by the sample, as indicated by equation 1. The percent space charge volume is calculated assuming a constant thickness of the space charge of 10 nm [14, 16], and using a particle size as estimated by TEM or REM [18, 19].

$$P_{spec} = \rho \cdot C_p \left(\frac{\delta T}{\delta t_{heating}} - \frac{\delta T}{\delta t_{cooling}} \right) \tag{1}$$

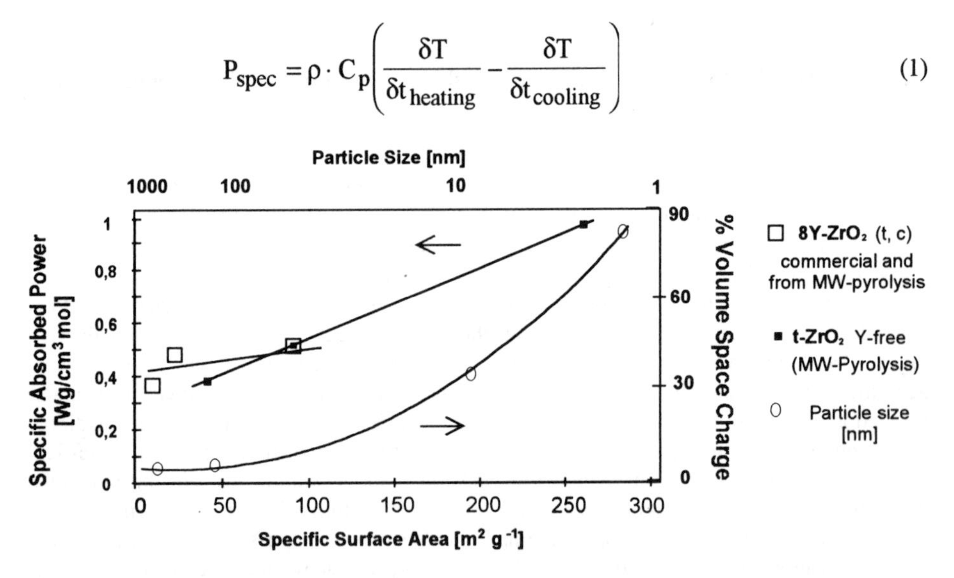

Figure 2. Microwave (2.45 GHz) heating behavior (T_{max} = 125 °C) of compacted t-ZrO_2-powders with different specific surface area and dopant content (Y-free and 8 mol% Y).

As shown in Figure 2, a difference exists between the heating efficiency of t-ZrO_2 powders from microwave pyrolysis as compared to commercial powders. For powders obtained from a precursor pyrolysis route, a pronounced correlation between space charge volume and heating efficiency can be assumed, regardless of the dopant content. The data are taken after subsequent heating to 125 °C, when the initially higher absorbed power reached a steady state value. From the waterfree microwave synthesis [18], contamination of the nanocrystylline powders with paraffin or xylene is possible. As indicated by dielectric loss measurements at 2.45 GHz, a similar H_2O-content due to atmospheric moisture is present in these powders as compared with those processed conventionally. For the two 8Y-ZrO_2-powders with specific surface area < $50 m^2 g^{-1}$, Ca^{2+}-contamination from milling media is assumed. Because Ca^{2+} is an aliovalent additive, it could strongly influence the space charge. It is known from grain growth kinetics and dopant segregation in t-ZrO_2 [17], that aliovalent cation additions change the sign and magnitude of the space charge.

As shown in Figure 3a and 3b, measurements of tan δ over the 10^4 to 10^7 Hz range of frequency indicate a particular loss mechanism existing in nanocrystalline t-ZrO_2 only. A detailed description of the influence of moisture, compaction pressure and thermal conditioning on the temperature- and frequency dependence, as well as the magnitude of the tan δ -maximum appearing at 150°C in Figure 3a, will be given elsewhere [21].

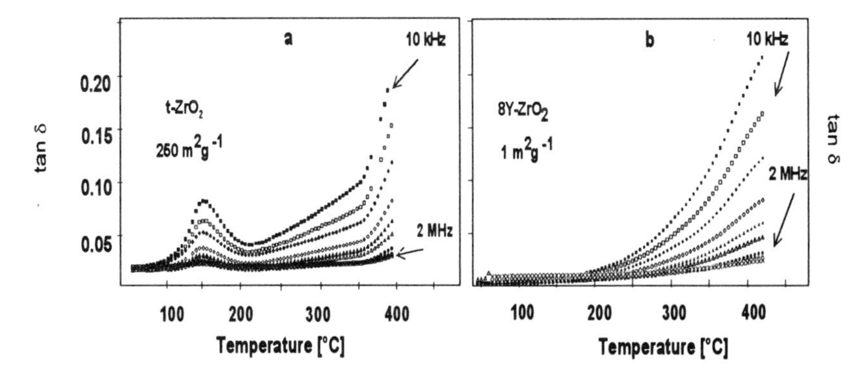

Figure 3. Impedance measurements on t-ZrO_2: a) nanocrystalline t-ZrO_2 without Y_2O_3, b) conventional 8Y-ZrO_2. Measurements taken on cooling from a 3h hold time at 430°C.

Because the tan δ is reduced with increased temperature and frequency, a mechanism based on space charge conductivity, as suggested by Lehovec [14] for ionic compounds with < 10^{-6} m grain size is assumed to become detectable in the green parts prepared from nanocrystalline ZrO_2 powder. Due to the potential difference between the surface and the interior of a crystal under the action of an ac-field, ion migration my occur as well as polarization of the space charge cloud, as indicated in Figure 1a. The space charge potential decreases with increasing temperature and changes its sign, as is known for alkalihalides and t-ZrO_2 [14,17], passing through 0 at a particular temperature. Therefore, a loss mechanism based on the space charge should be detectable in certain temperature and frequency regimes only.

The second tan δ-maximum at T>400°C is related to dipole relaxation of ion-vacancy clusters, as indicated in Figure 1b. [20].

Assuming a steady dependence of dielectric loss on frequency, as shown in Figure 3a and 3b, at microwave frequencies (10^9 Hz) only low tan δ-values can be expected from the process which is responsible for high tan δ at kHz frequencies. Therefore, a major contribution of space charge to the loss behavior at 2.45 GHz is not expected. Cavity Q-measurements on powder compacts at 2.45 GHz indicate an tan δ of 0.1-0.2, independent of the specific surface area of the powder. This value is very susceptible to erroneous interpretation, because within the cavity measurement no external heating was provided, and therefore, moisture desorption from the powder was less effective as was observed with the impedance measurements.

However, as shown in Figure 4, the mechanical loss caused by relaxation of volume defects (see Figure 1b) is different for ceramics prepared from nanocrystalline 8Y-ZrO_2 when compared to ceramics prepared using conventional 8Y-ZrO_2 powder. Both materials were sintered conventionally to a comparable grain size at 97% TD. The lower activation energy for volume defect mobility obtained in mechanical loss measurements could indicate that a lower activation energy is necessary to release oxygen vacancies from the vacancy-cation cluster [20] as it was for the ceramics sintered from nanocrystalline powders. This material property is very important for the development of oxygen ionic conductors with increased ionic conductivity at lower temperatures (e.g., fuel cell applications).

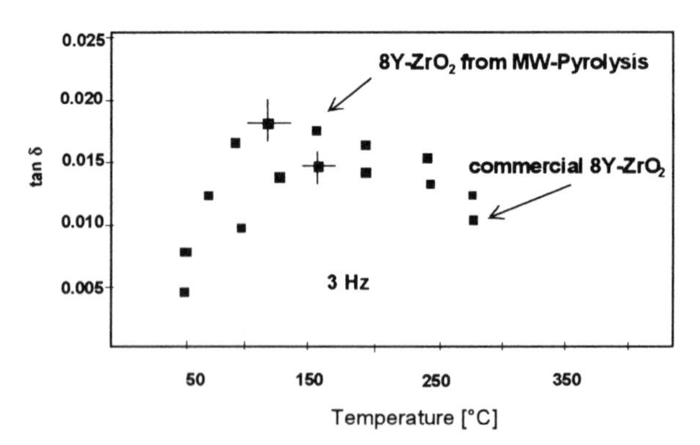

Figure 4. Mechanical loss behavior of t,c-ZrO_2 ceramics sintered conventionally from nanocrystalline and from conventional 8Y-ZrO_2 powders.

Space Charge Effects in Isothermal Microwave Sintering of Al_2O_3-Ceramics

As shown in Figure 5 there is a pronounced difference in grain morphology and densification behavior when >99.6% Al_2O_3-ceramics are sintered isothermally by microwaves as compared to conventional sintering. However, this difference is detectable only in an intermediate temperature region. After prolonged sintering at 1650°C, alumina with an identical grain size/density ratio is obtained from both conventional and microwave sintering experiments [7,9]. Therefore, it is assumed that again space charge effects may be responsible for a temporary change in neck growth, densification and grain faceting.

This assumption is supported by experiments in the late stage of Al_2O_3-ceramic sintering using additives with comparable elastic distortion of the lattice but different charge [8,9], e.g. Mg^{2+} (ionic radius in 8-fold coordination 0.089 nm) and Sc^{3+} (ionic radius in 8-fold coordination 0.087 nm). The Mg^{2+} addition to alumina or zirconia is known to decrease the grain boundary mobility by changing the space charge [17]. Upon microwave sintering, the highest density is reached at 50-100°C lower temperatures and at a smaller grain size when compared with conventional sintering for Mg^{2+}-additions only. Using Sc^{3+} as an additive, no difference in grain growth is observed between microwave and conventional sintering [8,9].

From this and the qualitative microstructure analysis shown in Figure 5, it is concluded that the alternating microwave field is supporting the densification process by a periodic polarization of the space charge region which represents an electrical double layer, as indicated in Figure 6.

Because densification requires the transport of uncharged vacancies from the pore region through the neck region to the exterior of the grain, and the transport of ions in the opposite direction, space charge polarization at the grain surface could significantly affect this process in a medium temperature range, where the space charge has a significant value.

Figure 5. Grain morphology and density in microwave isothermal sintering experiments (top) as compared to conventional sintering (bottom) for >99.6% Al_2O_3 doped with MgO

The intrinsic disorder in alumina may be of either the Schottky or the Frenkel type [15]. In analogy to TiO_2 [16], Schottky defect formation is assumed for Al_2O_3, with $g_{V_O} > g_{V_{Al}}$. According to equation (2) - (4) [16, 22], an excess of negative charges originating from Al-vacancies, V'''_{Al} , is present at the surface, equal to the space charge potential $e\phi$. However, in order to maintain the equilibrium according to equation (5), removal of vacancies from the *pore region* requires transport of both sorts of atoms.

When the external electric field present upon microwave sintering *inside the ceramic* is implemented into equation (2a) and (2b) as an additional local potential, ϕ_{ac} , as shown in equation (2'), the driving force for accumulation of V'''_{Al} would be reduced, according to equation (4'). The difference in formation energy for cation and anion vacancies would than be equilibrated by the sum of the ac- and the intrinsic potential. Therefore, the intrinsic potential would be reduced directly proportional to the external ac-electric field strength [22]. In Figure 6 this is indicated by $U_1 < U_0$.

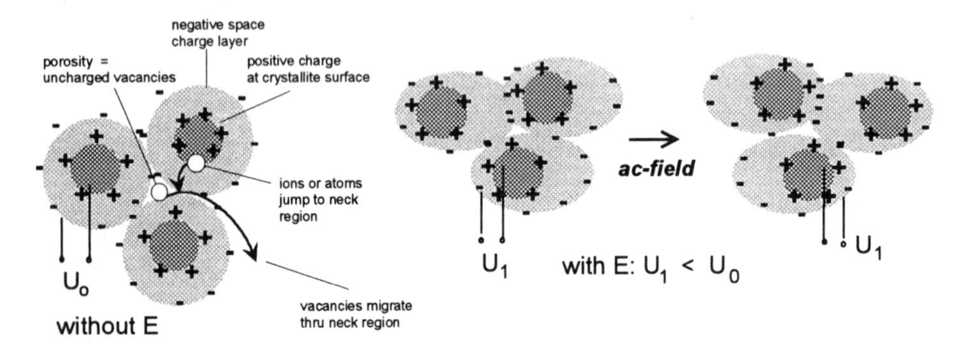

Figure 6. Periodic reduction of the potential barrier for vacancy flow from the pore region thru the neck region by polarization of the space charge layer by an alternating microwave field. *Note: the time scale of lattice vibrations and individual atom or ion jumps is 10^{-13}s, the time scale of the microwave alternating field is 10^{-9} s. Therefore, on the time scale of individual jump processes the microwave field is assumed to act as a dc-current.*

$$[V'''_{Al}]_\infty = 2e^{-\left[\frac{g_{V_{Al}} - 3e\phi_\infty}{kT}\right]} \quad (2a) \qquad [V^{\cdot\cdot}_O]_\infty = 3e^{-\left[\frac{g_{V_O} + 2e\phi_\infty}{kT}\right]} \quad (2b) \qquad (2)$$

$$2[V^{\cdot\cdot}_O]_\infty = 3[V'''_{Al}]_\infty \quad (3) \qquad e \cdot \phi_\infty = \frac{1}{5}\left[g_{V_{Al}} - g_{V_O}\right] \quad (4)$$

$$\text{null} \rightarrow Al^{3+} + V'''_{Al} \qquad \text{null} \rightarrow O^{2-} + V^{\cdot\cdot}_O$$
$$5\text{null} \rightarrow Al_2O_3 + 2V'''_{Al} + 3V^{\cdot\cdot}_O \qquad (5)$$

Owing to the fact, that an ac-field is used, the reduction of the space charge would be averaged with time. However, as pointed out earlier, reduction of porosity during sintering is an irreversible process. Therefore, the net effect on vacancy transport is not averaged with time. A periodic reduction of the intrinsic space charge would occur more frequently, when the frequency of the microwave radiation is increased. Therefore, with increasing frequency of the microwaves, the efficiency of the mechanism depicted above would increase. Furthermore, the temperature dependence of dopant solubility and segregation would strongly affect the efficiency of such an electrochemical transport mechanism. Consistent with this, the difference between microwave and conventional sintering of Mg-doped alumina starts at 1200°C, a temperature known for the beginning of solution of MgO in Al_2O_3.

$$[V_{Al}'''\,]_\infty = 2e^{-\left[\frac{g_{V_{Al}} - 3e(\phi_\infty + \phi_{ac})}{kT}\right]} \qquad [V_O^{\cdot\cdot}]_\infty = 3e^{-\left[\frac{g_{V_O} + 2e(\phi_\infty + \phi_{ac})}{kT}\right]} \tag{2'}$$

$$e \cdot (\phi_\infty + \phi_{ac}) = \frac{1}{5}\left[g_{V_{Al}} - g_{V_O}\right] \tag{4'}$$

In summary, the action of the ac-field could be compared with the action of an aliovalent acceptor dopant, like e.g., Mg^{2+}. When Mg^{2+} is used as a dopant in conventional sintering, the grain boundary mobility is reduced and a more effective reduction of porosity is possible [23]. Numerous mechanism are discussed, with no definite conclusion. When a space charge controlled mechanism is assumed [22], striking similarities become evident between both, the effect of a microwave field and of Mg^{2+} on the grain growth and densification of alumina. In addition, codoping with donor additives, known to counteract the effect of Mg^{2+} in conventional sintering [23], acts in the same way upon microwave sintering [7, 22, 23]. The benefit of microwave sintering for microstructure evolution would result from the fact, that ϕ_{ac} could be effective at lower temperatures, and not limited to the solubility, as dopants usually are.

Therefore, for the analysis of microwave sintering kinetics, the use of the electrochemical potential given in equation (6) is suggested, because, due to the space charge, the flux, j_i , may be affected by the microwave field.

$$\eta_i = \mu_i + Z_i F_i (\phi_i + \phi_{ac}) \qquad j_i = c_i v_i = -c_i B_i \frac{\delta \eta_i}{\delta x} \tag{6}$$

with Z_i: electrical charge; F_i: Faraday constant; ϕ_i: intrinsic electrical potential; ϕ_{ac}: local electrical potential due to the external microwave field; η_i: electrochemical potential; μ_i : chemical potential of the i-th component; v_i: velocity of the atom; B_i absolute mobility of an atom under the action of a unit force; c_i: concentration per unit volume; j_i: flux of the i-th component; x: direction of diffusion

However, facilitation of vacancy and ion transport through the neck region due to a periodic polarization of the space charge cloud should be difficult, if possible at all, to detect by diffusion coefficient measurements on single crystals. On polycrystalline samples the contribution of the neck region may be depleted by the overall volume diffusion process. Therefore, in order to obtain the information necessary for a quantitative analysis of a space charge layer contribution to enhanced densification or unique microstructures in microwave sintering, classical neck growth-grain growth kinetic measurements on pure and doped oxide ceramics in isothermal microwave sintering experiments as well as impurity segregation measurements should be performed.

CONCLUSION

Evidence for a microwave effect connected to the intrinsic space charge of binary ionic compounds is drawn from the following experimental results:

- the reduction of activation energy for ion jump relaxation in t-ZrO_2 ceramics processed from powders obtained from microwave pyrolysis
- the increased densification and reduced grain growth of isothermally microwave sintered Mg-doped alumina ceramics

Calculations based on Schottky defect chemistry suggest an additional potential acting upon microwave processing, and reducing the intrinsic space charge.

ACKNOWLEDGMENT

Thanks are due to Dr. Roland Heidinger (KFA Karlsruhe) for impedance measurements on t-ZrO_2-powders. Financial support of DFG (Wi 856/4-1 and SFB 316, Project A6) is gratefully acknowledged. I am also thankful to D. Clark and J. Katz for valuable discussions.

REFERENCES

1. S.J. Rothman, *MRS Symp. Proc.*, **347**, (M.F. Iskander, R.J. Lauf, and W.H. Sutton, eds.), Mater. Research Soc., Pittsburgh, PA, 9-18 (1994).
2. R.M. Young, and R. McPherson, *J. Am. Ceram. Soc.*, **72**, 1080 (1989)
3. D.L. Johnson, *J. Am. Ceram. Soc.*, **74**, 849-850 (1991)
4. T.T. Meek, X. Zhang, and M. Rader, *Ceram. Trans.*, **21**, (D.E. Clark, F.D. Gac, and W.H. Sutton, eds.), Amer. Ceram. Soc. Westerville, OH, 81-93 (1991).
5. K.I. Rybakov, V.E. Semenov, *Russ. Acad. of Sciences Preprint Nr. 357*, Inst. of Applied Physics, Nizhny Novgorod, Russia (1994).
6. S. Freeman, J. Booske, R. Cooper and B. Meng, *MRS Symp. Proc.*, **347**, (M.F. Iskander, R.J. Lauf, and W.H. Sutton, eds.), Mater. Research Soc., Pittsburgh, PA, 479-485 (1994).
7. S. Vodegel, "Microwave Sintering of Alumina", PhD Thesis, University Dortmund (1993).
8. S. Vodegel, C. Hannappel, M. Willert-Porada, *Metall* **48** [3], 206-210 (1994).
9. M. Willert-Porada, *MRS Symp. Proc.*, **347**, (M.F. Iskander, R.J. Lauf, and W.H. Sutton, eds.), Mater. Research Soc., Pittsburgh, PA, 31-43 (1994).
10. M. Willert-Porada, S. Vodegel, *Ceram. Trans.* **21**, /D.E. Clark, F.D. Gac, and W.H. Sutton, eds.), Amer. Ceram. Soc. Westerville, OH, 631-638 (1991)
11. S. Vodegel, M. Willert-Porada, MRS Spring Meeting, San Francisco, April 4-8, 1994, Symposium O, Microwave Processing of Materials IV, Poster O 14.20
12. S. Vodegel, M. Willert-Porada, to be published in J. Am. Ceram. Soc.
13. M. Willert-Porada, T. Gerdes, *MRS Symp. Proc.*, **347**, (M.F. Iskander, R.J. Lauf, and W.H. Sutton, eds.), Mater. Research Soc., Pittsburgh, PA, 563-570 (1994).
14. K. Lehovec, *J. Chem. Phys.*, **21** [7] 1123 (1953).

15. P.W-M- Jacobs, E.A. Kotomin, *Phil. Mag., A 1993*, **68** [4] 695-709 (1993).
16. J.A.S. Ikeda, Y.-M- Chiang, *J. Am. Ceram. Soc.*, **76** [10] 2437-46 (1993).
17. S.-L. Hwang and I-Wei Chen, *J. Am. Ceram. Soc.*, **73** [11] 3269-77 (1990).
18. M. Willert-Porada, S. Dennhöfer, D. Hachmeister, *MRS Symp. Proc.*, **347**, (M.F. Iskander, R.J. Lauf, and W.H. Sutton, eds.) Mater. Research Soc., Pittsburgh, PA, 571-578 (1994).
19. S. Dennhöfer, Masters Thesis, Dortmund (1994).
20. M. Weller, H. Schubert, and P. Kountouros, in "Science and Technology of Zirconia", (S.P.S. Badwal, M.J. Bannister, and R.H.J. Hannink, eds.), Technomic Publ. Co., Lancaster and Basel, 546-554 (1993).
21. C.-W. Schmidt, R. Heidinger, M. Willert-Porada, to be published in J. of Mater. Sci.
22. M. Willert-Porada, Habilitationsschrift, University of Dortmund, June 1995
23. S.J. Bennison and M.P. Harmer, *J. Am. Ceram. Soc.*, **68** [1] C 22- C 24 (1985).

MECHANICAL PROPERTIES OF ALUMINA SINTERED IN A MICRO-WAVE INDUCED OXYGEN PLASMA

Hunghai Su, Guy E. Salinas, Steve Kim, Vinayak P. Dravid and D.Lynn Johnson
Department of Materials Science and Engineering
Northwestern University, Evanston, IL 60208-3108, USA

ABSTRACT

Tubular specimens of Sumitomo alumina were sintered in both a microwave induced oxygen plasma and a rapid heating conventional furnace under identical conditions other than plasma. Surface morphologies and microstructures of the fracture surfaces of specimens sintered in both system were investigated. Surface etching was observed on the surface of samples exposed to plasma. Mechanical properties of specimens sintered in both systems were also compared using "C"-ring and "O"-ring diametrical compression tests. A higher average fracture strength and a larger Weibull modulus were found for plasma sintered specimens in the "C"-ring tests.

INTRODUCTION

Plasma sintering of ceramics has been under investigation for about two decades. Besides the significantly enhanced densification reported in all related studies [1][2][3], surface modifications induced by exposure to plasma were noticed by several researchers. Kemer observed changes in surface morphology of alumina specimens sintered in a microwave induced nitrogen plasma [4]. Hrdina studied surface modifications of densified alumina rods exposed to an induction coupled argon plasma with mixtures of H_2O, H_2, N_2, or O_2 gases [5]. He found that argon plasmas with different gas dopants resulted in distinct surface morphologies of the specimens, which indicated that chemical effects in addition to thermal effects may be involved in the plasma sintering process. The exact physical and chemical nature of this effect is not yet fully explored.

It is interesting to see whether the strength of ceramics is enhanced or degraded in plasma sintering through surface modification due to the exposure to plasma. This can be characterized using diametrical compression of "O"-rings (ID in tension) and "C"-rings (OD in tension) cut from tubes. The fracture strength of "O"-rings is strongly influenced by internal surface flaws, while the fracture strength of "C"-rings is sensitive to external surface flaws [6], as well as to the surface morphology change induced by environmental factors. Since the external surface is more likely to be affected than the internal surface of a small tube sintered in a plasma, a difference in surface morphology between the inside and outside surfaces is expected. The effect of this difference on the strength of sintered specimens may be found by using "O"-ring and "C"-ring tests.

EXPERIMENTAL

Sumitomo AKP-50 alumina powder with added binder was isostatic pressed at 270 MPa into thin wall cylindrical tubes with height and OD about 10 mm. Presintering was conducted at 650°C for 1/2 h. in air and the resulting green density was about 56%. Plasma sintering was carried on inside a single mode (TM_{012}) cylindrical microwave cavity [7]. The specimen was supported on a single crystal sapphire tube and was placed in the hot zone of the plasma tube which was insulated with an alumina fiber cylinder to minimize radial temperature gradients. The black body sensor was placed inside the specimen tube. High purity oxygen was used as the plasma gas. The pressure inside the plasma tube was monitored by a convectron vacuum gauge. A similar specimen setup was built inside an alumina tube furnace heated by graphite foil to study sintering behavior under identical conditions of plasma sintering (thermal histories, oxygen pressure) using conventional conductive/radiative heating.

For sintering studies, specimens were heated to 1400°C at constant heating rates ranging from 7.5 °C/min to 45 °C/min either in the plasma or in the conventional furnace at a pressure of 1.7 kPa. Some specimens tested here were sintered in other conditions both in the plasma and in the conventional furnace. The bulk density of the sintered samples was measured by the Archimedes method. The microstructures of fractured cross sections were observed using the SEM.

The unnotched samples for mechanical testing were prepared by sectioning tubular specimens sintered to various densities into either an "O"-ring or a "C"-

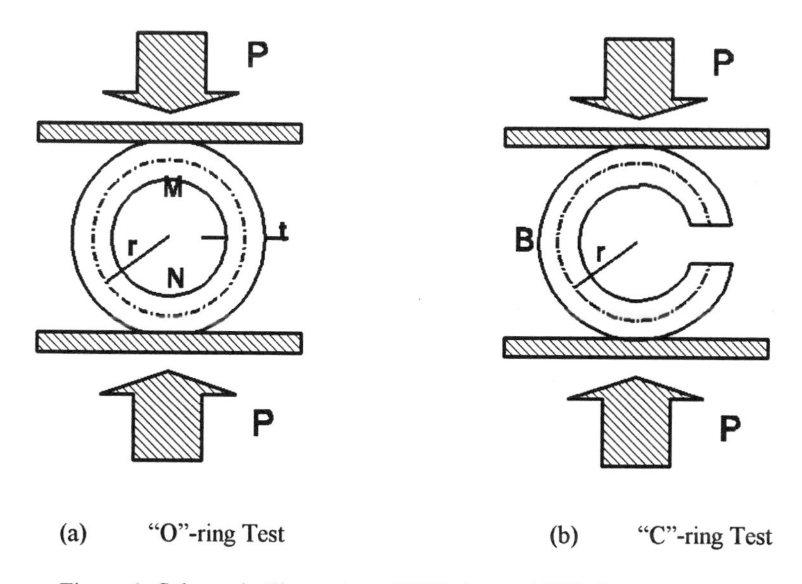

(a) "O"-ring Test (b) "C"-ring Test

Figure 1: Schematic Illustration of "O"-ring and "C"-ring test.

ring shape. Figure 1 is a schematic illustration for both the "O"-ring and "C"-ring tests. The samples were fractured at room temperature in a Sintech machine operating at a cross-head speed of 8.33 μm/s. For the "O"-ring test, the maximum tension appears on the inner surface of load points (M,N points in figure 1(a)) while for the "C"-ring test, the maximum tension occurs on the external surface at the midpoint of the "C" equidistant from the loading points (B point in figure 1(b)). The fracture strength was calculated from geometry parameters (r, t, and b are the mean radius, thickness, and width of the ring, respectively) and fracture load (P) using the following formulas [8]:

for o-ring test: $$\sigma_f = P\frac{r-e}{\pi}\frac{t/2-e}{bte(r-t/2)}; \qquad (1)$$

for c-ring test: $$\sigma_f = P\frac{r-e}{2}\frac{t}{bte(r+t/2)} \text{ in which } e = r - t(\ln\frac{r+t/2}{r-t/2})^{-1}. \qquad (2)$$

RESULTS AND DISCUSSIONS

The appearance of the fracture surfaces of plasma sintered specimens was not discernibly different from those of specimens conventionally sintered to comparable densities. Also the density dependence of both the closed pore volume fraction and the grain size was similar for both processes. This indicated that microstructure evolution in both the plasma and conventionally sintered specimens was quantitatively the same [9]. The final densities of specimens sintered in the plasma or in the conventional furnace with constant heating rate to 1400°C with no hold at temperature are shown in Figure 2. Although at a given heating rate specimens experienced the same temperature-time excursion in both plasma and conventional sintering, the plasma sintered samples reached higher density. This clearly demonstrates the athermal effect due to the plasma. Detailed discussions about sintering results and mechanisms are given in reference [9].

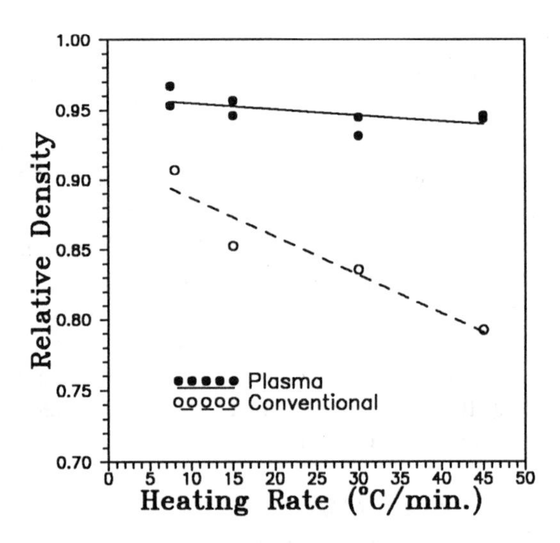

Figure 2: Comparison of Final Densities of Specimens Sintered to 1400 °C Without Holding at the Maximum Temperature in the Plasma and the Conventional Furnace.

Plotted in Figure 3 are the fracture strength vs. density data for both plasma and conventionally sintered specimens in "O"-ring and "C"-ring tests. The "O"-ring data (Figure 3(a)) for both sintering methods followed closely the same strength-density trajectory, while the "C"-ring data (Figure 3(b)) are more scattered, with the plasma sintered specimens showing a higher overall strength.

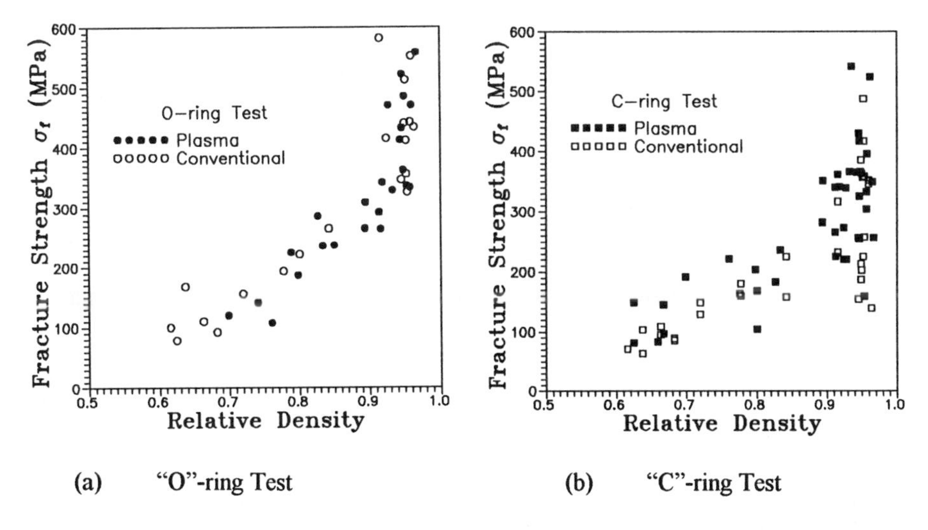

(a) "O"-ring Test (b) "C"-ring Test

Figure 3: Fracture Strength vs. Relative Density for "O"-ring and "C"- ring Tests.

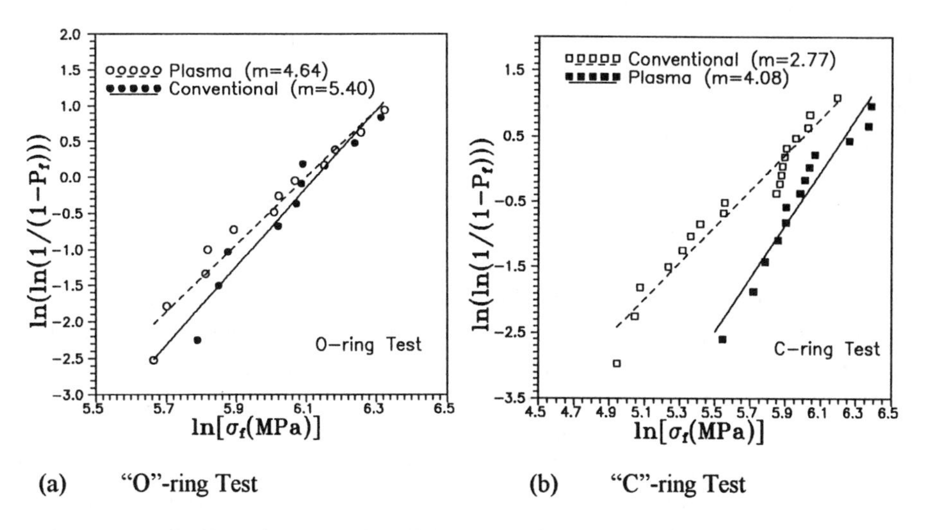

(a) "O"-ring Test (b) "C"-ring Test

Figure 4: Weibull Modulus Plot for High Density Specimens in Both "O"-ring and "C"-ring Tests.

The data for high density specimens (94.5% ~ 96.5% of theoretical density) were analyzed using failure statistics. The Weibull plots for "O"-ring and "C"-ring tests of specimens sintered in the plasma and the conventional furnace are shown in figure 4(a),(b). The statistic results are listed in Table I.

Table I: Mechanical Properties of Sintered Specimens.

	"O"-ring Test		"C"-ring Test	
	Conventional	Plasma	Conventional	Plasma
Average Fracture Strength (MPa)	425 ± 75	409 ± 88	297 ± 104	409 ± 103
Weibull Modulus	5.40	4.64	2.77	4.08

To explain the results shown in Table I, microstructures, especially surface morphology, of specimens sintered either in the plasma or in the conventional furnace were studied under the SEM. The obvious difference in strength and Weibull modulus of conventionally sintered specimens between "O"-ring and "C"-ring tests is due to the different morphology of the internal and external surfaces. When powder was pressed into tubes, the inner surface was pressed against a smooth steel mandrel while the external surface was squeezed by a polyurethane tube which led to bumps 20 ~ 100 μm in diameter, that are remnants of the granules in the powder before pressing. As a result, the internal surface of the sintered short tube was much smoother than the external surface, which may cause the difference in fracture strength.

The most interesting phenomenon was the difference in strength and Weibull modulus between specimens sintered in the plasma and in the conventional furnace for the "C"-ring test. A clear increase of the strength was achieved in plasma sintered specimens. Etching was observed on the specimen surfaces exposed to the oxygen plasma, though not as severe as that observed in nitrogen plasma [4]. Figure 5 is an SEM photograph taken of the surface of a specimen sintered in the plasma, which shows surface pores. This was only a surface effect, and the fracture surfaces of these specimens were indistinguishable from those of samples sintered conventionally. Furthermore, a fracture cross section near the external surface reveals that the pores extended only 1~2 grain layers deep, as shown in Figure 6. Etching may blunt or even remove critical flaws on the surface. As a result, the average fracture stress is increased and a higher Weibull modulus is observed for "C"-ring test which are sensitive to external surface flaws. Etching was also observed on the internal surfaces, but much less severe than that on the

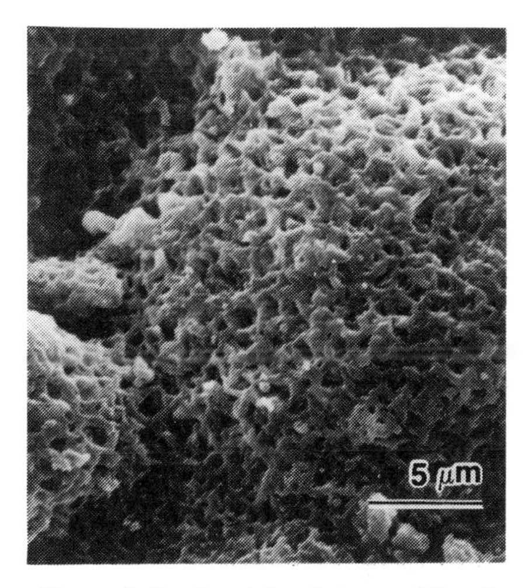

Figure 5: Surface Morphology of Specimen Sintered in Plasma.

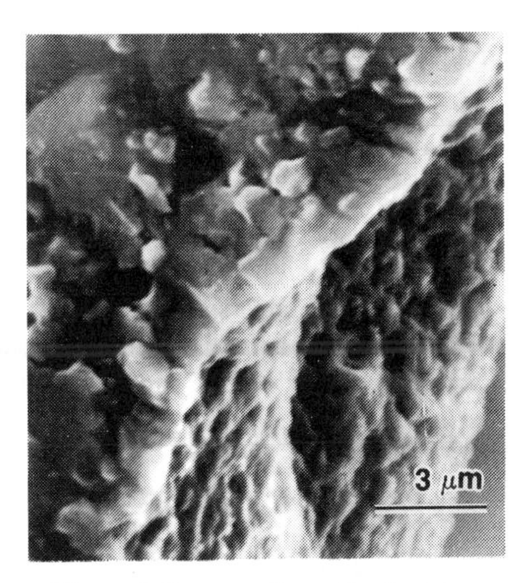

Figure 6: Fracture Cross Section Near the Surface of Specimen.

external surfaces. Also, the internal surfaces were very smooth as mentioned above. As a result, the fracture strength properties for the "O"-ring test were similar in samples sintered in both methods.

Bennett et al. attributed the increase of the strength of specimens sintered in a plasma to the smaller grain size achieved in plasma sintering and to clean grain boundaries with fewer mechanical imperfections [1]. In our sintering experiments, no evidence was seen for the depression of grain growth in the plasma sintering since the dominant diffusion mechanism was the same as conventional sintering [9].

CONCLUSIONS

Diametrical compression tests of "O"-ring and "C"-ring were employed here to characterize the effect of plasma exposure on the strength of ceramics. No distinguishable differences in fracture strength and Weibull modulus were observed in "O"-rings between plasma and conventionally sintered specimens. A higher fracture strength and larger Weibull modulus were found in "C"-rings from plasma sintered specimens than those from conventionally sintered specimens, which demonstrated the strengthening of ceramic tubes sintered in plasma. Plasma etching resulted in the formation of pores on the surface of specimens, and this surface modification induced by exposure to the plasma is considered to cause the increase in strength.

ACKNOWLEDGMENTS

This work was supported by the National Science Foundation Grant DMR-9200128. One of the authors, G. Salinas was a materials research intern at Northwestern University in summer 1994. This work made use of central Facilities supported by the National Science Foundation at the Materials Research Center of Northwestern University under Award No. DMR-9120521.

REFERENCES

1. C.E.G. Bennett, N.A. McKinnon, and L.S. Williams, "Sintering in Gas Discharges", Nature(London), 217, 1287-88 (1968).
2. D.L. Johnson and R.A. Rizzo, "Plasma Sintering of β"-alumina", Am.Ceram. Soc.Bull., 59[4] 467-68,472 (1980).
3. E.L. Kemer and D.L. Johnson, "Microwave Plasma Sintering of Alumina", Am.Ceram.Soc.Bull., 64[8] 1132-36 (1985).
4. E.L. Kemer, "Microwave-Induced Plasma Sintering of Alumina", M.S. Thesis, Northwestern University, Evanston, IL, June 1984.
5. K.E. Hrdina, "Thermal and Chemical Effects During Ultra-Rapid Sintering of Alumina in an Induction Coupled RF Plasma", M.S. Thesis, Northwester University, Evanston, IL, June, 1989.
6. J.R.G. Evans and R. Stevens, "The 'C'-Ring Test for the Strength of Brittle Materials", Br.Ceram.Trans.J., 83,14-18 (1984).
7. M.P. Sweeney and D.L. Johnson, "Microwave Plasma Sintering of Alumina", pp. 365-372 in Microwave: Theory and Application in Materials Processing, edited by D.E. Clark, F.D. Gac and W.H. Sutton, American Ceramic Society, Inc. (1991).
8. S. Timoschenko, Strength of Materials, part 1, 2nd ed. Van Nostrand, New York (1956).
9. H. Su and D.L. Johnson, "Sintering of Alumina in a Microwave Induced Oxygen Plasma", submitted to J.Am.Ceram.Soc. (1995)

Dielectric Properties

WHAT IS THE BEST MICROWAVE ABSORBER?

R.M. HUTCHEON, F.P. ADAMS AND M.S. DE JONG
AECL, Chalk River Laboratories, Chalk River, Ontario, Canada K0J 1J0

ABSTRACT

The absorption of microwave power by samples with different shapes and dielectric properties was studied for single-mode excitation of a microwave applicator. Exact calculations were done of the quality factor, Q and the frequency shift of a cylindrically-symmetric cavity with a small, cylindrically-symmetric sample on-axis in the cavity, demonstrating the influence of sample shape and dielectric properties on the power absorption per unit volume of the sample.

An approximate, but very general, analytical formalism was developed for calculating the power absorbed by the sample and thus the contribution of the sample to the Q of the cavity. This simple formalism is shown, by comparison with the exact calculations, to be quite accurate over a very broad range of properties. Thus, it can be used to calculate the power absorbed by samples which are smaller than the internal wavelength, and also to gain insight into the factors that influence the microwave absorption.

INTRODUCTION

Often, those who are learning about the use of microwaves for processing materials will ask, "What is the best microwave absorber?" The traditional reply is that the rate of microwave absorption depends on the magnitude of the dielectric loss factor and, if the dielectric constant is large, on the shape of the material. Such a response is not very satisfying to the novice, and generally does not produce much understanding or insight. Any attempt to provide a more complete answer for a multimode cavity, especially giving quantitative numbers, is foiled by the complexity and variability of the mode distributions. However, for a single mode electric field distribution, the problem is readily solved. The results are quantitatively applicable to a single-mode industrial applicator, and give some insight into multimode power absorption.

In this paper we demonstrate and discuss results for power absorption in a single-mode cavity resonator, and also outline a simple analytic formalism that allows one to easily calculate quantitative results, at least for small samples. To study the problem, one would ideally insert **many** samples of different shapes and dielectric constants into the peak electric- field region of a single-mode cavity, where the electric field is relatively uniform. As this is not practical, we have done the equivalent "computer experiments" in which the cavity wall losses were set to zero, leaving only the influence of the samples. The code SEAFISH was used, which has been extensively verified and shown to accurately model axially-symmetric, "material-in-cavity" problems[1,2]. It was desirable to have results for the two industrial frequencies, 915 and 2450 MHz, so the code was used to design a cavity (Fig. 1) which is resonant at both frequencies.

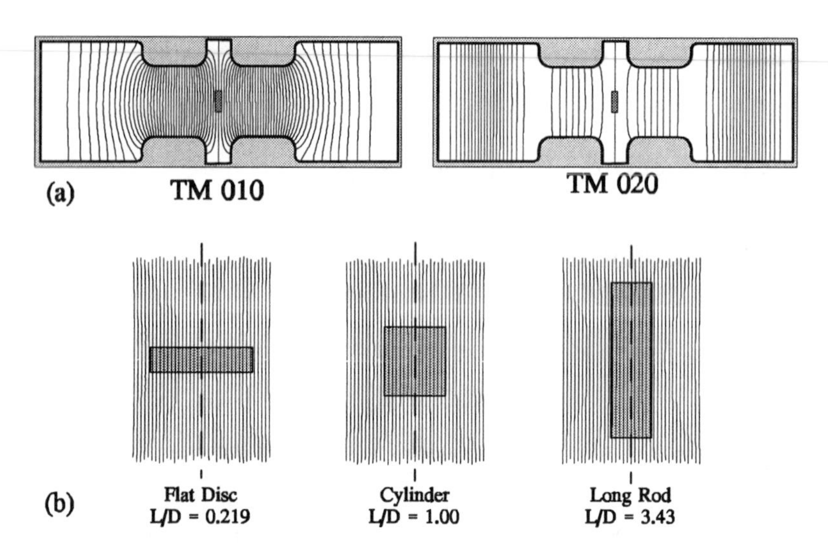

Figure 1. (a) A cylindrical cavity with TM resonances at both 915 and 2450 Mhz. (b) The three sample shapes used for calculations.

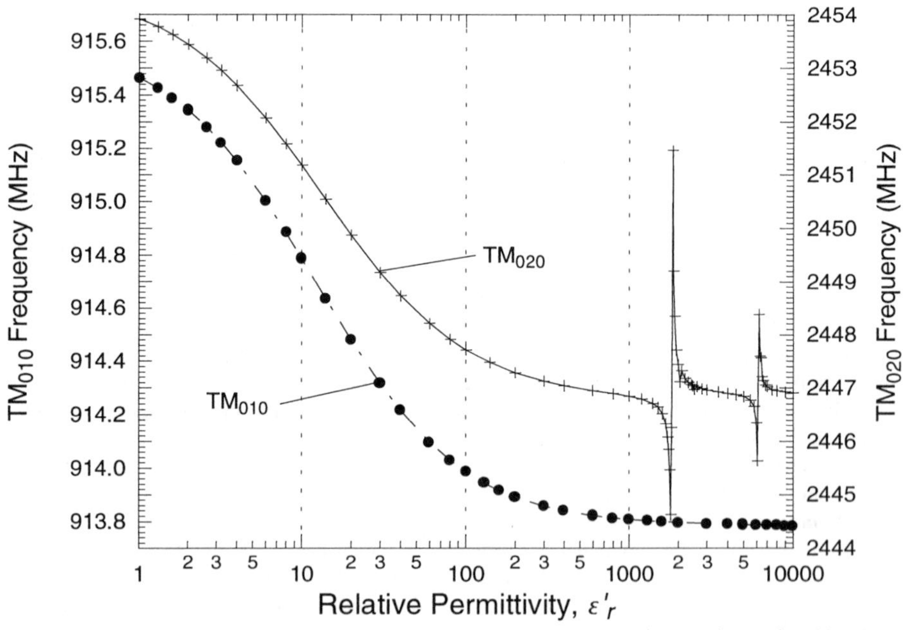

Figure 2. The calculated resonant frequencies of the cavity with a long rod sample of lossless material (ε''_r=0) as a function of ε'_r .

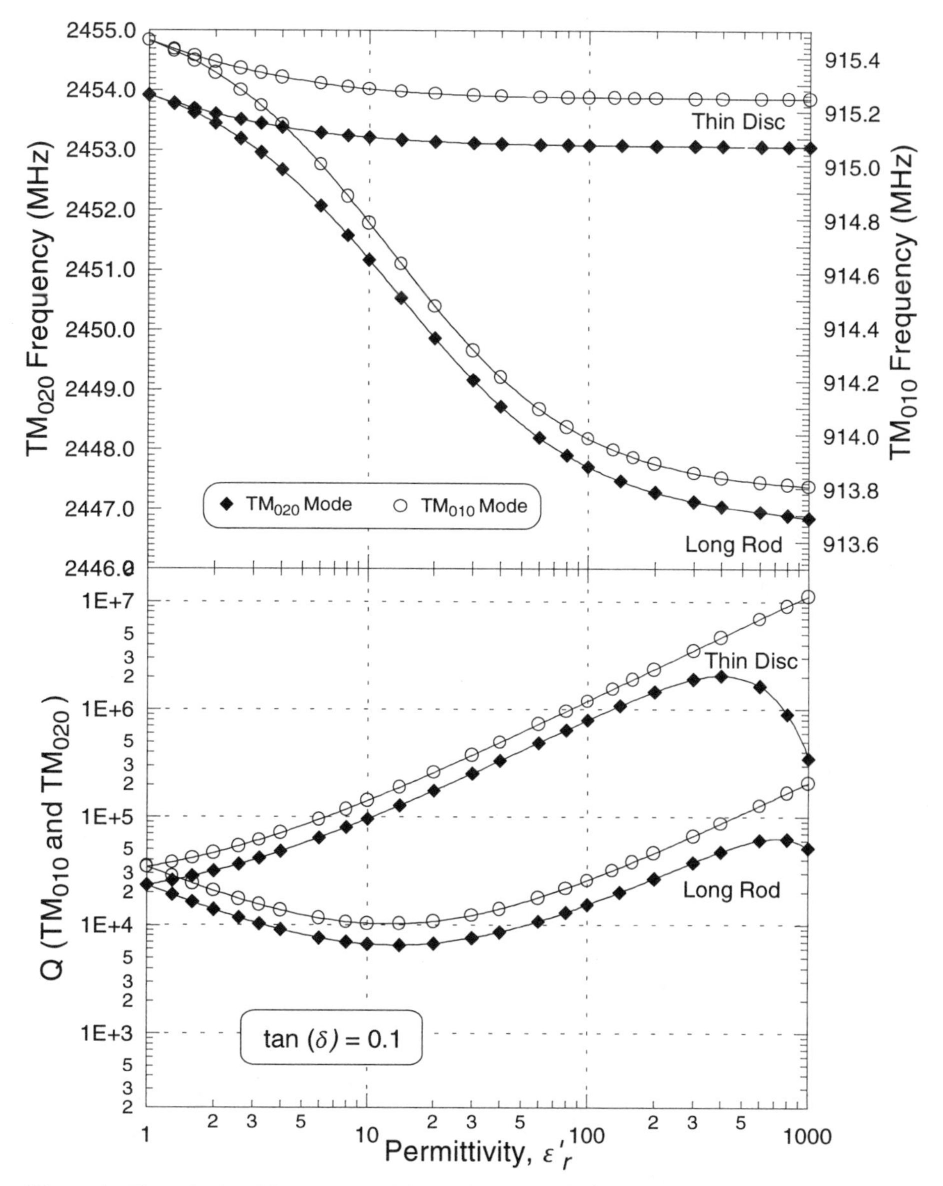

Figure 3. The calculated frequency and Q as a function of ε_r' for the extremes of sample shape.

Accurate computer calculations were done for three different shaped samples, of the same volume (Fig. 1), positioned on-axis in the centre of the cavity. These shapes are representative of those often encountered in practical problems. In the following three sections, the results and interpretation of these "computer" experiments are presented, a simple analytic dielectric loss formalism is developed, and finally the analytic expression is compared with the accurate results.

CYLINDRICAL DIELECTRIC SAMPLES IN A CYLINDRICAL CAVITY - A "COMPUTER" EXPERIMENT

The SEAFISH code was used to exactly model the "sample in a lossless cavity" problem, using samples of different shapes but the **same** fixed volume (115.45 mm^3) (Fig. 1) and for a wide range of values of complex relative dielectric constant, ε_r' and ε_r''. For example, Fig. 2 shows the frequency of each mode as a function of ε_r' for the 12 mm long, 3.5 mm diameter rod-shaped sample. Note that the frequency shift approaches a constant value for large relative dielectric constant, ε_r', although resonances appear at high values of ε_r'. These are the "dielectric resonator" modes, where the fields in the sample are large and very non-uniform, and occur when the wavelength **in** the sample, $\lambda / \sqrt{\varepsilon'_r}$, approaches the sample diameter.

The influence of sample shape is demonstrated in Fig. 3, where the Q and the frequency are shown for both the long rod and flat disc sample configurations (tan δ=0.10). The maximum power absorption (minimum Q) occurs at an intermediate value of $\varepsilon_r' \approx 15$ for the long rod, while for the thin disc the maximum power absorption occurs at the lowest possible value of $\varepsilon_r' \approx 1$. Clearly the shape of the sample has a strong influence on which value of the real part of dielectric constant produces maximum absorption! This is due to the influence of sample **shape** on the magnitude of the field **inside** the sample.

The effect of very large values of ε'' is demonstrated by a series of calculations for the short (right) cylindrical sample for $\varepsilon_r' \approx 10$. In Fig. 4 are shown plots of electric field lines in the region of the sample. The internal electric fields are relatively uniform for values up to $\varepsilon_r'' \approx 10^3$, (tan$\delta$=100) at which point exclusion of the fields from the centre of the sample is obvious. A value of $\varepsilon_r'' \approx 10^4$ (tanδ=1000) at a frequency of 915 MHz gives a skin depth or attenuation length of ≈ 1 mm, which is an appreciable fraction of the radius of the right cylinder (r=2.64 mm). Fields are increasingly excluded from the interior of the sample, and at larger values of ε_r'' the sample behaves increasingly like a metal.

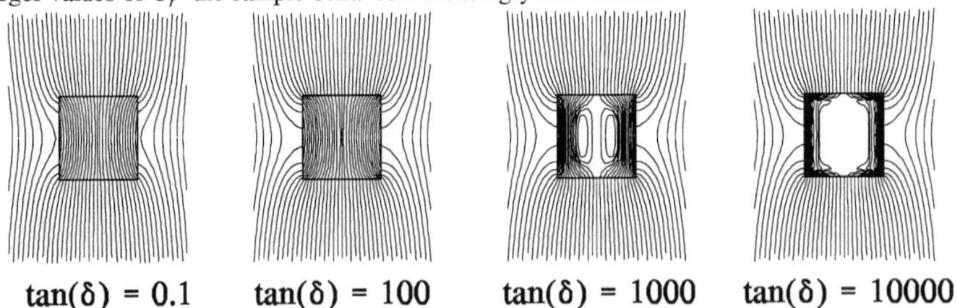

Figure 4. Calculated (SEAFISH) electric field lines in and around the sample for ε'=10, demonstrating more uniform fields for low loss and field exclusion from the sample core for high loss.

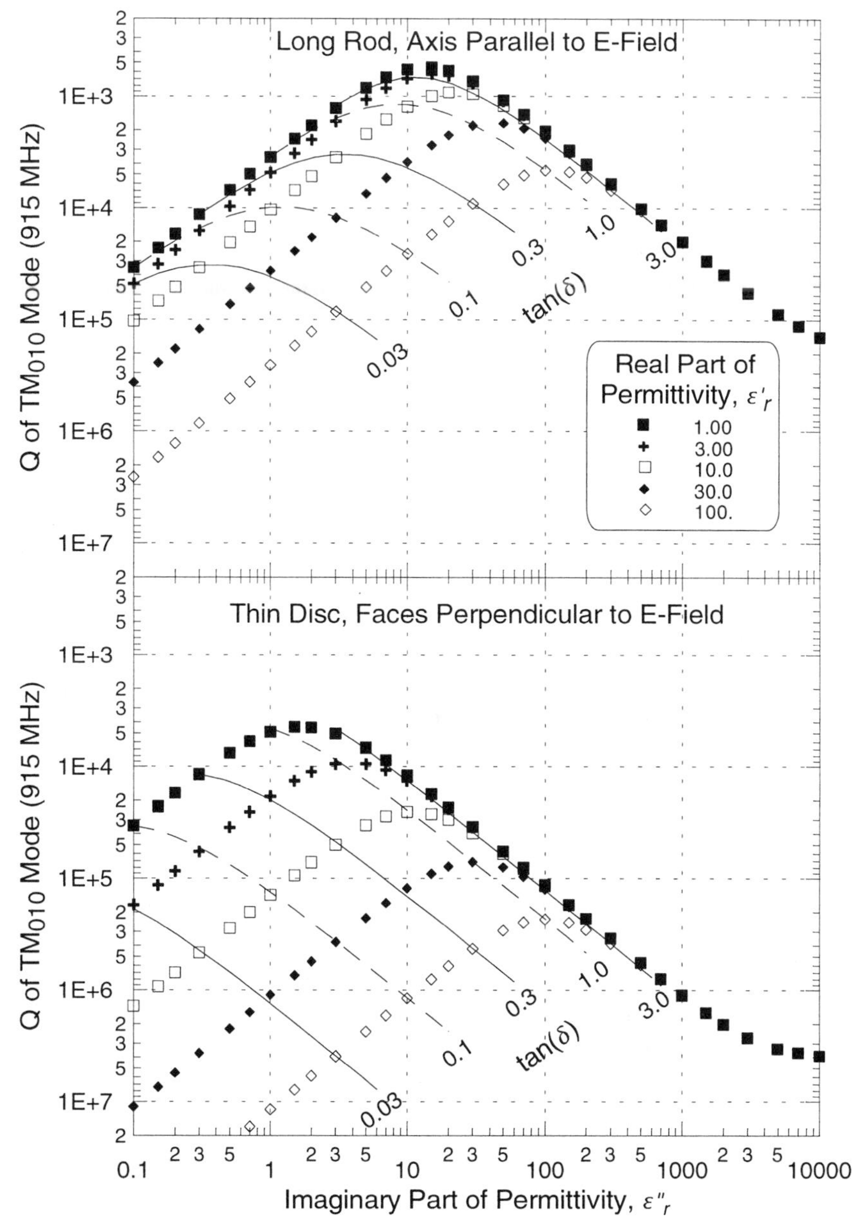

Figure 5. Calculated Q values of the 915 MHz resonance (with lossless cavity walls) as a function of ε_r' and ε_r'', for the extremes of sample shape.

The computer-generated Q values for samples of the same volume but with the extremes of shape (*i.e.*, the long rod and the flat disc), are shown in Fig. 5 for a wide range of values of the real part of the dielectric constant, ε_r'. Note that the Q value is plotted on an inverted logarithmic scale, so that a minimum value of Q, which represents a peak in the power absorption, is seen as a visual "peak".

The maximum power absorption for the two configurations clearly occurs at quite different values of ε", and overall, the power absorption of the long rod is an order of magnitude greater than that of the flat disc, in spite of their equal volumes. For the long rod, the "best absorber" (Q=550) has $\varepsilon_r'' \approx 15$ and tan $\delta \geq 3$ (*i.e.*, $\varepsilon_r' \leq 5$), while the flat disc "best absorber" (Q≈4400) is with $\varepsilon_r'' \approx 2$ and tan≈2 (*i.e.*, $\varepsilon_r' \approx 1$). For example, with an alumina ceramic of dielectric constant, $\varepsilon_r' \approx 10$ and a modest value of tan $\delta \approx 0.03$, approximately forty times more material would be required for the thin disc shape to produce the same Q or power absorption as the long rod. The key to understanding this effect is the influence of sample **shape** on the magnitude of the electric field in the material relative to that outside the material. The next section outlines an analytic formalism that addresses this problem in calculating the losses in the sample material.

A SIMPLE ANALYTIC EXPRESSION FOR MICROWAVE POWER ABSORPTION

The present "computer experiments" are intended to demonstrate the dependence of the microwave power absorption on the properties of a sample located in a region of approximately uniform electric field in a cavity excited in a single mode. The power absorbed on the **walls** of the cavity is of no interest in this study, and so was set to zero in the computer calculations. In fact, only the **ratio** of absorbed power in the sample to "circulating" power in the cavity is important in assessing the best absorber material, since the absolute power absorbed depends on the amount of power fed to the cavity. The quality factor, Q, is inversely related to the average power absorbed in the sample, P_{AV}, by

$$\left(\frac{1}{Q}\right) = \frac{P_{AV}}{\omega \cdot S}, \tag{1}$$

where $\omega \cdot S$ is the "circulating" power, $\omega = 2\pi f$, and S is the electromagnetic energy stored in the cavity. For a given frequency, the maximum absorption by the sample will be indicated by a minimum in Q.

An approximate, analytic expression for Q may be derived by using approximations for the energy stored in the cavity and for the average power absorbed in the sample. Both may be expressed as functions of the approximately-uniform electric field $E_{ext}(r \approx 0)$ **outside** of, but in the region of, the small sample. The stored energy in the cavity, for small samples (*i.e.*, neglecting the extra energy stored in the sample), is

$$S \approx \int_{V_c} \frac{1}{2} \varepsilon_0 |E_{ext}(r)|^2 \, dv_c \approx \frac{1}{2} \cdot \frac{A}{4} \cdot V_c |E_{ext}(r \approx 0)|^2 \cdot \varepsilon_0 , \tag{2}$$

where V_c is the volume of the cavity, ε_o is the permitivity of free space and A is the cavity electric-field shape factor, a unitless constant of order unity which depends **only** on the spatial distribution of electric field in the cavity. For a rectangular waveguide cavity, A=1 for the

TE_{10n} modes. For a pure cylindrical cavity, A=1.078 for the TM_{010} mode and A=0.463 for the TM_{020} mode. It should be noted that, throughout this paper, all electric fields are the peak temporal values.

The problem of calculating the energy absorbed by the sample for a given stored energy in the cavity (*i.e.*, for a given electric field **outside** the sample) has two parts: *(1)* determine the field inside the sample for a given **external** field, and *(2)* calculate the absorption in the sample for the determined value of the **internal** field.

In the situation of a traveling wave incident on a sample, part (1) of the calculation is the equivalent of solving for the reflection and transmission at the sample boundary. However, the present, standing-wave situation is the equivalent of equal and opposite traveling waves incident on opposite sides of the sample with specific phase relationships, and so is quite a different problem.

If the sample shape were an ellipsoid of rotation and the fields were uniform in the region of the samples, then the relative electric fields inside and just outside the sample would be given **exactly**[3] by the expression

$$\frac{E_{\text{int}}}{E_{\text{ext}}} = \frac{1}{1+F_{\text{sh}}\chi}, \tag{3}$$

where χ is the complex dielectric susceptibility of the material. The fields inside the sample are of constant magnitude and direction, but are phase-shifted in time relative to the external field. The **sample** shape factor, F_{sh}, is a real function of only the ratios of the lengths of the ellipsoid axis[4]. For a sphere, $F_{sh} = (1/3)$.

We have made the assumption that the functional form of Eq. (3) is still a good approximation, even for cylindrical samples, when considering the rms average field over the sample volume. For a particular cylindrical shape, the appropriate value of F_{sh} can be determined by measurement or by computation with SEAFISH. The agreement between measurement and SEAFISH computation for cylindrical samples has been confirmed[2] as long as the numerical approximations associated with the mesh size are valid.

The average energy absorbed per unit volume by a material is given by the usual formula

$$\frac{dP_{Av}}{dv_s} = \frac{1}{2}\omega\varepsilon_o\varepsilon_r'' |E_{\text{int}}|^2, \tag{4}$$

where E_{int} is the field **inside** the sample, and the relative complex dielectric constant, ε_r and the complex dielectric susceptibility, χ, are related by

$$\varepsilon_r - 1 = \varepsilon_r' - 1 - j\varepsilon_r'' = \chi = \chi' - j\chi''. \tag{5}$$

The average power absorbed by the sample for an approximately-uniform internal field is

$$P_{Av} \approx \frac{V_s\omega\varepsilon_o\chi''}{2}|E_{\text{int}}|^2, \tag{6}$$

where E_{int} is the spacial rms average, peak-temporal field over the sample. Relating this to the field outside the sample by using Eq. (3), the power absorbed is

$$P_{Av} \approx \frac{V_s \omega \varepsilon_o \chi''}{2} * \frac{|E_{ext}(r \approx 0)|^2}{|1+F_{sh}\chi|^2}, \tag{7}$$

where $E_{ext}(r \approx 0)$ is the external field in the region of the sample.

The contribution of the sample to the Q of the cavity may then be calculated using Eq. (1), (2) and (7), to give

$$Q \approx \frac{A}{4} \cdot \frac{V_c}{V_s} \cdot \frac{1}{\chi''} \cdot |1+F_{sh}\chi|^2 = \frac{A}{4} \cdot \frac{V_c}{V_s} \cdot \frac{[(1+F_{sh}\chi')^2 + (F_{sh}\chi'')^2]}{\chi''}. \tag{8}$$

Note that a large value of χ'', such that $\chi'' >> \chi'$ (or tan $\delta >> 1$), does not predict maximum absorption (*i.e.,* a minimum Q).

Equations (7) and (8) are significant refinements of the standard formula that are quoted for power absorption or sample contribution to Q. The inclusion of the sample shape factor dependence makes these useful for making engineering estimates of actual applicator designs. In the next section, the expression for the sample Q (Eq. (8)) is compared in detail with the exact computer calculations, and shape-factor values are determined which allow one to apply Eq. (3) to determine the mean, internal, sample field for a wide range of cylindrical sample shapes.

A COMPARISON OF THE SIMPLE ANALYTIC FORMALISM WITH EXACT CALCULATIONS

The contribution of the sample to the Q of the specified cavity (Fig. 1) was calculated for each of the three samples for a wide range of ε_r' and ε_r'' values. As an example, Fig. 6 shows

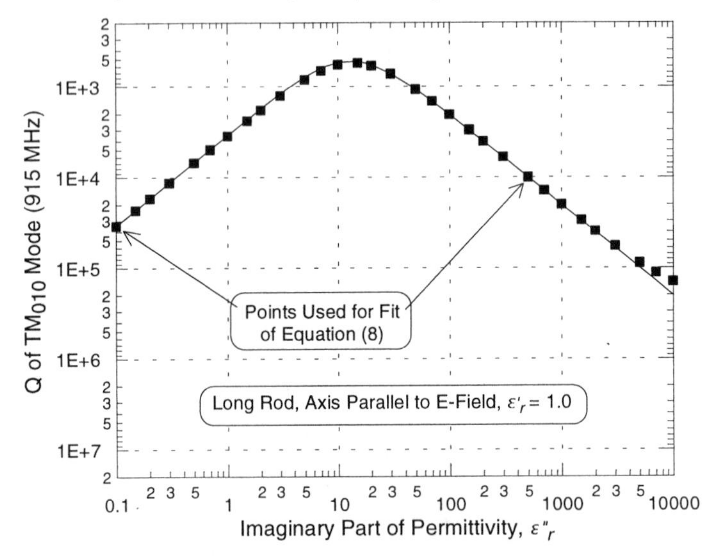

Figure 6. SEAFISH-calculated Q for the disc absorber (squares). The solid line is from Eq. (8), with A and F_{sh} determined by fitting to the calculations at the indicated points.

the exact solutions for Q for the disc absorber with $\varepsilon_r' = 1$, plotted against ε_r'', the relative dielectric loss factor. The analytic expression for Q (Eq. (8)) was fitted at the two specific points shown (Fig. 6), and the cavity electric field "A" values and sample shape factors were determined for the present re-entrant azimuthally symmetric cavity (Fig. 1). As the values of ε_r' and ε_r'' decrease, the influence of the sample shape factor in Eq. (8) decreases, so that the three different shaped samples have the same asymptotic dependence of Q on ε_r'' for $\varepsilon_r' = 1$. Hence, the value of Q at $\varepsilon_r'' = 0.1$ and $\varepsilon_r' = 1$ was used to determine the value of A for all samples at each frequency.

In the simple analytic theory, the value of the cavity electric field shape factor, A, is independent of sample shape. This is confirmed to good accuracy by the fit to the exact calculations. The present re-entrant cavity is not a simple cylinder in shape, and thus the A values are not those quoted for the theoretical cylinder cavity. The shape factor, F_{sh}, was determined by the fit at large ε_r'', avoiding however the very high ε_r'' region where the skin depth becomes significant. Obviously, the functional dependence of the analytic expression (Eq. (8)) reproduces the computer "data" very well except where skin depth or attenuation-length effects become important.

If one knows the value of the sample shape factor, F_{sh}, for a specific (ℓ/d) ratio, then one can calculate the field inside a sample and thus the frequency and Q shift generated by the sample in a cavity. A series of SEAFISH runs was done for cylinders with a range of (ℓ/d) ratios and the shape factors were determined (Fig. 7) and a fit interpolated. These shape factors are valid for simple cylinders with axis oriented along the electric-field direction in a field that is uniform over the sample dimensions. The degree of external field uniformity (without the sample) will determine the degree of approximation. For reference, the shape factors for ellipsoids of rotation are shown on Fig. 7, plotted against the ratio of the ellipsoid axes, (a/b). The difference between the shape factors is greatest just in the range of most practical samples.

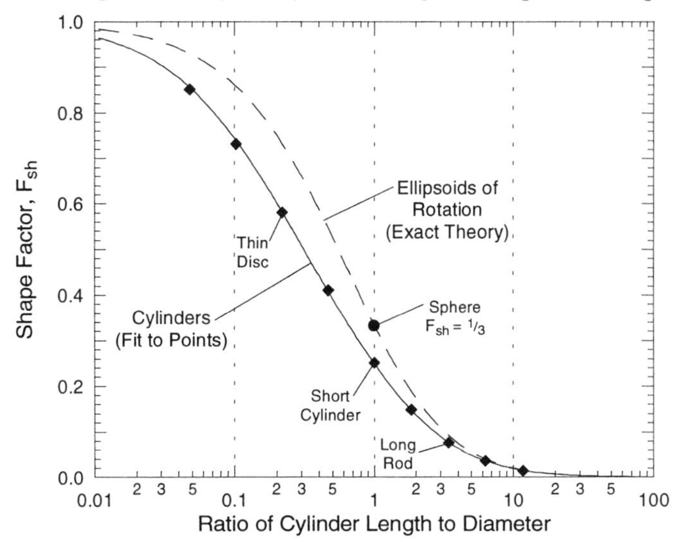

Figure 7. The sample shape factor, F_{sh}, for cylinders, determined by fitting Eq. 8 to SEAFISH calculations. The theoretical curve for ellipsoids of rotation is also shown (dotted line).

TABLE 1. The sample shape factors, F_{sh} and electric-field shape factors, A, for three cylindrical samples of the same volume but different length to diameter ratios in the re-entrant cavity (Fig. 1).

		Long Rod	**Right Cylinder**	**Thin Disc**
	(l/d)	3.43	1.000	0.219
915 MHz TM_{010}-like	A	0.703	0.697	0.696
	F_{sh}	0.0765	0.251	0.581
2450 MHz TM_{020}-like	A	0.477	0.473	0.474
	F_{sh}	0.0712	0.245	0.574

The SEAFISH-generated "data" for the three samples for a wide range of values of ε'_r and for both frequencies are shown in Fig. 8. For each frequency, the analytic curves were fitted using only the values of A and F_{sh} given in Table 1. The small systematic change of F_{sh} with

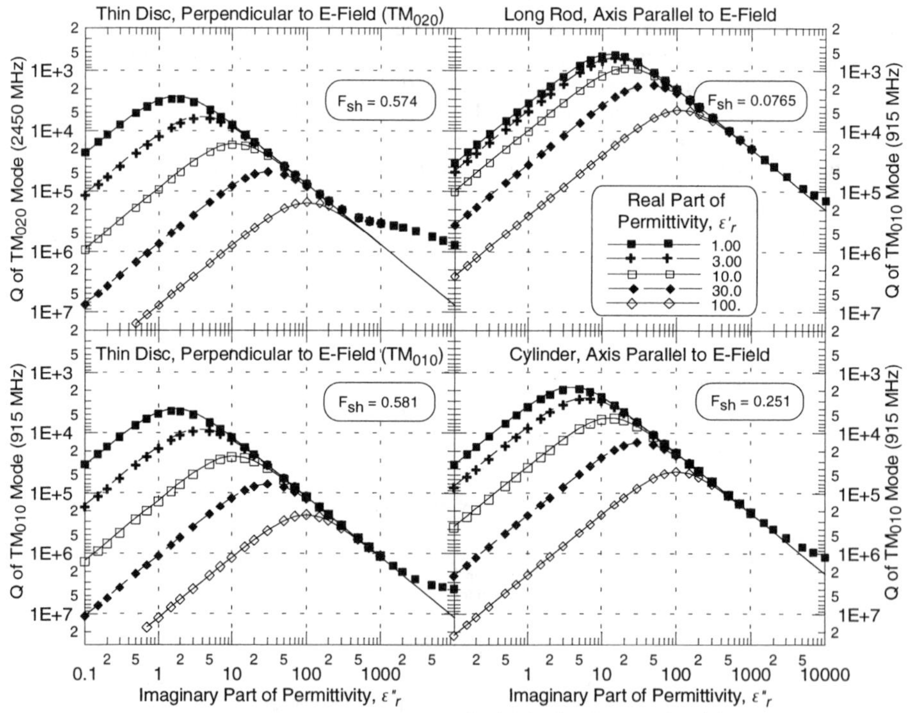

Figure 8. SEAFISH-calculated Q for the samples in the lossless cavity (Fig. 1). Note the inversion of the ordinate scale, to show maximum absorption in the sample as a peak.

increasing frequency was observed previously during experimental calibrations of a multi-mode, TM_{ono}, cylindrical cavity used for dielectric-perturbation measurements[5].

DISCUSSION

The calculations and analytic theory presented here apply accurately only to the situation of **small** samples of simple cylindrical shape in a relatively uniform, external electric field. In this case, the question of what is the best absorber material depends on the sample shape: a long rod aligned along the field absorbs best for $\tan\delta \geq 3$ and $\varepsilon' \leq 5$, while a flat disc aligned similarly has maximum absorption for $\tan\delta \approx 2$ and $\varepsilon' \approx 1$. As well, even the best "flat disc" absorber has almost an order of magnitude lower absorption per unit volume than the best long rod.

The practical application of microwave heating in industrial processes usually has constraints that exclude using the above-mentioned maximum absorption situation. Most materials have $\tan\delta$ much less than unity at room temperature, and only at elevated temperatures ($\approx$600°C for glasses, $\approx$1200°C for ceramics) does $\tan\delta$ exceed unity. Also, with the exception of some polymers and low atomic number ceramics, most materials have an intrinsic value of ε' greater than 5. Thus, achieving maximum absorption over a broad temperature range is rarely possible.

If one looks for the maximum absorption for an increasing value of ε', then the optimum value of $\tan\delta$ decreases to unity for both the disc and rod sample shapes. Thus, for example, a water solution ($\varepsilon'>80$) should achieve its maximum power absorption per unit volume when its ionic impurity content is increased such that $\tan\delta \approx 1$ at the desired frequency. In this case, a small rod-shaped sample will absorb $\approx$50 times more power per unit volume than a thin disc sample. In a multimode-cavity heating situation the oven dimensions are at least a few wavelengths, and thus strong, internal electric-field variations occur for each mode, although the usual existence of many modes produces some field averaging. In this case, for small samples, the long rod shape will still absorb better, but with a reduced advantage due to the effective averaging over field orientations.

In summary, we have demonstrated that with a simple refinement of the standard formula for microwave power absorption one can obtain reasonable quantitative estimates of absorbed power and Q for various sample shapes and dielectric properties, and, as well, gain some understanding of the process.

REFERENCES

1. M. de Jong, F. Adams and R. Hutcheon, Computation of RF Fields for Application Design, *Journal of Microwave Power & Electromagnetic Energy*, **27**, #3, 136 (1992).
2. F.P. Adams, M. de Jong and R. Hutcheon, Sample Shape Correction Factors for Cavity Perturbation Measurements, *Journal of Microwave Power & Electromagnetic Energy*, **27**, #3, 131 (1992).
3. J.A. Stratton, Electromagnetic Theory, McGraw-Hill, New York, 1941, Sec. 3.27, 211-215.
4. R.M. Hutcheon, M.S. deJong and F.P. Adams, A System for Rapid Measurements of RF and Microwave Properties up to 1400°C, *Journal of Microwave Power & Electromagnetic Energy*, **27**, #2, 87 (1992).
5. R.M. Hutcheon, M.S. deJong, F.P. Adams, P.G. Lucuta, J.E. McGregor and L. Bahen, RF and Microwave Dielectric Loss Mechanisms, in *Microwave Processing of Materials III*, eds. R.L. Beatty, W.H. Sutton and M.F. Iskander, *Mat. Res. Soc. Symp.* **269**, 541-551, 1992.

DIELECTRIC PROPERTY MEASUREMENT OF ZIRCONIA FIBERS AT HIGH TEMPERATURES

Gerald J. Vogt
Los Alamos National Laboratory, P.O. Box 1663
Los Alamos, New Mexico 87545

Wayne R. Tinga
Dept. of Electrical Engineering, University of Alberta
Edmonton, Canada T6G 2G7

Ross H. Plovnick
3M Ceramic Technology Center, 3M Center, Building 201-4N-01
St. Paul, Minnesota 55144

ABSTRACT

Using a self-heating, electronically tunable microwave dielectrometer, the complex dielectric constant of zirconia-based filaments was measured at 915 MHz from 350° to 1100°C. When exposed to a low temperature environment, this fibrous material cools rapidly within several seconds due to a large surface area to volume ratio. Such rapid sample cooling necessitates the use of a self-heating technique to measure the complex dielectric constant at temperatures up to 1100°C. Sample temperature was measured with optical fiber thermometry. The effect of sample temperature measurement on data accuracy is discussed.

INTRODUCTION

Successful application of microwave heating for ceramic fibers will require knowledge of the temperature-dependent dielectric properties of the desired ceramic filament. These dielectric properties can greatly affect the design and performance of the microwave process system, since they can change by several orders of magnitude with increasing temperature. Furthermore, the dielectric

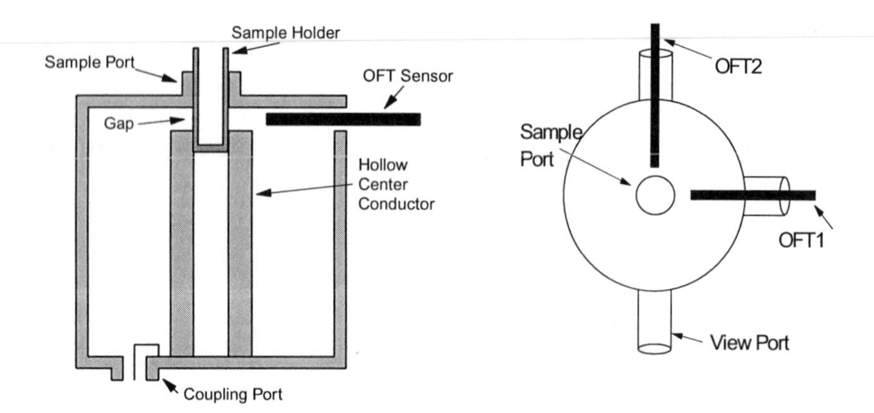

Figure 1a. Dielectrometer schematic showing position of the OFT light-pipe sensors in side view.

Figure 1b. Dielectrometer schematic showing position of the OFT1 and OFT2 lightpipe sensors in top view. OFT1 is the control OFT.

properties can significantly change due to chemical or structural changes in the ceramic during processing. Process modeling will naturally need the same data to produce realistic numerical results.

Because these properties are generally unknown, the dielectric properties must be measured over the range of possible process temperatures. Typically, a filament tow cools rapidly from a high temperature (>1000°C) to the temperature of its surrounding within several seconds due to a large surface area to volume ratio. Also, test samples cannot be consolidated into high-density bulk because this will significantly alter the dielectric properties of the ceramic fibers. Dielectric measurements under these sample constraints were performed for zirconia-based filaments with the self-heating, electronically tunable microwave dielectrometer developed at the University of Alberta [1-3].

EXPERIMENTAL

The theory and operation of the self-heating microwave dielectrometer has been described in earlier work [1-3]. Dielectric properties were measured at 915 MHz over a 350°-1100°C temperature range. The sample temperature was measured by two Accufiber optical fiber thermometers (OFT). Temperature was controlled by unit OFT1, operating at a single-wavelength band of 0.3-1.12 μm to measure temperature over a 300°-1200°C range. A second thermometer OFT2, operating

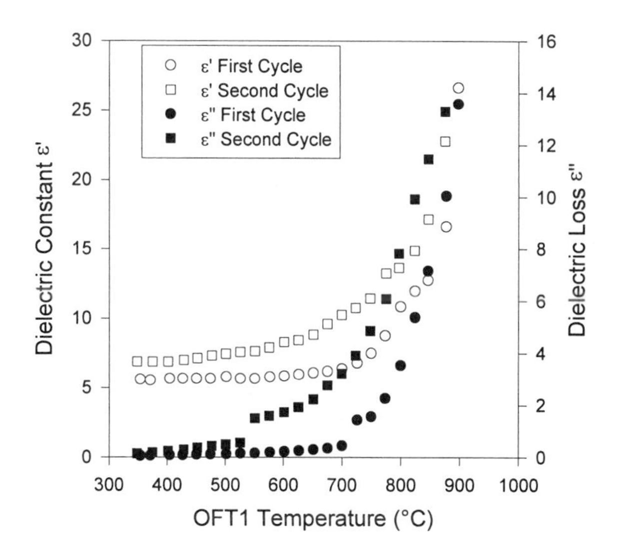

Figure 2. Dielectric constants measured at 915 MHz for a zirconia-based powder sample versus temperature measured by the control OFT. Mean sample density is 2.78 g/ml. Data for two measurement cycles are given.

with wavelength bands near 800 nm and 950 nm over a 550°-2000°C range, was added to the system for dual-wavelength (two-color) temperature measurements. Each OFT unit used an open-ended sapphire lightpipe to monitor the sample temperature. The lightpipe tips were positioned, as shown schematically in Figure 1, outside the measuring gap in the dielectrometer, ~15 mm from the sample for control OFT1 and ~12 mm away for OFT2. The spectral emissivity of the zirconia-based filaments was 0.17 at 800 nm and 0.18 at 950 nm, as measured by the ripple technique [4]. Temperatures from dual-wavelength OFT2 were measured as single-wavelength temperature values and converted to ratio (two-color) temperatures by taking the ratio of the measured light intensity (e.g. photodiode current) at the two wavelengths.

Test samples of 3M zirconia-based filaments were packed into a 7-mm OD x 5-mm ID x 10-cm long fused quartz tube with one closed end. Organic sizing on the filaments was first removed by firing at 600°C in air for 1 hour. Two sample groups with significantly different packing densities were prepared from chopped fibers in one group and from coarse powder in the second group. Filaments were chopped in a 1 wt% aqueous solution of poly(ethylene oxide) (MW≈5 million) in a food blender. Chopped filaments were dried and fired again in air at 600°C for

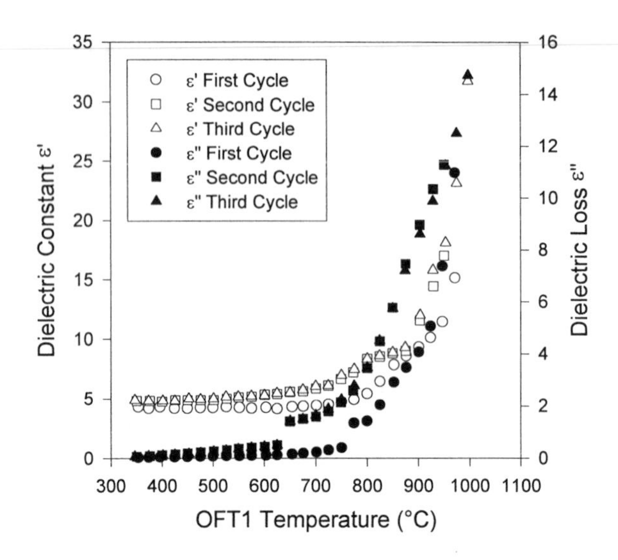

Figure 3. Dielectric constants measured at 915 MHz for a zirconia-based chopped fiber sample versus temperature measured by the control OFT. Mean sample density is 1.82 g/ml. Data for three measurement cycles are given.

1 hour to burn off the poly(ethylene oxide). The coarse powder was prepared by lightly grinding filaments with a porcelain mortar and pestle. The mean packing densities of the chopped filament samples and the powder samples were typically 1.5 to 1.8 g/ml and 2.7 to 2.9 g/ml, respectively. Dielectric properties were measured at two different locations along each tubular sample. The two measured segments along a sample were sufficiently spaced apart to avoid overlapping heating between the first and second measurements. Each location was measured either one, two or three times.

RESULTS AND DISCUSSION

Dielectric Properties

Figures 2 and 3 show temperature-dependent dielectric data measured for a powder sample and for a chopped fiber sample. Above 600°C the dielectric constant ε' for each sample increased rapidly. In the first measurement for the chopped sample, ε' at 975°C is approximately three times greater than at 350°C. Similarly for the powder sample, ε' at 900°C in the first measurement is approximately five

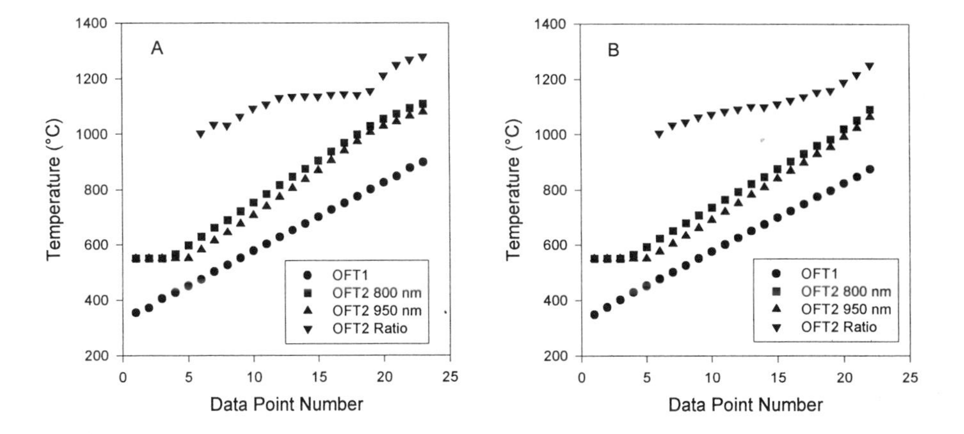

Figure 4. Measured temperatures for the zirconia-based powder sample in Figure 2 for the first (A) and second (B) measurement cycles at the same position. Data is plotted against its data point number in each measurement cycle.

times the value measured at 350°C. Even larger increases are seen in the dielectric loss data, approximately 100 times or more.

Interestingly, in Figures 2 and 3, the dielectric properties from a second measurement on the same sample volume yield even greater values than those from the first measurement, indicating a change in the material caused by the first measurement. Dielectric data from a third measurement in Figure 3 closely agree with those in the second measurement. The heated volumes for all samples were partially sintered, as evident from the presence of a weakly sintered piece found in the core of each heated volume. A sintered core was found in a sample that was measured only once. Each sample was probably sintered during each heating and measurement cycle. The increase in dielectric properties in a second measurement in Figures 2 and 3 is likely caused by partial sintering of the sample during the first measurement.

Temperature Measurement

Figure 4 shows the temperature values for the powder sample data in Figure 2 as measured by control OFT1 and dual-wavelength OFT2. The ratio temperatures in Figure 4 were calculated with the single-wavelength temperatures from OFT2. In both bands, the single-wavelength temperatures from OFT2 were significantly greater than the temperatures from OFT1, because the OFT2 sensor was closer to the sample. The OFT2-OFT1 difference is ~100°-150°C at the beginning of the test and ~200°C at the end of the test. The three single-wavelength temperatures show a continuous rise through the course of the measurement, where the slope of the OFT2 temperature curves is greater than that for the OFT1 temperature curve. From OFT2, the temperature measured in the 800 nm band was always greater than the temperature measured in the 950 nm band, although each was corrected with a measured emissivity.

Ratio temperature data from OFT2 display a very different behavior. In both tests, these ratio temperatures are substantially greater than the three single-wavelength temperatures. In fact, the ratio temperature is greater than the OFT1 temperature by ~500°C in the beginning and by ~400°C at the end. More surprisingly, the ratio temperature is nearly constant in the middle of both tests, although the single-wavelength temperatures show continuous increase. The ratio temperatures suggest that a hot spot formed in the samples from non-uniform heating. Whereas, because the sample's 2.5-mm gap target area only fills a fraction of the OFTs' conical field of view, the single-wavelength temperatures are probably quite low due to targeting error.

Large temperature variations in the measurements can be attributed to targeting error. The Accufiber lightpipe sensor tip collects emitted light over an optical cone angle of ~52°. If the emitting source does not fill the conical view of sensor tip, the measured temperature will be less than the source temperature in proportion to the fraction of view not filled by the source, producing a targeting error. The amount of light collected by an OFT sensor is measured as a photodiode current by a detector. If the source does not fill the conical view, the measured photodiode current will be lower than the expected current, producing a low temperature reading.

Control OFT1 was affected by targeting error because the sensor tip was located ~15 mm back from the heated sample surface. Single-wavelength temperatures from OFT2 also suffered from targeting error, as the sensor was ~12 mm back. The visible heated sample is ~2.5 mm high by ~5 mm wide surrounded by a 7-mm OD quartz tube. The 52° optical cone of the sensor tip allows the sensors to view a circular spot with a diameter approximately equal to the distance between the source and the sensor tip. Both control OFT1 and OFT2 were viewing part of the dielectrometer inner brass surface as well as the heated sample. The ratio temperatures are believed to be correct, as the ratio temperature measurement is not affected by targeting error. The ratio measurement is based on the light

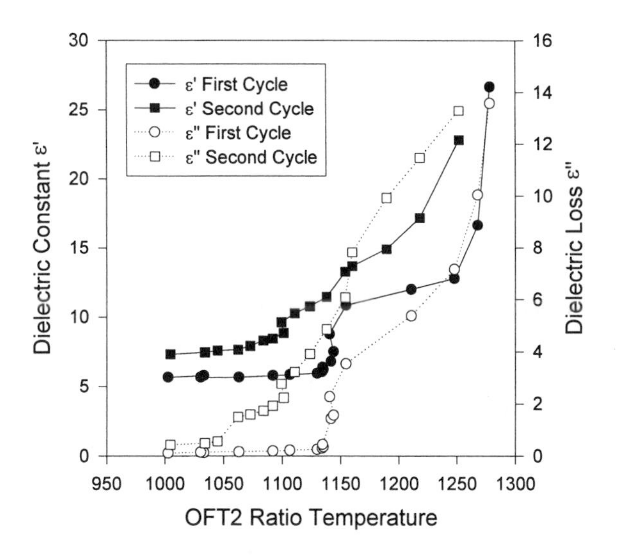

Figure 5. Dielectric constants measured at 915 MHz for the powder sample of Figure 2 versus the OFT2 ratio temperature.

intensity ratio, not intensity. Just how accurate the ratio temperatures are in this experiment is not known without further experimental study.

There are several other potential sources of experimental error that could effect the accuracy of the temperature measurements. Absorption error may result from the quartz sample tube absorbing and reflecting the emitted light from the sample. Error in the spectral emissivity can induce a large error in temperature [5-6] in both single-wavelength and dual-wavelength measurements.

Interpretation of Experimental Data

The dielectric measurements with temperatures as measured by the single-wavelength method, as shown in Figure 2 with OFT1 temperatures, exhibit the anticipated continuous increase in ε′ and ε″ with increasing temperature. However, the ratio temperatures measured by OFT2 are probably a better measure of the true sample temperature as the ratio temperature measurement was not affected by targeting error. In Figure 5, the dielectric data for the powder sample are plotted against the ratio temperature from OFT2. The ε′ and ε″ data in the first measurement contain an abrupt step change at ~1140°C, followed by gradual increases with increasing temperature. This step change is not seen in the ε' and ε'' curves for the repeat measurement. Presumably, this change in sample dielectric properties is associated with the sintering that occurred in the first measurement. Other discontinuities exist in the dielectric data in Figure 5, but can not be interrupted due to the uncertainty in the temperature measurements.

CONCLUSIONS

During the first dielectric measurement, the dielectric properties of the zirconia-based samples were changed due to heating and sintering of the sample. Sample sintering was confirmed during sample removal from the quartz tube. Significant experimental uncertainty exists in the temperature measurement by single-wavelength optical fiber thermometry due to targeting error. Temperature uncertainty could be significantly reduced by moving the lightpipe sensor closer to the sample, eliminating targeting error, or by using the dual-wavelength optical fiber thermometry. Interpretation of the dielectric data is not possible until the temperature measurement is improved.

ACKNOWLEDGMENTS

This work at Los Alamos was supported by the Department of Energy, Office of Industrial Technologies, Advanced Industrial Materials Program.

REFERENCES

1. W.R. Tinga, "Microwave System Design and Dielectric Property Measurements," **Microwave Processing of Materials IV**, Materials Research Society Symposium Proceedings, Vol. 347, pp. 427-440 (1994).
2. W.R. Tinga and W. Xi, "Design of a New High-Temperature Dielectrometer System," Journal of Microwave Power & Electromagnetic Energy, 28(2), pp. 93-103 (1993).
3. W. Xi and W.R. Tinga, "Error Analysis and Permittivity Measurements with Re-entrant High-Temperature Dielectrometer," Journal of Microwave Power & Electromagnetic Energy, 28(2), pp. 104-112 (1993).
4. Private communication with Chuck Schietinger, Luxtron Accufiber.
5. T.D. McGee, **Principles and Methods of Temperature Measurement**, John Wiley & Sons, New York, pp. 364-367 (1988).
6. **Microwave Processing of Materials**, Publication NMAB-473, National Academy Press, Washington, D.C., pp. 49-57 (1994).

HIGH TEMPERATURE DIELECTRIC CONSTANT MEASUREMENT - ANOTHER ANALYTICAL TOOL FOR CERAMIC STUDIES?

R.M. HUTCHEON, P. HAYWARD,[†] B.H. SMITH AND S.B. ALEXANDER
AECL, Chalk River Laboratories, Chalk River, Ontario, Canada K0J 1J0
[†] Whiteshell Laboratories, Pinawa, Manitoba, Canada

ABSTRACT

The automation of a high-temperature (1400°C), microwave-frequency, dielectric constant measurement system has dramatically increased the reproducibility and detail of data. One can now consider using the technique as a standard tool for analytical studies of low-conductivity ceramics and glasses. Simultaneous temperature and frequency scanning dielectric analyses (SDA) yield the temperature-dependent complex dielectric constant. The real part of the dielectric constant is especially sensitive to small changes in the distance and distribution of neighboring ions or atoms, while the absorptive part is strongly dependent on the position and population of electron/hole conduction bands, which are sensitive to impurity concentrations in the ceramic. SDA measurements on a few specific materials will be compared with standard differential thermal analysis (DTA) results and an attempt will be made to demonstrate the utility of both the common and complementary aspects of the techniques.

AN INTRODUCTION TO SCANNING DIELECTRIC ANALYSIS

Automation of the previously reported,[1] high-temperature, microwave-frequency dielectric properties measurement system has increased the reproducibility and reliability of the data, and in practical operation allows much smaller temperature steps to be used yielding more detailed information. The system measures both the real (ε') and the absorptive part (ε'') of the relative complex dielectric constant, $\varepsilon = \varepsilon' - j\varepsilon''$, as the temperature is stepped over a user-specified range. This refinement has opened the possibility of using scanning dielectric analysis (SDA) as an analytical tool in much the same way as one would use differential thermal analysis (DTA) to determine the chemical phase composition of a sample or to determine the response of a material to high-temperature processing.

Solid state phase transformations usually produce a change in electronic and ionic polarizability (and thus ε') of a material, since the change in relative atomic positions and spacings generally changes the deformability of outer-shell electron clouds as well as the phonon mode distribution. The change in spacing and arrangement also influences the electron and/or hole band gaps (changing the electronic conductivity) and can cause changes in the lattice relaxation time constants. Both effects produce observable changes in the microwave absorption of the material (i.e. in ε''). These effects have been well studied and the principles understood for many years.

Measurements of the frequency and temperature dependence of dielectric constants have been published, but usually only for a few materials whose specific electronic properties were of interest - for example, ferroelectric materials and high dielectric constant materials. These measurements have usually been restricted to temperatures below 450°C and frequencies below 1 MHz. In this region interfacial polarization at the electrodes and interconnectedness of the sample make measurements on powders and green ceramics difficult to do and interpret. With the exception of low-temperature polymers, the technique has rarely been used as an analytic tool to characterize the state of a material or the progress of a reaction. However, recent development of dielectric-measurement apparatus[1,2,3,4] have opened up the high-temperature range between 200°C and 1400°C for studies of the high-frequency response of ceramics.

In this paper we will show several SDA spectra for polycrystalline powder samples of materials that are traditionally used for temperature calibration of DTA systems, as well as spectra of well-known ferroelectric materials. An example will be given of monitoring the progress of a decomposition reaction (binder burnout in alumina compacts). A simple formalism will be outlined with which to start the process of interpreting the SDA spectra.

FREQUENCY DEPENDENCE OF DIELECTRIC CONSTANT AND RELAXATION MECHANISMS

The relative complex dielectric constant, ε, and the susceptibility, χ, of a single crystallite are represented by

$$\varepsilon = \varepsilon' - j\varepsilon'' = \chi' + 1 - j\chi'' , \tag{1}$$

where $\varepsilon' = \chi' + 1$ is the usual "dielectric constant" and depends on the material polarizability, whereas $\varepsilon''=\chi''$ is the "absorptive" part and is a measure of the rate at which energy is absorbed by the material from an ac electric field. At high frequencies (>10 MHz), the polarization contributions primarily are electronic polarizability (electron cloud deformation), ion-ion polarizability (change in mean spacing of oppositely-charged ions), and permanent dipole rotation (as with water). The primary energy loss mechanics are electron (and hole) conductivity, ion-ion relaxation, and permanent dipole relaxation. These yield

$$\varepsilon' - 1 = \chi' = \chi'_{\text{electronic}} + \chi'_{\text{ionic}} + \frac{\Delta\chi'_D}{(1+\omega^2\tau_D{}^2)} \tag{2}$$

and

$$\varepsilon'' = \chi'' \approx \frac{\sigma_{el}^{dc}}{\omega\varepsilon_o} + \chi''_{\text{ionic}} + \frac{\omega\tau_D\Delta\chi'_D}{(1+\omega^2\tau_D^2)} , \tag{3}$$

where σ_{el}^{dc} is the dc electron conductivity (Siemens/metre), ε_o is the permeability of free space, τ_D is the relaxation time of oscillating permanent dipole structures and $\Delta\chi'_D$ is the low-frequency contribution to the real part of the dielectric constant by the relaxing dipoles. Water, in loosely bound states and polymers are the usual sources of the permanent dipole terms. Although there are exceptions, the terms are introduced here in what is usually the decreasing order of importance at high frequencies (>10 MHz). The term $\chi'_{\text{electronic}}$ has no explicit temperature dependence, while the electronic conductivity usually has a standard thermal activation dependence with a specific activation energy for each phase. The ionic susceptibility term has only a weak temperature dependence, related to the thermal expansion coefficient[5], except in ferroelectric and antiferroelectric materials. In the latter cases, the value of ε'' is influenced by the presence of strong, very temperature dependent, "soft" ion-oscillation modes which can have either resonant or relaxational character. In the region of a ferroelectric transition, the relaxation frequency of the soft modes can drop down into the microwave range. The signature of this situation is a value of ε'' which **increases** with increasing frequency to frequencies well above 10^9 Hz. In contrast, the signature of electronic conduction is a value of ε'' that is inversely proportional to frequency.

Polycrystalline powder samples have "effective" dielectric constant values that are lower than the single-crystal values, as expected of a gas-solid mixture. If one wishes to extract a single-crystal value from the polycrystalline measurements, then the shape and dielectric anisotropy of the crystallites must be determined, and then the appropriate mixing formula chosen. However, the transition temperature and the character of individual phases may be determined without this step, since the single-crystal and "effective" dielectric constant values are monotonically related.

COMPARISON OF SDA AND DTA FOR STANDARD CALIBRATION MATERIALS

A DTA system provides a **continuous** measure of the temperature difference between a sample and a known reference material when both are placed near each other in an oven whose temperature is changing at a constant rate. The temperature difference is proportional to the latent heat (enthalpy change) associated with a phase transition in the sample or with the specific energy released by a chemical reaction, as well as being proportional to the ramp rate. Temperature calibration of DTA systems is often done by measuring the response of materials which have fast, reversible, first-order phase transitions. The first four materials listed in Table I are DTA calibrators, and their spectra are shown in Fig. 1. The DTA signal is truly differential, reflecting only the heat release during the transition. The signal polarity is

TABLE I. List of known properties of the phase transitions studied during this work.

Compound	Transition Onset Temperature (°C)	Transition Type and Description	Crystal Configuration Low Temp.→ High Temp.
$SrCO_3$	928	reversible, diffusionless	orthorhombic→hexagonal
K_2CrO_4	665	1st order (hysteresis), anisotropic conductivity	orthorhombic→hexagonal?
K_2SO_4	582	1st order, ordering of SO_4 group, semiconducting in high phase	orthorhombic→hexagonal
SiO_2	571	network puckering, displacement/rotation of SiO_6 tetrahedra	α(trigonal)→ β(hexagonal)
ZrO_2	≈1150	displacive martensitic, 9% volume contraction	monoclinic→tetragonal
$PbTiO_3$	T_c=490	ferro→paraelectric, soft modes underdamped	tetragonal→cubic
$KNbO_3$	T_c= 434	ferro→para electric soft modes overdamped	tetragonal→cubic

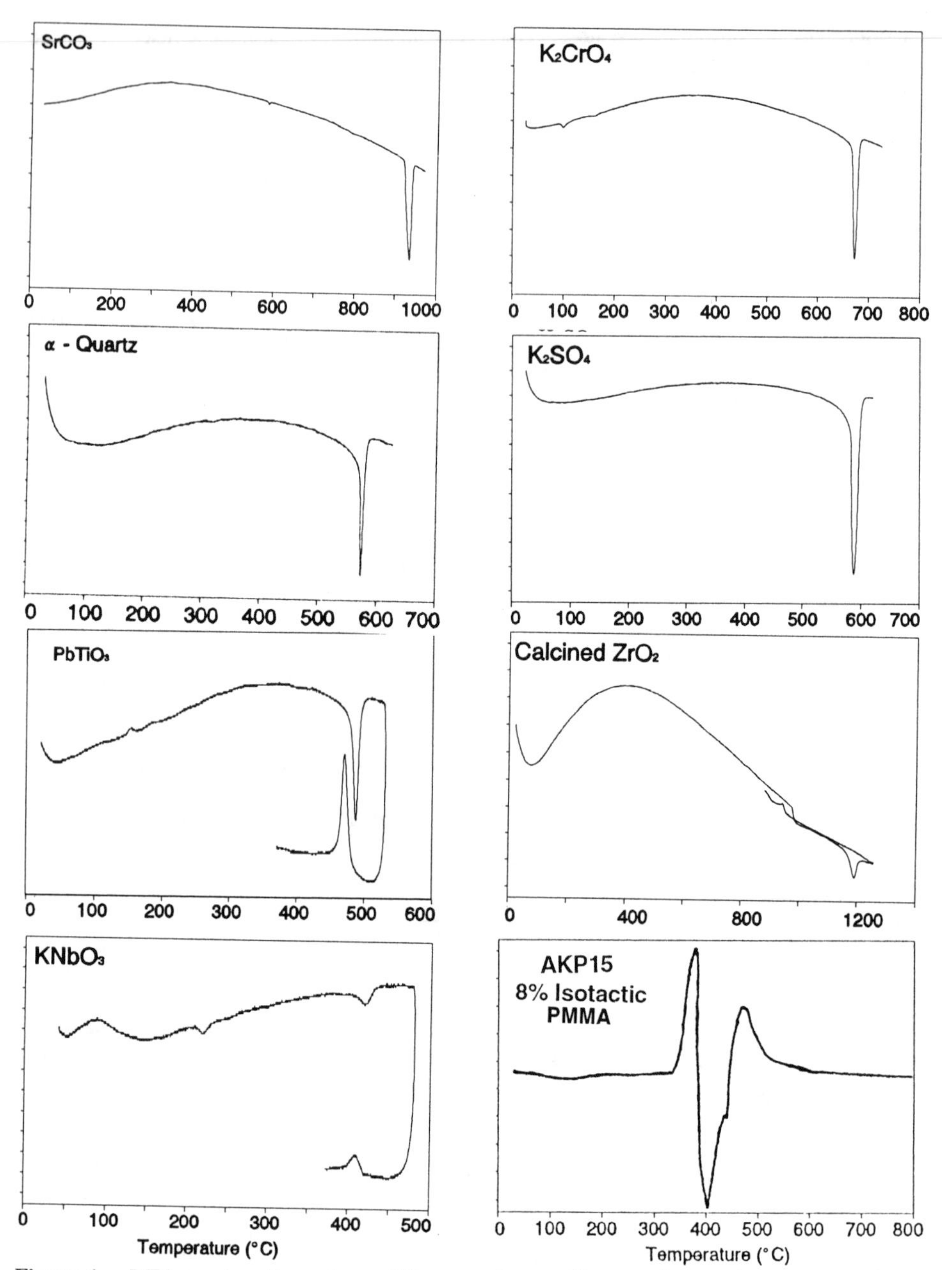

Figure 1. DTA spectra of samples. All were polycrystalline powder specimens except the alumina-isotatic PMMA sample which was a densly-pressed powder compact. Most spectra represent the second thermal cycle of a sample.

governed by the direction of the ramp; with increasing temperature, an endotherm gives a negative signal while an exotherm gives a positive one. These polarities are inverted with decreasing temperature (see $PbTiO_3$, Fig. 1). If a transition is slow (e.g. some reconstructive transitions are "sluggish"), the DTA signal can be small and diffuse, making the interpretation uncertain.

The SDA spectra reflect the state of the sample both before and after the transition, as demonstrated by the dielectric response around the reconstructive transition at 928°C in strontium carbonate, $SrCO_3$ (Fig. 2). The values of ε' before and after the transition reflect the polarizability in the two states; the usual increase in average lattice constant in going to the higher-temperature state produces an increased polarizability and larger ε'. The ε'' values have an inverse frequency dependence typical of electron conductivity, but only by fitting to an exponential thermal activation dependence can the small change in conductivity at the transition be seen.

The SDA spectra for potassium chromate (K_2CrO_4) (Fig. 2) provide a much better demonstration of how the rearrangement of atoms can change the electron conductivity. The step in ε' at ≈650°C (onset temperature) clearly indicates the existence of a transition. The dramatic increase in ε'' (which has electron conductivity character) suggests a factor of 9 increase in dc electron conductivity (0.0020→0.018 S/m) at the transition. A small range of hysteresis (≈20°C) is seen. The SDA spectra for potassium sulphate (K_2SO_4) (Fig. 2) is very similar to that of potassium chromate, with the same large increase in apparent electron conductivity appearing at the transition.

To make the SDA spectra appear more like the DTA spectra, one could differentiate the ε' curve with respect to temperature and produce a differential spectrum with a peak at each step change in dielectric constant. However, this process actually discards the information on the relative dielectric constants of the two phases, and is usually inapplicable to analysis of the ε'' values.

The final example, the α–β transition in quartz (SiO_2) at 571°C (onset) (Fig. 2), demonstrates the problem of using the SDA technique in isolation; the transition produces essentially no change in ε' (≤5%) and, within experimental error, has no readily apparent threshold in ε''. The reason for this lies in the nature of the transition. It has been termed a network puckering transition, one where the silicon-oxygen tetrahedra only rotate slightly, without change in coordination or oxygen bonding, and with only a 1% increase in volume. Clearly, the polarizability of the atoms is unaffected by this transition.

SDA AND DTA ON FERROELECTRIC MATERIALS

SDA and DTA spectra were run on 99.5% purity lead titanate ($PbTiO_3$) and high-purity (99.999%) potassium niobate ($KNbO_3$) powders obtained from Johnson-Matthey (Figs. 1 and 2). Both materials have ferroelectric to paraelectric (tetragonal to cubic) transitions, and both are known to have "soft" modes in the ferroelectric state. The dielectric responses of the two systems are clearly quite different.

The lead titanate shows large hysteresis (≈25°C) and has a very large paraelectric response just above the Curie point, decreasing with temperature with an approximate $(T-T_c)^{-1}$ response. The potassium niobate, however, shows only a small increase in dielectric constant over the transition (compatible with only small changes in electronic polarizability from phase to phase) although it is known to have a large, low- frequency paraelectric response just above the Curie point. This suggests strong damping of the soft modes. The dielectric-loss response, although different in magnitude for the two materials, shows the same inversion in frequency

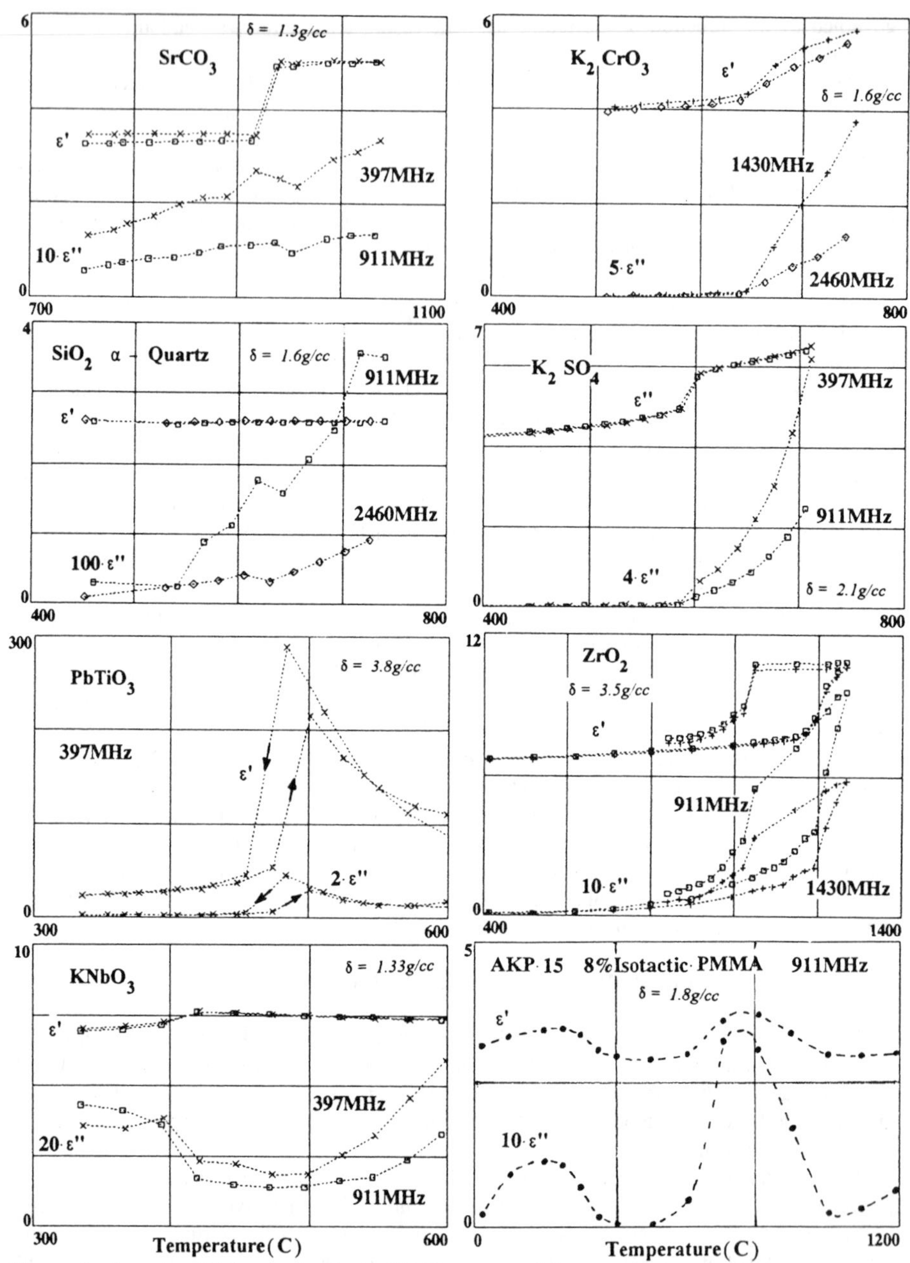

Figure 2. Scanning Dielectric Analysis (SDA) spectra of the permittivity of (ε',ε'') of polycrystalline powder samples. Most spectra were taken on the second thermal cycle. The AKP-15 alumina sample was a pressed-powder compact, ramped at ≅20°C/minute.

dependence over the transition. In both cases, the ε'' frequency dependence in the paraelectric cubic phase suggests a standard thermally activated electronic conductivity.

The ferroelectric phase transitions are readily seen in the SDA spectra and some information concerning the fundamental properties of any specific transition can be obtained.

SPECTRA FOR THE TETRAGONAL-MONOCLINIC TRANSITION IN ZIRCONIA

Zirconia has three phases: a high-temperature cubic phase which transforms at 2370°C to tetragonal phase, which in turn transforms to a monoclinic-phase via a well-studied transition at $\approx$1170°C. The latter transition is a diffusionless shear process (martensitic-like) with thermal hysteresis between heating and cooling cycles and an accompanying 9% increase in the atomic volume as the system cools and transforms to the lower-temperature monoclinic phase. This unusual situation of abrupt expansion on cooling made the manufacture of large sintered zirconia pieces impossible until the discovery that the addition of small amounts of some particular oxides (e.g. yttria) will stop the tetragonal phase from transforming to the monoclinic phase even at room temperature.

The DTA spectrum for a standard-purity, calcined, zirconia powder (Fig. 1) shows the tetragonal-to-monoclinic transition, with $\approx$160°C hysteresis between heating and cooling. The SDA spectrum (Fig. 2) shows the expected hysteresis in ε', but the increase of ε' (and thus of the polarizability) in the more dense tetragonal phase is not the usual trend. A sample of very high-purity (99.995%) zirconia had the same increased value of ε' in the tetragonal phase. However it exhibits a factor of twenty smaller value of ε'', although with the same typical electron conductivity. A sample of 7% yttria-stabilized zirconia showed the same inverse- frequency dependence of ε'', but with a dc electron conductivity of $\approx$0.4 ohm-metres at 800°C ($\varepsilon''\approx$100 at 400 MHz).

BINDER BURNOUT FROM ALUMINA COMPACTS

A previous study[6] used SDA to assist in understanding the process of polymethylmethacrylate (PMMA) binder burnout during the low-temperature stages of sintering of pressed alumina pellets. The DTA data (Fig. 1) show an initial exotherm, followed by an endotherm and at least one succeeding exotherm. The endotherm disappears and the exotherms coallesce into one peak if the temperature ramp rate is very low, suggesting that the endotherm is associated with oxygen starvation in the pellet core and resulting pyrrolysis of the binder. The second exotherm is believed to result from carbon burnout in the pellet core. The dielectric-loss portion of the SDA spectrum (Fig. 2) suggests the generation of a high-loss component in the material (relaxation of PMMA fragments?) whose disappearance is complete at 400°C and the appearance of another, thermally-activated, loss mechanism (carbon-based electron conductivity?) which slowly disappears as the carbon burns away at high temperatures.

The differences between DTA and SDA results are very evident here. The DTA signal is proportional to the rate of evolution of energy in the sample, while the SDA response reflects the state of the system at the time of the measurement. Thus, the thermal activation and dissociation of the PMMA seen in SDA at 200°C is hardly noted in DTA, while the large exothermic spike at 370°C in DTA corresponds with the rapidly decreasing SDA value, reflecting the decreasing amount of reactant.

GENERAL CONCLUSIONS

The basic premise that SDA can be used as a standard analytic tool, much as one uses DTA, seems well substantiated. It appears that a phase transition generally appears as a step in the real part of the dielectric constant, while the change in the absorptive part provides additional information on changes in the relaxation and conductivity mechanisms.

With the present apparatus, the dielectric constant is measured every four minutes. With 20°C measurement steps (temperature ramp rate ≈5°C/min) the measurements are made somewhat slower than for a usual DTA run. However, our experience has shown that when a specific temperature range is of interest, the furnace can be rapidly ramped to the range and then slowly stepped over it, thereby speeding up the data acquisition. The lapsed time for a full run is, at most, twice that of a DTA run. The usual problems of reproducibility in DTA spectra occur also in SDA. If a known, reversible-transition, sample is cycled over the transition a few times, the first heatup ramp yields somewhat different results from the rest of the cycles.

Work is now in progress to apply SDA as a tool to assist in understanding specific ceramic synthesis reactions[7] and polymer-composite processing.

REFERENCES

1. R.M. Hutcheon, M.S. de Jong, F.P. Adams, P.G. Lucuta, J.E. McGregor and L. Bahen, *"RF and Microwave Dielectric measurements to 1400°C and Dielectric Loss Mechanisms"*, MRS Symposium Proceedings, **269**, pp. 541-551; Microwave Processing of Materials III, edited by R.L. Beatty, W.H. Sutton, M.F. Iskander (1992).
2. W. Xi and W.R. Tinga, *"Microwave Heating and Characterization of Machinable Ceramics"*, ibid, pp. 569-577.
3. S. Bringhurst, O. Andrade and M.F. Iskander, *"High-Temperature Dielectric Properties Measurements of Ceramics"*, ibid ,pp. 561-568.
4. M. Arai, J.G.P. Binner, G.E. Carr and T.E. Cross, *"High Temperature Dielectric Property Measurements of Engineering Ceramics"*, Ceramic Transactions, **36**, pp. 483-492, Microwaves: Theory and Application in Materials Processing II, edited by D.E. Clark, W.R. Tinga and J.R. Laia, (1993).
5. W.W. Ho, *"High-Temperature Dielectric Properties of Polycrystalline Ceramics"*, MRS Symposium Proceedings, **124**, pp. 137-148; Microwave Processing of Materials, edited by W.H. Sutton, M.H. Brooks, I.J. Chabinsky, (1988).
6. E.H. Moore, D.E. Clark and R.M. Hutcheon, *"Removal of Polymethyl Methacrylate Binder from Alumina using Conventional and/or Microwave Heating"*, Ceramic Transactions, **36**, pp. 325-332 in ref. 4.
7. M.D. Mathis, D.K. Agrawal, R. Roy, R.H. Plovnick and R.M. Hutcheon, *"Microwave Synthesis of Aluminum Titanate in Air and Nitrogen"*, these proceedings.

A Parallel Measurement Programme in High Temperature Dielectric Property Measurements: An Update.

J Batt[†], J G P Binner[*], T E Cross[*], N R Greenacre[*], M G Hamlyn[¶], R M Hutcheon[‡], W H Sutton[†] and C M Weil[§]

† United Technologies Research Center, East Hartford, Connecticut, 06108, USA
* The University of Nottingham, University Park, Nottingham NG7 2RD, UK
¶ Staffordshire University, Beaconside, Stafford ST18 0AD, UK
‡ AECL, Chalk River Laboratories, Chalk River, Ontario K0J1J0, Canada
§ National Institute of Standards and Technology, Boulder, Colorado, 80303-3328, USA

ABSTRACT

Following the Materials Research Society Symposium on Microwave Processing of Materials held in San Francisco during April 1992 a Parallel Measurement Programme for high temperature dielectric properties was established. Initial results of this programme were presented at the ACerS symposium in Cincinnati in 1993[1] and preliminary results of the second stage at the MRS meeting in San Francisco in 1994. This paper will review the results obtained in the second stage of the programme since 1993 and give an inter-comparison of the applicability of the different measurement techniques.

INTRODUCTION

The interest in microwave energy for materials processing has been gathering momentum over recent years. It has found wide use in specific areas such as drying, rubber and food processing but its wider application in materials processing has been limited. One of the problems associated with the use of microwave energy is the calculation of the distribution of energy absorbed within a body undergoing heating. This requires knowledge of the variation of both the electrical and thermal properties of the material with temperature.

Computer modelling has been applied to the problem of the microwave heating of materials and this too requires knowledge of the variation of the electrical and thermal properties of the material to be heated with temperature.

The parameter which characterises the electrical properties of a material and hence describes the interaction between an applied electric field and a material is the permittivity. In general the permittivity is a complex quantity, the real part, ε', describing the energy stored within the material whilst the imaginary part, ε'', describes the energy dissipated. Both the real and imaginary parts of the permittivity are functions of temperature and ε", in particular, may change by several orders of magnitude between ambient temperature and 1000°C. Knowledge of the permittivity and its variation with temperature is therefore vital for the effective use of microwave energy.

A number of laboratories world-wide have developed or are developing techniques for the measurement of the microwave properties of materials at high temperature (>500°C). To be successful these techniques must either withstand the high temperature or the measurement time must be short enough so as not to allow significant thermal energy transfer from the sample to the measuring equipment.

The errors involved in the measurement of permittivity at high temperature are difficult to estimate. In order to establish confidence in the different measurement techniques it is necessary to quantify the errors involved and compare measurements made in different laboratories using different techniques. In order to do this a parallel measurement programme has been established where measurements may be made on a range of nominally identical samples by the different laboratories involved in this work.

PARALLEL MEASUREMENT PROGRAMME

Ideally a measurement programme of this sort should perform measurements on the same samples at each laboratory. In practice this was not possible due the differing requirements of the various measurement techniques. There was also some concern that repeated temperature cycling in potentially different atmospheres could change the properties of a single sample. A compromise was reached at a special discussion session at the Materials Research Society Symposium on the Microwave Processing of Materials in San Francisco, USA in April 1992. It was agreed that a series of measurements would be performed on samples cut from a single batch of alumina ceramic. The measurements should be performed at two frequencies and three temperatures; these conditions were 915MHz and 2.45GHz and 25, 500 and 1000°C. Results were received from three laboratories (AECL Research, Canada; Staffordshire University, UK and The University of Nottingham, UK) and were published at the symposium Microwaves: Theory and Application in Materials Processing II at the Annual Meeting of the American Ceramics Society held in Cincinnati during April 1993[1].

At the ACerS meeting in 1993 it was decided to hold a further round of the programme with a larger number of samples. In order to broaden the applicability of the measurement programme three materials were chosen for study with different dielectric properties. These were a low loss material (alumina), a medium loss material (zirconia) and a high loss material (silicon carbide). The alumina samples were Dynallox A100 and the zirconia Techno 2000 both supplied by Dynamic Ceramic, Stoke on Trent, UK as tiles of 90mm x 90mm x 15mm. The measurement samples were then cut from a single tile of these materials. The silicon carbide samples were again cut from tiles (4" x 2" x 1") of "Type N" silicon carbide supplied by Cercom Inc., Vista, Ca, USA. The tiles were in turn cut from a larger plate in order to ensure uniformity of properties across the tiles.

At the time of writing this paper results had been obtained from the following laboratories:

- AECL Research, Chalk River, Ontario, Canada
- NIST, Antenna & Materials Metrology Group, Boulder, Co, USA
- The University of Nottingham, UK
- Staffordshire University, UK
- United Technologies Research Centre (UTRC), East Hartford, Ct, USA

The ranges of temperature and frequency of the measurements submitted by the different laboratories are summarised in Table 1.

Table 1 Detail of measurement temperatures and frequencies

Organisation	Frequency Range	Temperature Range
AECL	912 MHz and 2.46 GHz	23 to 1100 ºC
NIST	50 MHz to 3 GHz	Room temperature
Nottingham (coax probe)	200 MHz to 3 GHz	20 to 1200 ºC
Nottingham (cavity)	600 MHz to 4 GHz	20 to 1200 ºC
Staffordshire	2.45 GHz	20 to 1360 ºC
UTRC	2.5 GHz to 4 GHz	20 to 1100 ºC

Cavity Based Techniques

The techniques used by AECL, Staffordshire University and one of the those used at The University of Nottingham are based on the perturbation of a TM_{ono} cavity. Multi-mode cavities are used to give several spot frequencies within the range 500 MHz to 3 GHz. The technique moves a hot sample, heated in a conventional furnace, rapidly into a region of high electric field within a TM_{ono} cavity. The resonant frequency and Q factor of the cavity are then measured using a vector network analyser and the permittivity of the sample calculated. The technique has been extensively described elsewhere[2 & 3].

Transmission Line Techniques

The methods used at NIST and The University of Nottingham are based on a coaxial transmission line while UTRC uses a waveguide based technique.

At NIST the sample of about 10mm length is accurately machined to fit as the dielectric in a standard 14mm, 50Ω, air-line. The transmitted and reflected signals are measured using a calibrated HP8510 vector network analyser. The dielectric data is then calculated using the NIST EPS_MU_3 reduction algorithm. Precision measurements of the sample dimensions are made and corrections for the gaps between the sample and the air-line made[4-6].

At Nottingham the sample is used to terminate a 50Ω co-axial line made from material which will withstand the high temperature. An HP8510 vector network analyser is then used to measure the reflection coefficient of the sample from which the dielectric properties are calculated[7].

UTRC uses a dual waveguide with a sample in the form of a slab or post placed in one of the waveguides. An HP8510 is then used to measure the difference in transmission and reflection between the sample and reference arms. From this data the permittivity of the sample is calculated[8].

RESULTS

The samples were measured by NIST at room temperature and the results of these measurements are presented in Table 2.

Table 2 Room temperature results from NIST over the frequency range 1 to 3 GHz

Material	ε'	ε"
Alumina	9.95±0.05	0.1±0.03
Zirconia	33.7±0.2	0.2±0.1
SiC	240±30	300±50

The results on the alumina samples are presented in Figures 1 and 2. From these results it may be seen that there is good agreement between the different laboratories on the value of ε', all being within about ±5%. However there is some disagreement between the different techniques for the value of ε".

In the case of the zirconia, the results for ε', shown in Figure 3, are in good agreement up to about 800°C. Above this temperature the agreement becomes less good with differences in ε' of about ±10% at 1200°C. In the case of ε", shown in Figure 4, the agreement is poor with a difference of about ±50% between the different techniques. Data from the Nottingham probe is not shown as recent investigations have shown that the sample used was too small to produce accurate results.

The agreement on the silicon carbide results, Figures 5 and 6, is also poor with a value of ε' ranging from around 100 up to around 500 with some laboratories having difficulty obtaining values at all. The range in ε" is also large ranging from about 100 up to about 400.

CONCLUSIONS

The major conclusion to be drawn from this measurement programme is that the level of agreement on the values of ε' between the different laboratories for materials with ε'≤40 is good. Materials of higher permittivity, particularly at the higher temperatures, however produce less good agreement. There is also poor agreement between the different techniques for ε" even in the case of the alumina sample.

This indicates that much further work needs to be done on the current techniques particularly for measurements of ε". It also highlights the value of this type of measurement programme in establishing the limits of agreement between the different techniques.

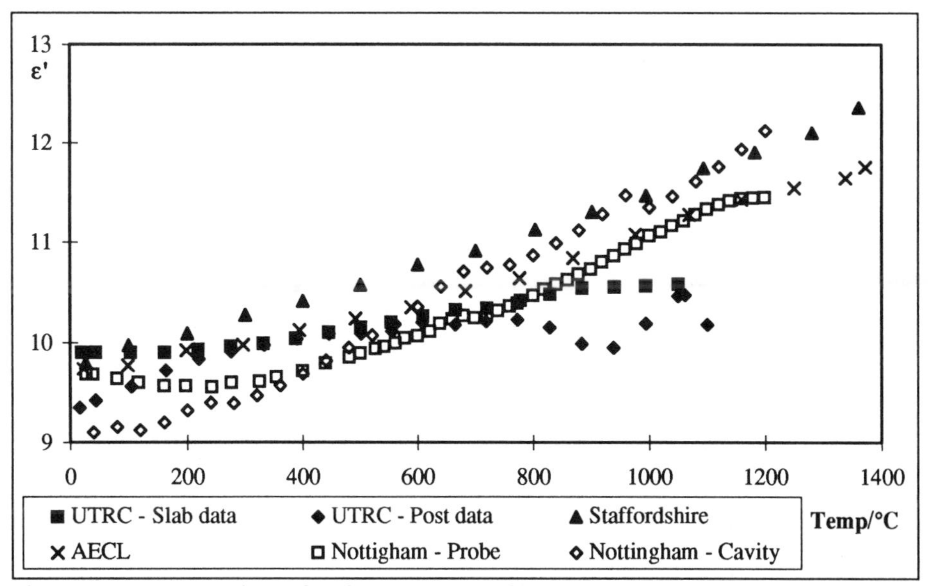

Figure 1 Inter-comparison of ε' for Dynallox A100 alumina vs temperature at a nominal frequency of 2.45GHz

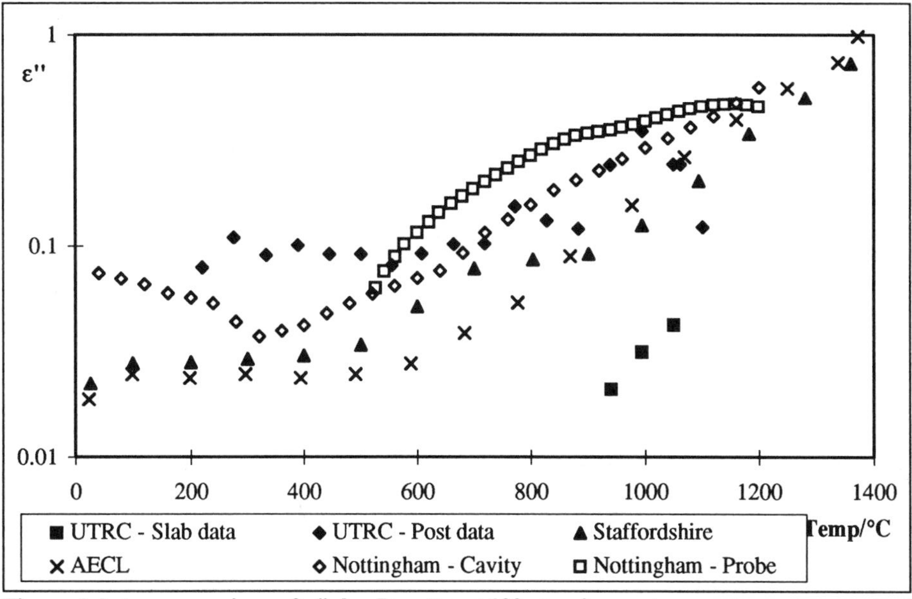

Figure 2 Inter-comparison of ε" for Dynallox A100 alumina vs temperature at a nominal frequency of 2.45GHz

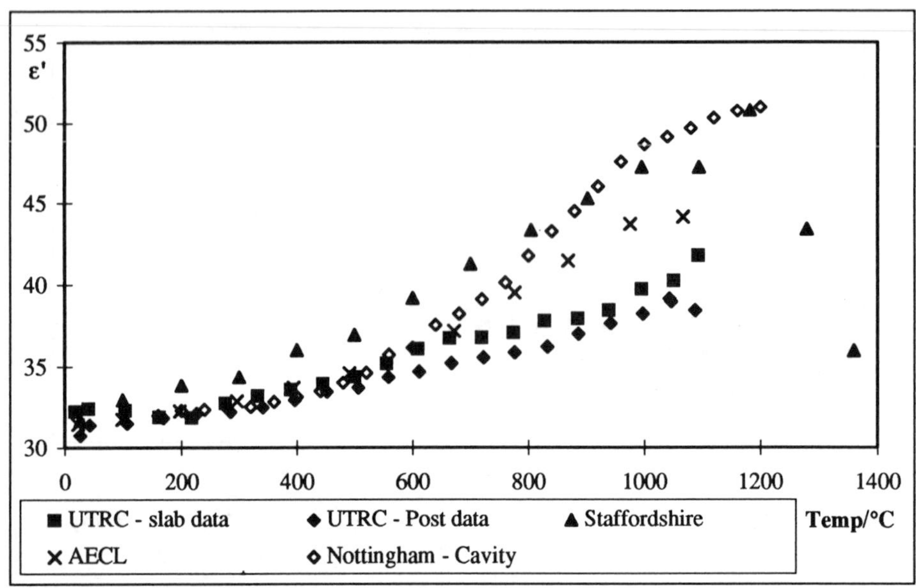

Figure 3 Inter-comparison of ε' for Techno 2000 zirconia vs temperature at a nominal frequency of 2.45GHz

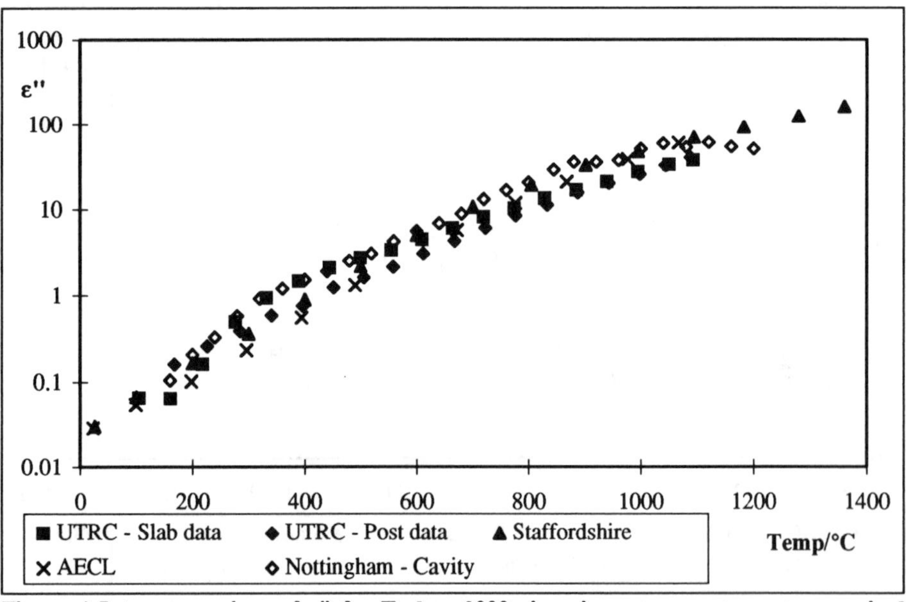

Figure 4 Inter-comparison of ε'' for Techno 2000 zirconia vs temperature at a nominal frequency of 2.45GHz

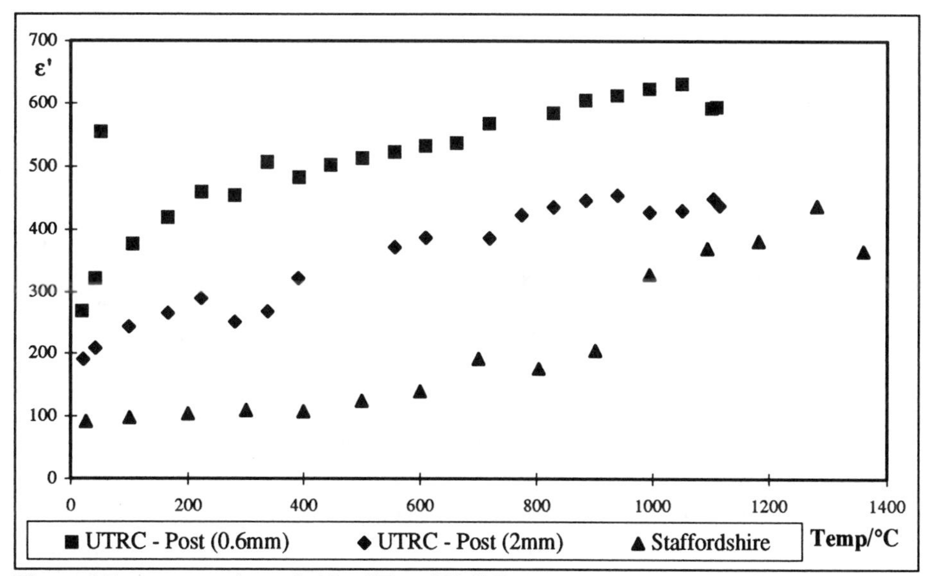

Figure 5 Inter-comparison of ε' for "Type N" SiC vs temperature at a nominal frequency of 2.45GHz

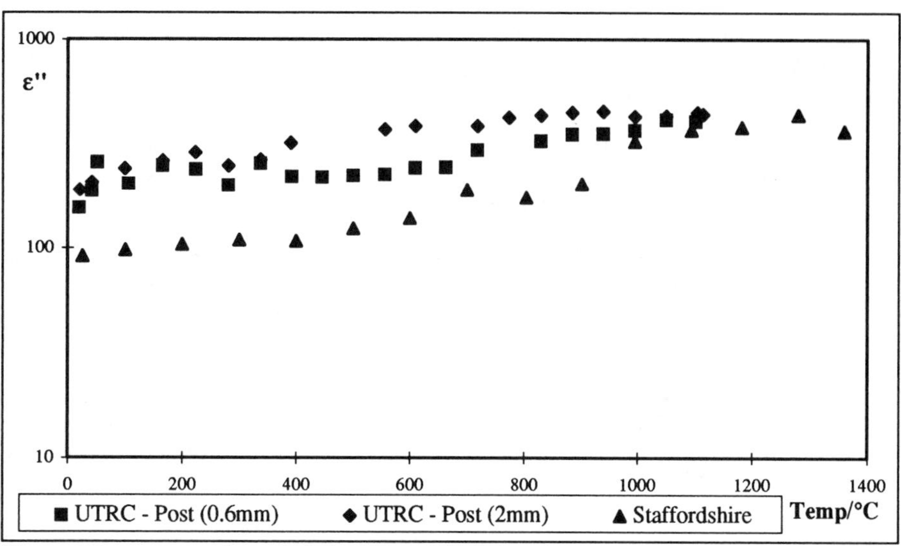

Figure 6 Inter-comparison of ε" for "Type N" SiC vs temperature at a nominal frequency of 2.45GHz

ACKNOWLEDGEMENTS

The authors would like to thank Cercom Inc. for the donation of the silicon carbide samples used in this programme.

REFERENCES

1 M Arai, J G P Binner, A L Bowden, T E Cross, N G Evans, M G Hamlyn, R Hutcheon, G Morin and B Smith, "Elevated Temperature Dielectric Property Measurements: Results of a Parallel Measurement Programme" in Microwaves: Theory and Applications in Materials Processing II, Ceramic Transactions, Vol 36, pp539-546, 1993.

2 R M Hutcheon, M S De Jong and F P Adams, "A System for Rapid Measurements of RF and Microwave Properties up to 1400ºC", J. Microwave Power and Electromagnetic Energy, Vol 27, No 2, pp 87-103, 1992

3 R M Hutcheon, M S De Jong, F P Adams, P G Lucuta, J E McGregor and L Bahan, "RF and Microwave Dielectric Measurements to 1400°C and Dielectric Loss Mechanisms", in Microwave Processing of Materials III, Ed., R L Beatty, W H Sutton and M F Iskander, MRS Symposium Proceedings, Vol 269, pp 541- 551, 1992

4 J Baker-Jarvis, E J Vanzura and W A Kissick, "Improved Technique for the Determining Complex Permittivity with the Transmission/Reflection Method", IEEE Trans Microwave Theory & Tech., MTT-38, p 1096, 1990

5 J Baker-Jarvis, R G Geyer, and P D Domich, "A Nonlinear Least-Squares Solution with Causality Constraints Applied to Transmission Line Permittivity and Permeability Determination", IEEE Trans I & M, Vol 42, p 646, 1992

6 E J Venzura, "Intercomparison of Permittivity Measurements using the Transmission/Reflection Method in 7 mm Coaxial Transmission Lines", IEEE Trans Micrwave Theory and Tech., MTT-42, pp 2063-2070, 1994

7 M Arai, J G P Binner, G E Carr and T E Cross, "High Temperature Dielectric Measurements on Ceramics", in Microwave Processing of Materials III, Ed., R L Beatty, W H Sutton, and M F Iskander, MRS Symposium Proceedings, Vol 269, pp 611-616, 1992

8 J A Batt, R Rukus and M Gilden, "General Purpose High Temperature Microwave Measurement of Electromagnetic Properties", in Microwave Processing of Materials III, Ed., R L Beatty, W H Sutton, and M F Iskander, MRS Symposium Proceedings, Vol 269, pp 553-559, 1992

MEASUREMENTS OF THE COMPLEX PERMITTIVITY OF LOW LOSS CERAMICS AT MICROWAVE FREQUENCIES AND OVER LARGE TEMPERATURE RANGES

Binshen Meng[a], John Booske[a,c], Reid Cooper[b,c], and Benjamin Klein[a]
[a] Department of Electrical and Computer Engineering
[b] Department of Materials Science and Engineering
[c] Materials Science Program
University of Wisconsin, Madison, WI 53706

ABSTRACT

A system has been developed for measuring the complex permittivities of low loss ceramic materials at frequencies from $2\,GHz$ to $20\,GHz$ and over a temperature range $20-1000^0C$. The measurement technique involves a modified version of the conventional cavity perturbation method. This extended cavity perturbation technique is presented. Details of the design and fabrication of the circular cylindrical cavity and the input and output coupling transmission lines are discussed. Data are presented for an illustrative measurement of the complex microwave dielectric properties of $NaCl$ single crystals between $20-400^0C$. The experimental results are in excellent agreement with theoretical models.

INTRODUCTION

An understanding of the fundamental physical mechanisms of interaction between microwave frequency electromagnetic radiation and crystalline solids is crucial to the development of predictive capabilities having an impact on several materials applications of growing interest and importance, namely: (1) microwave processing of ceramics (e.g., sintering, bonding), (2) complex dielectric properties of ceramic substrate materials for microwave and high speed digital circuits (microwave circuit substrates and integrated circuit packaging), and (3) materials selection and fabrication for window and insulator structures in high power coherent microwave sources employed in conventional power tube markets as well as research enterprises such as high energy physics accelerators and fusion plasma experiments. To accomplish these goals, it is necessary to develop a measurement system to accurately determine the microwave dielectric properties of low loss ceramic materials over a broad range of temperatures and frequencies.

In this paper, we first present an extended cavity perturbation method. The extended method is more accurate for larger samples and higher frequencies. We then proceed to describe the design and performance of a measurement system which makes use of this extended cavity perturbation method to determine the complex dielectric constant of low loss ceramics in $2-20\,GHz$ microwave regime. The specific demonstrated capabilities include: a minimum measurable loss tangent $tan\delta \sim 10^{-5}$ and a temperature range of $20-400^0C$.

EXTENDED CAVITY PERTURBATION TECHNIQUE

The cavity perturbation technique has been extensively and successfully employed to measure the complex dielectric constants of materials [1]-[4]. These measurements are performed by inserting a small sample with a certain shape into a microwave resonant cavity and determining the real part and the imaginary part of the complex permittivity from the shift of the resonance frequency and the change of the cavity Q factor, respectively.

The conventional cavity perturbation method may be difficult to perform in some practical situations, however, due to the requirement that the sample volume must be very small to produce a negligible perturbation to the electromagnetic field distribution inside the cavity. For the case of a circular cylindrical cavity operating in a TM_{0n0} mode, the sample is usually a thin circular cylindrical rod. In some applications, the sample materials may be fragile and extremely difficult to fashion into rods thin enough to satisfy accuracy requirements for higher frequencies or higher order modes. Considerable errors may result from the use of the conventional cavity perturbation technique with thicker rods.

Review of Conventional Cavity Perturbation Theory

The change of a complex eigenfrequency $\tilde{f}$ caused by a small sample having volume V_s with the dielectric constant ϵ and permeability μ in a cavity with volume V_c is [5]

$$\frac{\Delta \tilde{f}}{f_0} = -\frac{\int_{V_s}(\Delta\epsilon \mathbf{E}\cdot\mathbf{E}_0^* + \Delta\mu \mathbf{H}\cdot\mathbf{H}_0^*)\,d\tau}{\int_{V_c}(\epsilon \mathbf{E}\cdot\mathbf{E}_0^* + \mu \mathbf{H}\cdot\mathbf{H}_0^*)\,d\tau}, \tag{1}$$

where $\mathbf{E}_0$ and $\mathbf{H}_0$ represent electric and magnetic fields respectively in the empty cavity, and $\mathbf{E}$ and $\mathbf{H}$ represent the corresponding quantities in the cavity with the small sample. For the purposes of further discussion, we will restrict our attention to materials for which the permeability μ is a constant, hence $\Delta\mu = 0$. When the sample is small, it is reasonable to approximate $\mathbf{E}$ and $\mathbf{H}$ by $\mathbf{E}_0$ and $\mathbf{H}_0$ in the empty region of the cavity with the sample. Denoting the value of $\mathbf{E}$ inside the sample by $\mathbf{E}_{int}$ and writing the complex dielectric constant as $\epsilon = \epsilon_0(\epsilon' - j\epsilon'')$, one conventionally obtains [6]

$$\epsilon' \simeq 1 - 2C_{conv}\frac{\Delta f}{f_0}, \tag{2}$$

and

$$\epsilon'' \simeq C_{conv}\left(\frac{1}{Q} - \frac{1}{Q_0}\right), \tag{3}$$

where

$$C_{conv} = \frac{\int_{V_c}|E_0|^2\,d\tau}{\int_{V_s}\mathbf{E}_{int}\cdot\mathbf{E}_0^*\,d\tau} \tag{4}$$

and Q and Q_0 are the quality factors of the cavity with and without the sample, respectively.

For a circular cylindrical cavity resonating in TM_{0n0} modes with cavity radius a, cylindrical rod-shaped sample radius b, and cavity height d, use of the quasi-static approximation leads to [6]

$$C_{conv} = \frac{\int_0^a J_0^2(k_n r) r dr}{\int_0^b J_0^2(k_n r) r dr}, \tag{5}$$

where $k_n = 2\pi f_n\sqrt{\epsilon_0\mu_0}$, $f_n = c x_{0n}/2\pi a$, and x_{0n} is the nth zero of the Bessel function J_0.

Eqs.(2) and (3) are the basic expressions on which conventional cavity perturbation measurement methods are based. The most restrictive assumptions in this derivation are that the empty cavity fields are unchanged by the presence of the dielectric sample, and that the sample is so thin that the fields can be assumed to be uniform over the sample's diameter.

Extended Perturbation Theory

This method involves solving the eigenvalue problem of a dielectric sample in a resonant cavity. For the circular cylindrical cavity mentioned earlier, we solve Maxwell's equations subject to the appropriate boundary conditions at the walls and the dielectric sample surface. The final result is [7]

$$\epsilon'' \approx C_{exact}(\frac{1}{Q} - \frac{1}{Q_0}), \tag{6}$$

where

$$C_{exact} = \frac{\epsilon' \int_{r<b} |E_{z1}|^2 d\tau + \int_{b \geq r \geq a} |E_{z2}|^2 d\tau}{\int_{r<b} |E_z|^2 d\tau}. \tag{7}$$

and Q is the total (measured) quality factor (with sample in the cavity), whereas Q_0 is the empty cavity quality factor. $E_z(\mathbf{r})$ (E_{z1} or E_{z2}) is obtained as

$$E_{z1} = AJ_0(k_n r), \qquad r \leq b \tag{8}$$

$$E_{z2} = AJ_0(k_n b) \frac{J_0(k_{0n}a)N_0(k_{0n}r) - N_0(k_{0n}a)J_0(k_{on}r)}{J_0(k_{0n}a)N_0(k_{0n}) - N_0(k_{0n}a)J_0(k_{on}b)}, \qquad b < r \leq a \tag{9}$$

where

$$k_n^2 = \omega_n^2 \mu_0 \epsilon_0 \epsilon', \tag{10}$$

and

$$k_{0n}^2 = \omega_n^2 \mu_0 \epsilon_0 \tag{11}$$

and ω_n is obtained as the discrete set of roots when Eq.s (10) and (11) are substituted into:

$$\frac{k_{0n}}{k_n} \frac{J_0(k_n b)}{J_1(k_n b)} = \frac{J_0(k_{0n}b)N_0(k_{0n}a) - J_0(k_{0n}a)N_0(k_{0n}b)}{J_1(k_{0n}b)N_0(k_{0n}a) - J_0(k_{0n}a)N_1(k_{0n}b)}. \tag{12}$$

The constant C_{exact} is considered to represent an "exact" calculation as it employs the general field solutions of Eqs. (8) and (9), whose only a priori assumption was that the cavity walls were made of a good conductor (i.e., $\frac{\sigma}{\omega\epsilon_0} \gg 1$) material.

In practice, one first measures the shift of the resonant frequency due to the presence of a sample. By numerically solving the dispersion relation represented by Eq. (12), one can then determine ϵ' (this is considerably facilitated for materials where one can assume that $\epsilon'' \ll \epsilon'$). Thus the field pattern described by Eqs. (8) and (9) can be accurately known. Measuring the change of the Q factor resulting from the insertion of the sample, one then obtains ϵ'' from Eq. (6).

To illustrate the advantages of the extended method, we describe its application to a specific experimental system and compare its accuracy with that of the conventional method. The experimental configuration employs a circular cylindrical cavity operating in the first three TM_{0n0} ($n \leq 3$) modes. With an inner radius of $a = 5.22\,cm$ and a height of $2\,cm$, the resonant frequencies of these three eigenmodes are 2.20, 5.05, and 7.91 GHz. A cylindrical sample rod is inserted through a small hole on one of two flat walls. For current research purposes, we have been investigating microwave absorption mechanisms in $NaCl$ single crystals. Therefore, our interest in these higher order modes is based on a desire to characterize the sample's complex permittivity at several different frequencies without removing the sample from the cavity. However, the brittle ceramic properties of $NaCl$ make it difficult to obtain thin single crystal rods with a sample radius less than $2\,mm$. The combination of these interests and constraints necessitated the development of the extended cavity perturbation method.

Comparing the predictions for ϵ' from resonant frequency shift by using Eq.(12) for the extended method with those by Eq.(2) for the conventional approach we found that their difference for the lowest order mode (TM_{010}) is probably negligible but that the conventional determination of ϵ' errs by 10% for the TM_{020} mode and more than 20% for the TM_{030} mode.

The limitations of the conventional method for larger sample radii and higher order modes is further illustrated in Fig. 1. In this figure we have plotted the relative error $(C_{conv} - C_{exact})/C_{exact}$ in the conventional calculation of the integrated-field-distribution constant as a function of sample radius b in our cavity (we assume $\epsilon' = 5.6$ for this illustration). For the lowest order mode, the conventional method provides reasonable accuracy (less than 10% error) for a sample radius up to approximately $3\,mm$. However, similar accuracy for the second and third order modes would require sample radii less than 1.0 and $0.5\,mm$, respectively. These small dimensions are prohibitively difficult to achieve with single crystal specimens of most (brittle) ionic solids.

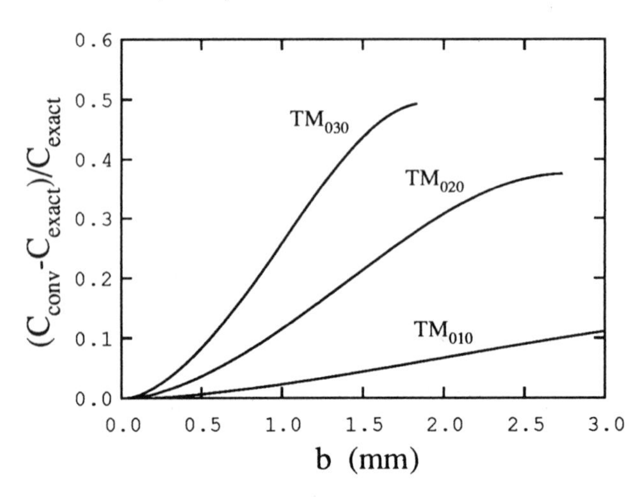

Figure 1: The relative difference between constant C's for conventional perturbation technique and extend method versus the radius of the sample.

DESCRIPTION OF THE MEASUREMENT SYSTEM

A schematic of the system configuration is illustrated in Fig.2. To span a frequency range of $2 - 18\,GHz$, two circular cavities resonating in several TM_{0n0} modes are employed. Two

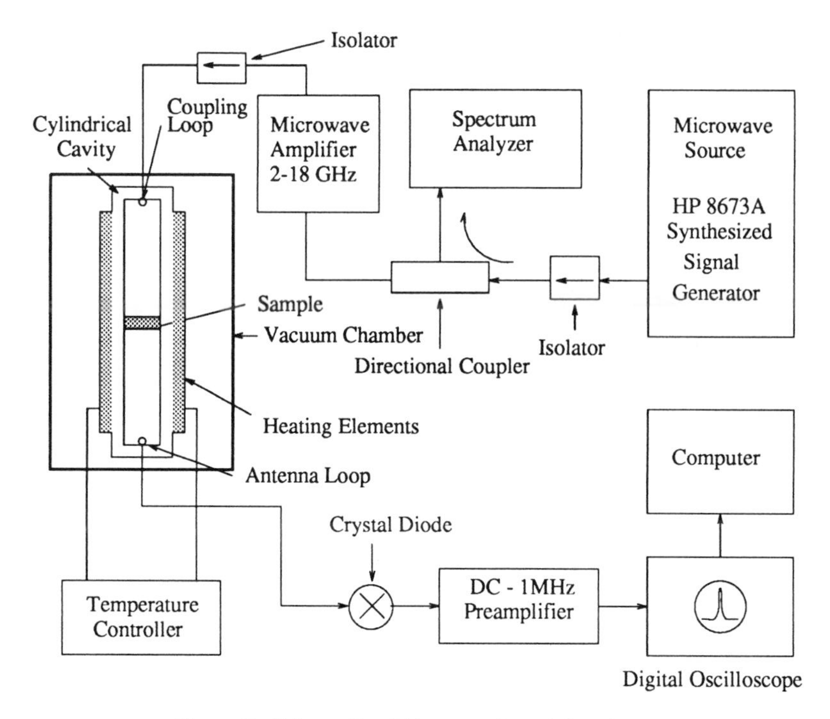

Figure 2: Schematic of the experimental system.

cavities are required since it was discovered that only the first three modes (TM_{010}, TM_{020} and TM_{030}) were usable in each cavity. Higher order modes (i.e., TM_{0n0}, $n \geq 4$) were found to overlap with other nearby, densely-spaced modes rendering it difficult to clearly ascertain their resonance lineshape properties. Each cavity is a right circular cylinder, one with an internal radius of $5.22\,cm$ and a height of $2\,cm$ the other having an internal radius of $2.3\,cm$ and a height of $1\,cm$. The larger cavity's first three eigenfrequencies are approximately 2, 5 and $8\,GHz$, while the smaller cavity's first three eigenfrequencies are approximately 5, 11 and $18\,GHz$. The redundancy at $5\,GHz$ was intentional to cross-check that both cavities yield identical data at $5\,GHz$. For initial measurements, prototype cavities were fabricated from copper.

Cavity Q is determined from the linewidth (full-width-half-maximum) of the 2-port transmission coefficient between two coupling loop antennas. The two antennas are inserted through diametrically-opposed small holes in the sides of the cavity and weakly couple the cavity to two shielded, two-wire transmission lines. The diameter of each transmission line is $31.75\,mm$ and the diameters of the loops are approximately $2\,mm$ for the larger cavity and $1.5\,mm$ for the smaller cavity.

In working with the prototype cavities, it was determined that a precise positioning of the loop antennas for extremely weak (but nonzero) coupling is necessary to ensure that the measured cavity Q is as close to the unloaded Q as practical. Even very small eigenmode field perturbations caused by insertion of a dielectric sample (through a small hole in the lid) can unacceptably distort the eigenfrequency lineshapes if the antenna-cavity coupling is not sufficiently weak. The combination of the $30\,dB$ ($1\,W$ maximum output) microwave amplifier(s) and a low-noise preamplifier (10^5 gain, DC-$1\,MHz$) after the crystal diode detector were found adequate to compensate for the weak signals associated with the very weak cavity coupling.

Observing variations in ϵ' and ϵ'' over large dynamic temperature ranges represents an important tool for determining the physical mechanisms that are responsible for the complex dielectric behavior of solid materials[10]. Consequently, considerable attention was focused on system design features that ensured repeatable, reliable operation at high temperatures and under conditions of regular thermal cycling. For example, the rigid two-wire transmission lines feeding the coupling loops were custom fabricated for high temperature operation from copper wire for the transmission line, double-bore alumina tubing for the dielectric, and thin copper tubing for electromagnetic shielding.

To ensure that the specimen is isothermal, the entire cavity is resistively heated by ohmic heating rings on the top and bottom (Fig. 2), and the entire assembly is enclosed in pressed ceramic fiber board insulation. To avoid oxidation of the hot cavity, the assembly is also placed in controlled gas conditions (typically, a $10\,mTorr$ vacuum is found adequate).

A good test of whether the cavity design is free from mechanical expansion and contraction effects during thermal cycling is to see whether the temperature dependence of the empty cavity quality factor Q_0 can be solely accounted for by the temperature dependence of the wall material's resistivity. In verification of a good cavity design and construction, measurements of the empty cavity Q_0 versus temperature agreed with the theoretical predictions to within a few percent for $300^0\,K \leq T \leq 700^0\,K$ [9].

TYPICAL EXPERIMENT RESULTS

To illustrate the capabilities of this new microwave dielectric spectrometer, we have measured the complex dielectric permittivity of nearly-pure and impurity-doped $NaCl$ single crystal specimens. Comparing experimental measurements of the imaginary part ϵ'' with theoretical models provides an excellent method for determining the impact of point defects on microwave absorption in ionic crystalline solids [10]. In Fig. 3 we provide such a comparison for the microwave absorption in both a nearly-pure $NaCl$ single crystal sample and a second sample doped with $250\,ppm$ Ca^{++} cation impurity concentration. This data specifically corresponds to the TM_{030} ($n = 3$) eigenmode (the one most sensitive to perturbation theory errors). Results from the second (higher frequency) cavity are similar to the data shown here. Note that the theory and experimental measurements are in excellent agreement, even at values of ϵ'' as small as $\sim 10^{-4}$ (corresponding to a loss tangent, $tan\delta \sim 10^{-5}$). A detailed discussion of the theory of microwave absorption in ionic crystalline solids is beyond the scope of this article but will be provided in a future publication. An outline of the basic physics can be found in Ref.[10].

SUMMARY

A system has been developed to measure the complex dielectric properties of low loss ceramic materials over large temperature ranges. A prototype measurement system has demonstrated reliable operation for a temperature range of $20-400^0C$ and at five frequencies spanning the frequency range of $2-20\,GHz$. The extended version of the cavity perturbation method allows the use of higher order mode measurements (TM_{0n0}, $n \geq 2$) with dielectric

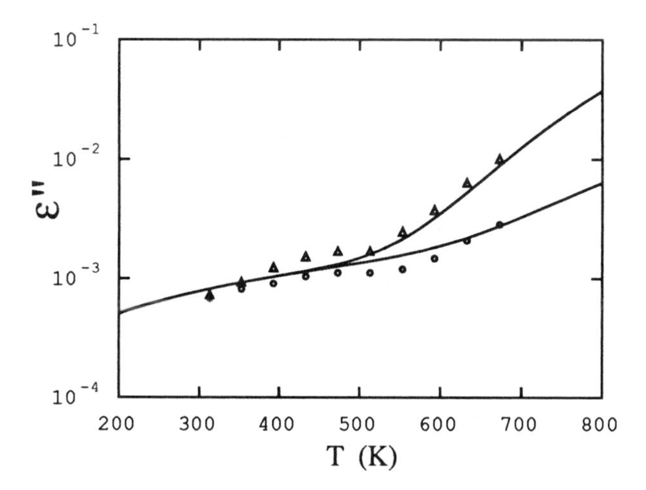

Figure 3: The experimentally measured ϵ'' and theoretically predicted ϵ'' in $NaCl$ single crystals at the TM_{030} eigenfrequency of $7.4 GHz$ versus temperature. In the plot, o= undoped crystal ("pure" crystal), $\triangle$= doped crystal, —= theoretical prediction.

samples that are difficult to fabricate to the small dimensions that would be required by the conventional cavity perturbation approach. Finally, illustrative measurements of the complex permittivity of $NaCl$ crystals with different levels of cation impurity concentrations are in excellent agreement with theoretical predictions over a large range of temperature, at several different frequencies, and at very small loss tangent values, $tan\delta \sim 10^{-5}$.

ACKNOWLEDGEMENTS

The authors gratefully acknowledge the support of the Electric Power Research Institute, the Wisconsin Alumni Research Foundation, the National Science Foundation through a Presidential Young Investigator Award, and Nicolet Corp. for donation of the digital oscilloscope used in the experiments. Sam Freeman and Tom Wadzinski contributed valuable assistance to the conduct of this research.

REFERENCES

[1] B. Birnbaum and J. Franeau, "Measurement of Microwave Dielectric Constant and Loss of Solids and Liquid by a Cavity Perturbation Method", *J. Appl. Phys.* **20,** 817(1949).

[2] K.S. Champlin and R.R. Krongard, "The Measurement of Conductivity and Permittivity of Semi-Conductors Spheres by an Extension of the Cavity-Perturbation Method", *IRE Trans. Microwave Theory and Tech.* **MTT-9**, 545(1961).

[3] J. Verweel, "On the Determination Microwave permeability and Permittivity in Cylindrical Cavities", in *Philips Res. Rep.*, **20**, 404(1965).

[4] A. Parkash, J.K. Vaid, and A. Mansingh, "Measurement of Dielectric Parameters at Microwave Frequencies by Cavity Perturbation Technique", *IEEE Trans. Microwave. Theory. Tech.* **MTT-27**, 791(1979).

[5] R.A. Waldron, "Perturbation Theory of Resonant Cavities", *Proc. Inst. Elec. Eng.*, **107C**, 272 (1960).

[6] R.F. Harrington, *Time-harmonic electromagnetic fields*, New York, McGraw-Hill, 317 (1961).

[7] B. Meng, Ph.D. Thesis, University of Wisconsin, 1995.

[8] C.A. Wert and R.M. Thomson, *Physics of Solids*, New York, McGraw-Hill, 205 (1964).

[9] B. Meng, J.H. Booske, and R.F. Cooper, *Rev. Sci. Instrum.* **66**(2), 1068-1071(1995).

[10] B. Meng, J.H. Booske, R.F. Cooper, and S.A. Freeman, "Microwave Absorption in NaCl Crystals with Various Controlled Defect Conditions", to be published in *MRS Symposium Proceedings: Microwave Processing of Materials IV*, (1994).

Process Modeling

DYNAMIC MODEL FOR ELECTROMAGNETIC FIELD AND HEATING PATTERNS IN LOADED CYLINDRICAL CAVITIES

Y. L. Tian, W.M. Black, H.S. Sa'adaldin, Dept. of Electrical and Computer Engineering, George Mason University, Fairfax, VA. I. Ahmad and R. Silberglitt, FM Technologies, Inc., Fairfax, VA

ABSTRACT

An analytical solution for the electromagnetic fields in a cylindrical cavity, partially filled with a cylindrical dielectric has been recently reported [1]. A program based on this solution has been developed and combined with the authors' previous program for heat transfer analysis. The new software has been used to simulate the dynamic temperature profiles of microwave heating and to investigate the role of electromagnetic field in heating uniformity and stability. The effects of cavity mode, cavity dimension, the dielectric properties of loads on electromagnetic field and heating patterns can be predicted using this software.

INTRODUCTION

Nonuniform heating and thermal runaway are two main problems frequently encountered in microwave processing of materials. They are caused by a number of factors related to both electromagnetic energy and thermal physical process [2]. A computer simulation was previously reported to study the effects of the thermal physical aspects on dynamic temperature profiles of microwave heating such as thermal conductivity, temperature dependence of dielectric loss factor, heating rate and insulation [3]. However no attempts were made in the previous study to investigate the effect of electromagnetic aspects on the microwave heating pattern. The electromagnetic field was simply assumed as a uniform distribution due to the lack of the knowledge of the real distribution of the electromagnetic filed at that time. Obviously, such a simplification might cause error and underestimate the problems.

An analytical solution for the electromagnetic fields in a cylindrical cavity coaxially loaded with a cylindrical dielectric has been recently reported by

Manring [1]. This progress motivated us to extend our computer simulation into electromagnetic fields. Based on Manring's approach a program has been developed to find solutions for resonant frequency, quality factor, electric field pattern and energy distribution for various types of cavity mode, cavity size, sample size and dielectric material. This program was combined with the authors' previous program for thermal analysis. The new software was then applied to simulate the dynamic temperature profiles during microwave heating and to investigate the effects of the electromagnetic field on the heating patterns. Some preliminary results of our modelling will be presented in this paper.

CONFIGURATION OF THE LOADED CYLINDRICAL CAVITY

The configuration of a coaxially loaded cylindrical cavity used in this study is shown in Figure 1. This configuration is referred to as cavity-short type where the load length is equal to the cavity length. The origin, 0, of a cylindrical coordinate system (ρ,ϕ,z) is located at the center of the bottom plate of the cavity. The dimension of the cavity is selected as close to that of the commercial cylindrical cavity, Wavemat model CMPR 250. The cavity length can vary from 6 to 16 cm, which allows eight resonant modes to be excited at 2.45 GHz. In this study, only two types of low index mode, namely, TE_{011} and TM_{011} modes, will be discussed. The TE_{011} mode is especially interesting because it exhibits an electric field pattern which may lead to a high energy density at the surface of the dielectric load and a low energy density in the center. This type of energy distribution may be beneficial in compensating for the heat loss at the surface and improving the temperature uniformity during microwave heating. Alumina (92%) is selected as the dielectric load in the cavity. The data for the dielectric and thermal properties of this material is collected from references [4,5].

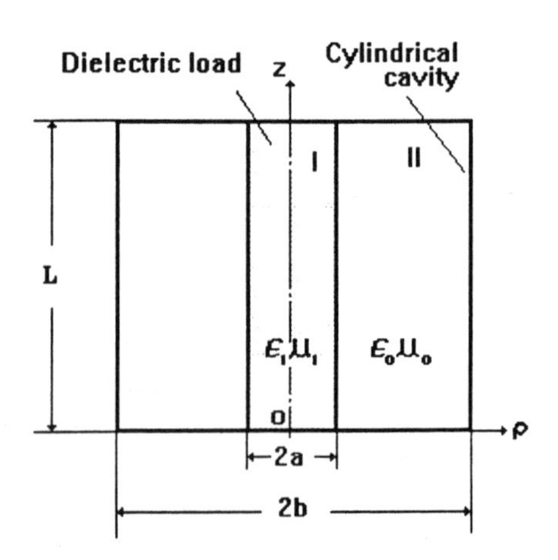

Fig. 1 Configuration of a dielectric loaded cylindrical cavity.

RESONANT FREQUENCY

Following Manring's approach, the complex radian resonant frequency ω and the wave numbers of $k_{\rho 1}$ in region I and $k_{\rho 2}$ in region II for the TE_{npq} or TM_{npq} modes are the roots of the following three simultaneous equations:[1]

$$V_n W_n - U_n^{\ 2} = 0 \tag{1}$$

$$\omega^2 \mu_0 \epsilon_1 = k_{\rho 1}^{\ 2} + (q\pi/L)^2 \tag{2}$$

$$\omega^2 \mu_0 \epsilon_0 = k_{\rho 2}^{\ 2} + (q\pi/L)^2 \tag{3}$$

where, $V_n = k_{\rho 1} k_{\rho 2} \mu_0 \left(k_{\rho 2} F_2' F_4 - k_{\rho 1} F_2 F_4' \right) \quad (V_n = 0 \text{ for } TE_{0pq} \text{ mode})$

$W_n = k_{\rho 1} k_{\rho 2} \left(\epsilon_1 k_{\rho 2} F_1' F_3 - \epsilon_1 k_{\rho 1} F_1 F_3' \right) \quad (W_n = 0 \text{ for } TM_{0pq} \text{ mode})$

$U_n^{\ 2} = [nq\pi/\omega a L]^2 F_1 F_2 F_3 F_4 \left(k_{\rho 1}^{\ 2} - k_{\rho 2}^{\ 2} \right)^2$

and ϵ_1 is the complex dielectric constant of the load in region I, which can be expressed as $\epsilon_1 = \epsilon_1' - j\epsilon_1''$. ϵ_0 and μ_0 are the electric permittivity and magnetic permeability of free space in region II, respectively. F's are combinations of Bessel's functions of the first and second kind using Harrington's notation [6]. The detailed information for F's can be found in Reference [1].

Equation (1) is an implicit expression containing complex terms of Bessel function for ω, $k_{\rho 1}$ and $k_{\rho 2}$. A complex root finding subroutine, called ROOT , developed by Tijhuis [1] was adopted in the computer program to solve for the complex ω, $k_{\rho 1}$ and $k_{\rho 2}$. Variation of the resonant frequency f_0 with sample size and cavity geometry for $Al_2 0_3$ (92%) loaded TE_{011} and TM_{011} modes is shown in Figure 2. The resonant frequency f_0 decreases with the increase of filling ratio (a/b) and decrease of the aspect ratio of the cavity (2b/L).

ELECTROMAGNETIC FIELD PATTERNS

If the complex roots of ω, $k_{\rho 1}$ and $k_{\rho 2}$ are known, the three components of the electric fields for n=0 case can be solved using the following equations :[1]

For TE_{0pq} mode: $\quad E_{\rho 1} = 0; \quad E_{z1} = 0; \quad E_{\rho 2} = 0; \quad E_{z2} = 0;$

$$E_{\phi 1} = B\, k_{\rho 1} F'_2 (k_{\rho 1} \rho) \sin (q\pi z/L) \tag{4}$$

$$E_{\phi 2} = D k_{\rho 2} F'_4 (k_{\rho 2} \rho) \sin (q\pi z/L) \quad (5)$$

For TM_{0pq} mode: $E_{\phi 1}=0;$ $E_{\phi 2} =0;$

$$E_{\rho 1} = - (A q \pi k_{\rho 1} / \acute{\omega} \epsilon_1 L) F'_2 (k_{\rho 1} \rho) \sin (q\pi z/L) \quad (6)$$

$$E_{z1} = (A k_{\rho 1}^2 / \acute{\omega} \epsilon_1) F_1 (k_{\rho 1} \rho) \cos (q\pi z/L) \quad (7)$$

$$E_{\rho 2} = - (C q \pi k_{\rho 2} / \acute{\omega} \epsilon_0 L) F'_2 (k_{\rho 2} \rho) \sin (q\pi z/L) \quad (8)$$

$$E_{z2} = (C k_{\rho 2}^2 / \acute{\omega} \epsilon_0) F_1 (k_{\rho 2} \rho) \cos (q\pi z/L) \quad (9)$$

where A,B,C,D are constants representing the amplitude of each field.

Typical examples of the computed electromagnetic field patterns for TE_{011} mode are shown in Figure 3 and Figure 4. The average power density, PD_{av}, defined as the total absorbed power divided by the volume of the sample, is assumed to be constant during the microwave process. Figure 3 plots the variation of the E_ϕ with the dielectric properties of the materials along the radial direction. As dielectric constant ϵ_1' increases, the peak of the electric field shifts from the center of the empty region II to thecenter of the load region I. Increase of the dielectric loss factor ϵ_1'' reduces the magnitude of the electromagnetic field.

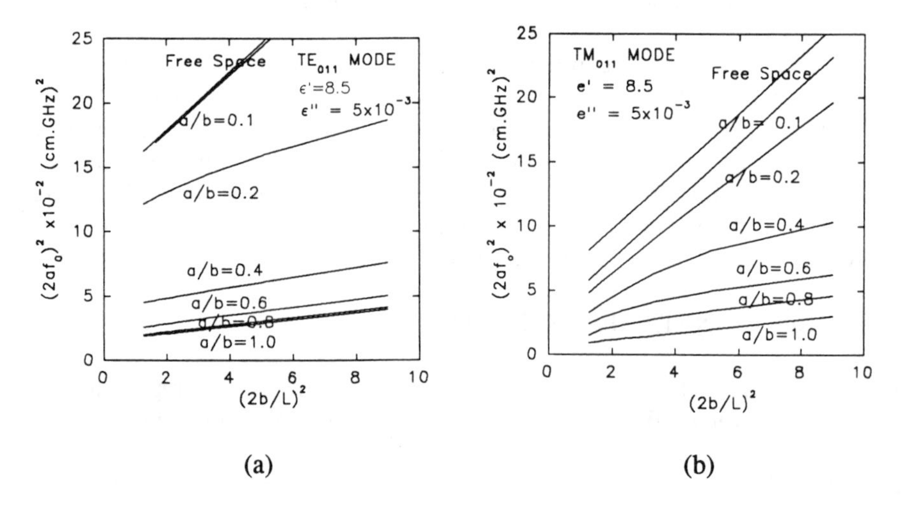

(a) (b)

Fig.2 Resonant frequency f_0 vs. cavity aspect ratio (2b/L) for various sample filling ratio (a/b). (a) TE_{011} mode. (b) TM_{011} mode.

Since both ϵ_1' and ϵ_1'' increase with the increase of the temperature, it can be seen from Figure 4 that the E_ϕ peaks shift to the center of region I and the magnitudes of the fields reduce as the temperature rises. Therefore, there is less chance of arcing in air (region II) when temperature is high or the material has a high dielectric constant or a high loss factor.

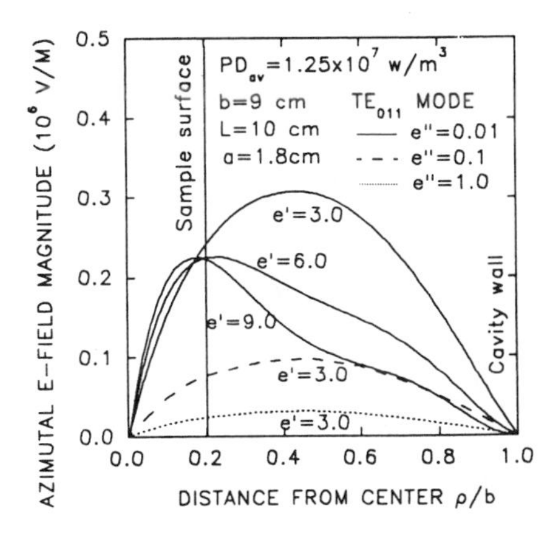

Fig.3 Variation of electric field E_ϕ distribution at z=L/2 with dielectric properties for TE_{011} mode.

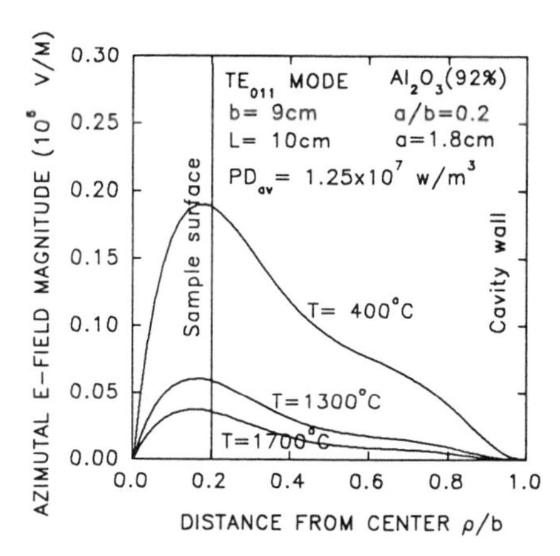

Fig.4 Variation of electric field E_ϕ distribution at z=L/2 with temperature for TE_{011} mode.

ENERGY DISTRIBUTIONS IN THE DIELECTRIC LOAD

The dissipated power density, PD, in a load can be calculated from:

$$PD = 2\pi f_0 \epsilon_0 \epsilon_1'' E E^* \qquad (10)$$

where E is the sum of three components of the electric field E_ρ, E_ϕ and E_z calculated from Eq.(4) to Eq.(9). Figure 5(a) and (b) show the variation of energy distribution patterns with temperature for TE_{011} and TM_{011} modes, respectively. The power density patterns of the two modes are just opposite. For the TM_{011} mode, the peak density position is in the center, whereas that of the TE_{011} mode is close to the surface. Although the magnitude of the energy density of both modes increases with the increase of temperature, the change of the peak density in TM_{011} mode is considerably higher than that of TE_{011} mode at temperatures over 1000 °C. This may have a negative impact on the thermal runaway problem as will be discussed in the following section.

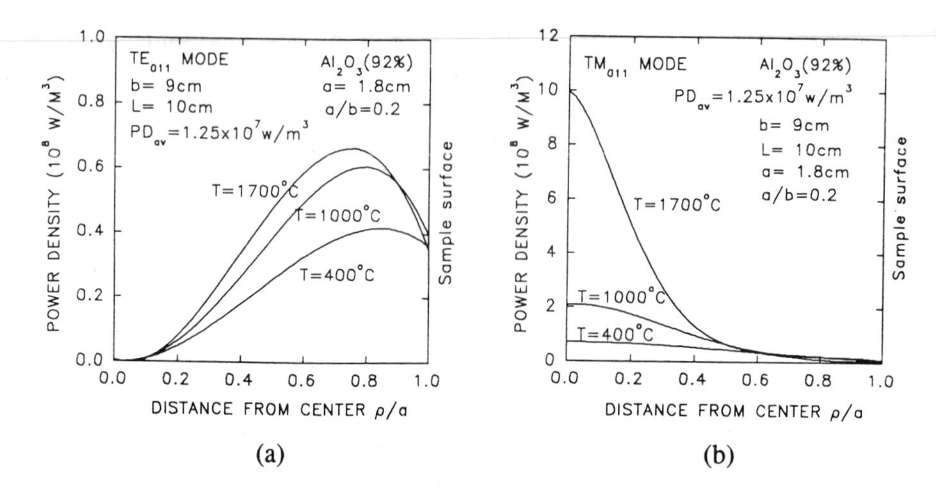

Fig.5 Variation of energy density distribution at z=L/2 with temperature. (a) TE_{011} mode. (b) TM_{011} mode.

DYNAMIC TEMPERATURE PROFILES

The temperature distribution inside the sample is governed by the following general heat transfer equation [3]:

$$d_n C_p(\partial T/\partial t)=(1/\rho)\{\partial[\ k\rho(\partial T/\partial \rho)\ ]/\partial \rho\} + \partial[\ k(\partial T/\partial z)\ ]/\partial z + PD \quad (11)$$

where d_n, C_p and k are the density, specific heat and thermal conductivity. The power density dissipated in the sample, PD, is given by Eq(10). For a given time step, the temperature rise in the load is computed using a finite difference algorithm. The resonant frequency, f_0, electric field E, dielectric properties ϵ_1' and ϵ_1'', and thermal properties of k and C_p are recalculated at next time step to accommodate the effects of the temperature increment on these parameters. This process is repeated until a steady state reached or a thermal instability happens. The details of our numerical approach can be found in References [2,3].

Figure 6 plots the radial temperature profiles for heating of an alumina sample in a TE_{011} mode with a low PD_{av}. A steady state of temperature distribution was reached after 1200 seconds because of the slow heating rate. The temperature distribution is relatively uniform in the center part of the sample but shows high gradient near the surface area due to the large radiation loss at high temperatures. Figure 7 shows the temperature profiles of TM_{011} mode under the identical conditions. Thermal runaway occurs at the center within about 120

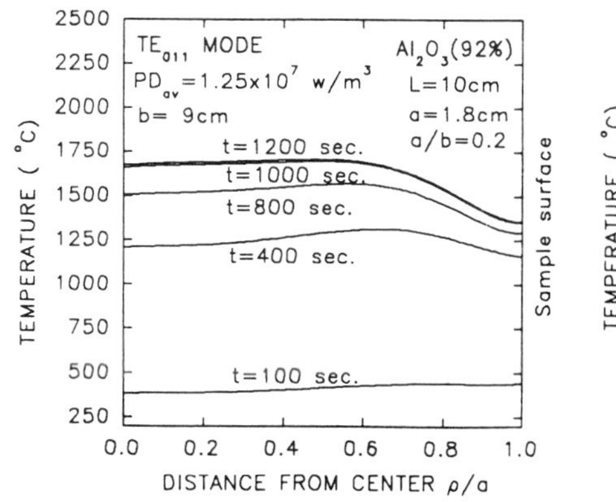

Fig.6 Dynamic temperature profile of Al_2O_3 sample in TE_{011} mode.

Fig.7 Dynamic temperature profile of Al_2O_3 sample in TM_{011} mode.

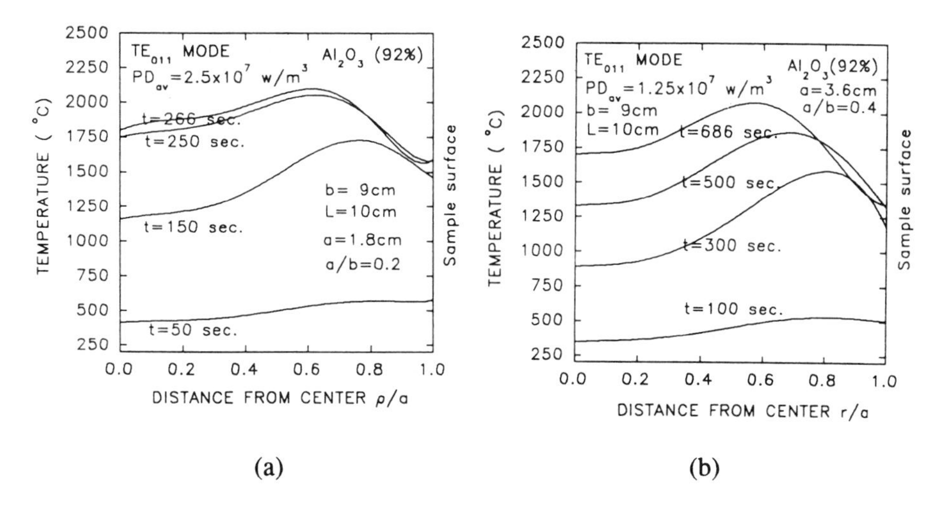

(a) (b)

Fig.8 Thermal runaway of Al_2O_3 samples in TE_{011} mode caused by (a) high heating rate and (b) large sample diameter.

seconds which can be attributed to the rapid rising of energy density in the center for T>1000°C as shown in Figure 5 . Obviously TE_{011} mode is in favor of improving the temperature uniformity and stability. Nevertheless, thermal runaway may still occur for TE_{011} mode as illustrated in Figure 8 where all the conditions are almost the same as Figure 6 except for a doubled average energy density PD_{av} in Figure 8(a), and a doubled sample diameter in Figure 8(b).

CONCLUSIONS

The electromagnetic field pattern plays an important role in determining the temperature distribution during microwave heating. The TE_{011} mode improves heating uniformity and stability, but limited to a certain extent. The resonant frequency, f_0, decreases with the increases of dielectric constant, ϵ_1', sample filling ratio, a/b, and the aspect ratio, L/2b. High heating rate and large sample size will enhance thermal instability and nonuniformity.

ACKNOWLEDGMENTS

The authors thank Dr. Ben Manring of Michigan State University and Dr. Anton G. Tijhuis of Delft University of Technology for their help in supplying computer programs. This work was sponsored in part by the U.S. Dept. of Energy Advanced Industrial Materials Program (AIM) through a subcontract from Los Alamos National Lab. and by the Virginia Center for Innovative Technology.

REFERENCES

1. E.B. Manring, "Electromagnetic Field Solution for the Natural Modes of a Cylindrical Cavity Loaded with Lossy Materials", Ph.D. Dissertation, Michigan State University, (1992).
2. Y.L. Tian, "Practices of Ultra-rapid Sintering of Ceramics using Single Mode Applicators", Ceramic Transactions, vol.21, p283-300, (1991).
3. Y.L. Tian, J.H. Fang and C.J. Tu, "Computer Modeling of Two Dimensional Temperature Distributions in Microwave Heated Ceramics', MRS Proceedings, vol.269, p41-46, (1992).
4. H.Fukushima, T.Yamanaka, and M.Matsui, "Measurement of Dielectric Properties of Ceramic at Microwave Frequency", J. Japan Soc. of Precision Engr., 53 [5], p1-6, (1987).
5. W.D. Kingery, H.K. Bowen and D.R. Uhlmann, Introduction to Ceramics, 2nd edition, John Wiley & Sons, Inc., (1976).
6. R.F. Harrington, Time-Harmonic Electromagnetic Fields, McGraw-Hill, (1961).

MATHEMATICAL MODELS OF MICROWAVE HEATING: BISTABILITY AND THERMAL RUNAWAY

G. A. Kriegsmann
Department of Mathematics
Center for Applied Mathematics and Statistics
New Jersey Institute of Technology
Newark, NJ 07102

ABSTRACT

The heating of a ceramic slab in a TE_{103} waveguide applicator is modeled in the small Biot number limit. If E_0 is the strength of the incident mode, then the relationship between the temperature and E_0^2 is shown to be bistable, or s-shaped in this limit. The shape of this response curve is intimately dependent upon the coupling of the microwave to the cavity, and upon the effect of the subsequent heating on this coupling. Several examples are presented to illustrate this dependence and two optimal coupling strategies are described.

INTRODUCTION

Recent modeling in microwave processing of ceramics has evolved in two directions. The first has focused on full numerical simulations of the heating process which includes applicator and sample geometry [1-2]. These codes require extensive computer resources, especially when parameter studies are required to deduce trends and functional relationships. The other approach has focused on idealized heating problems where the applicator is ignored, the driving microwave is completely specified, and the sample geometry is simple. The heating of ceramic slabs has been studied in the small Biot number limit [3]; these asymptotic results describe a bistable mechanism which gives a plausible explanation of the runaway phenomenon. This bistable character persists even when the Biot number is not small, as was shown by numerical simulations on a spherical sample [4]. The problem considered in this paper is a compromise between these somewhat divergent approaches. It deals with the heating of a ceramic slab in a TE_{103} waveguide applicator. The analysis shows how the cavity affects the bistable character of the heating process and how one can exploit it.

FORMULATION

A ceramic sample occupies a portion of a TE_{103} waveguide applicator, whose dimensions are shown in Figure 1. The cavity portion of the applicator is comprised of a shorted waveguide and a symmetric iris. The dimensions of the guide are such that only one mode propagates and L is chosen to yield a TE_{103} cavity which is large enough to neglect all evanescent modes.

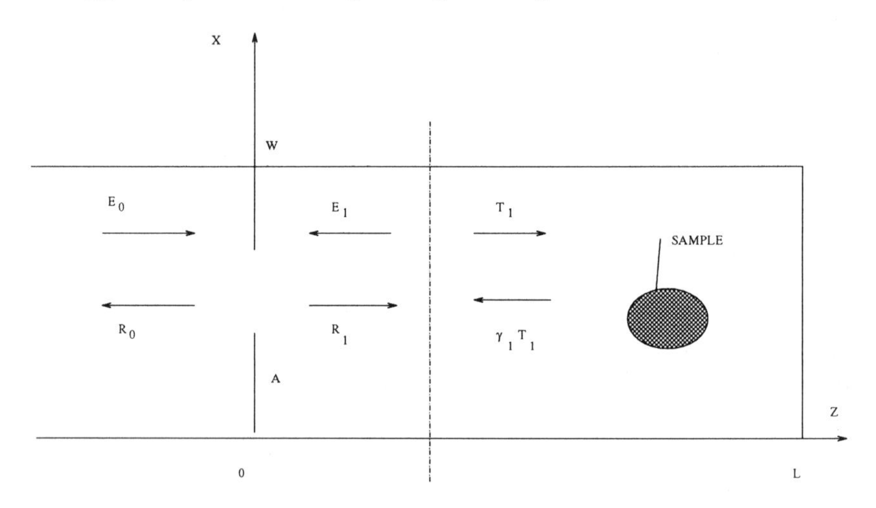

Figure 1.

The interactions of the iris with the incident mode and the mode reflected by the ceramic load are best described using scattering matrix theory. Consider the portion of the applicator to the left of the dashed vertical line in Figure 1. If E_j and R_j denote the amplitudes of incident and reflected modes, respectively for $j = 0, 1$, then they are related by the linear relationships

$$R_0 = r_1(A)E_0 + [1 + r_1(A)]E_1 \tag{1a}$$

$$R_1 = [1 + r_1(A)]E_0 + r_1(A)E_1 \tag{1b}$$

where $r_1(A)$ is the reflection coefficient for the iris alone in an infinite and homogeneous waveguide. It depends upon the iris height A and the frequency ω for a fixed W and H, where H is the height of the guide. Approximations for r_1 are given in many standard texts [5] and are always of the form

$r_1 = -iS(A)/[1 + iS(A)]$ where, for example, $S(A) = \frac{\pi}{2k_1} tan^2(\pi A)$ and k_1 is the propagation constant of the propagating mode.

The amplitude of the mode T_1 impinging upon the ceramic sample is needed in the thermal analysis to follow. Using the connection equations at the dashed interface $T_1 = R_1$ and $E_1 = \gamma_1 T_1$, where γ_1 is the reflection coefficient of the ceramic load, along with (1b), yields after a little algebra

$$T_1 = \frac{1 + r_1(A)}{1 - r_1(A)\gamma_1} E_0. \tag{2}$$

This shows the explicit dependence of T_1 upon the iris height, the amplitude of the incident mode E_0, and the reflection coefficient of the load γ_1.

The temperature in the ceramic satisfies

$$\rho C_P \frac{\partial}{\partial t} T = K \nabla^2 T + \frac{1}{2} \sigma(T) |\mathbf{E}|^2 \tag{3a}$$

and the initial condition

$$T(X, Y, Z, 0) = T_A \tag{3b}$$

where T_A is the ambient temperature in the applicator, K is the thermal conductivity of the ceramic, ρ is its density, C_P its thermal capacity, and $\mathbf{E}$ is the electric field in the sample which is proportional to T_1. The thermal boundary condition on the surface S of the ceramic sample is

$$K \frac{\partial}{\partial n} T + h(T - T_A) + se(T^4 - T_A^4) = 0, \tag{3c}$$

where h is the effective heat transfer coefficient, s is Stefan-Boltzman constant, and e is the thermal emissivity of the ceramic surface.

The actual determination of the electric field in the ceramic and its dependence upon the temperature are quite involved problems. They will be addressed approximately in the next section.

THE SMALL BIOT NUMBER THEORY

The method of analysis presented here is quite similar to the one-dimensional slab case given in Reference 3. It is based upon the assumption that the Biot number $B = hD/K$, where D is a typical dimension of the sample, is small which is true in many applications. This can be exploited in a systematic fashion [6] to deduce that

$$T = T_0(t)[1 + O(B)] \tag{4a}$$

where T_0 is the leading term in an asymptotic expansion and the $O(B)$ terms represent corrections which are the same order as B. It is important to observe here that the leading order temperature is spatially homogeneous in this limit. The correction term in this expansion contains spatial dependence which gives rise to nonuniform heating. It will not be described here.

By integrating equation (3a) over the volume of the sample R, using the divergence theorem, and applying the boundary condition (3c), the function T_0 satisfies

$$V\rho C_P \frac{dT_0}{dt} = -2S\{h(T_0 - T_A) + se(T_0^4 - T_A^4)\} + \frac{1}{2}\sigma(T_0) \iiint_R |\mathbf{E}|^2 \, dx \, dy \, dz. \tag{4b}$$

where V is the volume of the ceramic sample and S is its surface area. The model incorporates both convective and radiative heat loses through the first two terms on the right hand side of (4b), respectively. The term involving the triple integral is the microwave power absorbed by the ceramic and it can be related to the reflection coefficient using a simple power balance. Inserting this relationship into(4b) yields

$$V\rho C_P \frac{dT_0}{dt} = -2S\{h(T_0 - T_A) + se(T_0^4 - T_A^4)\} + \frac{k_1 S}{2\omega\mu_0}|T_1|^2[1 - |\gamma_1|^2] \tag{5}$$

where k_1 is the propagation number, μ_0 is the magnetic permeability, and T_1 is given by (2a). Thus, the effect of the electromagnetic field upon the heating process is contained in the reflection coefficient γ_1.

The determination of γ_1 in the context of the present theory is straightforward; it is computed as follows. First, an incident mode of unit amplitude is sent down the guide and impinges upon the sample. The guide is still shorted at $Z = L$, but has no iris. Since the leading order electric field satisfies Maxwell's equations with $\sigma(T)$ replaced by $\sigma(T_0)$, the equations are spatially homogeneous and may be solved by a variety of numerical methods. Finally, the amplitude of the reflected field γ_1 can easily be deduced from such a calculation.

RESULTS

When the ceramic sample is a slab that fills the entire cross-section of the guide, separation of variables can be used to deduce an analytical formula for γ_1 [6]. The function is exhibited in the Appendix for reference. Its insertion into (5) along with the initial condition $T_0(0) = T_A$ yields an autonomous initial value problem for T_0 which can be solved explicitly. The steady state solution of (5) is given by

$$E_0^2 = \frac{4\omega\mu_0}{k_1}\{h(T_0 - T_A) + se(T_0^4 - T_A^4)\}\frac{|1 - r_1\gamma_1|^2}{(1 - |\gamma_1|^2)|1 + r_1|^2} \tag{6}$$

where E_0 is again the known amplitude of the incident mode. This gives T_0 as a function of E_0 and the various cavity parameters that enter through γ_1.

In the examples that follow several parameters are held constant. They are $W = 109.22$mm, $H = 54.61$mm, and $\omega \sim 2.45$Ghz. The thickness of the slab is $D = 50$mm and it is always positioned midway in the cavity. The ceramic is taken to be alumina, and the exponential conductivity model [7] $\sigma = \omega\epsilon'' = 0.002e^{.78(\frac{T-T_A}{T_A})}$ is used. The Biot number is approximately 0.01.

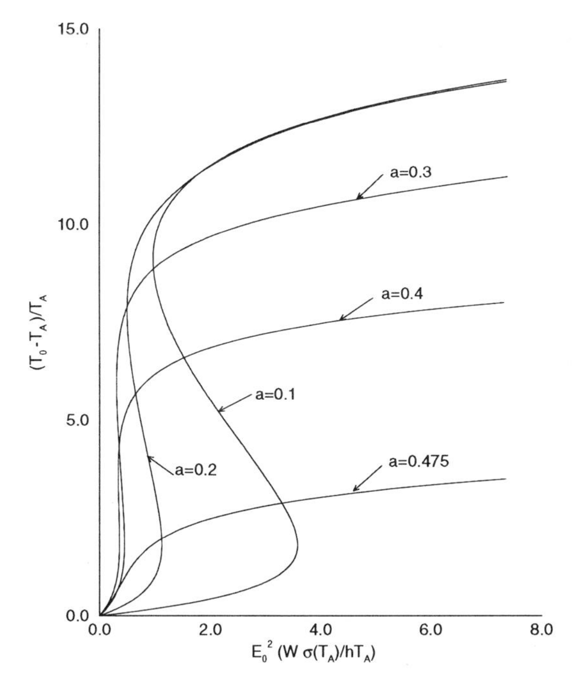

Figure 2.

Several curves corresponding to (6) are shown in Figure 2. Each is labeled by a value of $a = A/W$ which corresponds to an iris height; the bigger A, the smaller the aperture. The length of the cavity L was chosen to make the cavity resonate, for a given A, at the ambient temperature, where the ceramic is lossless. The response curves are s-shaped for large apertures and are essentially the same as those derived for one-dimensional slabs [3]. The upper branches are caused by the skin effect which shields the interior of the sample, thus limiting the power deposited. The dynamics of (5) are precisely those described in [3]. What is important here is to note how these curves deform as the aperture is decreased; gradually the s-shape becomes monotonic and the upper branch decreases significantly. These are the effects of cavity detuning. Finally, observe that when $a = 0.1$ the maximal temperature the sample can achieve occurs on the lower branch and is $\sim 630^oC$ for a dimensionless power of ~ 5. Increasing the power past this point will drive the sample to destruction on the upper branch where $T_0 \sim 3600^oC$. With a larger aperture corresponding to $a = .475$, the same amount of power will yield $T_0 \sim 1200^oC$ which is less than the desired sintering temperature of $\sim 1800^oC$.

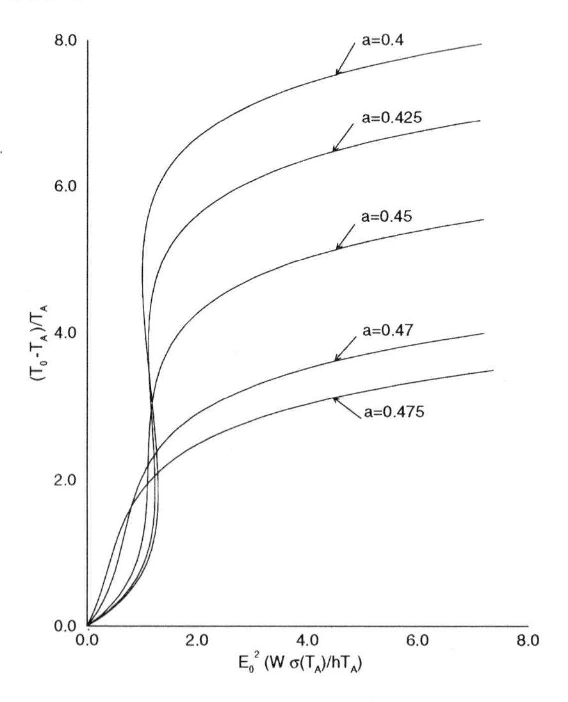

Figure 3.

Figure 3 contains several graphs which correspond to a laboratory strategy for sintering a ceramic slab. Each curve corresponds to a different aperture opening. The length of the cavity was fixed at $L = 317.91$mm; this length caused the cavity to resonate at T_A with $a = 0.475$. The upper branch of each curve increases as the aperture becomes bigger. This suggests the following scenario; choose a dimensionless power corresponding to 3 and take $a = 0.475$, The temperature will evolve according to (5) until it reaches the value $\sim 690^oC$ shown in the figure. Then, open the aperture (at a rate faster than either the diffusive or convective time scales) until $a \sim .425$. The temperature will track this change and end up on the corresponding response curve at $\sim 1800^oC$.

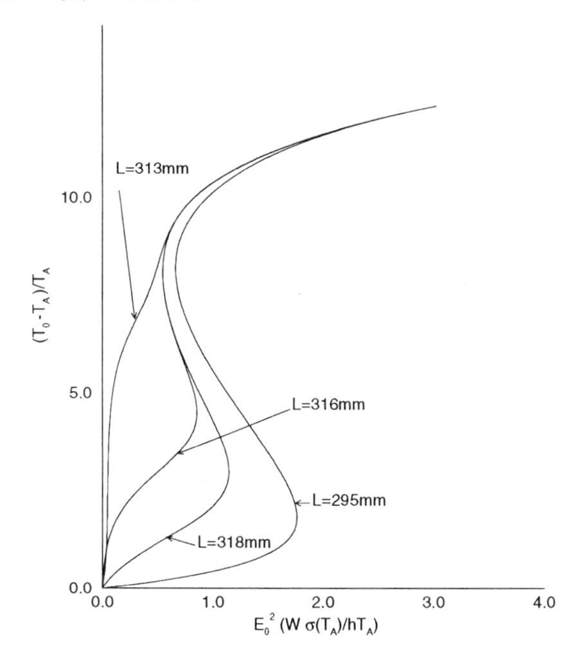

Figure 4.

The final example illustrates another optimal heating strategy. The idea is to pick an initial value of A and choose the L which makes the cavity resonate at T_A. Then hold L fixed and adjust A dynamically (as described above) to maximize $|T_1|^2$ as the sample heats. It is easy to deduce from (2) that this occurs when $r_1 = -iS_M/[1+iS_M]$ where $S_M = \gamma_I/|1+\gamma_1|^2$ and γ_I is the imaginary part of γ_1. Inserting this value of r_1 into (6) yields an optimal response curve for the given cavity length. Figure 4 contains several graphs, each corresponding to a value of L. It is clear from this figure that $L = 313$mm gives the optimal heating strategy for sintering; the temperature is an increasing function of power and runaway does not occur, and the sintering temperature is achieved with a minimal amount of power.

CONCLUSIONS

The heating of a ceramic slab in a TE_{103} waveguide applicator was modeled in the small Biot number limit. The relationship between the temperature and E_0^2, the strength of the incident mode, was shown to be bistable, or s-shaped in this limit. The shape of this response curve was shown to depend intimately upon the coupling of the microwave to the cavity, and upon the effect of the subsequent heating on this coupling. Several examples were presented to illustrate this dependence and two optimal coupling strategies were described.

ACKNOWLEDGMENTS

This work was supported by grants from the AFOSR, DOE, and NSF under grant numbers AFOSR 91-0252, DE-FG02-94ER25196, and DMS 9305828, respectively.

REFERENCES

1. Iskander M. F., Andrade, O., Virkar, A., Kimrey, H., Smith, R., Lamoreaux, S., Cheng, C., Tanner, C., Knowlton, R., and Metha, K., "Microwave Processing of Ceramics at the University of Utah", **Microwaves: Theory and Applications in Materials Processing**, Ceramic Transactions, Vol. 21, pp. 35-48 (1991).
2. Iskander, M. F., "Computer Modeling and Numerical Simulation of Microwave Heating Systems", *MRS Bulletin*, **27**[11], pp. 30-36(1993).
3. Kriegsmann, G. A., "Thermal Runaway in Microwave Heated Ceramics: A One-Dimensional Model", *Journal of Applied Physics*, **71** [9], pp. 1960-1966(1992).
4. M. Barmatz, et. al., "Transient Temperature Behavior of a Sphere Heated by Microwaves", **Microwaves: Theory and Applications in Material Processing II**, Ceramic Transactions Vol. 36, pp. 189-199 (1993).
5. Lewin, L., **Theory of Waveguides**, John Wiley and Sons, New York, pp. 120-130 (1975).
6. Kriegsmann, G. A., "Cavity Effects in Microwave Heating of Ceramics", *SIAM Journal of Applied Mathematics*, submitted.
7. D. G. Watters, "An Advanced Study of Microwave Sintering", Ph.D. Dissertation, Northwestern University (1988).

APPENDIX

The reflection coefficient for a ceramic slab that fills the entire cross-section of the guide can be obtained by separation of variables (see, e.g., Reference 6). Let the slab occupy the region $0 < Q < Z < Q + D$ where D is the thickness of the slab. Then omitting the details of the derivation, the reflection coefficient is given by

$$\gamma_1 = e^{2ik_1 Q}[\frac{2k_1 \tilde{Z} e^{ik_1\theta_1 D} - 2k_1 Z e^{-ik_1\theta_1 D} - \Delta}{\Delta}].$$

Here the propagation constants in the guide and the slab are given by

$$k_1 = \sqrt{\omega^2 \mu_0 \epsilon_0 - \pi^2/W^2}$$

and

$$\theta_1 = \sqrt{\omega^2 \mu_0 \epsilon_1 + i\omega\mu_0\sigma - \pi^2/W^2},$$

respectively. The modal impedance Z and the related $\tilde{Z}$ are given by

$$Z = k_1 cot[k_1(L - Q - D)] \quad \text{and} \quad \tilde{Z} = Z + 2i\theta_1\,,$$

respectively. And finally, the determinant Δ is

$$\Delta = \tilde{Z}(k_1 - \theta_1)e^{i\theta_1 D} - Z(k_1 + \theta_1)e^{-i\theta_1 D}.$$

It is important to note here that the temperature of the ceramic enters γ_1 through the propagation constant of the slab, θ_1.

STEADY STATE TEMPERATURE PROFILE IN A CYLINDER HEATED BY MICROWAVES

H.W. Jackson, M. Barmatz, and P. Wagner
Jet Propulsion Laboratory, California Institute of Technology
Pasadena, CA 91109

ABSTRACT

A new theory has been developed to calculate the steady state temperature profile in a cylindrical sample positioned along the entire axis of a cylindrical microwave cavity. Temperature profiles were computed for alumina rods of various radii contained in a cavity excited in one of the TM_{0n0} modes with n = 1, 2 or 3. Calculations were also performed with a concentric outer cylindrical tube surrounding the rod to investigate hybrid heating. The parametric studies of the sample center and surface temperatures were performed as a function of the total power transmitted into the cavity. Also, the total hemispherical emissivity was varied at boundaries of the rod, surrounding tube, and cavity walls. The results are discussed in the context of controlling the average rod temperature and the temperature distribution in the rod during microwave processing.

INTRODUCTION

A realistic model of microwave cavity reactors that provides accurate and rapid predictions of thermal and electromagnetic characteristics for materials processing has been developed recently. Some features of the model will be described in this paper, and calculated results that demonstrate usefulness of the model will be presented. The results can be applied in exploring new configurations for microwave reactors, predicting and controlling processing conditions, and designing and optimizing new reactors.

In previous work [1-4] we have treated a spherical sample in a rectangular cavity using a spherical shell model for the sample. Electromagnetic fields were calculated using methods that were improvements over ordinary cavity perturbation theory, but which still were not exact. Results of the electromagnetic shell model were combined with thermal equations in calculating properties of both the sample and the reactor as a whole. Those properties included the Q of the cavity, and power absorption distributions and temperature profiles in the sample under both steady state and transient conditions.

In later work [5], we treated a cylindrical sample in a cylindrical cavity using a cylindrical shell model. Exact formulas for normal mode frequencies, quality factors, electromagnetic fields and for power absorption distributions in the sample were derived under idealized conditions in which the cavity walls were assumed to be perfect electrical conductors.

This paper addresses the problem of extending the cylindrical shell model to calculate temperature profiles inside microwave reactors. This new theory contains several noteworthy features. First, the finite electrical conductivity of the cavity walls is taken into account accurately. Second, realistic values of thermal emissivity are taken into account at solid boundaries inside the cavity and at the cavity walls. Third, effects of both direct microwave heating of the sample and radiant heating of the sample by a microwave-heated tube that surrounds it are included in the model. That is, hybrid heating of the sample can be accommodated in the model.

As in our earlier work a combination of analytic and numerical approaches is used in solving the problem. The main elements of the theory will be described next. Then calculated results will be presented and discussed in the context of materials processing. Our conclusions will be presented in the final section.

THEORY

The geometry of the cylindrical shell model is shown in Fig. 1 for the case where a microwave-heated tube surrounds the cylindrical sample. Interior regions of both the sample and tube are partitioned into radially thin zones, but vacuum spaces are not subdivided. In each zone the complex dielectric constant is assumed to be uniform, but it may vary from one zone to another. The N+1th zone is the curved cavity wall, which is infinitely thick in the model.

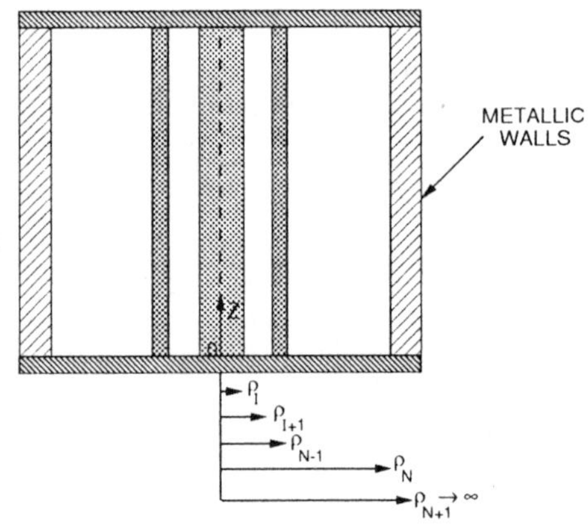

Fig. 1. Geometry of cylindrical shell model.

Solution of the electromagnetic problem was divided into two steps. In the first step, the flat end plates of the cavity were treated as perfect electrical conductors. Then by applying a technique described in our earlier work, we found exact solutions of Maxwell's equations with harmonically varying fields $e^{-i\omega t}$ and subject to appropriate boundary conditions. The method of solution involved expanding the fields in each zone in terms of cylindrical Bessel functions J_ℓ and cylindrical Neumann functions Y_ℓ. In the curved cavity walls, zone N+l, the solutions can contain only outgoing waves corresponding to cylindrical Hankel functions of the first kind, $H_\ell^{(1)} = J_\ell + iY_\ell$, since no energy propagates inward from infinity. Using this observation, we were able to include the curved cavity wall in the part of the model that could be treated exactly.

Normal mode frequencies, which are in general complex-valued, were calculated by finding the zeros of a complex-valued determinant of a certain 4x4 matrix, having 4 rows and 4 columns. Subsequently, the eigenfunctions of the 4x4 matrix were calculated, and these eigenfunctions provided input for applying a stepwise procedure to evaluate the expansion coefficients that occur in formulas for electromagnetic fields in each zone. For a TM0n0 mode the electric and magnetic fields in zone j are given by:

$$\boldsymbol{E}_n^j(\boldsymbol{r}) = 2\left[c_n^j J_0\left(\lambda_n^j \rho\right) + d_n^j Y_0\left(\lambda_n^j \rho\right)\right]\hat{z} \tag{1}$$

$$\boldsymbol{H}_n^j(\boldsymbol{r}) = \frac{2ik_n^j}{\omega\mu_0}\left[c_n^j J_0'\left(\lambda_n^j \rho\right) + d_n^j Y_0'\left(\lambda_n^j \rho\right)\right]\hat{\theta}. \tag{2}$$

Here c_n^j and d_n^j are expansion coefficients, and J_0' and Y_0' are derivatives of functions with respect to their arguments. Also, (ρ, θ, z) are cylindrical coordinates with origin at the center of the bottom end plate of the cavity. These formulas for the fields can be used to find formulas that can be used to evaluate the power absorbed in each zone and the time average energy stored in each zone.

In the second step of the electromagnetic problem, the finite conductivity of the cavity end plates was taken into account using the surface resistance approximation. In this approximation, the time average microwave power W absorbed at a surface S_0 is given by

$$W = \frac{1}{2} R_S \int_{S_0} \boldsymbol{H}_t(\boldsymbol{r}) \bullet \boldsymbol{H}_t^*(\boldsymbol{r}) ds. \tag{3}$$

$\boldsymbol{H}_t(\boldsymbol{r})$ is the tangential component of the magnetic field found in Step 1, and R_S is the surface resistance given by

$$R_S = \frac{1}{\sigma \delta_S}, \tag{4}$$

where σ is the electrical conductivity of the end plate material and δ_S is the penetration depth,

$$\delta_s = \left[\frac{2}{\omega' \mu_0 \sigma} \right]^{\frac{1}{2}}. \quad (5)$$

The integral in Eq. (3) can be evaluated analytically for the case we are considering, and this provided an explicit formula for power absorption in the two end plates.

The formulas described so far can be used to accurately calculate total time average power absorbed inside the cavity and in the cavity walls. The formulas can also be used to calculate the total Q of the cavity containing the sample and the contributions to the total Q and total absorbed power made by the sample, by a tube that may surround the sample, and by the cavity walls.

Next the solution of the electromagnetic problem was combined with a thermal model to calculate temperature profiles. Under steady state conditions, the general forms of the thermal equations are

$$\nabla \cdot \boldsymbol{q}(\boldsymbol{r}) = \mathrm{P}(\boldsymbol{r}) \quad (6)$$

$$\boldsymbol{q}(\boldsymbol{r}) = -\kappa \nabla T(\boldsymbol{r}). \quad (7)$$

In each zone the heat source $P(r)$ was taken to be a constant equal to the microwave power absorbed per unit volume averaged over the volume of the zone and averaged over time. The thermal conductivity κ was approximated as a linear function of temperature in each zone. The continuity of κ at every zone boundary interior to the sample or tube was enforced. In the model, heat is transferred across the surfaces of the sample, tube, and cavity walls by thermal radiation only, and that is calculated in a gray body approximation. Narrow gaps and low conductance supports between the end plates and the rod or tube can make conductive heat transfer negligible there.

The interior faces of the end plates were assumed to be perfect reflectors of thermal radiation, a condition that can be closely approximated in a real reactor by using highly polished, shiny surfaces. This assumption greatly simplified the problem of solving the thermal equations. More specifically, it implies that the thermal radiation formula [5] for two infinitely long cylinders, viz.,

$$Q = \frac{\sigma_{SB} A_1 \left(T_1^4 - T_2^4\right)}{\frac{1}{\varepsilon_1} + \frac{\rho_1}{\rho_2}\left(\frac{1}{\varepsilon_2} - 1\right)}, \quad (8)$$

can be applied here.

The steady state thermal equations have been solved analytically for each zone, and these solutions must be used with appropriate boundary conditions on temperature and heat current at zone interfaces. The formal solution of the thermal problem must be combined self-consistently with the electromagnetic equations. This can be accomplished by using an iteration procedure. The thermal and electromagnetic equations are coupled in two ways: (i) the complex dielectric constant is in general a function of temperature; and (ii) power absorbed in each zone is a source of heat in the thermal equations. Some results calculated with this self-consistent procedure will be discussed next.

DISCUSSION

The new theory was used in conducting a parametric study of the temperature within a cylindrical cavity containing an axially aligned cylindrical rod. In some of the cases studied, an axially aligned tube surrounded the rod to provide hybrid heating conditions. The cavity radius, ρ_c = 4.69 cm, and length, L = 6.63 cm, correspond to a 2.45 GHz resonant frequency for an empty cavity TM_{010} mode. These dimensions are the same as in our previous microwave power absorption study [6]. In this investigation, the temperature dependent properties of alumina were used for all rods and tubes [7,8]. The calculations were performed on a Y-MP2E Cray computer generally using 8 zones for the rod and 2 to 8 zones for the tube. A preliminary study for a number of cases indicated that essentially the same numerical results were obtained when a larger number of zones were used. The quantities computed included the electric field strength at the surface of the rod, the temperature profile within the rod and tube, the power absorbed in the cavity walls, rod and tube, and the resonant frequency and Q of the cavity.

A nominal set of experimental parameters was used in the present calculations, unless stated otherwise. The nominal alumina rod radius and emissivity values were a = 0.2 cm (a/ρ_c = 0.043) and ε_s = 0.31 respectively. The input power was P = 100 W into the cavity for a TM_{010} mode and the cavity walls were copper with an emissivity of ε_w = 0.025. The tube thickness, mid-radius and inner and outer surface emissivities were d = 0.1 cm, r_{mid} = 0.4 cm and $\varepsilon_{t_i} = \varepsilon_{t_o}$ = 0.31 respectively.

The temperature profile within the rod using the nominal parameter set is shown in Fig. 2 for the first three TM_{0n0} modes. For this small diameter rod, the average temperature and the temperature gradient within the rod both increase with mode number. The average rod temperature for the TM_{030} was calculated to be ≈ 137 C higher than the TM_{010} while the temperature difference between the rod's center and surface was 39 C for the TM_{030} mode compared to only 18 C for the TM_{010} mode. These differences become significantly larger as the rod diameter is increased.

The cavity wall emissivity plays an important role in determining the final temperature of the rod as shown in Fig. 3. If the emissivity were zero at any interface, then no heat would be transferred across that boundary and the temperature of the enclosed material would increase continuously as its energy is increased by microwave absorption. A steady state would never be reached before

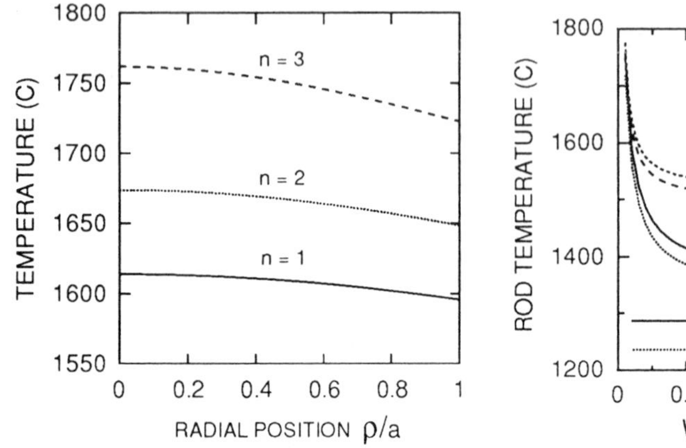

Fig. 2. Rod temperature profile for three lowest TM_{0n0} modes.

Fig. 3. Rod center and surface temperatures versus wall emissivity with and without the tube.

the enclosed material melted. This gives insight into the higher steady state temperatures attained as the emissivity is decreased. In the case of the cavity walls, the presence of the tube significantly increases the rod temperature for larger wall emissivities (> 0.05).

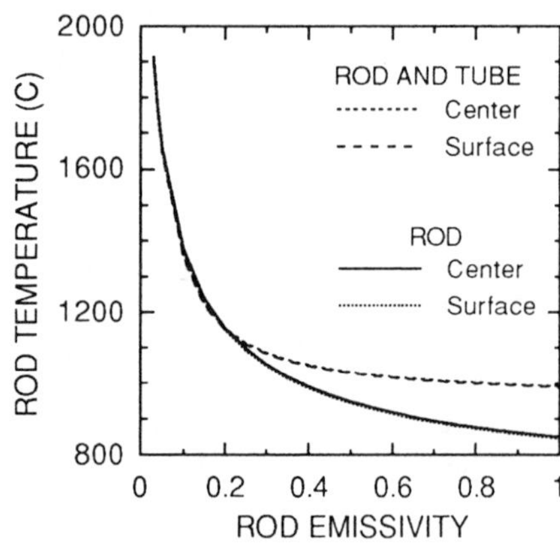

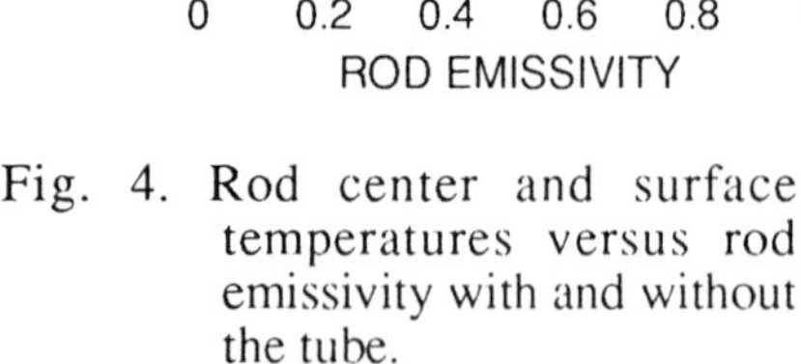

Fig. 4. Rod center and surface temperatures versus rod emissivity with and without the tube.

Fig. 5. Rod center and surface temperatures versus inner and outer tube surface emissivity.

The rod emissivity affects the rod temperature in a manner similar to the cavity wall emissivity. Figure 4 shows the strong dependence of the rod center and surface temperatures on the rod emissivity with and without the tube present. A power level of 30 watts was used for this calculation in order to observe the high temperatures attained at low rod emissivities without melting the rod. Again the rod temperature rises sharply as the emissivity is reduced. However, in contrast to the cavity wall emissivity, the tube increases the rod temperature only at higher emissivity values (> 0.25).

The effect of the emissivity of the inner and outer surfaces of the tube on the rod temperature was also investigated. Figure 5 shows the strong dependence of the rod center and surface temperatures on the tube emissivities. The tube inner surface emissivity has a slightly stronger effect on the rod temperatures than the tube outer surface. These calculations indicate that the rod temperature can be increased by almost 300 C by reducing the emissivity of the alumina tube inner surface to 0.1. This reduction could possibly be accomplished by coating the tube with a low emissivity material that adheres to alumina at high temperatures.

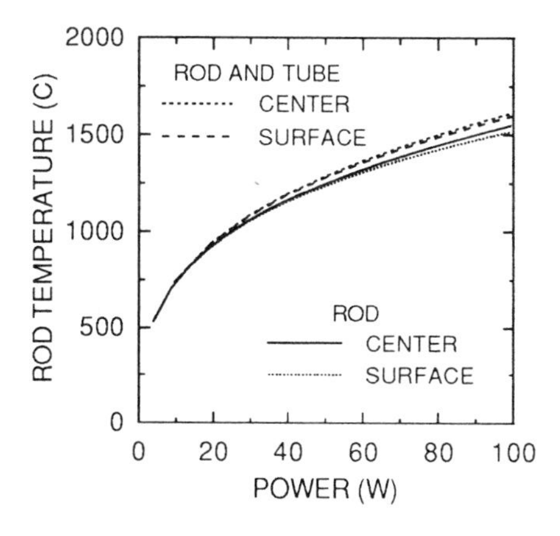

Fig. 6. Rod center and surface temperatures versus power with and without tube.

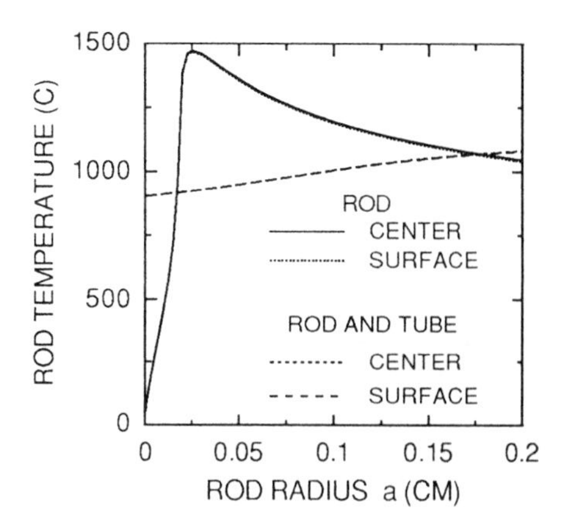

Fig. 7. Rod center and surface temperatures versus rod radius with and without tube.

Figure 6 shows the center and surface temperatures of the rod with and without the tube present as a function of the power introduced into the microwave cavity. When the input power to the cavity is 100 W and the tube is present, the average temperature of the rod is about 74 C higher than when the tube is absent. The fact that the tube has little effect on the rod temperature is associated with the chosen nominal parameter set. The effect of the tube becomes more significant as the rod radius is reduced as shown in Fig. 7 where the power has been fixed at 30 W. The presence of the tube actually reduces the temperature of the rod for radii in the range $0.017 \text{ cm} < a < 0.18 \text{ cm}$. For smaller or larger rods, the tube increases the temperature of the rod. There is a critical rod radius ($a \approx 0.025$ cm) that leads to a maximum isolated rod temperature ($\approx$ 1470 C). Of particular interest is the small rod region, below the critical radius, which is relevant to the microwave processing of fibers. In this region the temperature of an isolated fiber is significantly reduced from the peak temperature. However, the fiber temperature remains relatively constant (900 C < T < 950 C) with a tube surrounding the fiber.

The behavior exhibited in Fig. 7 for small rod radii can be understood as follows. When the tube is absent and the rod radius approaches zero, the field strength at the rod asymptotically approaches that of the empty cavity. The power absorbed in the rod is directly proportional to the rod volume, i.e., it is proportional to the square of the rod radius, a^2, and it is almost negligible compared to the power absorbed in the walls. On the other hand, the power radiated at the rod surface is proportional to the surface area of the rod, i.e., it is proportional to the rod radius, a. As the rod radius $a \rightarrow 0$, the surface to volume ratio increases as $1/a$ and the radiative cooling dominates over microwave heating. Therefore as $a \rightarrow 0$, the rod temperature decreases. On the other hand, when the rod radius approaches zero with the tube present, the radiative heating of the rod by the tube changes very little, and so the rod temperature remains high.

From the results of the nominal parameter set study, we chose an optimum parameter set for heating a rod. This parameter set only differed from the nominal set in that the TM030 mode was used and the emissivity of the tube surfaces was set at 0.1, which is a technically achievable value. Figure 8 shows the rod center and surface temperatures as functions of power into the cavity for the nominal 0.2 cm radius rod and a 20 μm radius fiber that is 100 times smaller. This calculation indicates that with this optimum parameter set the rod can be heated to melting with less than 80 watts. Furthermore, the fiber can reach a temperature above 1500 C using this same power level.

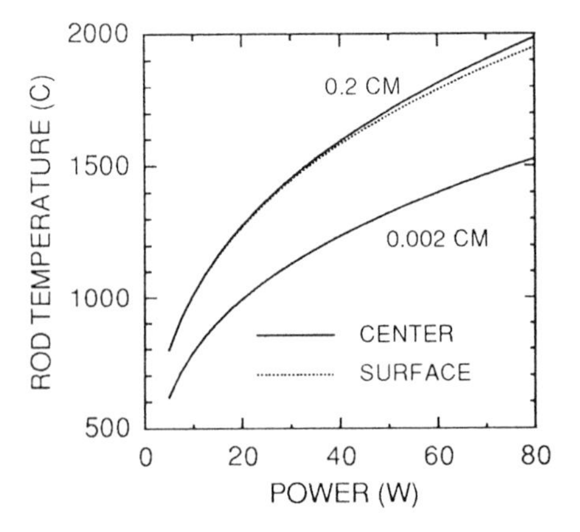

Fig. 8. Rod and fiber temperatures versus power for an optimum parameter set.

CONCLUSIONS

We have developed a steady state microwave model for a cylindrical rod situated along the entire axis of a cylindrical cavity. The model was applied to the <u>microwave</u> heating of an isolated rod as well as the <u>hybrid</u> heating of a rod surrounded by a microwave-heated tube. A parametric heating study was performed for samples ranging from fibers to large rods. Enhanced heating of the rod was obtained by reducing the emissivity of the cavity walls, tube and rod. In most cases, the presence of the tube increased the average temperature of the rod, however for the smaller size rods investigated, the tube did not significantly reduce the temperature gradient within the rod. For all the rods studied, a monotonic increase in power led to the melting of the rod without the presence of thermal runaway. This study demonstrated that by optimizing the experimental parameters, efficient microwave/hybrid heating of fibers and rods can be achieved. These results should support future economical development of commercial microwave processing applications.

ACKNOWLEDGMENT

The research described in this article was carried out at the Jet Propulsion Laboratory, California Institute of Technology, under contract with the National Aeronautics and Space Administration.

REFERENCES

1. H. W. Jackson and M. Barmatz, "Microwave Absorption by a Lossy Dielectric Sphere in a Rectangular Cavity," J. Appl. Phys. **70**, pp. 5193-5204 (1991).
2. H.W. Jackson and M. Barmatz, "Method for Calculating and Observing Microwave Absorption by a Sphere in a Single Mode Rectangular Cavity," Ceramic Transactions **21**, pp. 261-269 (1991).
3. M. Barmatz and H. W. Jackson, "Steady State Temperature Profile in a Sphere Heated by Microwaves," MRS Symp. Proc., **269**, pp. 97-103 (1992).
4. H. W. Jackson, M. Barmatz and P. Wagner, "Transient Temperature Behavior of a Sphere Heated by Microwaves," Ceramic Transactions, **36**, pp. 189-199 (1993).
5. J.P. Holman, *Heat Transfer*, (McGraw-Hill, New York), p. 193 (1972).
6. H.W. Jackson, M. Barmatz, and P. Wagner, "Microwave Power Absorption Profile in a Cylindrical Sample Contained in a Resonant Cylindrical Cavity," MRS Symp. Proc., **347**, pp. 317-323 (1994).
7. H. Fukushima, T. Yamanaka and M. Matsui, "Measurement of Dielectric Properties of Ceramics at Microwave Frequency," J. of Japan Soc. of Prec. Eng. **53**, pp. 743-748 (1987).
8. Y. S. Touloukian, *Thermophysical Properties of Matter*, (IFI/Plenum, New York - Washington), Vol. 8, p. 98 (1972).

MICROWAVE HEATING OF A CHANGING MATERIAL

Daniel J. Skamser, Jeffrey J. Thomas, Hamlin M. Jennings, D. Lynn Johnson
Department of Materials Science and Engineering, Northwestern University,
Evanston, IL 60208-3108

ABSTRACT

Some ceramic systems involve changes in composition during high temperature processing. If microwave heating is to be employed, it is important to understand how microwave heating parameters are affected by the property changes that occur in these systems. The present study uses a numerical model to simulate both the power density and the temperature distribution established inside a compositionally changing specimen heated with microwave energy. Simulation of the microwave processing of reaction–bonded silicon nitride shows the development of temperature spikes, which lead to reaction fronts that propagate through the specimen. Material systems that evolve from *lossy to low loss* exhibit more uniform and controlled heating.

INTRODUCTION

Microwave heating is a useful way of creating temperature gradients for ceramic processing. Processes which require that reactants be transported from outside the specimen into the interior through a diminishing pore structure can benefit from microwave heating. Examples are reaction-bonding and chemical vapor infiltration. Isothermal heating of these processes often results in unreacted material or incomplete densification in the specimen interior. Steady-state temperature gradients inside the specimen create an inside-to-outside chemical reaction, leading to increased conversion or densification of the final product before the outer pores close.[1,2]

Many of the processes which benefit from the temperature gradients created inside the specimen also involve changes in composition via a chemical reaction. Both density and composition gradients, which affect the microwave heating characteristics, will develop inside the specimen because the rate of reaction is dependent on both the temperature and the gas composition. Some regions of the specimen will react faster than others, resulting in a partially-reacted material that has spatially varying dielectric properties and thermal conductivity, causing non-uniform microwave power absorption and heat flow. For example, in the chemical vapor infiltration (CVI) process the matrix material is deposited inside a fiber preform which increases the density and changes the properties.[1] Another example is reaction-bonded silicon nitride (RBSN), in which a silicon powder compact is reacted in a nitrogen atmosphere to form Si_3N_4, causing a large change in the electrical conductivity of the specimen.[2,3] Thus, a knowledge of the changing properties of the material is necessary before microwave heating can be used effectively in a process.

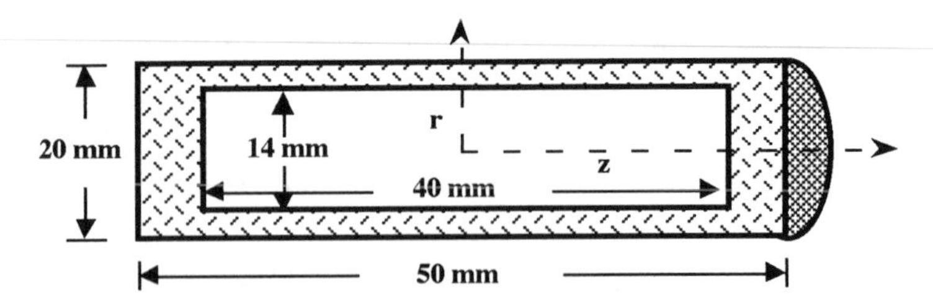

Figure 1 Cutaway view of the model configuration for the ceramic specimen surrounded by insulation. The model simulates heating conditions for one 2-D quadrant of the cylindrical specimen, shown in the upper right corner. The physical properties of the insulation remain constant while the properties of the specimen change with composition during the course of a simulation.

Another consideration for microwave heating of a compositionally changing system is that control of the microwave power distribution produced inside a specimen can be difficult. Variations in the dielectric property distribution inside a specimen can lead to a local instability for the absorbed power, and possibly result in thermal runaway and an undesirable final microstructure. Modeling can provide an insight on how to control microwave heating, as previously demonstrated in a model for thermal stability of ceramics. [4]

DESCRIPTION OF MODEL

A model is used to predict the temperature, absorbed power, and composition inside a microwave heated specimen during processing. Compositional changes in the specimen, leading to varying dielectric properties, are simulated for two cases likely to be encountered during microwave processing of ceramics. For both cases the RBSN system was chosen as the model process because this material undergoes large changes in dielectric properties as the composition changes from conducting silicon to insulating Si_3N_4 The model was used successfully to predict temperature and composition distributions during microwave RBSN processing[5,6]. The model provides insight into the microwave heating of other compositionally changing processes, such as CVI.

Simulations of microwave heating were conducted for (1) a conductor converting into a lossy dielectric and then into an insulator (e.g., RBSN process), and for (2) a lossy dielectric converting into an insulator. In both cases the lossy dielectric was modeled as a conductor-loaded dielectric (CLD) composite, which consists of electrically conducting particles randomly dispersed within an insulating host. In the RBSN process this occurs above 70% conversion when the conducting silicon particles are separated by Si_3N_4 reaction product.

A finite difference model in cylindrical coordinates was constructed to simulate the microwave heating of the RBSN process using a control volume methodology.[7] The configuration consists of a cylindrical silicon powder compact surrounded by thin layer of alumina insulation which is transparent to microwave energy, as shown in Fig. 1. The insulation reduces but does not eliminate the temperature gradients produced in the specimen.

A uniform microwave power density was assumed to be incident on the surface of the specimen. Simplified equations were used to describe the microwave-material interactions, which are adequate for calculating composition-based differences in the microwave heating response of the specimen. A generalized equation[8] for the absorbed microwave power is

$$P_{mw} = P_o\left(1 - p^2\right)\left(1 - e^{-2\alpha d}\right) \quad (1)$$

where P_o is the incident power, p is the reflection coefficient, α is the attenuation constant, and d is the specimen size. Both p and α are functions of the relative dielectric constant, ε_r', and the electrical conductivity, σ, of the specimen, and are functions of the specimen composition.[9,10] Insulators (e.g., Si_3N_4) have low values of p and α, indicating that the incident power is neither highly reflected nor strongly attenuated within the specimen. Conductors (e.g., silicon) have high values of p and α, producing a high amount of reflected power and rapid attenuation of the microwave field near the surface of the specimen. Neither of these heats as well as a CLD since it has intermediate p and α that cause low reflection and high attenuation of the microwave field, and result in efficient power absorption inside the specimen.

The variation in the microwave penetration depth ($D_p=1/2\alpha$), the reflection coefficient, and the resulting absorbed power calculated using Eq. 1 are shown in Fig. 2 for the two cases described above. The penetration depth is the distance into the specimen at which the microwave power drops to 1/e of its surface value. The magnitude of D_p indicates the amount of volumetric heating that will occur inside the specimen. If D_p is much smaller than the specimen size then the power will be deposited mostly at the surface, similar to conventional heating. When D_p is greater than or equal to the radius of the specimen then the heating will be volumetric, and significant radial temperature gradients can form.

The equation governing heat transfer within the specimen is

$$\rho c_p \frac{\partial T}{\partial t} = \frac{1}{r}\frac{\partial}{\partial r}\left(rK\frac{\partial T}{\partial r}\right) + \frac{\partial}{\partial z}\left(K\frac{\partial T}{\partial z}\right) + P_{mw} + \frac{\left(dX/dt\right)\rho_g \Delta H}{3M_w} \quad (2)$$

where ρ is the density, c_p is the heat capacity, K is the effective thermal conductivity, (dX/dt) is the rate of conversion from silicon to Si_3N_4, ρ_g is the green density of the silicon compact, ΔH is the heat of reaction, and M_w is the molecular weight of the reactant gas. The heat generated inside the specimen consists of the absorbed microwave power, P_{mw}, and the heat released by the chemical reaction, given by the last term in Eq. 2. The c_p and K are weighted averages of the values for silicon, Si_3N_4, and porosity. The porosity varies linearly with X, starting at 36% for a silicon compact and decreasing to 22% for pure Si_3N_4. Heat is dissipated at the surface of the insulation by convection and radiation, so that the heat dissipated must be equal to the heat absorbed for the specimen to be in thermal equilibrium.[11]

The equation for mass transfer and molar conservation of the reactant gas is

$$\phi\frac{\partial m}{\partial t} = \frac{1}{r}\frac{\partial}{\partial r}\left(rD\frac{\partial m}{\partial r}\right) + \frac{\partial}{\partial z}\left(D\frac{\partial m}{\partial z}\right) - \frac{\left(dX/dt\right)\rho_{inc}}{M_w} \quad (3)$$

where m is the molar density of the reactant gas, D is the effective diffusion coefficient, and the last term in Eq. 3 represents the consumption of the reactant by chemical reaction. For a green density of 64%, the bulk density increases from 1.49 to 2.48 g cm^{-3} (i.e., ρ_g=0.99 g cm^{-3}). The effective diffusion coefficient takes the following form for the Knudsen regime

$$D = \frac{4}{3}\left(\frac{8R_gT}{\pi M_w}\right)^{1/2}\eta \quad (4)$$

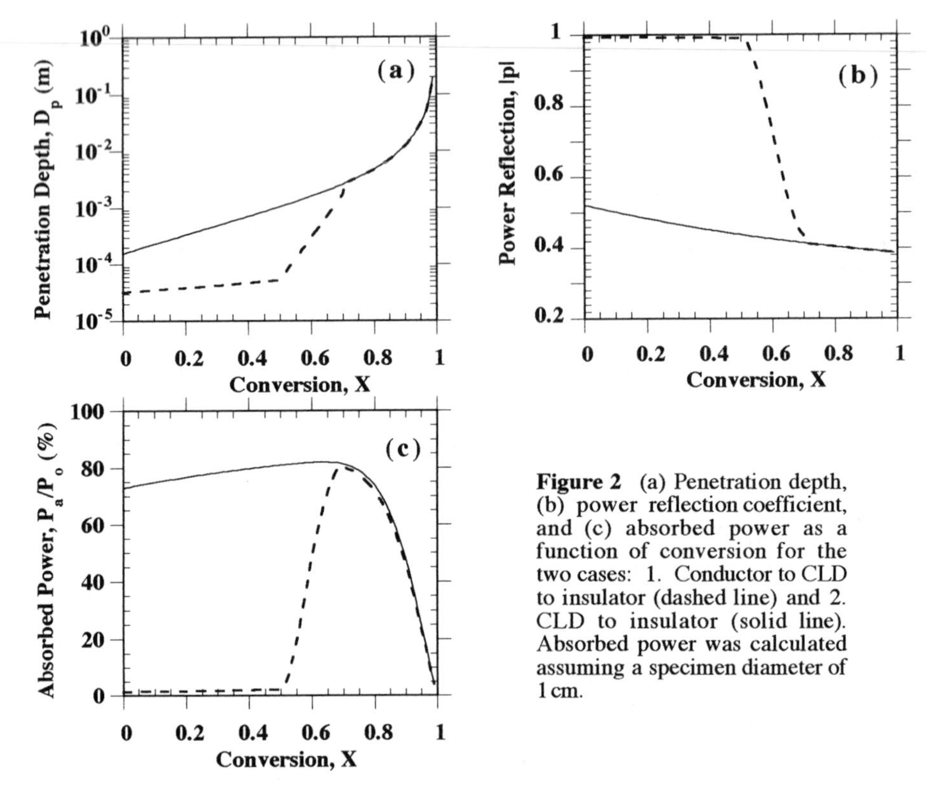

Figure 2 (a) Penetration depth, (b) power reflection coefficient, and (c) absorbed power as a function of conversion for the two cases: 1. Conductor to CLD to insulator (dashed line) and 2. CLD to insulator (solid line). Absorbed power was calculated assuming a specimen diameter of 1 cm.

where η is the permeability coefficient, which decreases exponentially with reaction from 50 nm for a green silicon compact to 0.1 nm for fully reacted Si_3N_4 product.[12]

The kinetic rate expression for the RBSN chemical reaction has been discussed at length in review articles.[13,14] and follows

$$\left({dX}/{dt}\right) = CP_{N_2} \exp\left(\frac{-E_a}{RT}\right)(1-X)^n \tag{5}$$

where X is the fractional conversion from silicon to either α- or β-Si_3N_4, P_{N2} is the partial pressure of nitrogen, C is a pre-exponential constant, E_a is the activation energy, and n is the order of the reaction (n=1 or 2/3 for α- or β-Si_3N_4). During every simulation the overall fractional conversion rate was constant at 3% of the starting silicon per hour.

RESULTS AND DISCUSSION

Case 1, which represents the RBSN system, has three distinct regimes of dielectric properties based on composition, as shown in Fig. 2. Below 50% conversion the silicon phase is continuous, causing a low D_p and a high p which remain fairly constant with conversion. This results in a small amount of absorbed power. Between 50 and 70% conversion there is a rapid transition in the heating

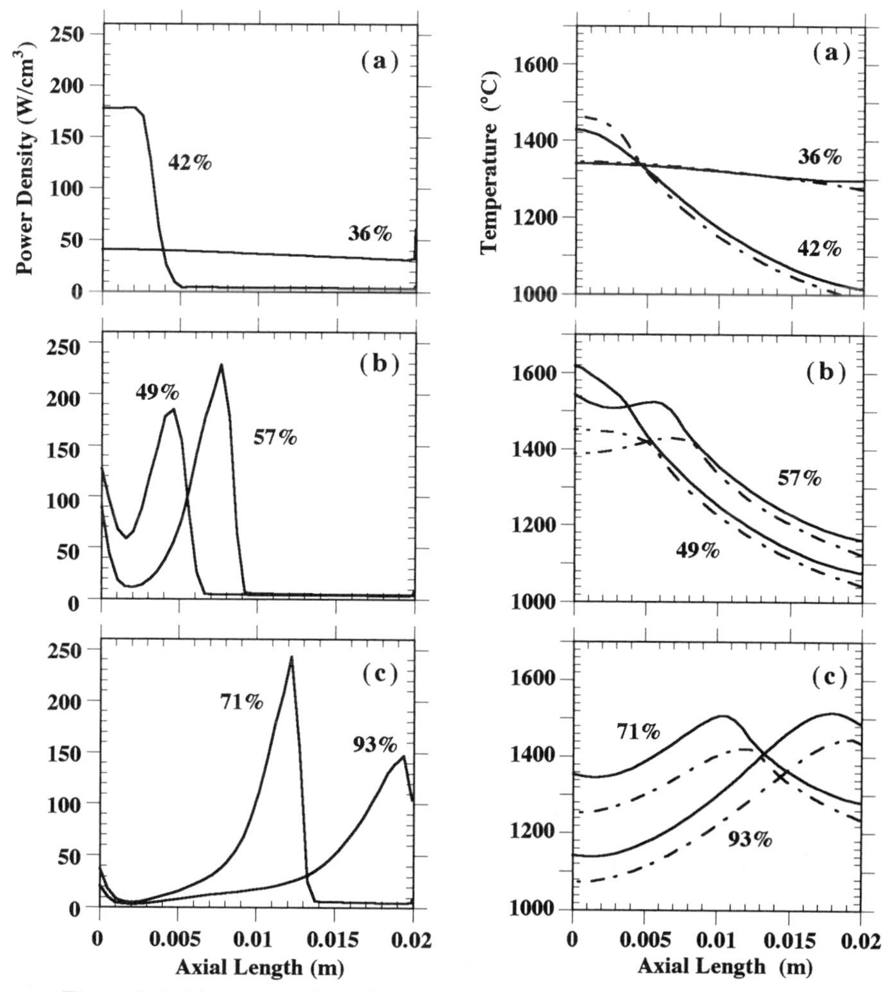

Figure 3 Axial power density and temperature profiles, from the middle to the end of the specimen, for the conductor to CLD to insulator process. The temperature is shown at both the centerline (solid line) and at the surface of the specimen (dashed line). a) 36 and 42% overall conversion. b) 49 and 57% overall conversion. c) 71 and 93% overall conversion.

behavior as the silicon phase depercolates into silicon particles separated by the Si_3N_4 reaction product. The p decreases and D_p increases causing the power absorption to increase. Optimum heating conditions are achieved near 70% conversion because p is low and D_p is near the radius of the specimen. Above 70% conversion the silicon particles are separated by the insulating Si_3N_4, and a CLD material is formed. The CLD material changes from a lossy dielectric to an insulator as the reaction proceeds, because p remains low and D_p increases from near the specimen size to a few meters. As a result, the absorbed power decreases

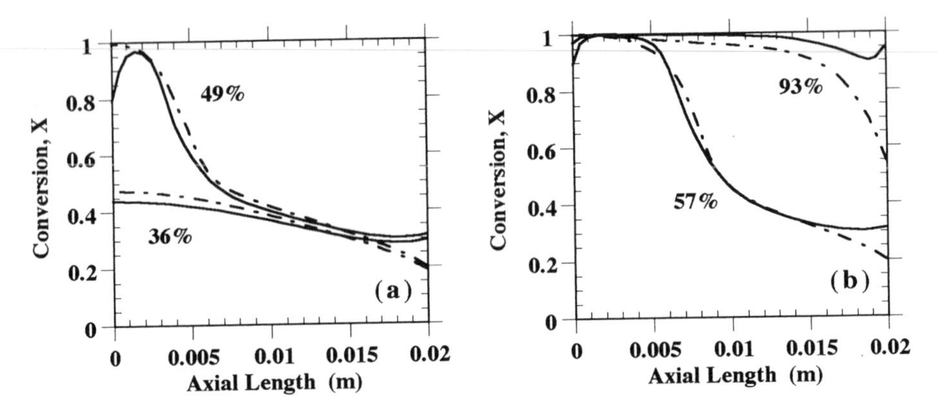

Figure 4 Axial conversion profiles, from the middle to the end of the specimen, for the conductor to CLD to insulator process. The conversion is shown at both the centerline (solid line) and at the surface of the specimen (dashed line). a) 36 and 49% overall conversion. b) 57 and 93% overall conversion.

from 80% to less than 5% of the incident power.

Simulation results for case 1 are shown in Figs. 3 and 4 from the middle to the end of the specimen. After 36% overall conversion the axial temperature and composition profiles are relatively flat because the absorbed power is only weakly dependent on composition. Radial temperature and composition gradients are not present since the power is absorbed only near the surface. At 42% overall conversion some material in the middle of the specimen reaches 50% conversion and absorbs microwave power more efficiently, causing a rapid temperature increase. By 49% overall conversion the temperature in the middle of the specimen has reached 1600°C because the conversion in the middle is nearly complete while the ends are unreacted. A transition zone is formed between the reacted and unreacted material. The power is absorbed most efficiently in the transition region, causing spikes for both the absorbed power and the temperature to form. At 57% overall conversion the absorbed power and temperature spikes have moved away from the middle of the specimen, following the transition zone. The transition zone continues to propagate along the z-axis until the specimen is fully reacted. Radial temperature gradients develop in the low-loss wake of the transition zone. The conversion in the center of the specimen is incomplete because the amount of reactant is low. This causes the power absorbed in the middle to be higher than surrounding regions and results in a local maximum for the temperature in the center.

Case 2 represents a CLD material which changes from a lossy dielectric to an insulator. The p remains low throughout the process, resulting in only a small amount of power being reflected from the surface of the specimen. The D_p rises slowly until about 70% conversion, when it begins to increase rapidly. The result is a material that absorbs a high amount of microwave power up to nearly full conversion, at which point it becomes a low loss insulator and the absorbed power rapidly decreases.

Simulation results for case 2 are shown in Fig. 5. At 10% overall conversion the temperature profile is flat because the D_p is low and the power is absorbed at the surface. The conversion profile is also flat following the temperature. The absorbed power remains highest at the ends throughout the reaction because they

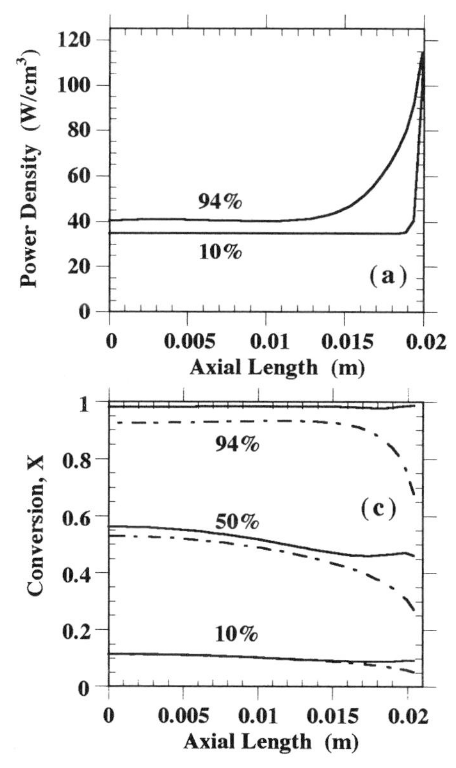

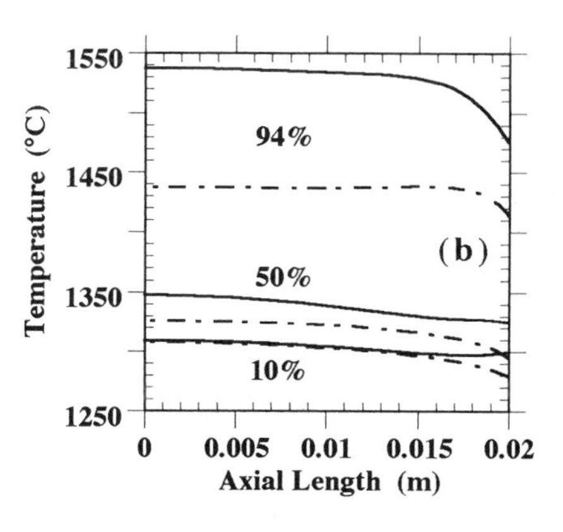

Figure 5 (a) Axial power density, (b) temperature, and (c) conversion profiles, from the middle to the end of the specimen, for the CLD to insulator process. The variables are shown at both the centerline (solid line) and at the surface of the specimen (dashed line) for 10, 50, and 94% overall conversion.

are the last region to become fully reacted and always have lower D_p than the rest of the specimen. By 50% conversion the D_p has increased sufficiently to allow radial temperature gradients to form. The conversion throughout the specimen shows the beginning of an inside-to-outside reaction profile. Finally at 94% overall conversion the temperature has risen dramatically as a result of maintaining a 3%/hr overall conversion rate. Large temperature gradients are present in the low loss material, and the conversion profile continues to follow an inside-to-outside progression. The absorbed power profile is more gradual as microwaves penetrate deeper into the specimen.

It is clear that microwave processing of materials that change from a conductor to an insulator will result in the creation of spikes for the temperature and the absorbed power. This tends to produce a sharp reaction front that propagates through the specimen, and may result in loss of control over the process. This is because the absorbed power passes through a maximum at intermediate conversion.

On the other hand, materials that change from a lossy dielectric to an insulator may be more applicable in reaction-bonding and CVI processes. Microwave processing of these systems produces radial temperature gradients which result in inside-to-outside reaction. More importantly, microwave heating of these types of materials can be controlled since excessive temperature gradients or temperature spikes are avoided. This is because the absorbed power remains high and nearly constant for most of the conversion before decreasing at nearly complete conversion, and avoids passing through a sharp maximum in the absorbed power as a function of conversion. Generally, this type of behavior is expected for any system in which the microwave heating characteristics are not greatly affected by the composition and the penetration depth increases with conversion.

CONCLUSIONS

A model has been used to simulate the microwave heating of a compositionally changing material. Microwave processing of material systems which change from a conductor into an insulator results in the creation of reaction fronts that may become unstable. Material systems which change from a lossy dielectric into an insulator should have better success in industrial applications. The creation of radial temperature gradients allows for an inside to outside reaction, and the process can be controlled.

ACKNOWLEDGMENT

Supported by NIST under Contract No. SC91209 and by ARPA/ONR under Contract No. N00014-90-J-4020. Special thanks to Barbara L. and Robert H. Martin for donating a computer in memory of their son, Robert S. Martin.

REFERENCES

1. D.J. Skamser, P.S. Day, H.M. Jennings, and D.L. Johnson, "Optimization of Microwave Assisted Chemical Vapor Infiltration"; pp. 143–54 in Ceramic Transactions, Vol. 46, *Advances in Ceramic Matrix Composites II*, Edited by J.P. Singh, and N.P. Bansal. American Ceramic Society, Westerville, OH (1994).
2. J.J. Thomas, "Microwave Processing of Reaction-Bonded Silicon Nitride"; Ph.D. Dissertation Northwestern University, Evanston, IL (1994).
3. J.J. Thomas, D.J. Skamser, H.M. Jennings, and D.L. Johnson, "Formation of Reaction-Bonded Silicon Nitride using Microwave Heating," *J. Am. Ceram. Soc.*, submitted.
4. M.S. Spotz, D.J. Skamser, and D.L. Johnson, "Thermal Stability of Ceramic Materials in Microwave Heating," *J. Am. Ceram. Soc.*, **78** [4] 1041–48 (1995).
5. J.J. Thomas, D.J. Skamser, and D.L. Johnson, "Microwave Processing of a Compositionally Changing System"; pp. 140–43 in Proceedings 29th Microwave Power Symposium, *A Forum on Electromagnetic Technology & Applications from Around the World*, Edited by D.A. Lewis, and M.D. McCutchan. International Microwave Power Institute, Manassas, VA (1994).
6. D.J. Skamser, J.J. Thomas, and D.L. Johnson, "A Model for Microwave Processing of Compositionally Changing Ceramic Systems," *J. Mater. Res.*, submitted.
7. S.V. Patankar, *Numerical Heat Transfer and Fluid Flow*, Hemisphere Publishing Corp., New York (1980).
8. D.G. Watters, "Advanced Study of Microwave Sintering"; Ph.D. Dissertation Northwestern University, Evanston, IL (1989).
9. A.C. Metaxas, and R.J. Meredith, *Industrial Microwave Heating*, Peter Peregrinus Ltd., London (1983).
10. P.S. Neelakanta, "Complex Permittivity of a Conductor-loaded Dielectric," *J. Phys.: Condens. Matter*, **2** 4935–4947 (1990).
11. H.S. Carslaw, and J.C. Jaeger, *Conduction of Heat in Solids*, 2nd Edition. Clarendon Press, Oxford (1989).
12. M.H. Abbasi, and J.W. Evans, "Diffusion of Gases in Porous Solids: Monte Carlo Simulations in the Knudsen and Ordinary Diffusion Regimes," *AIChE J.*, **29** [4] 617 (1983).
13. A.J. Moulson, "Reaction-Bonded Silicon Nitride: Its Formation and Properties," *J. Mater. Sci.*, **14** 1017 (1979).
14. G. Ziegler, J. Heinrich, and G. Wotting, "Relationships Between Processing Microstructures and Properties of Dense and Reaction-Bonded Silicon Nitride," *J. Mater. Sci.*, **22** 3041 (1987).

IMPORTANCE OF DIELECTRIC PROPERTIES IN MODELING OF MICROWAVE SINTERING OF CERAMICS

J. R. Thomas, Jr.
Mechanical Engineering Department
Virginia Polytechnic Institute and State University
Blacksburg, VA 24061

ABSTRACT

Sintering experiments at the Los Alamos National Laboratory involved small cylinders of high-purity alumina encased in a "casket" of low-density zirconia insulation and heated in a large multi-mode microwave oven. Optical fiber sensors were used to monitor the temperature at several locations in the system. It was found that the alumina samples apparently heat faster than the zirconia insulation at temperatures above 1000°C, and that the temperature distribution in the sample is essentially uniform during the heating process.

A previously reported two-dimensional mathematical model of the heat transfer process reproduced the essential features of the observed phenomena when artificially corrected for uncertainties in dielectric property data. Recent measurements of the dielectric properties now allows a much more correct model of the process, which leads to new information about the heat transfer processes involved and their interaction with the dielectric properties.

INTRODUCTION

Successful sintering of multiple small alumina cylinders in a large multi-mode microwave oven at 2.45 GHz has previously been reported [1]. In that work the samples were enclosed in an insulating "casket" of low-density alumina or zirconia insulating board. Temperatures were measured by optical-fiber sensors at a single location within the enclosure. In spite of the success of this work, several questions remained. In particular, nothing was known about the temperature distribution within the samples or insulation during the heating cycle. Thus it was not clear whether the samples were being heated primarily by absorption of the microwave energy or indirectly by thermal radiation from the insulating material. Even though the samples were sintered without cracking, the degree of uniformity of the temperature distribution during the heating cycle remained unknown. Answers to these questions were sought through more complete temperature measurements on single cylindrical alumina samples, and by development of a detailed mathematical model of these experiments. This work was reported previously [2]; it utilized literature data for the dielectric properties of zirconia and alumina of different composition from that used in the experiments [1]. Reasonable agreement between calculation and experiment could only be obtained by artificially adjusting the electric field strength in the sample. Moreover, use of two similar sets of dielectric property data for alumina of very similar composition lead to opposite conclusions concerning the relative microwave energy absorption of the alumina sample and zirconia insulation.

Recently, dielectric properties of samples of the materials used in the experiments have been measured [3] thereby reducing one important source of uncertainty in the numerical simulation. The present contribution reports results of using these new data in the mathematical model described in Ref. [2] along with some additional refinements. These calculations suggest some rethinking of the heat transfer process involved in microwave sintering of ceramics, and point up some of the limitations in modeling of this process.

MATHEMATICAL MODEL

The system being modeled consists of a small cubic container ($256 cm^3$) enclosed in a large ($5.66 \times 10^4 cm^3$) multimode cavity operating at 2.45 GHz [2]. Since the sample is cylindrical, the insulating container was modeled as a cylinder with the same volume as the actual cubic container. Thus the problem was formulated as a 2-dimensional $r - z$ cylindrical system. The mathematical problem may be stated as

$$\nabla \cdot (k \nabla T) - \nabla \cdot \vec{q_r} + \dot{q}(\vec{r}, T) = \rho C_p \frac{\partial T}{\partial t}, \tag{1}$$

where T represents the absolute temperature at location r at time t, ρ is the density of the material, c_p is its heat capacity at constant pressure, k is the thermal conductivity, and $\dot{q}(r, T)$ is the volumetric heat generation rate created by dissipation of microwave energy, computed from

$$\dot{q}(\vec{r}, T) = 2\pi f \epsilon''(T) \mid E_f \mid^2, \tag{2}$$

where ϵ'' represents the dielectric loss coefficient of the material at microwave frequency f, E_f is the local field strength, and $\vec{q_r}$ represents the radiative flux within the material. Eq. (1) describes the temperature in both the sample and the insulating container when appropriate properties are used for each. The sample surface and the interior wall of the insulating container exchange heat by radiation except where the sample rests on the container floor. Here, heat exchange occurs through a contact conductance h_c. The air within the cavity between the sample and the casket wall was treated as a vacuum since radiation heat transfer dominates over conduction and convection in this small space. At the outer surface, the container exchanges heat by radiation and convection with the cavity walls and the air within the cavity, respectively. All of these heat transfer mechanisms are accounted for in the model.

An additional heat transfer mechanism not accounted for in the previous study [2] is radiative transfer from within the insulating material, represented by the net radiative flux vector $\vec{q_r}$. This term was found necessary to prevent thermal runaway in the calculations when correct values for the dielectric properties of the specific materials used in the experiment were incorporated. Thermal runaway was never observed in the experiments, but the insulation casket is observed to "glow" and become semitransparent to visible radiation during the experiments. Since the zirconia insulation is 92% porous, radiation heat loss from the interior is a plausible explanation for the self-limiting temperature behavior observed in the experiments.

A rigorous determination of the radiative flux vector requires a solution of the equation of radiative transfer [4], which in turn requires knowledge of the wavelength-dependent absorption and scattering properties of the materials, which are not available. For an optically thick medium, the equation of transfer is well approximated by a diffusion approximation [5], which leads to the concept of a radiative conductivity k_r,

$$k_r = \frac{16}{3} \frac{n^2 \sigma T^3}{K_e}, \tag{3}$$

where n is the index of refraction and K_e is the monochromatic extinction coefficient. Thus Eq. (1) was modified as

$$\nabla \cdot [(k_c + k_r) \nabla T] + \dot{q}(\vec{r}, T) = \rho C_p \frac{\partial T}{\partial t}, \tag{4}$$

where now k_c represents the ordinary solid conductivity and k_r is given by Eq. (3).

The microwave field strength E_f is assumed uniform at the outer surface of the container. An initial attempt was made to correct the internal field by reflection at each interface and exponential attenuation with distance from the surface using the "skin depth" approach [6]. This led to unsatisfactory agreement with measured internal temperatures, so the field was assumed to be uniform throughout the system. The field strength at the outer surface was set at each time step to match the total microwave power absorption by the sample and casket to the power absorption measured in the experiments.

Eq. (4) is solved using a fully implicit finite-difference scheme. A typical solution utilizes a 30×30 $r - z$ mesh, for a total of 900 mesh points. The resulting algebraic equations are solved using the sparse-matrix routines Y12m [7]. The time step size is started at 1 s and decreased as necessary to ensure convergence of inner iterations. Simulation of a 40-minute experiment requires 1 - 2 hours of CPU time on a fast engineering workstation. In order to mimic physical reality as closely as possible, material properties were treated as temperature-dependent in all cases

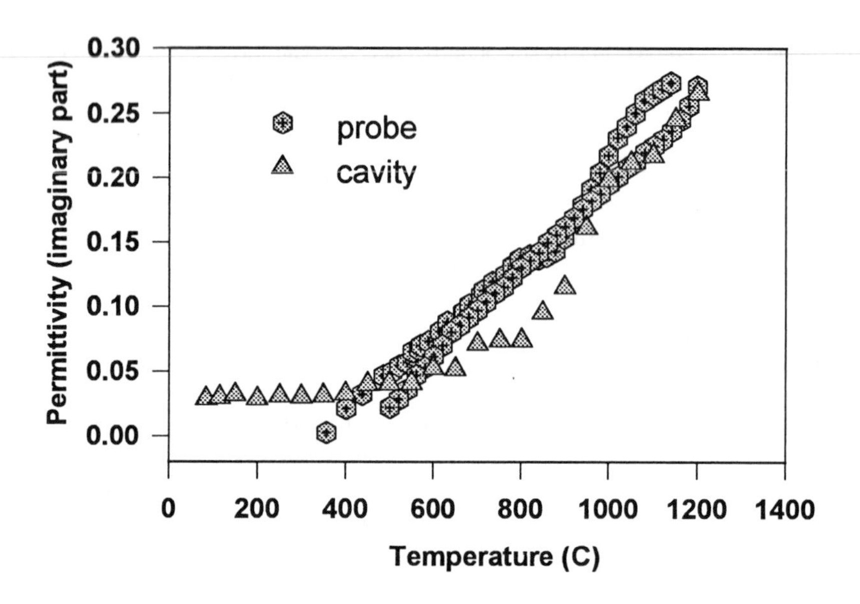

Figure 1: Dielectric loss coefficient of alumina samples measured by probe and cavity perturbation techniques [3].

where reasonably reliable data is available. In the experiments reported here, the samples were high-purity alumina, and the insulation was highly porous yttria-stabilized zirconia. Curve fits for thermal conductivity and specific heat for these materials were generated from literature data. The alumina samples were initially at approximately 55% theoretical density and reached 98% T.D. during sintering. The density of these samples was approximated as a linear function of temperature between 1200 and 1600°C.

The most important temperature-dependent property in this work is, of course, the dielectric loss coefficient ϵ''. In the previously reported work [2], we used the data of Hutcheon, et al. [8] and Arai, et al. [9]. The materials for which dielectric property data were given in Refs. [8,9] were high-purity alumina [8] and another high-purity alumina as well as zirconia of unspecified composition [9]. The two sets of alumina data are sufficiently different that they led to different conclusions regarding the extent of direct heating of the alumina sample by the microwave field. Recent measurements [3] of the dielectric properties of samples of the materials used in the experiments described here allow calculations to be made with more confidence.

Fig. 1 shows the dielectric loss (imaginary part of the permittivity) of alumina measured by two different techniques [3], a probe method and a cavity perturbation technique. These results are in quite good agreement at temperatures above

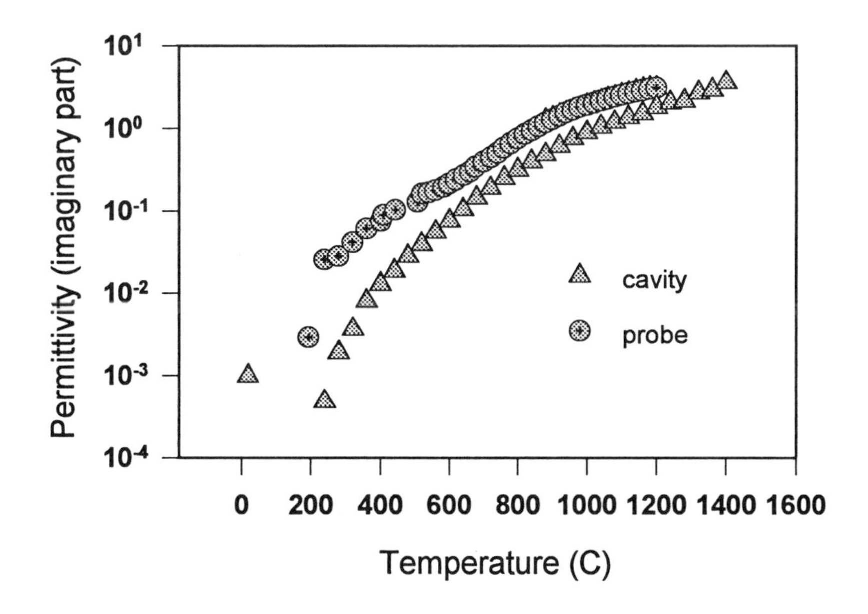

Figure 2: Dielectric loss coefficient of zirconia insulation [3].

about 400°C, but there is considerable uncertainty at lower temperatures where the dielectric loss is quite small.

Similar results are shown in Fig. 2 for the zirconia insulation, which is plotted on a logarithmic scale to highlight the uncertainty at low temperatures. These noisy low-temperature data pose a problem for the modeler, since the value of ϵ'' at low temperatures determines success or failure for a simulation. To achieve good agreement between simulation and measured temperatures from a sintering experiment requires some arbitrary adjustment of low-temperature values for ϵ'', within the bounds of experimental uncertainty. Piecewise polynomial curve fits were generated for these data, with linear extrapolation to temperatures above those for which data are available.

RESULTS

Data from a typical sintering experiment [2] is displayed in Fig. 3 along with simulation results. The curve labeled "sample" is the temperature measured by an optical fiber probe positioned at approximately the center of the alumina sample, while the curve marked "insulation" is the temperature measured by another probe positioned approximately halfway through the wall of the zirconia insulation "casket". The temperature monitoring system does not function for temperatures

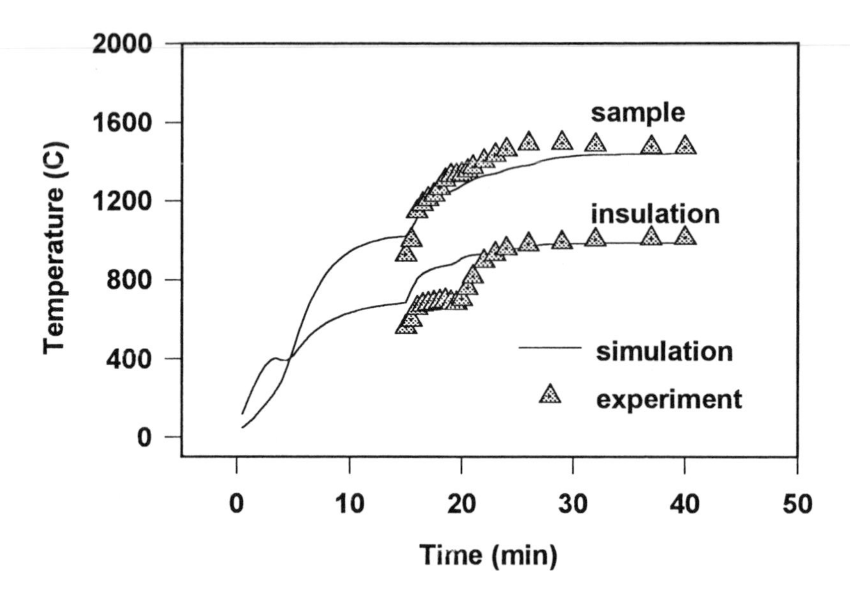

Figure 3: Measured and computed temperatures during sintering run with alumina workpiece and zirconia insulation. The temperatures were measured with optical fiber sensors in holes drilled into materials.

below 500°C. For this experiment, the magnetron power was set at 20% of maximum and maintained at that value throughout the experiment. The forward and reflected power from the cavity was measured, and the difference between these values was taken as the power absorbed by the insulation and sample. Since no data were taken until the temperatures exceeded 500°C, the absorbed power was not known during the time period 0 - 15 minutes. Consequently, excellent agreement between model and experiment is not expected. Nevertheless, the agreement is excellent past this initial transient.

To attempt to answer the fundamental question of whether or not the alumina samples were being heated directly by the microwave field or indirectly by thermal radiation from the sample, a simulation was run with the alumina permittivity ϵ'' set to zero. This allows for no direct heating of the sample by the microwave field so that any temperature rise that results is caused by indirect heating. Results of this simulation show that the alumina sample displays the same temperatures within 20°C throughout the heating cycle. The obvious conclusion is that the samples were heated principally by thermal radiation from the zirconia insulation.

The internal temperature distribution within the system during a typical experiment is of interest. This is illustrated in Fig. 4 at 5 minutes into the heating cycle. Note the hot spot that develops immediately above the sample during the

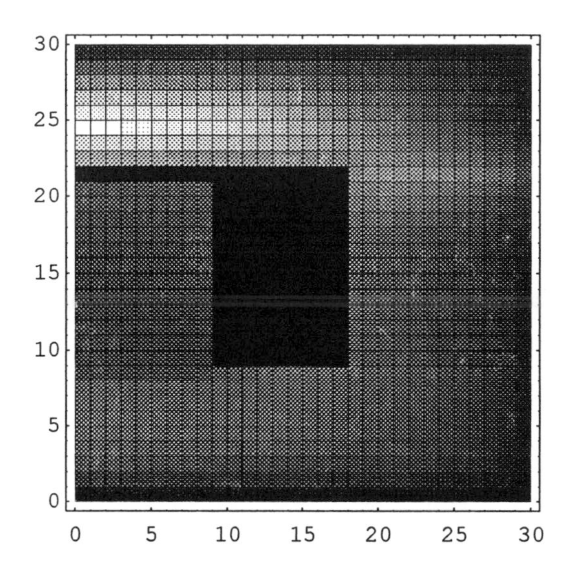

Figure 4: Density plot of the two dimensional temperature distribution in cross section of experimental system at 5.0 minutes. Centerline is to left of figure. The brightest area is at about 1840°C.

early portions of the heating cycle. The energy from this hot spot then slowly diffuses outward to the surrounding areas of the insulation casket. This hot spot is not caused by field concentration since this calculation assumed uniform field strength throughout. It results from the fact that a point far from the surface quickly heats up to the point that the dielectric loss ϵ'' is on the steep portion of the temperature curve (Fig. 2). Without accounting for internal radiation loss, this spot quickly exceeds the melting temperature of the zirconia insulation, a result not observed in any of the experiments.

CONCLUSIONS

Good qualitative information about temperature distributions and power absorption in microwave sintering of ceramic samples can be obtained through a simplified model of the microwave field distribution coupled to a detailed thermal model. These results indicate that alumina samples sintered in low-density zirconia insulation do not heat by direct absorption of microwave power but that temperatures nevertheless remain uniform during the heating cycle. Thermal runaway is predicted by the model unless an additional heat loss mechanism is introduced, specifically, thermal radiation emission from within the interior of the system. Hot spots can develop in such a system even with uniform microwave fields because of the non-linear chracter of the microwave heat source. Definitive quantitative results must await a coupled solution of the electromagnetic field and heat transfer equations.

ACKOWLEDGEMENTS

The author would like to express appreciation to Dr. Thomas Cross for communication of measured dielectric properties of the materials. Helpful discussions with Dr. W. A. Davis are also gratefully acknowledged.

REFERENCES

1. Joel D. Katz and Rodger D. Blake, *Am. Ceram. Soc. Bull.* **70**, 1304 (1991).
2. J. R. Thomas, Jr., Joel D. Katz, and Rodger D. Blake, Mat. Res. Symp. Proc. **347**, 311-316 (1994).
3. T. E. Cross, Univ. Nottingham UK, Personal Communication (1995).
4. J. R. Thomas, Jr., *Transp. Th. Stat. Physics* **19**, 405 (1990).
5. M. N. Özisik, **Radiative Transfer**, John Wiley, New York (1973).
6. Constantine A. Balanis, **Advanced Engineering Electromagnetics**, John Wiley & Sons, New York (1989).
7. Z. Zlatev, *SIAM J. Numer. Anal.* **17**, 18 (1980).
8. R. M. Hutcheon, M. S. De Jong, F. P. Adams, P. G. Lucuta, J. E. McGregor, and L. Bahen, in *Microwave Processing of Materials III*, edited by R. L. Beatty, W. H. Sutton, and M. F. Iskander Mater. Res. Soc. Proc. **269**, Pittsburgh, PA, pp. 541-551 (1992).
9. M. Arai, J. G. P. Binner, G. E. Carr, and T. E. Cross, in *Microwaves: Theory and Application in Materials Processing II*, edited by D. E. Clark, W. R. Tinga, and J. R. Laia, Jr. **Ceramic Transactions 36**, pp. 483-492, Am. Ceram. Soc., Westerville, OH (1993).

MODELING OF MULTI-FREQUENCY MICROWAVE SINTERING OF ZnO CERAMIC

A. Birman*, B. Levush*, Y. Carmel*, D. Gershon*, D. Dadon L.P. Martin*** and M. Rosen*****
***Institute for Plasma Research, University of Maryland, College Park, MD 20742**
****N.R.C.N, P.O. Box 9001, Beer Sheva, Israel.**
*****John Hopkins University, Baltimore, MD 21218**

ABSRACT

To gain better insight of microwave heating of ceramic materials and to optimize this process, we have developed a code modeling the heat propagation in the ceramic. The model takes into account the variation of the heat conduction coefficient, specific heat and permittivity with temperature. In addition, the heat conduction coefficient and permittivity dependence on the porosity is modeled using Maxwell Garnet mixing theory (MGT). The code was used to model experiments of ZnO microwave heating. These computations prove that the model used in the code is capable of reproducing the salient features of the heating process. The code was modified to allow modeling of simultaneous microwave heating at different frequencies. In particular, heating of ZnO sample with two microwave sources at widely separated frequencies (2.45GHz and 28GHz) showed that it is possible to keep a homogeneous temperature profile across the sample throughout the sintering process.

INTRODUCTION

Use of microwaves for ceramic sintering has the potential of producing high quality ceramic bodies of complex shape using less energy. This process however is extremely complex and optimizing it requires a large number of trial and error experiments. Therefore, without proper computational modeling further progress is hindered [1]. We have developed a one dimensional heat transport code to model heat propagation and microwave energy absorption during the sintering process. Microwave absorption is governed by the dielectric properties of the ceramic material which depend on the temperature, porosity and frequency. Specific heat and heat conduction coefficients depend on the temperature and heat conduction further depends on the porosity. All this is taken into account in our model. The code was first tested against the analytical results of Kriegsmann [2] and then used to model sintering of ZnO ceramic. The numerical results were found to be in good agreement with the data taken in our laboratory during ZnO sintering. The model was further expanded to allow multi frequency processing. In particular, it is shown that with two microwave sources (2.45GHz and 28GHz) applied simultaneously it is possible to achieve a homogeneous temperature profile across the sample throughout the sintering process.

THEORETICAL MODEL

To obtain the temperature distribution in the sample at each moment one has to solve the heat transport equation with the appropriate boundary conditions. For the 1D case this equation is given by:

$$c_p(T)\frac{dT}{dt} = \frac{d}{\rho(T)dx}[\kappa(T,\rho)\frac{dT}{dx}] + \frac{P(T,\rho,f,t)}{\rho(T)} \qquad (1)$$

where c_p is the heat-capacity, κ is the heat-conduction, ρ is the density and P is the power absorbed in the ceramic. The boundary conditions are either: 1) no thermal flux, 2) prescribed temperature temporal profile, 3) heat loss by radiation and convection. For microwave sintering the power P absorbed per unit volume is given by: $P=2\pi f\varepsilon_0\varepsilon'_r\tan\delta|E|^2$ where f is the frequency of the microwave source, ε_0 is the permittivity of free space, ε'_r is the relative dielectric constant, $\tan\delta$ is the loss tangent and $|E|$ is the local electric field. Strictly speaking the spatial distribution of the electric field should be found by solving the field equations in the ceramic. However, in this work the electric field is assumed to decay within the sample according to:

$$E = E_0 \exp(\frac{-x}{d}) \qquad (2)$$

where d is the skin depth and E_0 is the electric field in the unloaded cavity, which is proportional to the microwave power temporal profile g(t). The skin depth is further related to the frequency of the microwave source and to the loss tangent through:

$$d(T,\rho,f) = \frac{0.0675}{f(GHz)\sqrt{\varepsilon'_r}\sqrt{\sqrt{1+\tan^2\delta}-1}} \; [m] \qquad (3)$$

MODELING OF MICROWAVE SINTERING OF ZnO CERAMIC

To perform the computation knowledge of the dependence on temperature, porosity and frequency of all the relevant quantities is required. The specific heat temperature dependence on the temperature in the range relevant to the sintering process is shown in Fig. 1. The data was taken from Ref. [3] and fitted to a curve of the form $a+bT+cT^{-2}$ ($a=0.140$, $b=1.77*10^{-5}$, $c=2360$). The temperature dependence of the heat conduction coefficient [3] [4], is shown in Fig. 2.

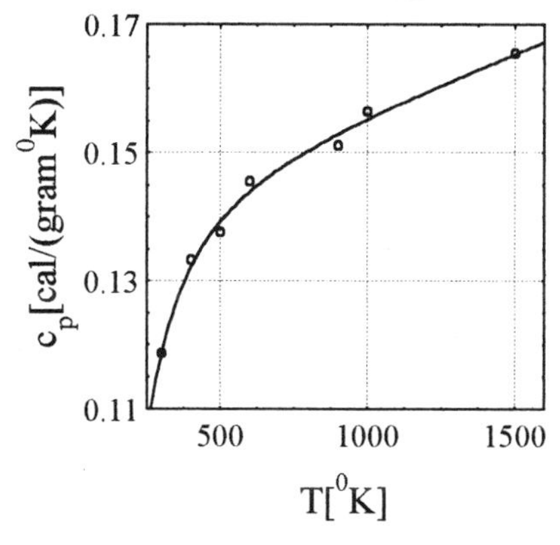

Figure 1. Specific heat of ZnO versus temperature

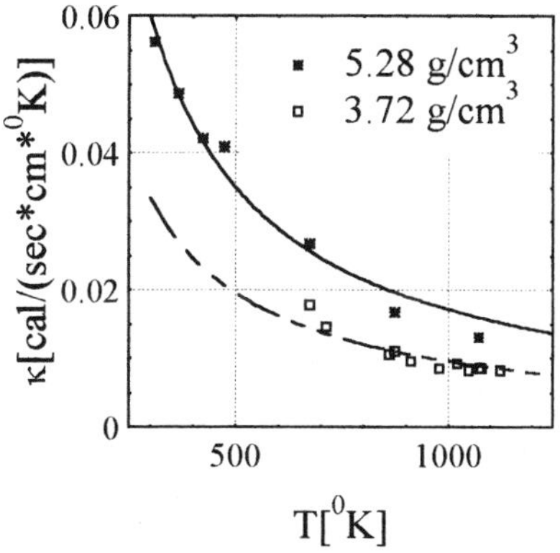

Figure 2. Heat conduction of ZnO versus temperature.

The two sets of experimental data correspond to two densities 5.28 g/cm^3 (95% of the theoretical density) and 3.72 g/cm^3 (67% of the theoretical density). Both sets of data were fitted to a curve of the form $a/(b+T)$. For the high density we found $a = 16.8$ and $b = -18.5$ and for the low density we found $a = 9.4$ and $b = -18.5$. According to Maxwell Garnet mixing theory [5] heat conductivity of porous materials κ_m, is related to that of the full dense material through: $\kappa_m = \kappa_c \frac{(d-1)\upsilon_c}{d-\upsilon_c}$ where κ_c is the heat conduction coefficient for fully dense ZnO ceramic, υ_c is the ratio of the density to that of theoretical density and d is either 2 or 3 for a 2D or 3D mixing model correspondingly. For $\upsilon_c = 0.95$ and $\upsilon_c = 0.67$ the ratio of the corresponding heat conduction coefficients is predicted to be 0.620 for d=3 and 0.557 for d=2. Note, that the ratio found from the fitted data is 0.559 which is in agreement with MGT.

The real and imaginary parts of the dielectric constant of fully sintered ZnO of theoretical density were measured at 2.2GHz [6] and are shown in Fig. 3 and Fig. 4 .

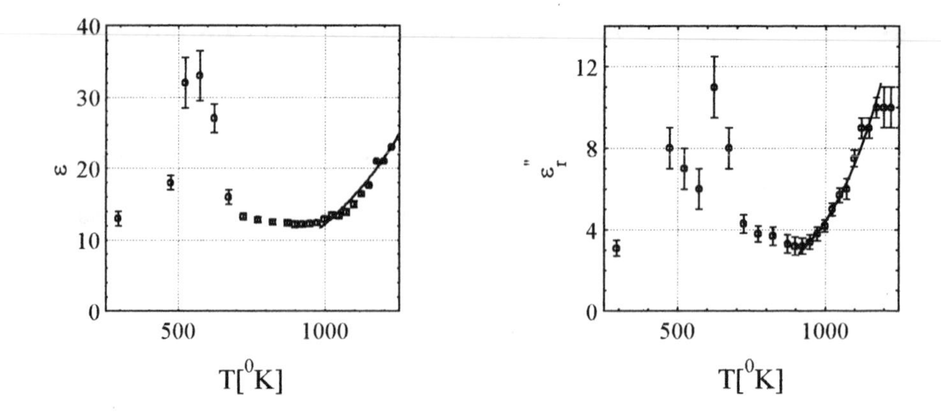

Figure 3. Relative real part of the dielectric constant of ZnO vs. temperature.

Figure 4. Relative imaginary part of the dielectric constant of ZnO vs. temp.

At temperatures above 900^{0}K we fitted the data to an exponential curve $ae^{(bT)}$. We found b = 2.82*10^{-3} for ε and b = 4.77*10^{-3} for ε$^{"}$. The exponential dependence on the temperature of ε$^{'}$ is corroborated by earlier data [7] where the exponent is found to be 2.65*10^{-3} (note that this data was taken at a low frequency of 1kHz). For the sample sintered in our laboratory the complex permittivity at room temperature of a fully sintered sample ε_{c0} was measured and found to be $\varepsilon_{c0}' = 19 \pm 2$ $\varepsilon_{c0}'' = 9 \pm 2$ (which is somewhat different from the data obtained in [6]). For our simulations we used the following:

$\varepsilon_c' = \varepsilon_{c0}'$ for T < 900^{0}K $\quad$ $\varepsilon_c' = \varepsilon_{c0}' \exp[0.00282*(T-900)]$ for T > 900^{0}K

$\varepsilon_c'' = \varepsilon_{c0}''$ for T < 900^{0}K $\quad$ $\varepsilon_c'' = \varepsilon_{c0}'' \exp[0.00477*(T-900)]$ for T > 900^{0}K

For the relative dielectric constant the dependence on the porosity we used MGT [5], namely: $\varepsilon_m = \varepsilon_c \dfrac{2\upsilon_c\varepsilon_c + (3-2\upsilon_c)}{\upsilon_c + (3-\upsilon_c)\varepsilon_c}$ where ε_c is the complex permittivity for fully dense ZnO. The permittivity at room temperature was also measured over the frequency range 0.2GHz to 20GHz [8] and the data fitted to a linear logarithmic curve:

$\varepsilon_{c0}' = -11.5 \log_{10} f(\text{GHz}) + 23.5$ and $\varepsilon_{c0}'' = -5 \log_{10} f(\text{GHz}) + 11.1$.

During the sintering process the density of the sample is continuously changing thus modifying the thermal and electrical properties throughout the sample. Hence, the heat transport equation should be supplemented by an equation relating the density change to the local temperature and the heating rate. In this work however the density was assumed to be a single valued function of the temperature. The density as function of the temperature was measured [8] and preliminary results are shown in Fig. 5. For temperatures above 650^{0}C and below 1100^{0}C a third order polynomial was fitted to the data. For temperatures above 1100^0 C we assumed the density to be linearly extrapolated to full density. Note that densification does not start before 650^{0}C. In solving the heat transport equation eq. (1), we have used the above fitted data for the density change with the temperature, supplemented with the condition that density is only increasing (solidifcation is irreversible).

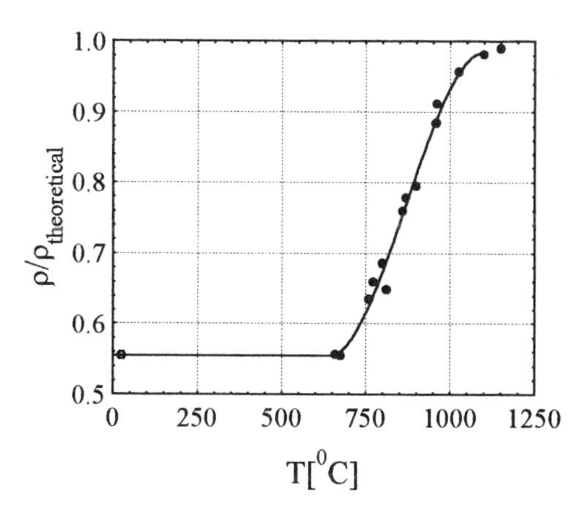

Figure 5. Relative density versus temperature.

SIMULATIONS AND EXPERIMENTAL RESULTS

Our simulation of the experiment set-up is shown schematically below.

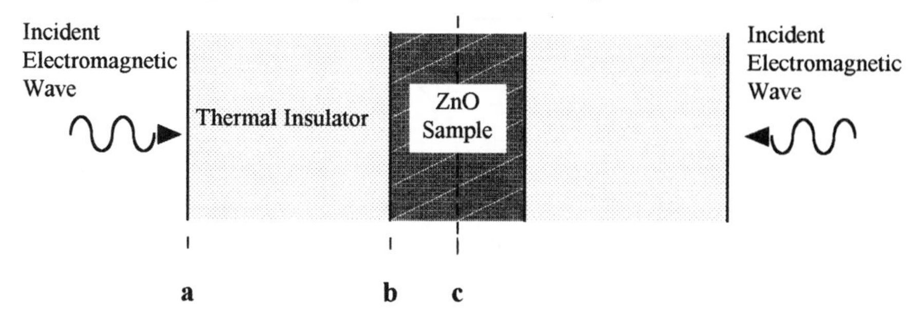

The cylindrical ZnO sample of height 1.65cm and diameter 3.3cm is represented by a 3.3cm thick slab sandwiched between 4cm thick slabs of low density porous ZnO ($\rho/\rho_{theoretical}$ =0.12) representing the thermal insulation (not undergoing sintering). Plane **a** is kept at room temperature and the simulation is symmetric with respect to plane **c**. The electric field is assumed to decay within the sample (from **b** to **c**) according to eq. (2). To compute the skin depth (eq. (3)) the value of $\tan\delta$ at **b** is used. In the experiment the input power g(t) had a step-like shape as shown in Fig. 6 and the surface and core temperatures were measured with two type K thermocouples. For this input power, eq. (1) was solved yielding the temperature distribution in the sample as function of time and, in particular, the temperature at the core (plane **c**) and at the surface (plane **b**). The measured and computed core temperatures are shown in Fig. 7, while the difference between the core and the surface temperature is shown in Fig. 8. The electric field E_0 was first estimated from the measured g(t) and then adjusted so that the maximum calculated core temperature (at 28800 sec) is approximately equal to the maximum measured temperature.

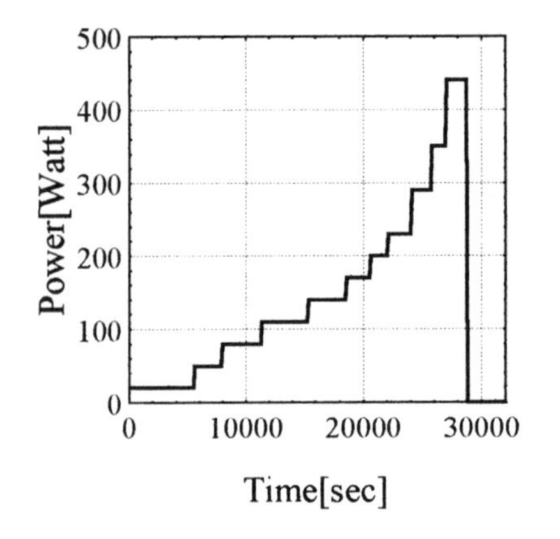

Figure 6. Microwave input power temporal profile.

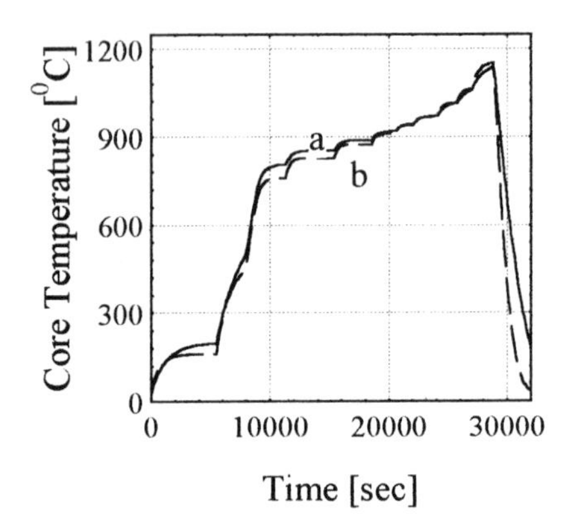

Figure 7. Core temperature a) measured b) computed (dashed line).

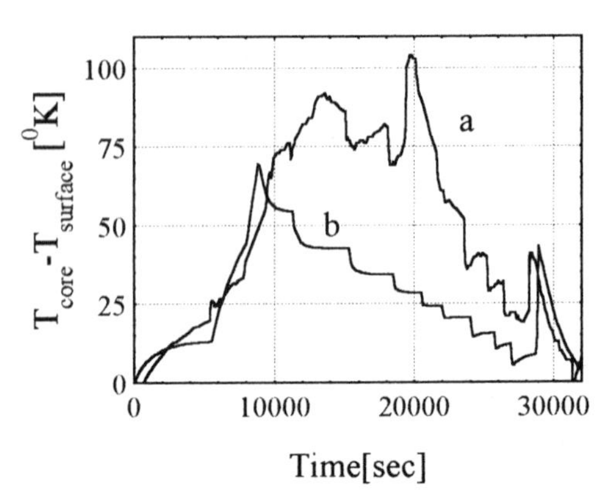

Figure 8. T_{core}-$T_{surface}$ vs. time a) measured and b) computed

We note that for the temperature difference further refinements may be needed to improve the modeling. This includes such issues as 1) more accurate spatial distribution of the electric field in the sample 2) improved modeling of the densification process 3) 2D computations (2D modeling of microwave sintering of Alumina had been reported in [9]).

APPLICATION OF TWO MICROWAVE FREQUENCIES IN THE SINTERING PROCESS.

One of the advantages of microwave sintering is higher heating rates. However, this might lead to nonuniform temperature distribution during processing. This nonuniformity may be fixed by applying microwave heating at two distinct frequencies simultaneously. Having shown that our code is capable of predicting behavioral trends of the sintering process we set forth to simulate an accelerated heating process. For example, a ZnO sample is heated for 300sec at a constant rate of about 200^{0}C per minute. The results of these numerical experiments are shown in Fig. 9. The curves in this graph are the difference between the core and surface temperature versus time for three cases: a) only a 2.45GHz microwave source is applied; b) only a 28GHz source is applied; c) both sources are applied.

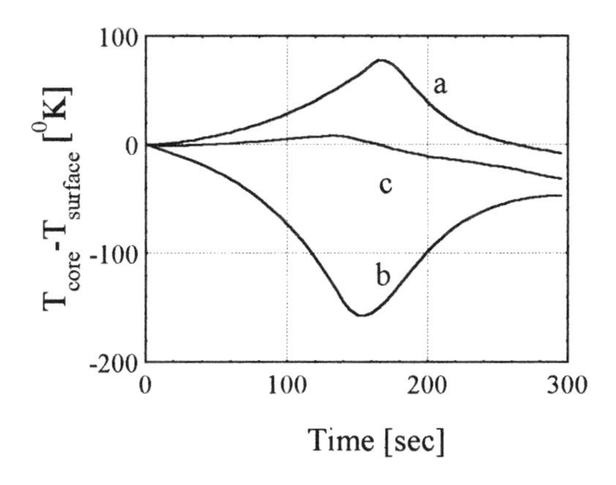

Figure 9. Calculated T_{core} - $T_{surface}$ vs. time a) 2.45GHz b) 28GHz c) 2.45GHz & 28GHz

For the low frequency and low temperatures the skin depth is 2.41cm which is larger than the sample's size, which therefore is heated almost uniformly. However, the conductive cooling of the surface causes the core temperature to exceed that of the surface. With the increase of the temperature, the skin depth along with the heat conduction decrease, thus increasing the temperature difference. For the high frequency the initial skin depth is 0.3cm which is smaller than the sample size. Therefore most of the power absorbed is close to the surface and the core is heated mainly by conduction, which causes the core's temperature to be lower. At higher temperatures the heat conduction increases due to the densification of the ceramic. Thus, the temperature gradient decreases in both cases. Using the two frequencies simultaneously, the temperature difference is reduced. In this example, the same temporal power profile was assumed for both frequencies; however, the temperature gradient can be further decreased by using different temporal power profiles.

CONCLUSION

Any attempt to simulate the sintering process requires detailed knowledge of the thermal and electrical physical parameters as function of the temperature, the porosity and the frequency applied, as well as the electrical field spatial distribution in the sample. This work is an attempt to model the sintering process of ZnO ceramic. The physical parameters dependence on ρ, T and f was in part based on measurements done in our laboratory and partly compiled from the literature. We found qualitative agreement with the experiment's results. We show that in the case of high rate sintering it is possible to achieve a history of uniform temperature distribution in the sample by applying two frequencies simultaneously.

ACKNOWLEDGEMENT

Work supported by U.S. Department of Energy Division of Advanced Energy Projects.
We are grateful to J. Binner and T. Cross for allowing us to use their facilities at the University of Nottingham. In particular, many thanks to J. Binner for making it possible to have D. Gershon perform the measurements. We would also like to thank V. Semenov for helpful discussions and comments.

REFERENCES

1. M. F. Iskander, "Computer modeling and numerical simulation of microwave heating systems" MRS Bulletin **18** pp 30-36 (1993), and references therein.
2. G. A. Kriegsmann, "Thermal runaway in microwave heated ceramics" J. Appl. Phys. **71** (4) pp 1960-1966 (1992).
3. Y.S. Touloukian, R.W. Powell, C.Y. Ho, P.G. Klemens, "Thermophysical Properties of Matter" The TPRC Data Series (Plenum New-York 1970).
4. M.P. Shascalski, "Akustichekie Kristally" Nauka, 1982.
5. W.D. Kingery, H.K. Bowen, D.R. Uhlmann, "Introduction to Ceramic" John Wiley & Sons, Inc, New York 1975.
6. D. Gershon, J. Binner, T. Cross, and N. Greenacre (Private communication)
7. E.F. Tokarev, I.B. Kobyakov, I.P. Kuzmina, A.N. Lobachev, G.S. Pado, "Dielectric and Piezoelectric Properties of Zincite in the 4.2-800K temperature Range" Sov. Phys. Solid State **17**, No. 4 (1975).
8. L.P. Martin, D. Dadon, D. Gershon, B. Levush, Y. Carmel and M. Rosen, "Characterization of microwave processed ZnO" Paper presented in this conference.
9. J.R. Thomas, J.D. Katz and R.D. Blake, "Temperatue distribution in microwave sintering of Alumina cylinders" Mat. Res. Symp. Proc. **347** p 311 (1994)

TEMPERATURE DISTRIBUTION CONSIDERATIONS DURING MICROWAVE HEATING OF CERAMICS

Michael T.K. Koh, Rajiv K. Singh and David E. Clark
Department of Materials Science & Engineering
University of Florida
Gainesville. FL 32611

ABSTRACT

A simulation model based on the accurate solution of Maxwell's equations as the power source term is used to investigate the microwave heating of ceramics. The computer program dynamically couples the power absorption calculated from Maxwell's equations with the heat transfer equation and includes temperature varying parameters. Simulations are done to study the optimal conditions in the microwave heating of alumina and silicon carbide. The effects of hybrid heating and the size of the work-piece are also considered.

INTRODUCTION

Microwave heating has been identified as an alternative to conventional heating in the processing of ceramics. Shorter processing times and its unique heating characteristics are some advantages of microwave heating. Problems faced in microwave heating of ceramics include poor coupling of microwave energy with many ceramics and the occurrence of thermal runaway. Because of these problems, there is a need for detailed studies on solutions for better microwave processing of ceramics. Computer simulations enable a close examination of sample temperature profiles heated under various processing conditions.

Alumina (Al_20_3) and silicon carbide (SiC) were selected for the present study as representing weak and strong absorbers of microwave radiation, respectively. In terms of heating efficiency, alumina and silicon carbide are almost the antithesis of each other. Silicon carbide's good absorption properties lead it to be used to line the interior of microwave heating chambers [1] or as radiating posts in picket fence arrangements [2] in the hybrid heating of alumina and other materials that are poor microwave absorbers.

SIMULATION MODEL

The simulation model employs the finite difference method which has proven to be an accurate and flexible technique. The temperature dependence of crucial input parameters such as dielectric permittivities, thermal conductivities and volumetric heat capacities [3-7] were considered in the model. Considering the thermal variations of these parameters was especially important as the permittivities generally vary rapidly at higher temperatures and have a significant influence on the heating behavior of the material.

Another important distinctive feature of the model was the accurate determination of the power distribution from Maxwell's equations. The accurate determination of the electric field (from which the power absorbed is calculated) should give confidence in using the simulation in studies of microwave/material interactions. It can be seen that the magnitude of the electric field within the sample varies greatly within the sample, a point which is sometimes neglected in other simulations. Furthermore, the use of Lambert's law approximation in assuming that the power absorption within the material follows an exponential decay relationship is accurate only for large samples, or where the penetration depth is small.

In a simulation cycle, all temperature dependent parameters are first initialized at the ambient temperature of 300K. The electric field was calculated from Maxwell's equations followed by the calculation of the absorbed power per unit volume through the formula determined from the Poynting theorem. This value was then dynamically passed to the heat conduction equation as the power source term in order to calculate the temperature within the sample. The heat conduction equation was solved through the use of an accurate implicit finite difference technique [8]. At regular time intervals, the temperature-dependent parameters are recalculated at the new temperatures at various positions within the sample and the heat absorbed per unit volume was reevaluated. The effects of heat transfer through radiation and convection were considered through the boundary conditions.

RESULTS AND DISCUSSION

From the simulation, temperature profiles were obtained for different sample lengths, incident power values and frequencies for alumina and silicon carbide. Variation of radiation frequency was considered as a solution for heating weakly absorbing materials. In addition, for alumina, heating at different elevated ambient temperatures was simulated to investigate the effect of having a secondary external heating source.

Sample Size

Thermal runaway occurred earlier, but at higher temperatures for the 6cm sample as compared to the 12 cm sample as shown in Figure 1. It was also observed that the rapid increase in surface temperature occurs within the range

1000K to 1500K. Relatively slow but uniform heating of alumina occurs in the early stages (Figure 2). However, at later stages heating becomes rapid and the center is preferentially heated at a much higher temperature with the development of significant temperature gradients, quickly leading to thermal runaway conditions, Figures 1 and 2. Heating within this temperature range creates regions of uneven heating with the formation of temperature peaks within the sample (Figure 2). The peak temperatures are often above alumina's melting point (~2050K) and frequently leads to serious material damage and thermal stresses. Thermal runaway is directly linked to a jump in the absorption of microwave energy. Low initial power absorption however means a large portion of incident energy is not utilized in heating the sample.

For silicon carbide, the temperature profile peaks near the face of the sample where radiation is incident (Figure 3). This is expected for the strongly absorbing SiC, with the strongest absorption near to where radiation is incident. The stronger absorption at the surface is different from the results generally seen for alumina where the highest temperature occurs close to the center.

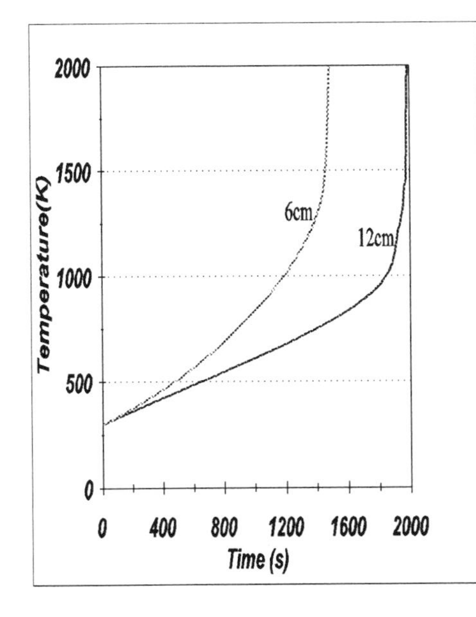

Figure 1. Surface temperatures for 2 alumina samples of 6 and 12 cm.

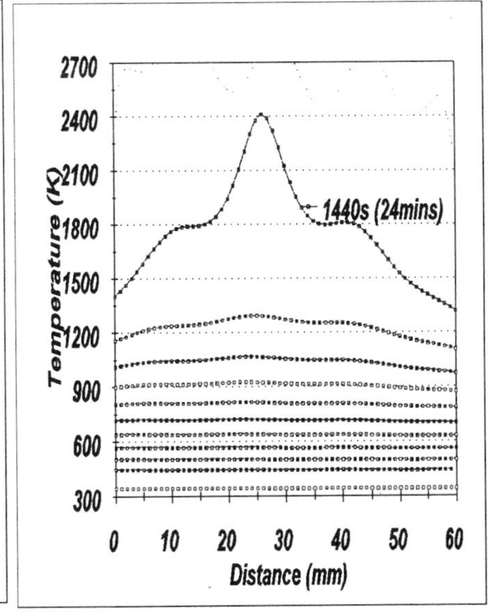

Figure 2. Temperature profiles at 2 min intervals.

The temperature peaks do not occur at the surface of the sample itself as heat is lost more rapidly through radiation and convection [5]. The heating profiles of four different sample lengths in Figure 3 show the peaks to be about the same distance (about 10cm) from the face of incident radiation. However the average temperature calculated at 1 500s is higher for the shorter sample due to the higher temperatures attained by the sample. The power absorption curves show a highly attenuated oscillatory behavior (Figure 4). The power absorption and its oscillatory nature are also increasingly damped with time. This is due to the decrease of penetration depth with increasing temperature. Depending on the thermal conductivity of the material, this can lead to an increasing heating peak near the surface.

The strong attenuation of the absorbed power within a silicon carbide sample cautions against assuming uniform electric field in cases of strongly absorbing materials, large samples or high frequency radiation. The power absorption for alumina also shows highly irregular profiles **within the sample.**

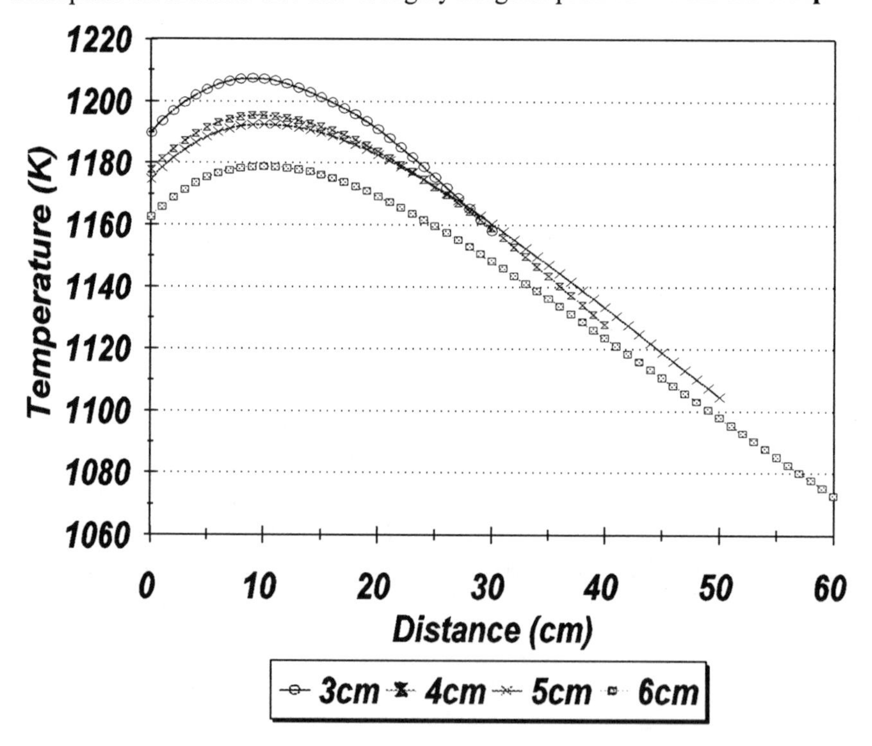

Figure 3 . Comparison of temperature profiles for different SiC sample lengths at 1500s.

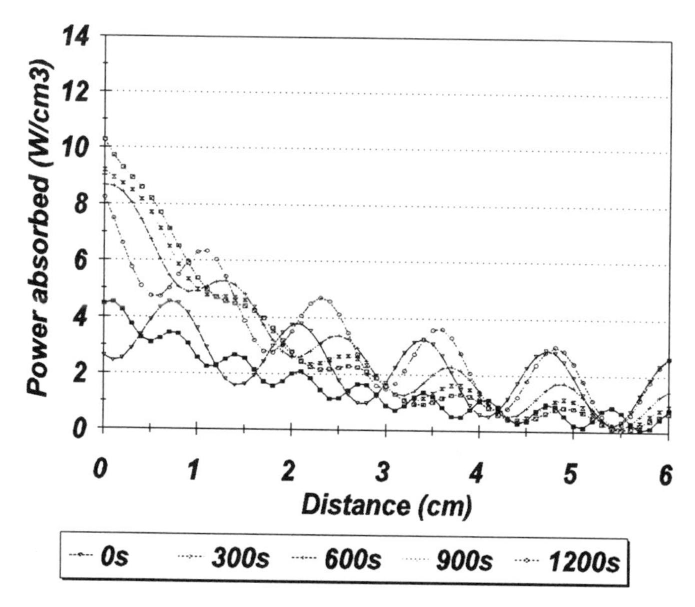

Figure 4. Power absorption for a 6cm SiC sample.

Incident Power

A comparison of sample temperature profiles for similar average temperatures for alumina (Figure 5) shows a slight improvement in temperature uniformity for lower power inputs. The trade off is longer heating times. The heating rate itself increases with time as shown in Figure 6 and is consistent with the increasing gradient seen for the rise of temperature with time (Figure 1). At 10 mins heating time, the heating rate varies almost linearly with the input power. This relationship changes to an exponential variation type after an additional 10 mins. A balance can thus be struck between having a high heating rate and yet avoiding thermal runaway and heating nonuniformity.

Higher incident power in silicon carbide causes a higher steady-state temperature to be reached in a shorter time as shown in Figure 7. With increasing incident power, the increase in temperature rate also decreases. Heating uniformity is sacrificed at higher powers. Figure 8 shows average temperatures for some input powers. The first example at about 1900K, where the heating profile at $400W/cm^2$ shows a steeper gradient, even though it is only 12K lower in its average temperature as compared to its $300W/cm^2$ counterpart (at 1918K). The nonuniformity of higher incident power appears to be greater initially as seen in the

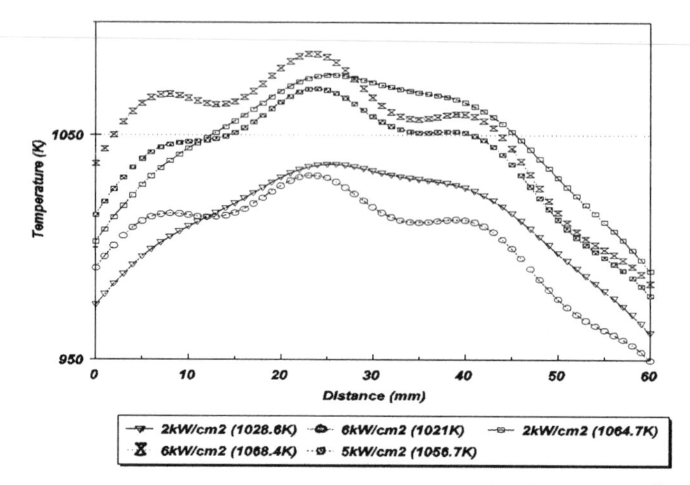

Figure 5. A comparison of heating uniformity in alumina at similar average temperatures.

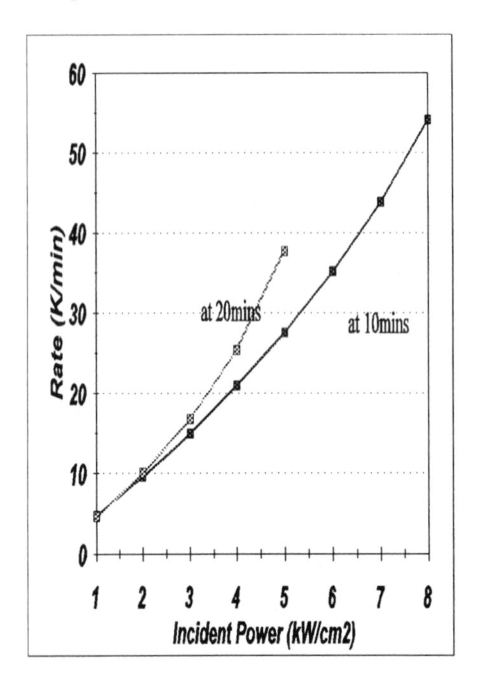

Figure 6. Heating rates for 6 cm alumina at 2.45 GHz.

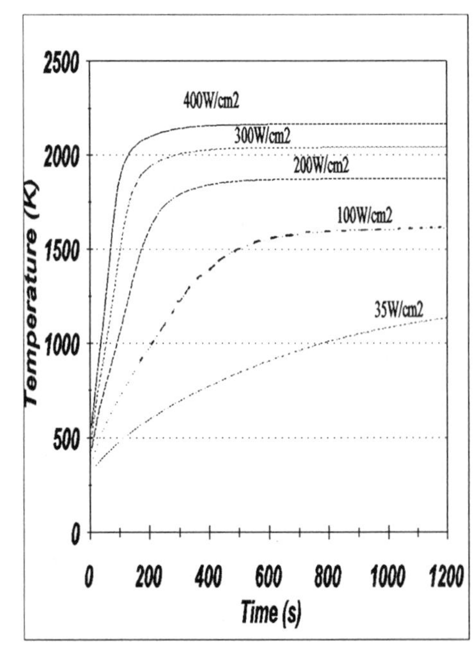

Figure 7. Variation of power input for a 4 cm SiC sample.

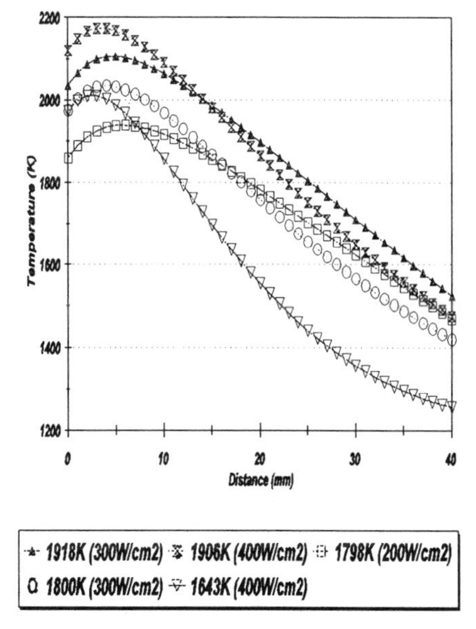

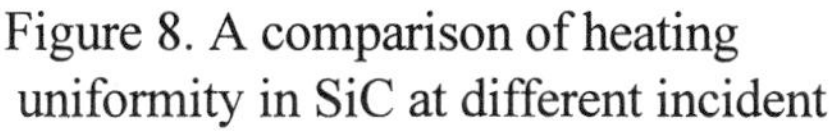

Figure 8. A comparison of heating uniformity in SiC at different incident

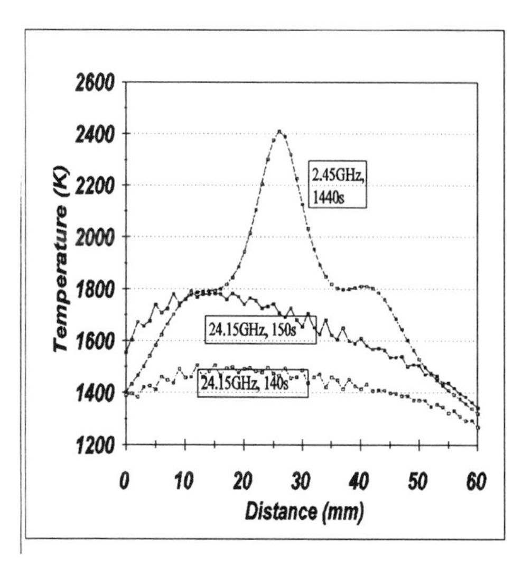

Figure 9. A comparison of heating uniformity of 2.45GHz and 24.15 GHz.

second situation at the average temperature of about 1800K. Here temperature uniformity is greatest at $200W/cm^2$ followed by $300W/cm^2$ and lastly by $400W/cm^2$. The temperature peak also shifts toward the center with decreasing power.

Frequency

For alumina, the penetration depth for 24.15GHz microwave radiation is about 10 times less than that of the 2.45GHz radiation. Even at the temperature of about 2000K, the penetration depth at 2.45GHz is about 30cm, which is considerably larger than the usual workpiece in a microwave heating run. Therefore though heat absorption is considerably uniform throughout the sample length, power absorption efficiency is very poor. The penetration depth however for at 24.15GHz is generally below 10 cm which enables more efficient absorption of microwave energy.

Comparing the respective profiles for 2.45GHz and 24.15GHz, Figure 9, shows that below runaway conditions at about the same temperature, higher temperature uniformity can be obtained for 24.15GHz radiation. Higher frequencies also result in higher heating rates as it takes only 2mins for the 24.1 5GHz radiation to achieve a temperature of 1000K compared with 20mins for the 2.45GHz radiation. Great care is needed to control the temperature at higher frequency however in view of the rapid increase in the rate of heating leading to thermal runaway.

Ambient Temperature

The effect of raising the ambient temperature is rather dramatic. With the ambient temperature at 700K, the temperature range before thermal runaway occurs spans the temperatures for solid state processing (Figure 10). The incident power input required for this temperature range varies from 2kW/cm^2 to 3kW/cm^2 In comparison, in the absence of hybrid heating, the same temperature range could be achieved only at incident power greater than 5kW/cm^2. Johnson [9] also reported the increase of the critical temperature with an increase in ambient temperature. This can also be seen from Figure 10. Thus higher ambient temperatures enable better control of thermal runaway. Results published in these proceedings also demonstrate that while insulation leads to earlier onset of thermal runaway, the addition of a susceptor actually impedes the occurrence of thermal runaway [10].

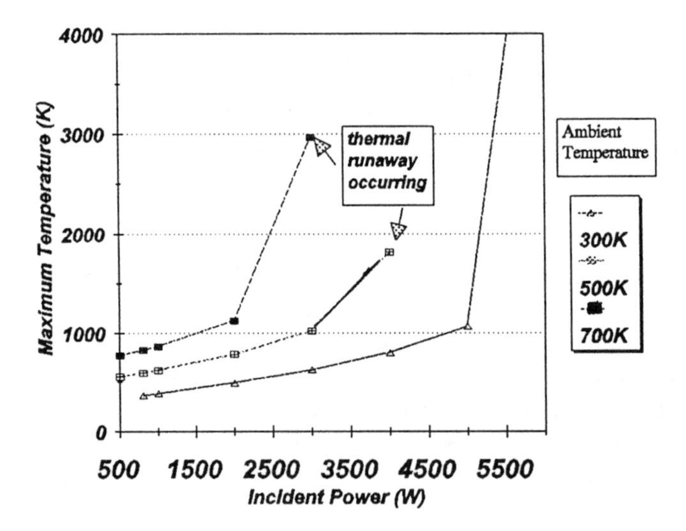

Figure 10. Effect of ambient temperature on maximum temperature reached in alumina sample.

CONCLUSIONS

The heating temperatures in the cases of SiC studied in the simulation characteristically reach a point of steady state [11]. Porada [12] also reports no thermal runaway observed for a SiC/ZrO_2 sample regardless of heating rates, in contrast to thermal runaway observed for a Al_2O_3/ZrO_2 sample.

In the present simulation, input data was obtained from various sources where the material was in some cases assumed to be dense. Future simulation could be done with input data of green material for more accurate investigation of ceramic sintering.

REFERENCES

1. A. De, I. Ahmad, E.D. Whitney, and D.E. Clark, Microwave Processing of Materials II (W.B. Snyder Jr., W.H. Sutton, M.F. Iskander, and D.L. Johnson, eds), MRS, Pittsburgh, PA, pp. 283-288 (1990).
2. M.A. Janney, C.L. Calhoun, and H.D. Kimrey. *J. Am. Cer. Soc.* 75, pp. 341-346 (1992).
3. Gitzen, W.H. Special Publication #4. American Ceramic Society (1970).
4. R.M. Hutcheon, M.S. De Jong, F.P. Adams, P.G. Lucuta, J.E. McGregor, and L. Bahen, Microwave Processing of Materials III (R.L. Beatty, W.H. Sutton, and M.F. Iskander, eds), MRS, Pittsburgh, PA, pp. 371-378 (1992).
5. R.K. Singh, J. Viatella, Z. Fathi, and D.E. Clark, Microwaves: Theory and Application in Materials Processing II (D.E. Clark, W.R. Tinga, and J.R. Laia, Jr., eds), American Ceramic Society, Westerville, OH, pp. 247-255 (1993).
6. G.W.C. Kaye and T.H. Laby, ed. Tables of Physical and Chemical Constants. 15th ed., Longman (1986).
7. W. Xi and W.R. Tinga, Microwaves: Theory and Application in Materials Processing (D.E. Clark, F.D. Gac, and W.H. Sutton, eds) Am. Ceramic Society, Westerville, OH, pp. 215-224 (1991).
8. R.K. Singh and J.Narayan, Materials Science and Engineering B3, pp. 217-230 (1989).
9. D.L. Johnson, D.J. Skamser, and M.S. Spotz , Microwaves: Theory and Application in Materials Processing II (D.E. Clark, W.R. Tinga, and J.R. Laia Jr., eds), American Ceramic Society, Westerville, OH, pp. 133-146 (1990).

10. D.J. Skamser, J.J. Thomas, D.L. Johnson. To be published in these proceedings (1995).
11. E.P. Bescher, U.Sarkar, and J.D.Mackenzie, Microwave Processing of Materials III, ed. by R.L. Beatty, W.H. Sutton, and M.F. Iskander, pp. 371-378 (1992).
12. M. Willert-Porada, Microwave Processing of Materials IV, edited by M.F. Iskander, R.J. Lauf, and W.H. Sutton, MRS, pp. 31-43 (1994).

FINITE ELEMENT COMPUTER MODEL OF MICROWAVE HEATED CERAMICS

Liqiu Zhou,* Gang Liu,* Jian Zhou* and Jiping Cheng**
*Wuhan University of Technology - Wuhan 430070 P. R. China
**The Pennsylvania State University - University Park, PA

ABSTRACT

In this paper, a 3-D finite element model to simulate the heating pattern during microwave sintering of ceramics in a TE_{10}^{n} single mode rectangular cavity is described. A series of transient temperature profiles and heating rates of the ceramic cylinder and cubic sample were calculated versus different parameters such as thermal conductivity, dielectric loss factor, microwave power level, and microwave energy distribution. These numerical solutions may provide a better understanding of thermal runaway and solutions to microwave sintering of ceramics.

INTRODUCTION

Recent work has addressed the simulation and modeling of microwave heating [1,2]. Several numerical methods have been proposed, e.g., Finite Element Method (FEM), Finite Difference Method (FDM), Methods of Moments (MOM) and Transmission Line Matrix (TLM). The Finite Element Method is more convenient to deal with a special of complex geometry ceramic body than the Finite Difference Method. This method was selected to simulate and calculate the heating pattern of microwave sintering.

The differential equation as well as the boundary and initial conditions which describe the heating pattern in a ceramic sample during microwave heating are given below:

$$K\nabla^2 + Q_V = \rho C_p \frac{\partial T}{\partial t} \quad (1)$$

$$-K \frac{\partial T}{\partial n}\Big|_s = H(T_s - T_f) \quad (2)$$

$$T\big|_{t=0} = T_0(x,y,z,0) \quad (3)$$

where, T is the temperature in the (x,y,z) location at time t, K is the thermal conductivity, ρ is density of the ceramic material, C_p is its heat capacity at constant pressure, S is the boundary domain in which connective and/or radiant heat exchange occurs, H is heat exchange coefficient for all heat exchange taking place on the boundary domain S, and Q_v is the volumetric heat created by dissipation of microwave energy.

The equation group (1-3) is considered to be equivalent to the corresponding extreme value function of functional J on the initial condition (3)[3]:

$$J[T(x,y,z,t)] = \iiint [[\frac{K}{2}(\frac{\partial T}{\partial x^2} + (\frac{\partial T}{\partial y})^2 + (\frac{\partial T}{\partial z})^2] - Q_v\, T + \rho C_p \frac{\partial T}{\partial t} T] dv +$$

$$\iint H(\frac{1}{2}T_s^2 - T_f T_s) ds \quad (4)$$

The solving domain is divided into E space elements with network nodes. In each of these space elements an element functional J is defined:

$$J = \sum_{e=1}^{E} J_e \quad (5)$$

A linear interpolation function with the unknown node temperatures can approximate the temperature in the elements, so the variation problem of J is transformed into the extremum problem of the interpolation function of several variables. Therefore, one can write (where k = 1,2,3....., N):

$$\frac{\partial J}{\partial T_k} = \frac{\sum_{e=1}^{E} J_e}{\partial T_k} = \sum_{e=1}^{E} \frac{\partial J_e}{\partial T_k} = 0 \qquad (6)$$

The time variable, t, is treated with the backward difference method that is where T^t and $T^{t-\Delta t}$ are temperatures at time t and t-Δt, respectively.

$$\frac{\partial T}{\partial t}\Big|_t = \frac{T^t - T^{t-\Delta t}}{\Delta T} \qquad (7)$$

In view of the above, equation (6) can be expressed:

$$([K] + \frac{[N]}{\Delta T})(T)^t = (P) + \frac{N}{\Delta T}(T)^{t - \Delta T} \qquad (8)$$

where, [K] is a temperature rigid matrix, [N] is a temperature changing matrix, $(T)^t$ is node temperature vector at time t and (P) is the vector embodying boundary conditions, surface temperature, and other parameters that vary with time.

Introducing the initial condition (3), equation (8) can be solved using the marching integration method.

COMPUTER SCHEME

The computer plan was carried out on the basis of the material parameters of alumina which possesses rather extensive valuing range both in dielectric loss factors and thermal conductivity.

Sintered samples with two different geometries were investigated. The first was a cylinder (see Figure 1). To simplify the problem, a 2-D axis--symmetric mathematical model was adopted. The electric field distribution was as shown below (a =60mm) :

$$E_y = E_{y0}\sin(\frac{\pi}{a}X + \frac{\pi}{2}) \qquad (9)$$

The second was a cubic sample, the 3-D mathematical model used in this case. The sample shape and the couple electric field pattern are shown in Figure 2.

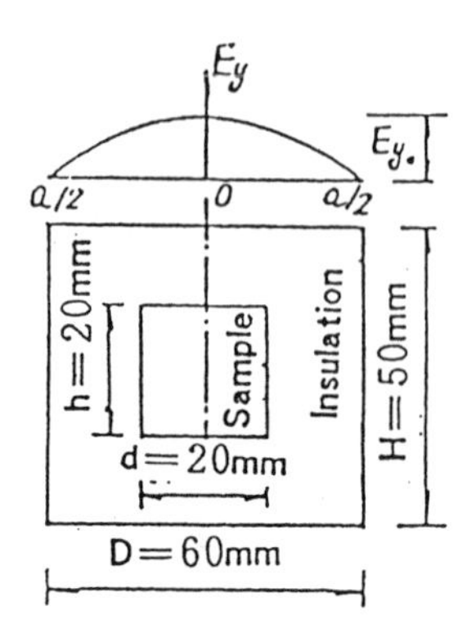

Figure 1. Schematic of 2-D axis-symmetric cylinder sample and coupled electric field pattern.

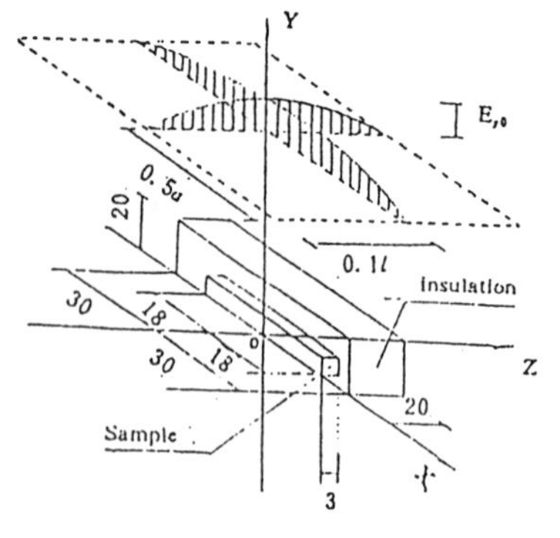

Figure 2. Schematic of 3-D cubic sample and coupled electric field pattern. Cube size: 6x6x36mm.

The electric field distribution is

$$E_y = E_{y0}(\frac{\pi}{a}x + \frac{\pi}{2})\sin(\frac{5\pi}{L}z + \frac{\pi}{2}) \tag{10}$$

where a = 150 mm, L = 275mm.

The dielectric loss factor, ϵ'', is an important parameter for computing intensity of the interior heat Q_v. The variation of ϵ'', which depends on temperature [4, 5], can be expressed as function of time. Figure 3 shows two ϵ'' t curves accepted in the computer program.

Thermal conductivity, as a temperature function, can be introduced into the program directly. The K-T curve [6], accepted in the computing is shown in Figure 4. The computing schemes are shown in Tables 1 and 2.

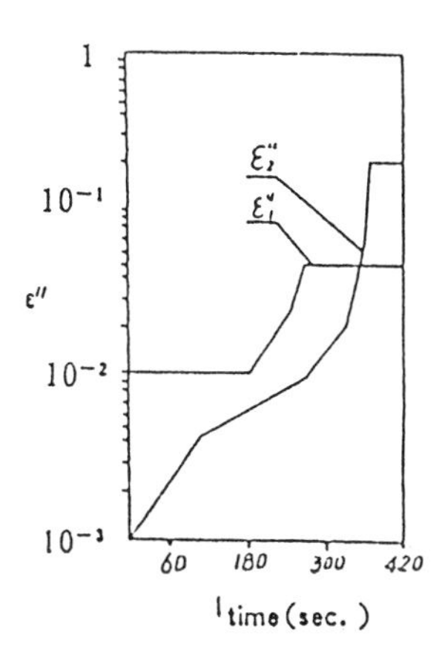

Figure 3. Relationship between dielectric loss factor, ϵ'', and time, t.

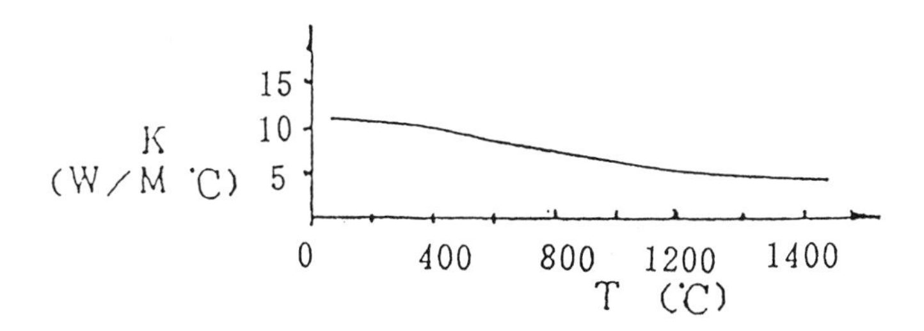

Figure 4. Relationship between dielectric loss factor,thermal conductivity, K, and temperature, T.

Table 1. The Computing Scheme for the Cylindrical Sample.

Sample #	**Microwave Power (W)**	**Dielectric Loss Factor, ϵ''**	**Thermal Conductivity, K, W/M°C**	**Microwave Field Pattern**
1	500	ϵ'' 1	10	Eqn. 8
2	500	ϵ''1	5	Eqn. 8
3	500	ϵ''1	10	$E_y = E_{y0}$ (constant)
4	500	ϵ''2	10	Eqn. 8
5	300	ϵ''1	10	$E_y = E_{y0}$ (constant)
6	300	ϵ''1	10	Eqn. 8
7	300	ϵ''1	5	Eqn. 8

Table 2. The Computing Scheme for the Cubic Sample.

Sample #	**Microwave Power (W)**	**Dielectric Loss Factor, ϵ''**	**Thermal Conductivity, K, W/M°C**	**Microwave Field Pattern**
8	500	ϵ'' 1	10	Eqn. 9
9	500	ϵ''1	K=K(T)	Eqn. 9
10	500	ϵ''2	K=K(T)	Eqn. 9
11	500	ϵ''1	K=K(T)	Eqn. 9

The thermal insulation was assumed to be transparent to microwaves, the thermal conductivity, K = 0. 1W/m°C, the heat capacity, C_p=0. 96232 x 10^3 J/kg°C and the density, ρ = 0. 4 x 10^3 kg/ m^3.

RESULTS AND DISCUSSION

The numerical solutions show that the heating course can be divided into two stages according to rate. In each one, the T--t curve is regarded as approximately linear. During the first stage, the heating rate isrelatively low and was affected by both the input power and the dielectric loss factor. During the second stage, the heating rate was high and determined primarily by the dielectric loss factor because of its dependence on temperature.

Elevation of dielectric loss factor during the heating process was the major reason for thermal runaway. It can be seen from Figure 5 (B) and (E) that thermal runaway occurred for both the cylinder sample no. 4 and the cubic sample no. 10 due to the fact that ϵ'' was equal to 2 for both samples. From the $\epsilon''2$ - t curve shown in Figure 3 and Figure 5 (B) and (E), it can be seen that the varying tendency of $\epsilon''2$ with time is similar to the those of samples no. 4 and no. 10 after 320 sec. This reflects the effect of ϵ'' on thermal runaway. In other words, when the samples reach a certain temperature range, the intensity of interior heat, Q_v, will increase in an erratic manner for a short period of time due to the increase in ϵ'' even if the microwave power remains constant. Thus thermal runaway occurs.

On the basis of the computer results for the cubic and cylindrical samples (see Figure 6 (A, B) and Figure 7 (A, B)), it is known that thermal conductivity exerts a basic influence on the temperature gradient inside the sample but has little influence on the heating rate. Since thermal conductivity varies with the temperature, it should be considered in the program for cases requiring accurate temperature distribution. Otherwise, it would be better to treat thermal conductivity as a constant (see Figure 5 (D)).

Homogeneity of the microwave electric field gives some positive affect on decreasing temperature difference inside the sample. Figure 6 (A,C, E, F) shows that a uniform electric field can change the shape of isothermal line and make variation of temperature gradient tend toward gentleness. In the view of the heat generating and scattering, sample shape and volume also have some effects on the temperature distribution. It can be seen from Figures 6 and 7, that there was a significant difference in the temperature distribution for the cylindrical and cubic samples even when all other computing parameters were kept constant.

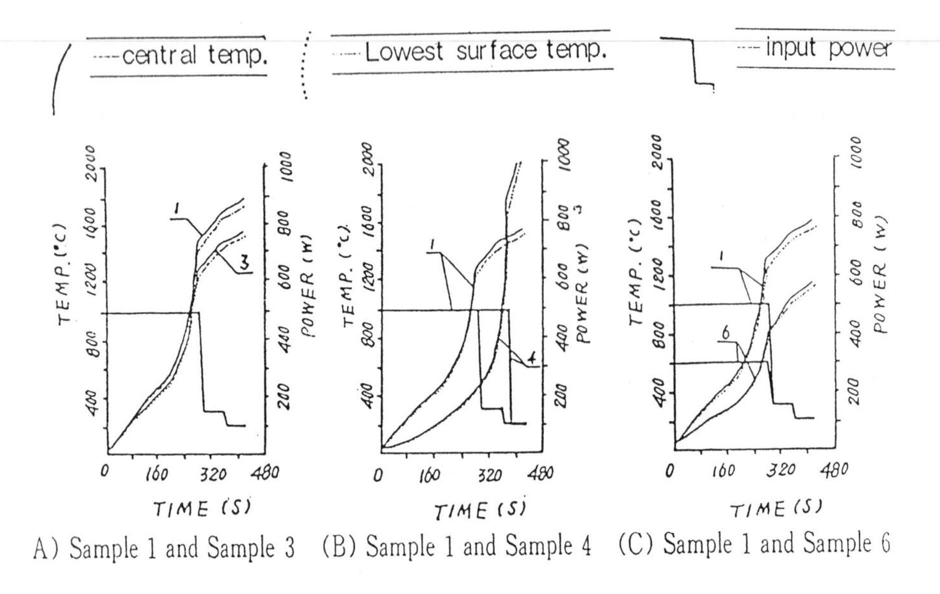

A) Sample 1 and Sample 3 (B) Sample 1 and Sample 4 (C) Sample 1 and Sample 6

(D) Sample 8 and Sample 9 (E) Sample 9 and Sample 10 (F) Sample 9 and Sample 11

Figure 5. Relationship between sample temperature, T, and sintering time, t. The computing condition coupled with the sample number can be seen in Tables 1 and 2.

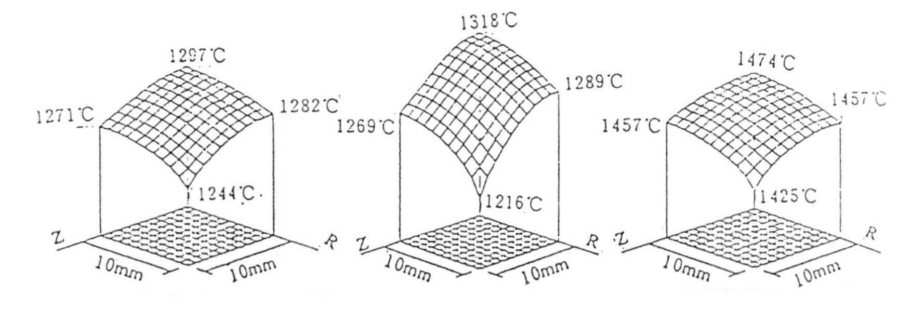

(A) Sample 1 for 280 sec. (B) Sample 2 for 280 sec. (C) Sample 3 for 280 sec.

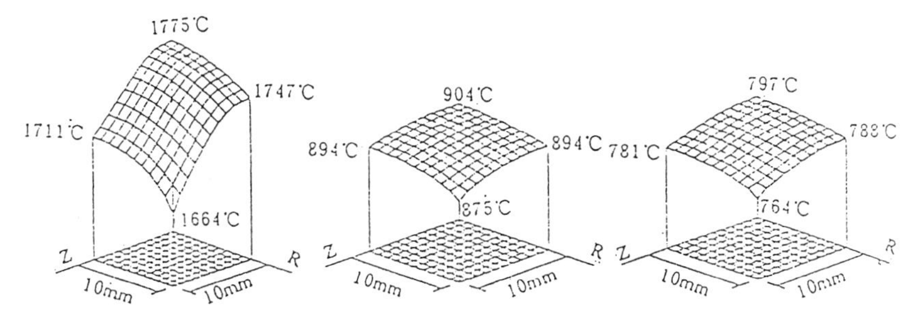

(D) Sample 4 for 370 sec. (E) Sample 5 for 280 sec. (F) Sample 6 for 280 sec.

Figure 6. Temperature distributions in the cylinder samples on the coordinate plane (r, 0, z).

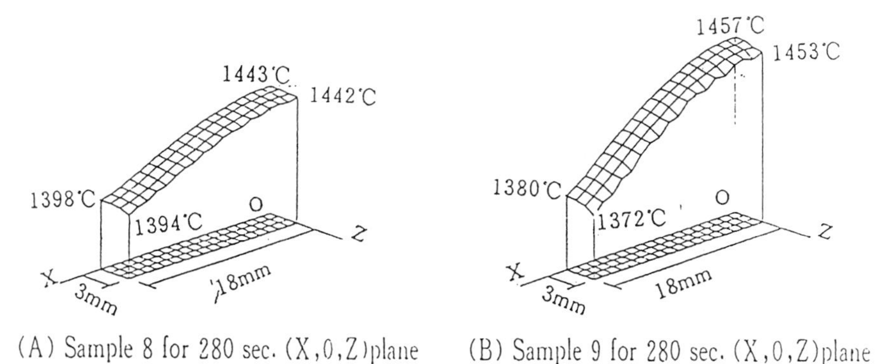

(A) Sample 8 for 280 sec. (X,0,Z)plane (B) Sample 9 for 280 sec. (X,0,Z)plane

Figure 7. Temperature distribution for the cubic sample on the coordinate plane (X,O,Z) (see also Figure 2).

CONCLUSIONS

The actual sintering process is more complex and deals with many nonlinear factors. Accounting for all these factors in a computer program is challengingand few alternatives are currently available to simplify the problems. Nevertheless, the numerical solutions given here can reflect the varying features of the microwave heating processes on the whole. The mathematical model and relevant computing methods described in this paper are not limited to single mode cavities. The program is adaptable to initial and boundary condition and to a variety of material parameters.

REFERENCES

1. *Y.* Tian, J. Feng, L. Sun and C. Tu in, **Microwave Processing of Materials III**, (R.I . Beatty, W.H. Sutton and M.F. Iskander, eds), Mater. Res. Soc. Proc., Vol. 269, pp. 41-46 (1992).
2. J. Tucker, M.F. Iskander and Z. Huang in, **Microwave Processing of Materials IV,** (M.F. Iskander, R.J. Lauf and W.H. Sutton, eds), Mater. Res. Soc. Proc., Vol. 347, pp. 353-362 (1994).
3. X. Kong, **Finite Element Method for Heat Transfer**, 2nd edition, Chinese Press House (1984).
4. J. Cheng, Ph.D. Dissertation, Wuhan University of Technology (1991).
5. J. Zhou and J. Cheng in, **Microwave Sintering Processing,** Proc. Of International Conference on Microwave Communications, P.R. China, pp. 698-701 (1992).
6. S. Youyichi, **Ceramic Handbook**, Chinese Version, Chinese Light Industry Press House, p. 715 (1980).

Joining

MICROWAVE JOINING OF ENGINEERING CERAMICS

JGP Binner[†], PA Davis[†], TE Cross[‡] and JA Fernie[¶]
† Department of Materials Engineering and Materials Design
‡ Department of Electrical and Electronic Engineering
The University of Nottingham, University Park, Nottingham, UK
¶ TWI, Abington Hall, Abington, Cambridge, UK

ABSTRACT

A TE_{102} single mode microwave applicator has been used to examine the potential for joining a range of engineering ceramics without the use of interlayers. Materials examined included silicon carbide, alumina and zirconia. Experimental variables included both temperature and pressure whilst joining time was kept to a minimum, typically ≤ 10 minutes. Successful joins were produced with reaction bonded SiC, medium and low purity Al_2O_3 and ZrO_2. Analysis of the results has led to the conclusion that it is possible to achieve successful microwave joins by both solid state and liquid phase mechanisms.

INTRODUCTION

The ability to form rapid, high quality joins between a range of ceramic materials using microwave energy has been demonstrated by a number of researchers[1-16] using two main approaches, viz direct and indirect joining. In the latter case, use is made of some form of interfacial layer to assist in achieving bonding. Ceramics which have been investigated to date include Al_2O_3[1-11], mullite[4], Si_3N_4[1,2,4,10] and SiC[11-16]. Rods, tubes and bars of varying size have all been joined.

Fukushima et al[1,2] and Davis et al[3] have investigated the microwave joining of different grades of alumina. In both cases, relatively low purity materials ($\leq 96\%$) could be joined readily whilst high purity ceramics ($\geq 99\%$) were insufficiently lossy to heat, making direct joining difficult. Characterisation of the joined samples showed little difference before and after joining; no grain growth or cracking was evident. Similar results were found by Palaith et al[4] and in all cases evidence was found that the grain boundary phases had been melted during the process. Strengths of the joined materials were at least as strong as, and often stronger, than the base material[1,2,4]. The inability to achieve direct joining in high purity alumina led to experiments involving lower purity[1,2] interlayers. Using this approach, successful joins were achieved with strengths approaching 90% of the base material. Sealing glasses have also been tried and found to be successful[4-6] as have alumina gels[7-9]. The latter have the advantage that, at the joining temperature, the gel transforms into colloidal α-alumina which subsequently sinters providing consistency across the join region. The use of alumina powder intermediaries[10] was less successful in that, although high strengths were achieved, results were more erratic due to shrinkage cracks caused during sintering within the intermediary. Attempts to simultaneously sinter and join green ceramics with a slip interlayer were more successful[11].

Successful microwave joins have also been produced in mullite[4] and silicon nitride[1,2,4,10]. Palaith et al[4] found that in the region of the join the hardness and toughness were both increased by 20-30%. This was attributed to a localised reduction in porosity caused by the melting of a glass phase in the mullite. Join strengths equivalent to those of the base material were found for a high dielectric loss, Fe-rich grade of Si_3N_4 whilst a lower loss Y/Al-containing grade only displayed strengths of ~20% of the base material[1,2]. The introduction of a thin Fe-rich Si_3N_4 sheet between the surfaces of the Y/Al-doped Si_3N_4 improved the quality of the resultant join and the strength increased up to ~70% of the base material[1]. Similarly, the use of powdered silicon nitride intermediaries[10] between hot pressed Si_3N_4 samples resulted in a continuous bond with the parent material.

Reaction bonded SiC (RBSC) samples have been directly microwave joined[12,13]; a Si-rich band being observed at the join line. The use of Al interlayers resulted in joins as strong as the base material itself[14]. Reaction bonded SiC has also been bonded to sintered SiC (SSC)[12,13]. The results suggested that although substantial cracking had occurred in the interfacial region, Si from the RBSC had wetted and bonded to the SSC. Although high strength microwave joins (400 MPa) have been produced between SSC samples by utilising conditions of 2050°C and 8 MPa[15], indirect approaches have resulted in joins being produced under substantially less demanding conditions. Dense and homogeneous interfacial regions could be achieved when Si interlayers were used[16]. Attempts at using in-situ combustion synthesis of SiC/TiC interlayers were less successful due to oxidation of the metals prior to reaction and the formation of a porous reaction product[16].

The above results imply that some form of phase which softens or melts at the joining temperature is a requirement for microwave joining of engineering ceramics. In materials where this occurs naturally as a grain boundary phase the ceramics can be directly joined. In materials where no such phase exists, then some intermediary is required which will provide a similar function. Often this can be a different grade of the base ceramic itself, but other materials such as glasses, gels and silicon have been found to be successful depending on the composition of the ceramics to be joined. Two questions are left unanswered by such a simple approach. First, does the softening phase fulfil any other role besides what could be crudely described as a 'glue' and second, is there no chance of achieving successful joins when softening phases are absent, ie by solid state processes? This paper, based on work carried out at the University of Nottingham over the past 3-4 years, attempts to address these questions.

EXPERIMENTAL

A range of ceramics were chosen to investigate the effects of various material characteristics on microwave joining. The materials included 2 grades of SiC, 3 grades of Al_2O_3 and 1 grade of ZrO_2. Details of the materials used and the sample configurations are provided in Table 1 below. All butting faces were ground using a diamond wheel to a surface finish of ~ 0.5 μm and then cleaned in a solution of methanol, soap and warm water using an ultrasonic bath before drying.

Table 1. Materials and dimensions

Key: Ø = diameter, l = length.

Material	Density / gcm^{-3}	Grain size / μm	Dimensions
Hot pressed SiC (HPSC)	3.15	4	5 mm x 5 mm x 50 mm
Reaction bonded SiC (RBSC)	3.10	10	Ø=15 mm, l=9 mm
85% alumina (AD85)	3.41	6	Ø=10 mm, l=50 mm
94% alumina (AD94)	3.70	12	Ø=10 mm, l=50 mm
99.8% alumina (AD99.8)	3.96	3	Ø=10 mm, l=50 mm
8 wt% yttria doped partially stabilised zirconia (YPSZ)	5.98	0.4	Ø= 7 mm, l=40 mm

A simplified diagram of the microwave system used to carry out the joining experiments is illustrated in Figure 1. Either a 300 W or 1 kW 2.45 GHz

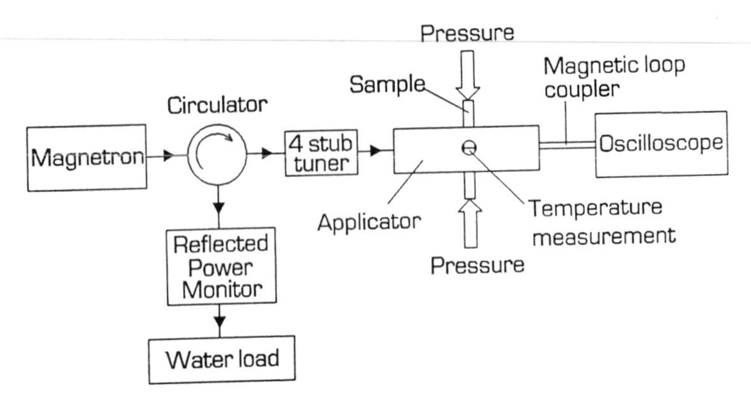

Figure 1. A simplified control diagram for the microwave heating system.

magnetron provided microwaves to a TE_{102} mode rectangular cavity. A matrix of experiments was performed for each material in which the joining temperature and pressure were varied. Heating rate, cooling rate and time at temperature were maintained as consistently as possible for any given material. To improve the uniformity of the temperature profile within the rods, they were encased in a foamed alumina insulation. A stainless steel support rig was constructed around the applicator to allow a compressive axial load to be applied to the ends of the rods during the joining cycle. The load was maintained at a constant level throughout the heating cycle by releasing the pressure as required by thermal expansion. The surface temperature of the sample was monitored using an Accufiber lightpipe system connected via a microprocessor to a personal computer. The sapphire lightpipe was inserted through a choke in the narrow wall of the cavity and clamped into position 1 mm above the butting faces. The manufacturers quote an accuracy of $\pm 5\,^{\circ}C$ for the system as used.

After joining, the rods were sectioned and polished using normal preparation procedures. An SEM fitted with an EDX detector was used to characterise the join region. When very successful joins were produced, the only method of determining the location of the join was via a $\sim 100\ \mu m$ misalignment of the rods. A number of the samples were etched in hydrofluoric acid to highlight the grain boundaries. The flexural strength of some of the joints made between alumina samples were measured using four point modulus of rupture tests.

The dielectric properties of the ceramics were measured at 2.45 GHz as a function of temperature using a surface probe technique[17]. Approximate values were calculated for the glassy grain boundary phase present in the lower purity alumina ceramics using the refractive index mixture equation[18]

$$\varepsilon_m^{1/2} = \upsilon_1 \varepsilon_1^{1/2} + \upsilon_2 \varepsilon_2^{1/2} \quad (1)$$

where, ε_m is the permittivity of a mixture comprising two phases of values ε_1 and ε_2 and volume fraction υ_1 and υ_2 respectively. Values for both AD85 and AD94 were substituted for ε_m and for AD99.8 for ε_2 in separate calculations and average values for ε_1, the glassy phase, calculated over the temperature range 100-1200°C. The calculations thus assume that the AD99.8 is representative of the composition of the polycrystalline phase in the two lower purity ceramics.

RESULTS AND DISCUSSIONS

Silicon Carbide

Table 2. Microwave joining parameters for silicon carbide.

Material	HPSC-HPSC	RBSC-RBSC
Maximum or joining temperature / °C	1312	1190 - 1376
Maximum power used / W	600	300
Axial pressure / MPa	0.5	0.5 - 2.0
Typical heating rate / °Cs^{-1}	0.6	2.0
Typical cooling rate / °Cs^{-1}	4.6	3.0
Holding time / minutes	-	up to 2
Typical total processing time / minutes	30	15

The processing parameters achieved are outlined in Table 2. Despite all attempts, including the introduction of an adjustable iris design, the HPSC could not be heated to a temperature much in excess of 1300°C. The RBSC ceramic could be joined under all conditions examined; however, the quality of join varied significantly. At low temperature and pressure, eg 1190°C, 0.5 MPa, a general lack of contact across the width of the sample was found. The butting surfaces were separated by a 10-20 μm gap which was only partially filled by free silicon (see Figure 2). As temperature and pressure were increased the width of the interfacial region reduced and became more completely filled with silicon. The gap was reduced to 5-10μm under conditions of 1235°C, 1 MPa, whilst by 1350°C, 1.5 MPa it had become 3-5 μm and was completely filled with silicon,

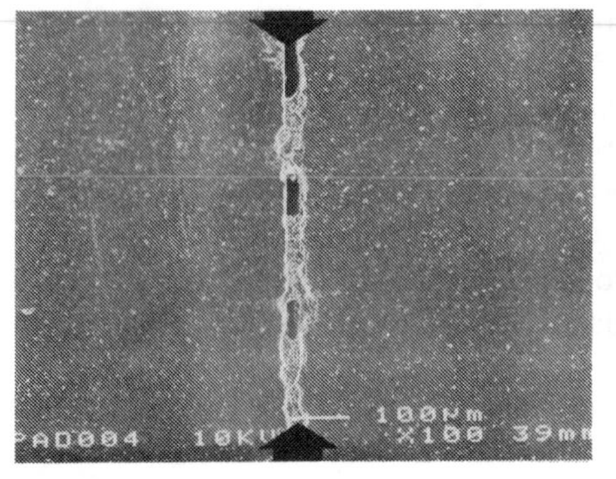

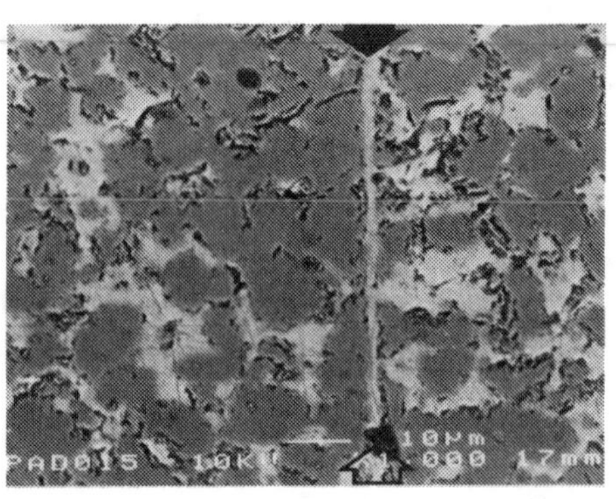

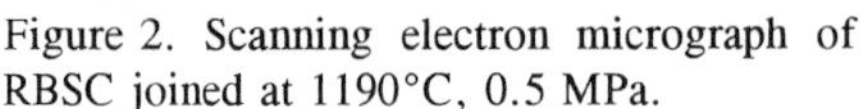

Figure 2. Scanning electron micrograph of RBSC joined at 1190°C, 0.5 MPa.

Figure 3. Scanning electron micrograph of RBSC joined at 1350°C, 1.5 MPa.

see Figure 3. Increasing the processing conditions to 1376°C, 2.0 MPa produced a join which suggested that they were excessive. A region of enhanced porosity spanning the width of the sample was observed and free silicon was found on the surfaces of the rods indicating it had flowed out of the join region.

These observations indicate that the silicon grain boundary phase is softening and providing a 'gluing' effect. The optimum temperature is that which causes the silicon phase to soften sufficiently to fill the void between the butting faces, but not so extensively as to flow out of the join region. Within reason, the higher the pressure the narrower the interfacial region. This will lead to higher strengths in the resulting join.

No evidence for local rearrangement of the SiC grains across the join interface was found in any of the samples examined.

Alumina

The processing parameters achieved are outlined in Table 3. Attempts to join the AD99.8 to itself were unsuccessful due to an inability to heat the material to a sufficiently high temperature. This occurred because the material displayed too low a dielectric loss to absorb sufficient microwaves. Both the AD85 and AD94, however, could be joined successfully, as could the AD85 to the AD99.8. The conditions required for successful direct joins between 85% and 94% pure alumina ceramics are shown in Table 4. These observations indicate that the greater the quantity of glassy grain boundary phase present the less extreme the temperature/pressure conditions required to effect a successful join. Figure 4,

Table 3. Microwave joining parameters for alumina.

Material	AD85-AD85	AD94-AD94	AD99.8-AD99.8	AD85-AD99.8
Maximum or joining temperature / °C	1400 - 1600	1500 - 1600	1400	1500 - 1600
Maximum power used / W	250	250	250	250
Axial pressure / MPa	0.25 - 1	0.5 - 1	1	0.5 - 1
Typical heating rate / $°Cs^{-1}$	3.0 @ $<$ 1200°C 0.4 @ $>$ 1200°C	2.5 @ $<$ 1000°C 1.0 @ $>$ 1000°C	3.0	3.6 @ $<$ 1200°C 0.5 @ $>$ 1200°C
Typical cooling rate / $°Cs^{-1}$	3.0	3.7	4.0	3.5
Holding time / minutes	10	10	-	10
Typical total processing time / minutes	30	40	25	30

Table 4. Degree of success achieved in joining alumina rods

	AD85 to AD85 alumina			AD94 to AD94 alumina	
Temp / °C ↓ Press / MPa →	0.25	0.50	1.00	0.50	1.00
1400	□	○	●		
1500	○	○	●	□	○
1600	●	●	●	○	●

□ = no/negligible join; ○ = joined but regions of poor contact; ● = successful join.

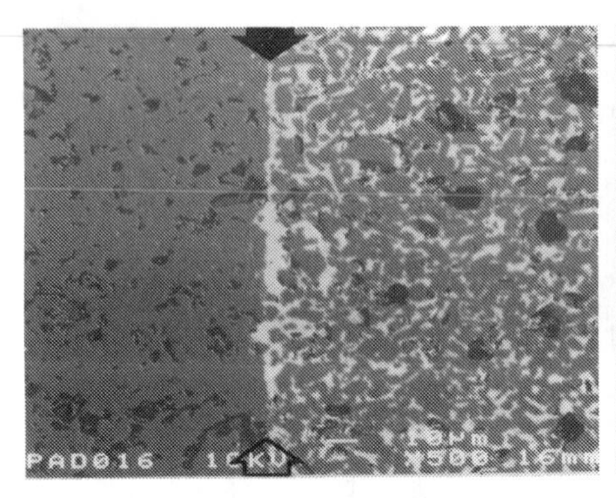

Figure 4. Scanning electron micrograph of AD85-AD99.8 joined at 1600°C, 0.5 MPa.

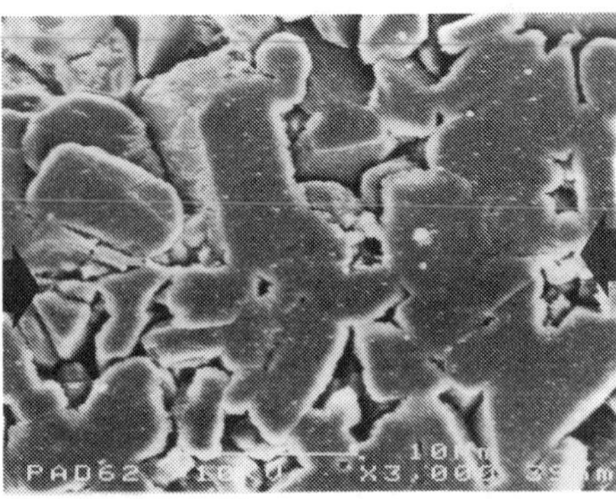

Figure 5. AD85-AD85 joined at 1500°C, 0.5 MPa and etched in hydrofluoric acid.

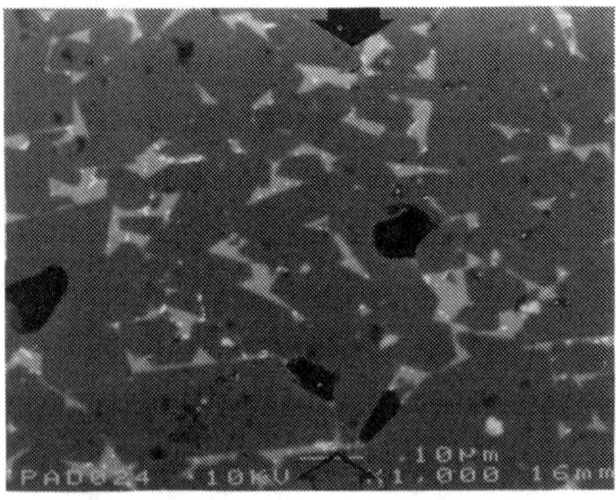

Figure 6. AD94-AD94 joined at 1600°C, 1.0 MPa showing homogeneous join region.

which shows a back scattered electron micrograph of the join region for a specimen of AD85 successfully attached to one of AD99.8 (1600°C, 0.5 MPa), supports this theory. A clear line of glass, squeezed from the AD85, can be seen along the join line.

Figure 5 is a highly magnified view of an AD85 sample joined at 1500°C, 0.25 MPa and etched in hydrofluoric acid. This figure shows evidence of two features. First, at locations where the α-Al_2O_3 grains have come into contact from the two rods they appear to have sintered together by atomic diffusion across the interface. Second, the join shows evidence of grains crossing the line and hence suggest they are able to move in the reduced viscosity of the glassy phase. Extensive examination of joins has indicated that the glass tends to flow out of the join towards the edges of the sample under the applied load. This can result in an homogeneous structure in which no obvious glassy line exists at the boundary (see Figure 6).

Dielectric property results (Figure 7) indicate that the presence of the glassy grain boundary phase decreases the real part and increases the imaginary part of the permittivity. Also plotted in the figure are the values for the glassy grain boundary phase obtained using equation 1. It can be seen that above ~700°C it

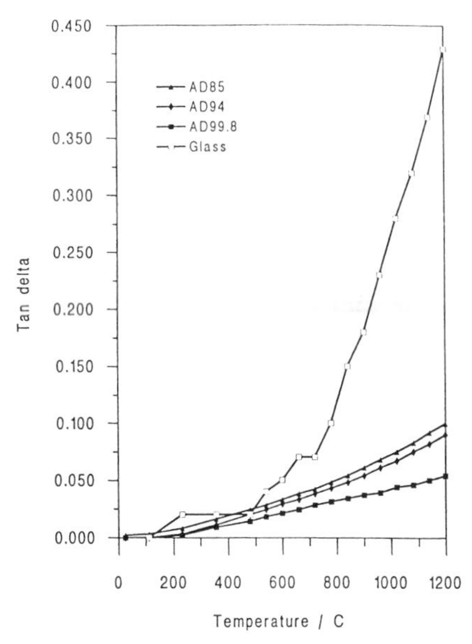

Figure 7. Dielectric properties of alumina ceramics as a function of temperature.

is the major contributor to the loss of the ceramics and its absence accounts for the inability to heat the AD99.8 from room temperature.

The flexural strength of the base and joined AD85 and AD94 materials are presented in Table 5. Examination of the fractured samples showed that all of the AD85 samples fractured away from the join, the cracks initiating from random locations on the tensile surface within the loading span. The strengths obtained suggest that fully homogenous joins have been produced. All of the AD94 samples, however, were observed to fail along the join line itself indicating that the temperature used was too low for this grade of alumina.

Table 5. 4-point flexural strength of alumina samples

Material	Strength / MPa	
	Base material	Joined @ 1600°C, 1 MPa, 10 mins
AD85	264 ± 53	312 ± 31
AD94	282 ± 68	144 ± 61

Zirconia

As discussed earlier, the feature apparently required for obtaining good quality joins is a grain boundary phase which softens at the joining temperature, effectively 'gluing' the butting surfaces together. However, Figure 5 appears to provide evidence of α-Al_2O_3 grains sintering together across the join interface. This suggests that, had the AD99.8 ceramic been sufficiently lossy to absorb microwaves, it is possible that it could have been joined by purely solid state diffusion processes. This tentative conclusion resulted in an investigation of zirconia, a ceramic which has no grain boundary phases likely to soften at the joining temperature but which will heat readily in a microwave environment.

The processing parameters achieved are outlined in Table 6. Zirconia proved to be a difficult material to join due to the large thermal expansion associated with zirconia ceramics (α = 10.6x10^{-6}°C^{-1}). This made control of the applied pressure difficult. The thermal expansion also yielded severe problems with thermal shock, limiting heating rates. Nevertheless, it proved possible to produce joins which were undetectable by scanning electron microscopy, except via the sample misalignment mentioned earlier. Figure 8 shows such a join obtained at 1560°C, 2.7 MPa. Joins could be produced at temperatures as low as 1400°C but regions of isolated elongated porosity remained along the join line.

Table 6. Microwave joining parameters for zirconia.

Material	YPSZ-YPSZ
Maximum or joining temperature / °C	1564
Maximum power used / W	370
Axial pressure / MPa	2.7
Typical heating rate / °Cs^{-1}	0.5@<900°C; 3.7@>1000°C
Typical cooling rate / °Cs^{-1}	3.5
Holding time / minutes	10
Typical total processing time / minutes	30

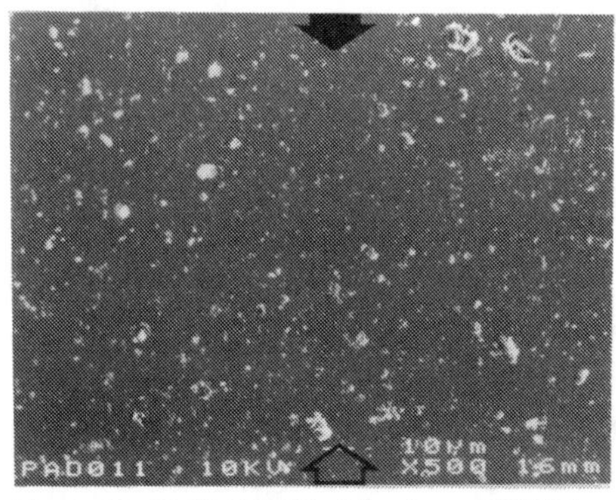

Figure 8. YPSZ-YPSZ joined at 1560°C, 2.7 MPa.

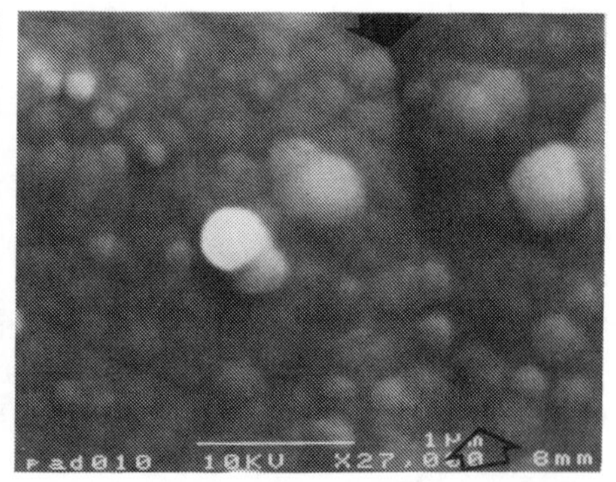

Figure 9. YPSZ-YPSZ join line etched using HF acid to reveal direct bonding of grains

Figure 9 is a higher magnification back scattered electron micrograph obtained on a sample etched using hydrofluoric acid to highlight the grain boundaries. Since no glassy grain boundary phase existed in this ceramic, solid state diffusion must be responsible for the bonding obtained. Microhardness measurements made in the join region were comparable to those of the bulk untreated material.

CONCLUSIONS

A wide range of engineering ceramic materials can be joined using 2.45 GHz microwave energy with joining times less than about 10 minutes. Joining is most easily affected when i) the ceramic is sufficiently absorbent of microwave energy and ii) grain boundary phases are present in the base material which will soften at the joining temperature allowing the butting faces to be 'glued' together. For materials without one or other (or both) characteristic, appropriate interlayers can act as a suitable substitute.

For low purity alumina ceramics, the glassy grain boundary phases present appear to allow motion of the individual grains across the join interface resulting in an homogeneous structure in which no obvious glassy line exists at the boundary. For reaction bonded silicon carbide ceramics, the equivalent silicon grain boundary phase does not allow similar grain motion but rather forms a discrete and continuous layer at the interface under optimum joining conditions. The reasons for these two types of behaviour are not yet known, but are probably due to the different viscosities and wetting characteristics of the two phases. The size of the grains is not thought to be relevant since both the AD94 and RBSC ceramics had similar sized grains.

Work with zirconia, a ceramic which absorbs microwaves readily but which does not possess glassy grain boundary phases, has confirmed that it is possible to join ceramics via purely solid state diffusion processes. This may open up a number of new and exciting joining possibilities previously thought unsuitable.

ACKNOWLEDGEMENTS

The authors would like to acknowledge the financial contribution made to this project by TWI and the Engineering and Physical Science Research Council, both in the UK.

REFERENCES

1. H. Fukushima, T. Yamanaka & M. Matsui, in *Microwave Processing of Materials*, MRS Proc **124**, ed. WH Sutton, MH Brooks and IJ Chabinsky, MRS, Pittsburgh, pp. 267-272 (1988).
2. H. Fukushima, T. Yamanaka & M. Matsui, J. Mater. Res. **5** [2] 397-405 (1990).

3. P.A. Davis, J.G.P. Binner, T.E. Cross & J.A. Fernie, in *Microwave Processing of Materials IV*, MRS Proc **347**, ed. WH Sutton, MH Brooks & IJ Chabinsky, MRS, Pittsburgh, pp. 655-660 (1994).
4. D. Palaith, R. Silberglitt, C.C.M. Wu, R. Kleiner & E.L. Libelo, in *Microwave Processing of Materials*, MRS Proc **124**, ed. , MRS, Pittsburgh, pp. 255-266 (1988).
5. T.T. Meek & R.D. Blake, US Patent No. 4,529,857 (1985).
6. T.T. Meek & R.D. Blake, J. Mat. Sci. Letts **5** 270-274 (1986).
7. S. Al-Assafi, I. Ahmad, Z. Fathi & D.E. Clark, in *Microwaves: Theory & Application in Materials Processing*, Ceramic Trans **21**, ed DE Clark, FD Gac & WH Sutton, ACerS, Westerville, pp. 515-521 (1991).
8. S. Al-Assafi & D.E. Clark, in *Microwave Processing of Materials III*, MRS Proc **269**, ed. RL Beatty, WH Sutton & MF Iskander, MRS, Pittsburgh, pp. 335-340 (1992).
9. S. Al-Assafi & D.E. Clark, in 16th Annual Conference on Composites & Advanced Ceramic Materials, Ceramic Engineering & Science Proceedings **2** 1073-1080 (1992).
10. P.S. Apté, R.M. Kimer & M.C.L. Patterson, Microwave heating for ceramic joining, Report prepared for the Ontario Ministry of Energy, Project #600049, Alcan International Ltd, Kingston Research & Development Centre, Kingston, Ontario, Canada (1988).
11. X.D. Xu, V.V. Varadan & V.K. Varadan, in *Microwaves: Theory & Application in Materials Processing*, Ceramic Trans **21**, ed DE Clark, FD Gac & WH Sutton, ACerS, Westerville, pp. 497-505 (1991).
12. I. Ahmad & W.M. Black, in 16th Annual Conference on Composites & Advanced Ceramic Materials, Ceramic Engineering & Science Proceedings **1** 520-527 (1992).
13. I. Ahmad, R. Silberglitt, W.M. Black, H.S. Sa'adaldin, J.D. Katz, in *Microwave Processing of Materials III*, MRS Proc **269**, ed. RL Beatty, WH Sutton & MF Iskander, MRS, Pittsburgh, pp. 271-276 (1992).
14. T.Y. Yiin, V.V. Varadan, V.K. Varadan & J.C. Conway, in *Microwaves: Theory & Application in Materials Processing*, Ceramic Trans **21**, ed DE Clark, FD Gac & WH Sutton, ACerS, Westerville, pp. 507-514 (1991).
15. T. Sato, T. Sugiura & K Shimakage, J. Ceram. Soc. Japan **101** 422-427 (1993).
16. R. Silberglitt, D. Palaith, H.S. Sa'adaldin, W.M. Black, J.D. Katz & R.D. Blake, idem, pp. 487-495.
17. M. Arai, J.G.P. Binner, G.E. Carr & T.E. Cross, in *Microwaves: Theory & Application in Materials Processing II*, Ceramic Trans **36**, ed DE Clark, WR Tinga & JR Laia Jr., ACerS, Westerville, pp. 483-492 (1993).
18. S.O. Nelson & T-S. You, J. Phys. D: Appl. Phys., **23** 346 (1990).

OPEN COAXIAL MICROWAVE SPOT JOINING APPLICATOR

W.R. Tinga, J.D. Xu, F.E. Vermeulen
University of Alberta
Edmonton, Alberta, Canada.

ABSTRACT

To allow intense heating on a large surface or along a joint region requires an open-ended applicator. Currently used closed resonant structures greatly constrain the practical application of microwave joining. Initial results of an open coaxial structure are presented which can be moved along a surface to raise its temperature in a very narrow region to over 1000°C. Microwave leakage is kept very low by the use of a dual frequency choke design. Dual frequency operation is possible with this approach since the applicator length is tunable also. Results obtained show promise that this design approach could be used for microwave joining of ceramics as well for many other applications. This design was modeled by a modified finite element technique giving insight into the effect of various design variables.

INTRODUCTION

Current microwave applicators for the joining of ceramics require the material to be fed into a closed applicator [1,2]. This greatly constrains the practical use of microwave joining. To allow intense point heating along a surface or in a joint region, an open structure design is presented here. Photographs of the structure are shown in Figure 1.

A magnetic loop coupled, resonant, symmetrical coaxial structure is adopted as shown schematically in Figure 2. Placing the applicator on or near the material to be processed will cause the structure to resonate and create an intense field pattern in the interface region between the center conductor and the material to be heated. This resonance, or field amplification, can be enhanced if an electrically conducting plate is placed under the process material. Two quarter-wave cylindrical choke sections surround the applicator and limit the microwave leakage at the two ISM frequencies, 915 and 2450 MHz. Adjustment of the cavity length is possible by moving the sliding short circuit. An electronically tunable 915 MHz oscillator and a 150 Watt solid state power amplifier are used to track the dynamically changing applicator resonant frequency.

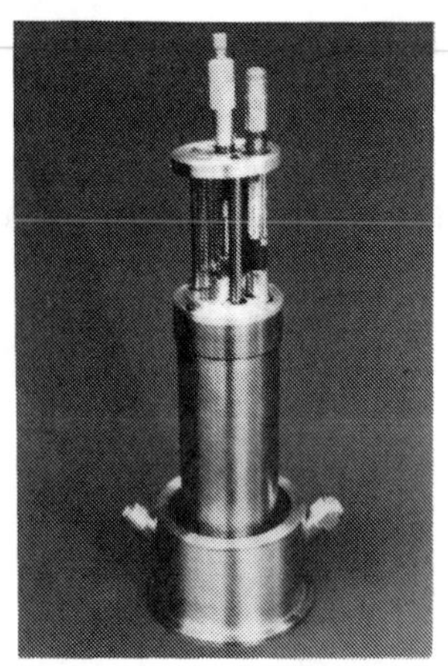

Figure 1. Photographs of the coaxial applicator showing the cylindrical quarter wave chokes. Resonator length is adjustable to allow for small frequency shifts during processing or to change frequency bands.

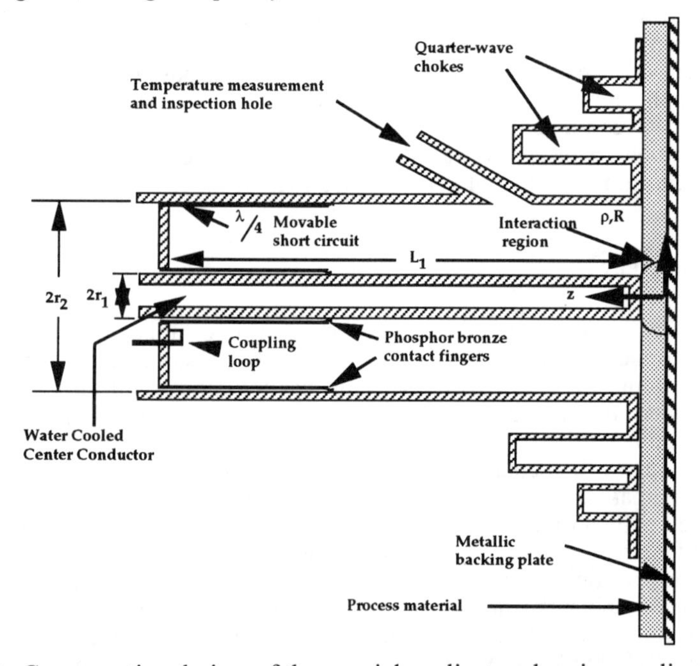

Figure 2. Cross-sectional view of the coaxial applicator showing applicator details and the coordinate system used in the analysis. Resonator length is adjustable to allow for small frequency shifts during processing or to change frequency bands. Resonator dimensions are: r_1 = 12mm, r_2 = 45mm, L_1 = 80-200mm.

To obtain the field distribution in the structure, the finite element modeling method (FEM) is chosen for this applicator because of its accuracy, flexibility and its suitability for inhomogeneous regions [3,4]. By using the finite element method, the field distribution in both the applicator and the medium to be heated are obtained. Different applicator dimensions are considered to find the set that optimizes the point concentration of the microwave energy at the applicator to material interface. Achieving the necessary intense energy zone, requires a resonance to be established and therefore the Q factor of the structure must also be found. Numerical results are presented for the field configurations in the interaction region. These are compared to experimentally obtained data. Good agreement is found between the two.

OBTAINING THE CAVITY PARAMETERS

Once the resulting eigen value equation is solved, using the FEM method, the resonant frequency and the magnetic field distribution can be calculated. From the calculated magnetic field and Maxwell's equations we obtain expressions for the radial and longitudinal electric fields [5]. Other quantities of interest to the applicator designer are the electric field distribution and the Q factor which must also be determined. By definition the Q factor is given by

$$Q_0 = \omega \frac{W}{P} = \omega \frac{W}{P_c + P_d} \qquad \textbf{(1)}$$

where W is energy stored in the cavity, P_a is the total time-average power absorbed in the resonator cavity, P_c *is* the time-average power loss in the cavity conducting walls and P_d is the time-average power loss in the dielectric material to be heated. Thus

$$Q_0 = \frac{Q_c\, Q_d}{Q_c + Q_d} \qquad \textbf{(2)}$$

where Q_c is

$$Q_c = \frac{\omega\mu \int_V |H|^2 dV}{R_s \int_S |H|^2 dS} = \frac{2 \int_V |H|^2 dV}{\delta \int_S |H|^2 dS} \qquad \textbf{(3)}$$

R_s *is* the surface resistivity and δ the skin depth. This integral equation is solved by the FEM method using previously determined discrete expressions for the field Hϕ.

The dielectric Q is given by

$$Q_d = \frac{W}{W_d} \frac{\epsilon'}{\epsilon''} \quad \textbf{(4)}$$

where W_d is the energy stored in the dielectric region. Inside the cavity, the absorbed power is

$$P_a = P_c + P_d = P_i - P_r \quad \textbf{(5)}$$

where P_i and P_r are the incident and reflected power respectively. Moreover, when the reflected power is normalized to unit incident power, it is related to the reflection coefficient, r, by

$$P_r = 1 - |\Gamma|^2 \quad \textbf{(6)}$$

Suppose the cavity is undercoupled to the external circuit and that the coupling is magnetic. Then it can be shown that the following relation holds [6]:

$$|\Gamma| = \frac{C-Q}{C+Q} \quad \textbf{(7)}$$

where C is the coupling coefficient. Now assume that this coupling coefficient remains unchanged with a slight variation in the cavity dimensions. One can then obtain the normalized absorbed power in the cavity, P_a, from $P_a = 1 - |\Gamma|2$ and Eq (7) since the cavity's Q factor is known already from Eq. (3). Then the electric field, E_n, normalized for excitation by unit power, can be written as

$$E_n = E_{FEM}\sqrt{\frac{P_a}{P_{FEM}}} \quad \textbf{(8)}$$

where P_{FEM} is the absorbed power calculated by the FEM method and E_{FEM} is the electric field found from the FEM solution.

EXPERIMENTAL AND NUMERICAL RESULTS

Based on the above problem formulation, a computer code has been developed. Results were calculated for a coaxial applicator of 132 mm length with an outer conductor inside radius of 45 mm and center conductor radii of 4, 6, 8,10 and 12 mm, respectively. This corresponds to a $5\lambda/4$ longitudinal resonant mode at 2450 MHz. In these calculations, it is assumed that the material in the gap region has a dielectric constant of 2.5 and is 5 mm thick. To reduce the calculation time, the FEM analysis assumes the material is lossless even though the materials used in the experiment had a non-zero loss factor. This standard perturbation assumption has no significant effect on the field pattern and magnitude calculation. It is clear from Figure 3 that the E_Z field component is dominant in the gap interaction region thereby ensuring good field penetration into the process material. Near the outer edge of the inner conductor, the electric field magnitude peaks sharply as a result of the large value of the $H\phi$ field at the center conductor surface.

Increasing the inner conductor radius decreases this effect. Peaking of the E-field in this region can cause arcing. In practice, rounding the end of the center conductor minimizes that problem.

An experimental prototype applicator, see photographs in Figure 1, has been constructed and tested. Experimental measurement results compare favorably with calculated numerical results for the resonant frequency, f_o, and Q-factor of the loaded applicator supporting the validity of the numerical model. Comparing the calculated and measured resonant frequency, in Figure 4, as a function of cavity length for an air gap of 5 mm, shows excellent agreement between the two.

Measured and calculated E-fields along the material surface next to the metallic backing plate are shown in Figure 5 for an inner conductor radius of 12 mm and for a $5\lambda/4$ resonance in the 2450 MHz band. Agreement in the region of interest under the center conductor is good considering the disturbance caused by the 2 mm long, 1.5 mm diameter field measuring probe used and any errors with placement.

For practical reasons, the field was measured and calculated at the lower sample surface next to the metallic backing plate. The sample test material was a 1/4 inch Plexiglas® sheet with $\epsilon' = 2.5 - j.025$. Perturbation caused by the probe resulted in a maximum frequency shift of 20 MHz over the 50 mm distance. It is clear from Figure 6 that the variation of the unloaded Q-factor is nearly independent of the center conductor radius but increases significantly for increasing resonant mode numbers. That is, the Q-value of the $7\lambda/4$ mode is twice that of the $3\lambda/4$ mode.

It is apparent that the highest field intensity in the gap is obtained for the thinnest inner conductor. In the gap region, the E_z field component value is about five times the value of the E_r component and decays rapidly with distance outside the gap region, ensuring effective and efficient heating in the central interaction region between the sample material and the center conductor. Leakage suppression by the quarter-wavelength chokes was very effective in keeping the leakage power well below 1mW/cm^2 as measured at 5 cm from any external applicator surface for an input power of 150 W at both 2450 and 915 MHz. This applicator can be used as a spot heater for various dielectric materials, including low loss materials such as polypropylene, glass and ceramics. Typical results, as shown in Figure 7, show a concentrated heating pattern in the gap interaction region corresponding, qualitatively, to the theoretical field density prediction.

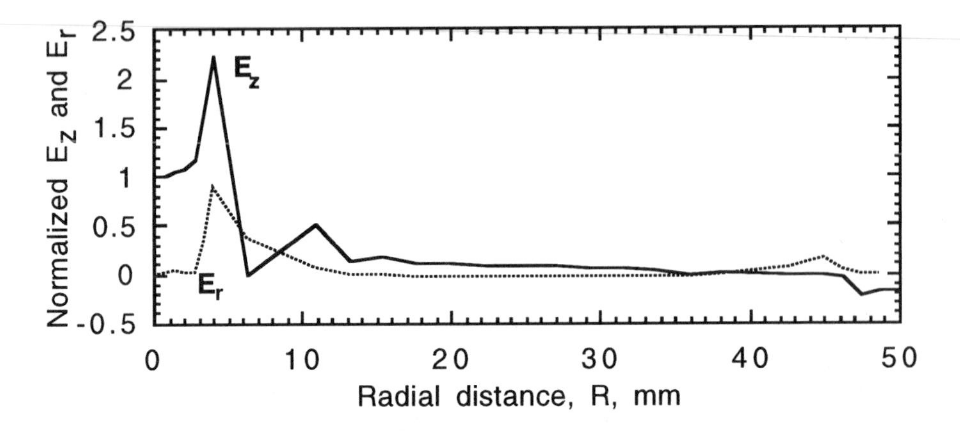

Figure 3 (a). Normalized electric fields E_z and E_r versus the radial distance, R, from the applicator center line for an inner conductor radius, r_1, of 4 mm and for resonance in the 5λ/4 mode in the 2450 MHz band. This is the field along the surface of the sample nearest the center conductor. The choke centerline lies at R=50 mm and represents a magnetic wall.

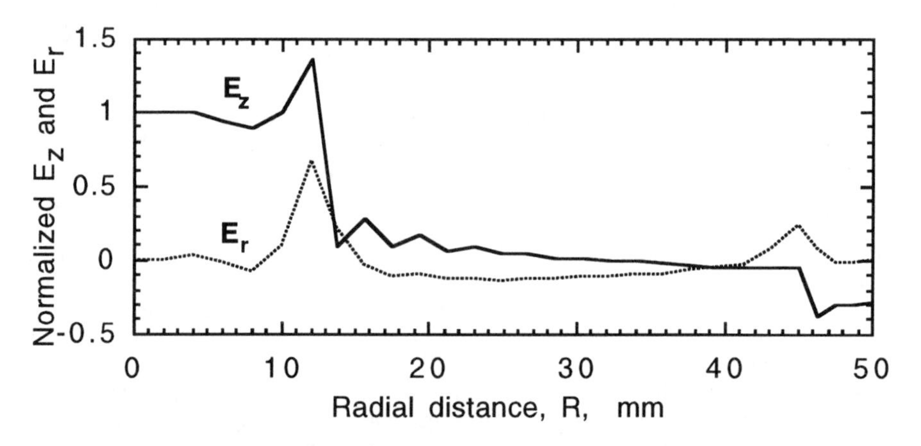

Figure 3 (b). Normalized electric fields E_z and E_r versus the radial distance, R, from the applicator center line for an inner conductor radius, r_1, of 12 mm and for resonance in the 5λ/4 mode in the 2450 MHz band. This is the field along the surface of the sample nearest the center conductor. The choke centerline lies at R=50 mm and represents a magnetic wall.

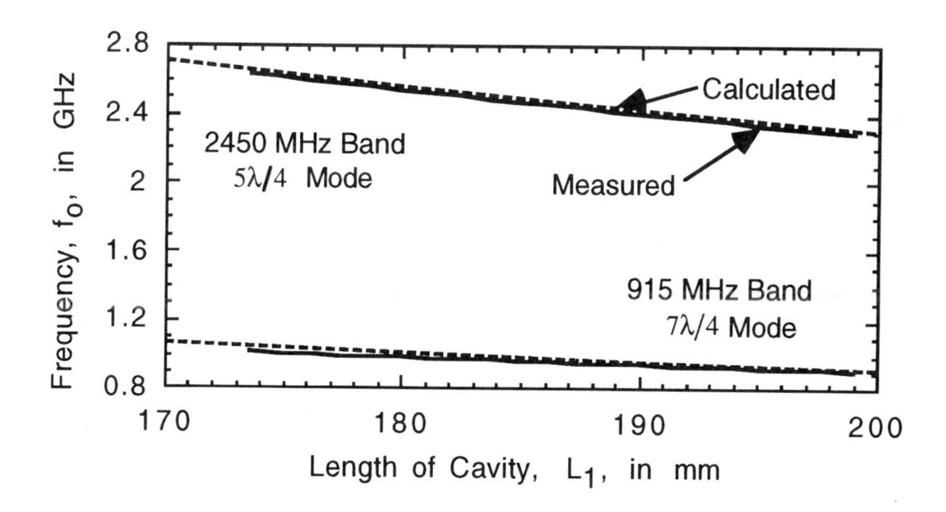

Figure 4 . Measured and calculated frequency variation in GHz as a function of the resonator length, L_l, in mm, for an air gap of 5 mm and for resonance in the ISM bands of 2450 MHz and 915 MHz. Note the different resonant modes used.

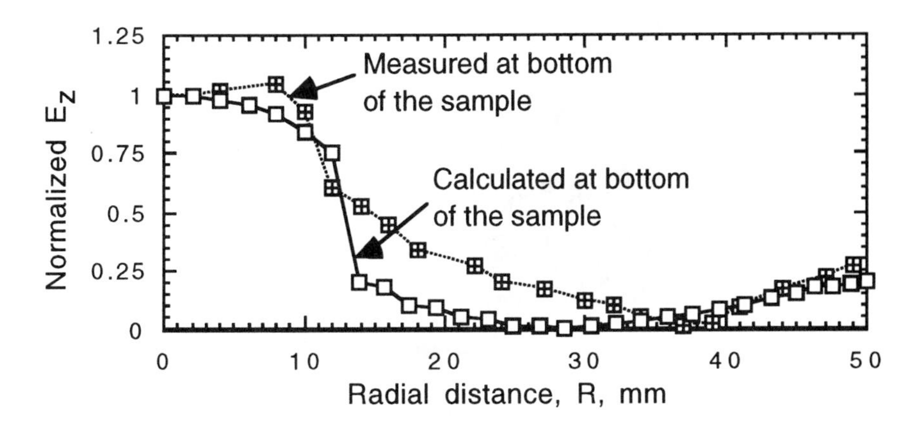

Figure 5. Measured and calculated normalized electric field, E_z, versus the radial distance, R, from the applicator center line for an inner conductor radius, r_l, of 12 mm and for a 5λ/4 resonance in the 2450 MHz band.

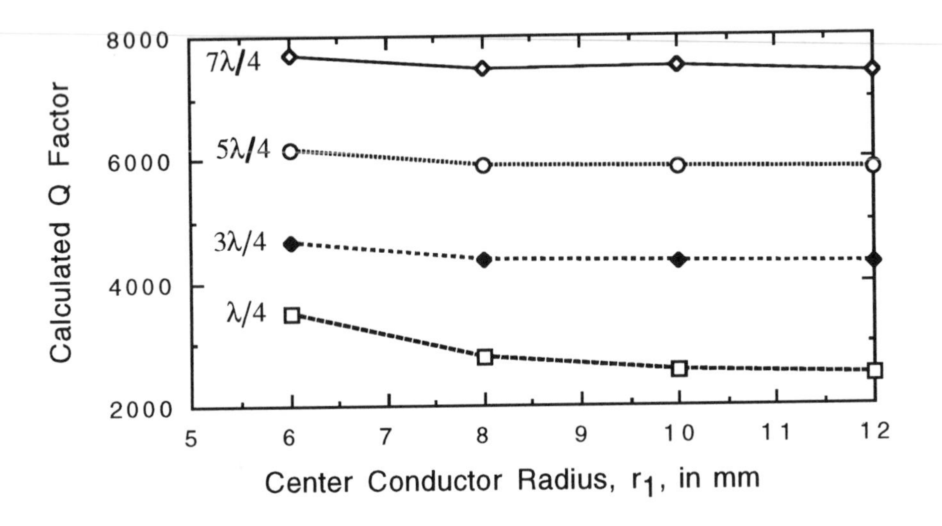

Figure 6. Calculated unloaded Q factor as a function of the inner conductor radius, r_1, for different resonant modes and an air gap of 5 mm. The cavity is constructed of brass with a $\sigma = 15 \times 10^6$ S.

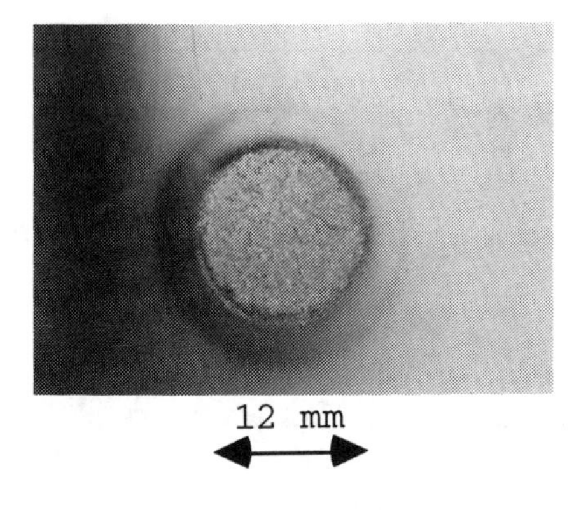

Figure 7. Heating pattern, at 2450 MHz, of the applicator as obtained on a Plexiglas® sample using an inner conductor radius of 6 mm qualitatively showing the effect of the intense E-field zone immediately below the center conductor.

CONCLUSIONS

It has been shown that an open coaxial applicator structure with a metal backed dielectric sheet load can be used to achieve high temperatures in a very focused region. This work has laid the foundation for the continuous joining of ceramic sheet materials at high temperatures. Further work is needed to adapt and optimize this applicator to solve practical problems in ceramic joining.

ACKNOWLEDGMENTS

This work was supported in part by grants from the Natural Science and Engineering Research Council of Canada.

REFERENCES

1. D. Palaith, R. Silberglitt, C.C.M. Wu, R. Kleiner and E.L. Libelo, "Microwave Joining of Ceramics," Mater. Res. Soc. Symp. Proc. Vol 124, pp 255-266 (1988).
2. H. Fukushima, T. Yamanaka and M. Matsui, "Microwave Heating of Ceramics and Its Application To Joining," Mater. Res. Soc. Symp. Proc. Vol 124, pp 267-272 (1988).
3. P. Silvester, "A General High Order Finite Element Waveguide Analysis Program," IEEE Trans. Microwave Theory Tech., MTT-17, pp 204-210 (1969).
4. J.B. Davies, F, Fernandes and G. Philippou, "Finite Element Analysis of All Modes in Cavities with Circular Symmetry," IEEE Trans. Microwave Theory Tech., MTT30, pp 1975-1980 (1987).
5. J.D. Xu, W.R. Tinga and F.E. Vermeulen, "Finite Element Analysis of a Coaxial Microwave Spot Joining Applicator," Submitted to IEEE Trans. Microwave Theory Tech. (1995).
6. B. Tian, "Single Frequency High Temperature Dielectrometers," PhD Thesis, University of Alberta, Dept. of Electrical Engineering (1993).

MICROWAVE-ASSISTED PYROLYSIS OF SiC AND ITS APPLICATION TO JOINING

I. Ahmad and R. Silberglitt, FM Technologies, Inc., Fairfax, VA; T.A. Shan, Y.L. Tian and R. Cozzens, George Mason University, Fairfax, VA.

ABSTRACT

Microwave energy has been used to pyrolyze silicon carbide from commercially available polycarbosilane precursor. The pyrolysis was performed on SiC surfaces having various surface treatments, to identify conditions which improve the wetting and adherence. Grinding and etching of the surfaces in hydrofluoric (HF) acid promotes the bonding of precursor derived ceramic to the SiC ceramic. Finally, the polycarbosilane precursor mixed with fine silicon carbide powder was used as the interlayer material to join silicon carbide specimens.

INTRODUCTION

Ceramic materials with superior strength at high temperatures and excellent thermal shock resistance are in demand for various applications such as advanced heat engines, heat exchangers, radiant burners and pump components. The outstanding candidate materials for such applications include silicon nitride and silicon carbide. Silicon nitride has found applications in some automobile engine components, whereas silicon carbide is being exploited for heat exchangers and radiant burner tubes. However, the high hardness of these materials makes it equally difficult to machine and join these ceramics to fabricate components of the desired shape.

Various methods have been explored to join these non-oxide ceramics to themselves and metals. These are either liquid-phase joining or solid-state bonding. Liquid-phase joining typically involves the use of an active metal brazing alloy. Active metal brazing materials containing a small percentage of an active metal are available for silicon-based ceramics. The common brazing alloy is silver-copper-titanium which could be used without any prior metallization. The brazing material sandwiched in between the specimens to be joined is held at temperatures ranging from 800 to 950°C in either vacuum or an inert gas. This method, although used for some applications employing ceramic inserts, is not suitable for high temperature applications. Furthermore, it does not provide the oxidation resistance of the joint.

Solid-state or diffusion bonding has been used for both ceramic/metal joining and ceramic/ceramic joining. This is accomplished with temperature, pressure and time to allow sufficient mobility and interdiffusion of species to form the bond. Processes have been developed for diffusion bonding of numerous polycrystalline and single crystal oxide ceramics. However, these joining techniques are not expected to be directly employed for nitrides, carbides or borides. A comprehensive description of the ceramic-ceramic and ceramic-metal joining issues can be found in references [1,2]. However, the important factors relevant to liquid-phase joining of two dissimilar materials are briefly described.

Basic Factors in Bonding

Surface free energy: Surfaces and interfaces can play an important role in the joining process. Surface free energy arises from the fact that some of the material lies on or close to the surface. Atoms at discontinuities- such as a jog or ledge on a surface, bonded to fewer neighboring atoms, have more surface energy than atoms on a smooth surface. When an atom has to attach itself to a solid, the smooth surfaces have the lowest probability for accommodation, because the interfacial energy will be increased. On the other hand, filling of scratches, cracks or pits is more probable, because it will reduce the interfacial energy of the system.

Wetting and spreading of a liquid on a solid surface depends on the short-range molecular forces of the surface. If the liquid does not wet the solid surface, it forms a bead. If the liquid wets the surface, it will spread and cover the surface. If the liquid wets the surface of two adjacent particles, it spreads over the surface and tends to concentrate at the contact region forming a neck.

Adhesion: This generally describes the molecular attraction between the surfaces in contact. Chemical bonding at the interface is obtained when chemical equilibrium exists at the interface. The interface may not initially be in the lowest energy state. With diffusion into the adjoining phases, the system tends to go toward equilibrium and reaction continues until equilibrium is maintained and the system is in the lowest energy state.

Thermal stress relaxation: When dissimilar materials are joined, thermal stresses can develop because of the mismatch in the coefficient of thermal expansion of the two materials. The greater the difference, the higher the thermal stress at the interface. Thermal stress increases as the sizes of the ceramics being joined increase. These thermal stresses can induce flaws in the joint area and weaken the joint. For best joint properties there should be no mismatch. However, if this is not possible, the existing thermal stress should be relaxed. Different methods have been developed using graded interlayers, soft metal interlayers and composite interlayers that help in minimizing the stress.

Surface roughness: The preparation of the surfaces to be joined can have a strong influence on the joint properties. Generally, a rough surface is expected to prevent full contact at the interface, thereby introducing severe residual stress at the regions of contact. On the other hand, a rough surface may have an anchoring effect which will promote mechanical interlocking. This is true for situations where surface reaction is extensive, so that damaged surface may disappear and will not influence the mechanical properties.

Joining of Silicon Carbide

Silicon carbide is a difficult material to sinter as well as to join by solid state methods. However, solid state bonding was achieved at 1950-2100°C with a one hour soak under a pressure of 27MPa. Bend strengths up to 300 MPa were achieved [3]. It has also been joined with microwave energy at 2050°C under a pressure of 8MPa for 5 min to give a strength of 71% of the original material [4].

Joining of silicon carbide has been achieved utilizing reaction sintering. A tape of silicon carbide and carbon is used as the joining interlayer, which is then siliconized at 1450°C to form the joint. The average room temperature strength of the joint was reported to be 327 MPa, which increased slightly with temperature up to 1200°C [5].

Polycarbosilane (PCS) is well known to convert to silicon carbide upon firing in an inert atmosphere [6]. Recently, PCS was used as the interlayer material to join SiC at 1500°C for 2 hrs, and an average flexural strength of 104 MPa was achieved. Investigation of the processing variables revealed that removal of the surface oxide on SiC with a pre-pyrolysis HF-dip improved the wetting of the surface by polycarbosilane [7]. Earlier, a brief communication reported the use of a mixture of this polymer and silicon carbide powder as the interlayer material. Joining was carried out at 1500°C in nitrogen without any pressure. The strength reported was only of the order of 40 MPa [8].

In this effort, microwave energy is used as the heating method with PCS as the interlayer material. The use of PCS and its conversion into SiC will eliminate the thermal stresses that may be introduced with other interlayer materials. Since wetting of the SiC is important in obtaining good joints, various surface treatments were explored.

MATERIALS AND METHODS

Commercial polycarbosilane having an average molecular weight of 1400 was purchased from Dow Corning (manufactured by Nippon Carbon Corporation, Tokyo, Japan). Silicon carbide (Hexoloy^{TM}) specimens approximately 0.5 cm in length were cut from rods 0.95 cm in diameter which were purchased from Carborundum Company (Niagara Falls, NY). Four sets of specimens were prepared for a comparative evaluation of the surface treatment before joining. The specimens were cut by a Buehler low speed saw using a high concentration blade . The first set of specimens were as-cut and PCS was dissolved in hexane and applied to the surface to be joined. On the second set, the surfaces to be joined were etched in 40% hydrofluoric acid before PCS was applied as the interlayer for joining. On the other two sets, the surfaces to be joined were ground on a diamond wheel. One set was etched in hydrofluoric acid and then the PCS was applied while on the other PCS was applied without etching. It is important to point out that the as-cut surface had a better surface finish (approx. 15μm) than the surfaces ground on a diamond wheel (30μm finish). All processing was conducted under a mixed reducing atmosphere of 95% nitrogen and 5% hydrogen at temperatures ranging between 1400-1450°C.

Results from the pyrolysis of polycarbosilane using microwave heating were published earlier [9]. Microwave curing of PCS at 1400°C resulted in sharp peaks for SiC in the XRD spectra. Thus temperatures in excess of 1400°C were used in a single mode TE_{103} rectangular microwave cavity to study the joining of silicon carbide with PCS as the interlayer material. Microwave power was coupled to the specimens by adjusting the plunger and an adjustable iris to minimize the reflected power [10]. The specimens heated in a few minutes after the cavity was initially tuned. The temperature was monitored by an optical pyrometer and was held in the 1400-1450°C range for 30 minutes. The entire cavity was sealed and purged with 95% N_2 and 5% H_2 prior to and during heating.

The specimens either did not join or were poorly joined and fell apart during handling. The surface of each of the four different types of specimens were scratched with a sharp metal point under a load of 820 grams to see the adherence of the SiC formed from PCS to the surface of the SiC specimens. The surfaces of these four different kinds of preparations were examined in the Hitachi Scanning Electron Microscope Model S530.

More specimens were cut, ground and etched for joining. On these specimens, a mixture of a fine SiC powder and PCS was applied. This set was also heated using microwave energy as discussed above. These specimens joined very well and it was possible to cut, grind and polish these specimens for SEM examination.

RESULTS AND DISCUSSIONS

Scanning electron micrographs of the surface of the four sets of specimens that did not join or had a very weak bond are shown in Figures 1-4. Figure 1 corresponds to the PCS applied to the as-cut surface. Figure 2 shows the PCS on the surface which was cut and then etched in hydrofluoric acid. As is apparent from these two micrographs, etching of the surface improved the wetting or spreading of PCS on the SiC specimen surface. Figure 3 shows the micrograph of the SiC pyrolyzed on a ground surface. The wetting of the SiC surface with the polymeric precursor is far better than the as-cut case (Figure 1). Figure 4 shows a micrograph of the wetting of the pre-ceramic precursor on a surface which was ground and etched. Indeed, this micrograph shows the maximum spreading of the polymeric precursor on the SiC specimen, and this joint was stronger than the other three.

The scratch test was performed on all four specimens. The white mark running horizontally on the micrographs indicates the scratched region. Figures 1-4 show that, where there was no SiC from the PCS there is no mark, because the SiC specimen was smooth enough not to show abrasion of the metal point. However, at places where PCS pyrolyzed into SiC, the scratch shows some abrasion on the flakes of the pyrolyzed SiC with SiC sanding some of the metal (like that on a sand paper). This is seen clearly in a higher magnification micrograph shown as Figure 5. The important point is that the metal point did not peel off the SiC layer pyrolyzed on the SiC surface. This is true only for the small flakes of the SiC. As seen in most of the micrographs, the larger flakes came off easily. In fact, some of the empty spaces may be so because the larger flakes had peeled off.

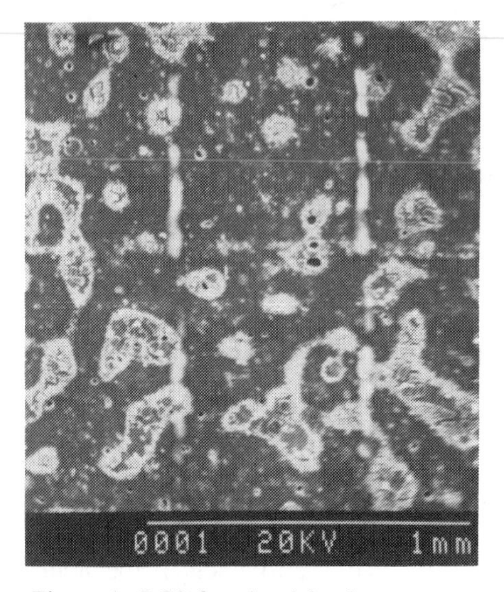

Figure 1. PCS forming islands on as-cut surface of SiC. (White vertical marks are artifacts of processing.)

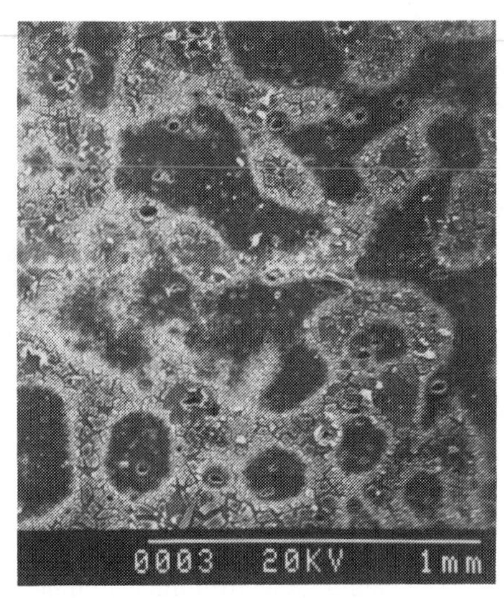

Figure 2. PCS on a surface which was cut and etched in HF.

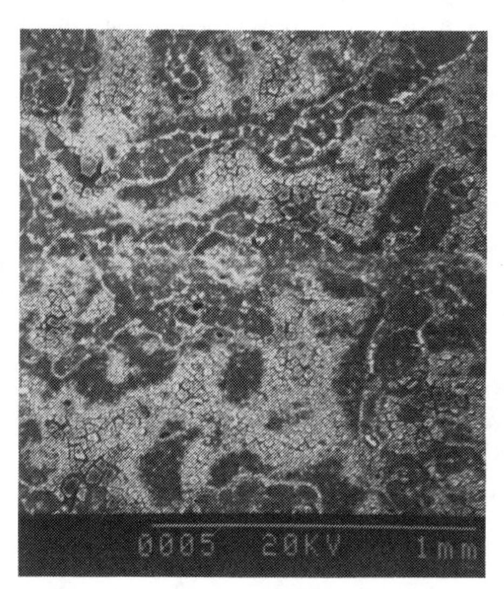

Figure 3. PCS wetting and spreading of PCS on SiC surface which was cut and ground.

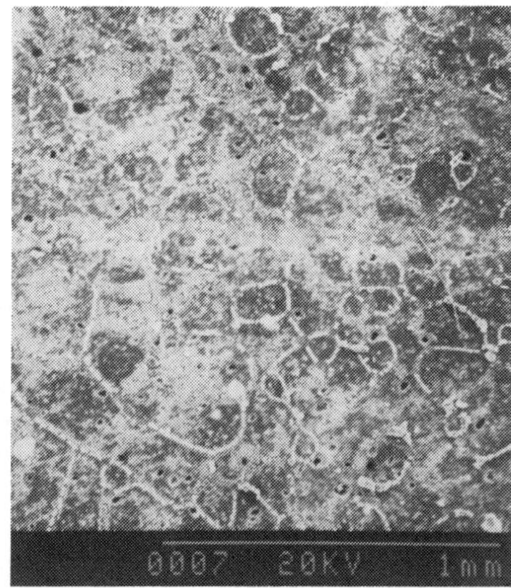

Figure 4. Maximum wetting and spreading of PCS on a surface which was cut, ground and etched in HF.

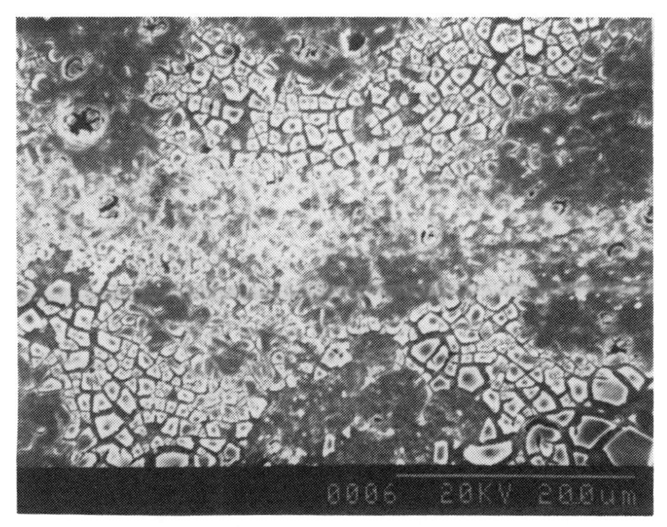

Figure 5. Larger magnification of micrograph in Figure 3, showing the adherence of the pyrolyzed silicon carbide to the SiC surface after the scratch test. The thin flakes deformed as probably did the metal point.

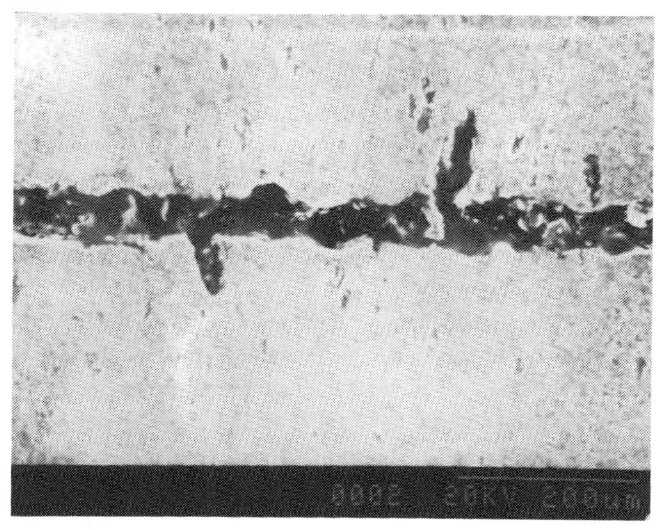

Figure 6. Cross-section of the specimen joined with a mixture of PCS and SiC powder as the interlayer material. The mixture filled the pores that existed in the SiC specimens to be joined.

A comparison of Figures 1-4 illustrates that when the surface was ground and etched, not only did it have the best coverage of PCS-derived SiC, it had the smallest flakes and the best joint among these four cases. This suggested mixing fine silicon carbide powder in PCS, which could provide a two-fold advantage. First, it could serve as a filler material and inhibit the flow of PCS when it melts, and second, it will provide much more surface area to which the PCS-derived SiC can adhere. Thus a fine SiC powder was mixed with PCS and applied to the surfaces that were ground and etched. The specimens were heated as described above. Figure 6 shows the cross-section of the joined specimen. The rough surface with some fairly large pores was almost completely filled by the mixture. However, a hole and some small cracks are visible in the micrograph, which is representative of the entire joint area.

SUMMARY AND CONCLUSIONS

Commercially available polycarbosilane precursor has been successfully pyrolyzed using microwave energy to form silicon carbide. Various surface treatments were investigated to identify the conditions which can improve the adherence of the pyrolyzed SiC to the SiC specimens. Polishing is usually performed to allow contact and hence joining of the specimens. It has been observed that grinding and etching of the surfaces in hydrofluoric acid can promote the bonding of precursor-derived ceramic to the SiC specimen. Silicon carbide specimens were joined using a mixture of fine silicon carbide powder and polycarbosilane precursor as the interlayer material.

ACKNOWLEDGMENTS

This work was sponsored in part by the US Department of Energy Advanced Industrial Materials (AIM) Program, through subcontract with Los Alamos National Laboratory. The Hitachi Scanning Electron Microscope used at the Shared Researched Instrumentation Facility (SRIF) at George Mason University was provided under a National Science Foundation Grant No. BSR-85 11148.

REFERENCES

1. R.E. Loehman, Chapter 7, Ceramic and Glasses, Engineered Material Handbook, Vol. 4, ASM International (1991).
2. A.H. Carim, D.S. Schwartz and R.S. Silberglitt, Eds. Joining and Adhesion of Advanced Inorganic Materials, MRS Proceedings, Vol. 314 (1993).
3. D. DeLeeuw, Effects of Joining Pressure and Deformation on the Strength and Microstructure of Diffusion-Bonded Silicon Carbide, Journal of the American Ceramic Society, Vol. 75, (No. 3), pp. 725-27 (1992).
4. T. Sato, T. Sugiura and K. Shimakage, Microwave Joining of Ceramics, Silicon Nitride and Silicon Carbide, Journal of the Ceramic Society of Japan Vol. 101, (No.4) pp. 422-427 (1993).
5. B.H. Rabin and G.A. Moore, Reaction Processing and Properties of SiC-to-SiC Joints, MRS Proceedings, Vol. 314, pp. 197-203 (1993).
6. G.D. Soraru, F. Babonneau and J.D. Mackenzie, Structural Evolution from Polycarbosilane to SiC Ceramic, Journal of Materials Science, Vol. 25 pp. 3886-3893 (1990).
7. Lynn M. Ewart, A Study of Process Variables and Bond Strength in the Use of Polycarbosilane to Join SiC, Proceedings of the 19th Army Science Conference, Orlando, Florida (1994).
8. S. Yajima, K. Okamura, T. Shishido, Y. Hasegawa and T. Matsuzawa, Joining of SiC to SiC using Polyborosiloxane, Communication American Ceramic Society, 60[2], p.253 (1981).
9. T.A. Shan, R. Cozzens, Y.L. Tian and I. Ahmad, Microwave Curing of Silicon Carbide Ceramics from a Polycarbosilane Precursor, MRS Proceedings, Vol. 347, pp.729-734 (1994).
10. I. Ahmad, R. Silberglitt, W.M. Black, H.S. Sa'adaldin and J.D. Katz, Microwave Joining of Silicon Carbide Using Several Different Techniques, MRS Proceedings, Vol. 269, pp.271-276 (1992).

MICROWAVE JOINING OF ALUMINA BOTH DIRECTLY AND WITH 2223 SUPERCONDUCTOR INTERLAYER

J.Cai, B.S.Li, H.R.Zhuang and J.K.Guo
Shanghai Institute of Ceramics, Shanghai 200050

Y. L. Tian
George Mason University, Fairfax, VA 22030

ABSTRACT

Microwave joining has many advantages, such as high joint strength, rapid increase of temperature in heating, local heating and ease of control. By adopting and modifying the latest technology of microwave sintering and joining, a low power joining system has been set up. The preliminary tests indicate that Al_2O_3-Al_2O_3 direct joining is achieved at 1400 °C with soaking time of 20 minutes. A comparison between microwave joining with and without 2223 superconductor interlayer has been made. Joining temperature and soaking time markedly decrease when 2223 superconductor is applied as interlayer. Scanning electron microscope(SEM) is used to study the microstructure of the different joints.

INTRODUCTION

High performance ceramics, which are excellent in resistance to heat and corrosion, are promising candidates for replacing metals under high temperature environment. These ceramics have been developed for various applications including aeronautical, automotive, and electronic materials. However, ceramics are hard to machine, which limits their applications. One possible solution is to develop techniques for fabricating a complex shaped article from small molded parts by joining.

Microwave joining is a new technique developed rapidly in recent years. Because microwave heating is generated internally within the material instead of originating from external heating sources[1], microwave joining has advantages, such as local heating and ease of control.

Ceramic-ceramic seals by microwave heating were reported by Meek and Blake[2]. They discovered that less energy and less time were required to form a glass-ceramic seal by microwave heating than by conventional heating.

Fukushima et al. joined lower purity(92%) alumina directly by microwave heating[3]. When joined at 1750 °C for 3 minutes with pressure of 0.6 MPa, the joint strength of the sample was equal to the strength of the bulk alumina. Different materials were tested as interlayers in microwave joining SiC to SiC by Palaith et al.[4] Al-Assafi et al. microwave joined alumina using alumina gel as interlayer; they found that the gel-assisted in lowering the joining temperature to less than 1400 °C[5].

In this paper, direct joining of alumina at different temperatures has been presented. In previous study of preparing 2223 superconductor samples[6], 2223 superconductors were observed to have a strong bonding with alumina and stainless steel, hence 2223 superconductor is used as interlayer for the microwave joining of alumina.

EXPERIMENTAL PROCEDURE

Alumina rods 8 mm in diameter and 15 mm long were polished before microwave joining. The composition of the alumina rod is shown in Table 1. Interlayers 2223 were prepared by the co-decomposing method. Microwave joining was carried out in a TE_{10n} rectangular single-mode applicator.

Table 1 Compositions of Alumina Rod

Al_2O_3 (wt%)	SiO_2 (wt%)	MgO (wt%)	CaO (wt%)	Na_2O (wt%)
92.02	4.82	0.93	0.48	1.06

The details of microwave joining system have been reported by Cai et al[7]. The joining region is located in the maximum electric field strength zone of the applicator. The joining of the samples is conducted by applying an axial compressive load to the end of each specimen.

Microwave joining of alumina was carried out at 1300 °C, 1400 °C and 1550 °C with different soaking times. Al_2O_3 with 2223 as interlayer was microwave joined at 1000 °C for 10 minutes and post annealing at 855 °C for 12 hours was developed to ensure the sufficient diffusion of the elements.

The bulk X-ray diffraction(XRD) pattern was taken(RAX-10 diffractometer) at room temperature. The microstructure studies were conducted using an electron probe microanalzer (EPMA) (Model EPMA-8705QH_{II},

Shimadzu Co., Kyoto, Japan) equipped with an energy dispersive spectrometer (Model TN-5502N, Shimadzu).

RESULTS AND DISCUSSION

Direct Microwave Joining of Alumina

Direct microwave joining of alumina-alumina was undertaken at 1300 °C, 1400 °C and 1550 °C, followed by fracture analysis and SEM observations. The results are discussed below.

1. Microwave joining at 1550 °C

When the specimen was heated at 1550 °C for only 3 minutes, local melting was observed. According to the phase diagram, there might be a small amount of $Na_2O \cdot Al_2O_3 \cdot 2SiO_2$(melting point is about 1526 °C) inside the sample. XRD analysis has proven the existence of such substance. When the specimen was heated above 1526 °C, ion conduction of Na_2O caused the loss factor to increase rapidly. The net result wass an exponential increase in temperature, which caused thermal runaway to happen in this region, and therefore the temperature of this region was much higher than other parts, and local melting of the alumina sample was observed.

2. Microwave joining at 1300 °C

Alumina samples were not able to join at 1300 °C, even for a 160 minute long period, as shown in Fig.1. It is hard for atoms of two samples to diffuse into each other at 1300 °C; therefore the samples cannot be joined together.

Figure 1. SEM observation of boundary of microwave joined Al_2O_3 at 1300 °C.

3. Microwave joining at 1400 °C

For the alumina specimen joined at 1400 °C for 20 minutes, SEM micrograph shows that the joined boundary line becomes invisible. Little difference in microstructure can be observed between before and after joining.

Figure 2. SEM observation of boundary of microwave joined alumina at 1400 °C for 20 min.

Microwave Joining of Alumina with 2223 Interlayer

When prepared Bi-2223 superconductors ($Bi_{1.6}Pb_{0.4}Sr_2Ca_2Cu_3O_x$) by conventional heating in previous experiments, it was found that melted 2223 can bond alumina and stainless steel plates very well. Hence 2223 was applied as interlayer for microwave joining of alumina.

After microwave joining at 1000 °C for 10 minutes, alumina samples were joined together. To ensure the elements of 2223 and alumina sample diffuse sufficiently, the joined samples were post annealed at 855 °C for 12 hours. Fig.3(a) shows the microstructural features of the joined region (the arrow is the mark to indicate the joined region), Fig.3(b), (c) and (d) are corresponding x-ray images of Al, Si and Bi. As shown in the microstructure of the joined region, joined surface has already disappeared and element Al, Si, Bi are distributed homogeneously.

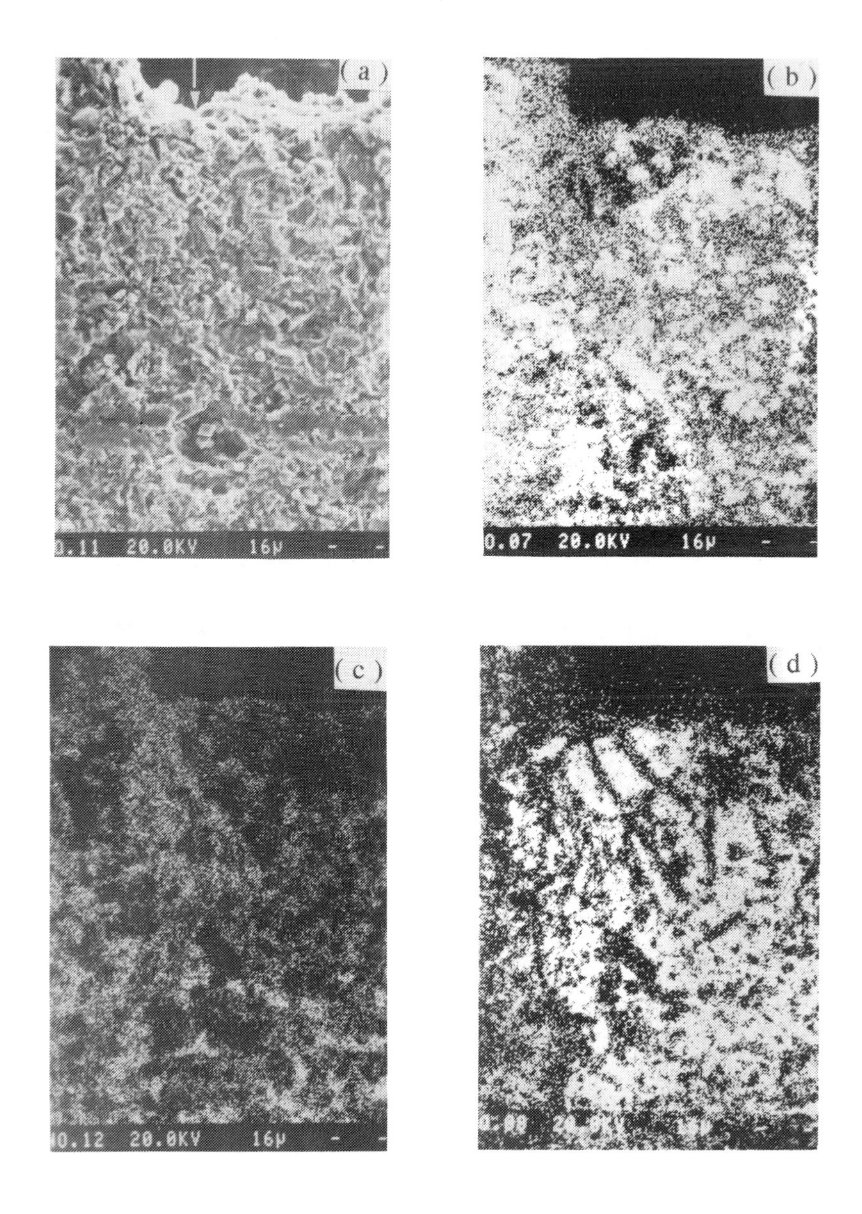

Figure 3. Micrography of joined region and its corresponding X-ray images after post annealing at 850 °C for 12 hr: (a) microstructure of the joined region;(b) Al K_{α} X-ray image;(c) Si K_{α} X-ray image;(d) Bi M_{α} X-ray image.

CONCLUSIONS

Microwave joining of alumina appears to be promising. After microwave joining at 1400°C for 20 minutes, the samples were joined and the SEM micrographs showed that the joined surface disappeared.

Indirect microwave joining of alumina with a 2223 interlayer is feasible. After microwave joining at 1000°C for 10 minutes and post annealing at 855°C for 12 hours, the samples were joined with elements distributed homogeously around the joined region.

REFERENCES

1. W.H. Sutton, *Am Cer. Soc. Bull.*, 68[2], pp. 376 (1989).
2. T.T Meek and R.D. Blake, *J. Mat. Sci.Letters, 5, pp. 270 (1986).*
3. H. Fukushima, T. Yamaka and M. Matsui, *J. Matls. Res.*, 5[2], pp. 397, (1990).
4. D. Palaith, R. Silberglitt, C.C.M. Wu, R. Kleiner and E. Libelo, *Am. Cer. Soc. Bull.*, 68[9], pp. 1601 (1989).
5. S. Al-Assafi and D.E. Clark, **Materials Research Society Symposium Proceedings**, Vol. 269, pp. 335 (1992).
6. J. Cai, X.M. Xie, B.S. Li, X.X Huang, T.G. Chen and J.K. Guo, *Supercond. Sci. Technology,* 5, pp. 599 (1992).
7. J. Cai, X-M. Xie, L-W. Zhang, X-T. Li, Y-L. Tian and J-K. Guo, *J. Am. Cer. Soc.*, 77[4], pp. 1090 (1994).

AUTOMATIC CONTROL DURING MICROWAVE HEATING OF CERAMICS

Mengli Li, Guy O. Beale, Yong Lai Tian
George Mason University
Fairfax, VA 22030

ABSTRACT

This paper presents results on the development of a feedback control system for regulating temperature in ceramic samples being heated by microwave energy for the purpose of joining. The main objective of the control system is to prevent thermal runaway in the ceramic samples during heating. Microwave power level and the positions of a movable iris and plunger are the variables being controlled. Results obtained from heating mullite and silicon carbide are presented. These results indicate the ability of the control system to prevent thermal runaway during microwave heating of the samples. This research has been funded by NSF Grant DDM-9115417.

INTRODUCTION

Advanced ceramics and composites based upon them can survive hostile chemical environments and temperatures as high as 1500-2000°C. Accordingly, there is great interest in using these materials in advanced automotive and aerospace engines, chemical process systems, high temperature heat exchangers, and advanced energy conversion systems. Since the brittleness of the ceramics, and the lack of reliable in-process methods to detect critical flaws make manufacturing of ceramic components difficult, there is a need for the capability of joining ceramic pieces to form larger pieces.

However, a major problem encountered when microwave heating certain ceramics, such as alumina, is thermal runaway. The exponential increase in loss factor [1] causes localized melting of the material at a temperature below that which allows joining of the components. The major goal of this research has been to develop a feedback control system which will prevent thermal runaway from occurring during the joining process. The cavity used in this research is a rectangular S-band waveguide with a sliding metal plunger and an adjustable inductive iris. The samples which have been tested are cylindrical rods and tubes of alumina, mullite, and silicon carbide.

Prevention of thermal runaway has been accomplished by sensing the temperature of the ceramic material, processing that measurement through a control algorithm stored in a computer, and using the results of that computation to control the level of the microwave power being applied to the cavity. A two-color pyrometer is currently used for temperature measurement, and the microwave source is a magnetron operating at 2.45 Ghz, in TE_{103} mode, with a maximum power level of 1000 watts. A detailed derivation of the control algorithm used for temperature regulation is given in Beale, et al. [2] and Beale, et al. [3].

Two other variables must also be controlled during the heating process, namely the coupling of microwave energy into the cavity and the resonant frequency in the cavity. If these variables are not carefully controlled, the ceramic material will not absorb enough microwave energy to reach the temperatures of interest. These variables are controlled by manipulating the size of the inductive iris and the position of the mechanical plunger. The plunger and iris are adjusted by controlling the positions of two stepping motors; the control of these motors is accomplished with the same computer that is used for the temperature control.

TEMPERATURE CONTROL

The core temperature of the ceramic sample being heated by microwave energy is governed by the following nonlinear differential equation [3]. The parameter $\kappa''(T)$ is the dielectric loss factor. The various k_i coefficients are functions of the material properties for the sample being heated, the material geometry, and the geometry of the microwave cavity. The loaded quality factor for the resonant cavity and sample is represented by Q(T).

$$\frac{dT}{dt} = k_1 \kappa''(T) Q(T) P_{IN} - k_2[T^4 - (296)^4] - k_3(T - 296)^{5/4} - k_4(T - 296) \tag{1}$$

Based on work presented by Fukushima, et al. [1], an analytic expression was developed for the loss factor of alumina. Since alumina was considered to be the material most subject to thermal runaway which would be considered during this research, the material parameter values for alumina were used for the design of the temperature control system. The expression for the loss factor for alumina is

$$\kappa''(T) = 3.7276 \times 10^{-3} \times 10^{(T-273)/700} \tag{2}$$

The quality factor for the cavity was modeled by equation (3), where V_{cav} and V_{samp} are the cavity and sample volumes, respectively, and κ' is the real part of the dielectric constant.

$$Q(T) = 1500\left(\frac{V_{cav}}{4V_{samp}} + \kappa'\right)\left(\frac{1}{3000\kappa'' + \frac{V_{cav}}{4V_{samp}} + \kappa'}\right) \quad (3)$$

A linearization of the nonlinear heating equations was developed to provide a linear system model which could be used for designing the temperature control system. The linearization was evaluated at a number of different microwave power and temperature values to produce a family of linear models. A worst case model from the perspective of control design was identified, and the design of the temperature control system was based on that model. The resulting controller is guaranteed to provide a stable response for all the other linear models as well. The temperature controller is modelled as

$$G_c(s) = \frac{60(s + 1.75)}{s} = \frac{P_{Commanded}}{T_{Meas}} \quad (4)$$

where s is the complex variable associated with the Laplace transform. Since the control algorithm is implemented on a digital computer, the above controller must be converted into a discrete-time representation through the Z-transform. The equation which is programmed in the computer is

$$P_{IN}(k) = P_{IN}(k-1) + K_{DE}T_{ERR}(k) - K_{DEP}T_{ERR}(k-1) \quad (5)$$

where k represents an arbitrary point in discrete time, and T_{ERR} is the difference between the desired temperature and the measured temperature. The values of the coefficients K_{DE} and K_{DEP} depend on the values in the continuous-time representation and on the value of the sample period, the time between consecutive measurements of the temperature. The sample period is 0.1 seconds. The value of the microwave power level computed from Eqn. (5) is compared with a user-supplied upper limit on value and on rate of change. If the computed value exceeds its maximum allowed value or rate of change, the computed value is limited accordingly. Although the temperature controller is simple in form, its guaranteed stabilization of the linearized models and the experimental results which have been obtained have justified its use. Figure 1 is a block diagram of the digital implementation of the temperature control algorithm.

EXPERIMENTAL CONTROL OF TEMPERATURE

Samples of silicon carbide rods and mullite tubes have been heated using this control system. Commanded temperatures for silicon carbide ranged from 1200°C to 1550°C. The temperature control algorithm was able to maintain the actual temperature within ±10°C [4]. Although silicon carbide does not suffer from thermal runaway, those experiments

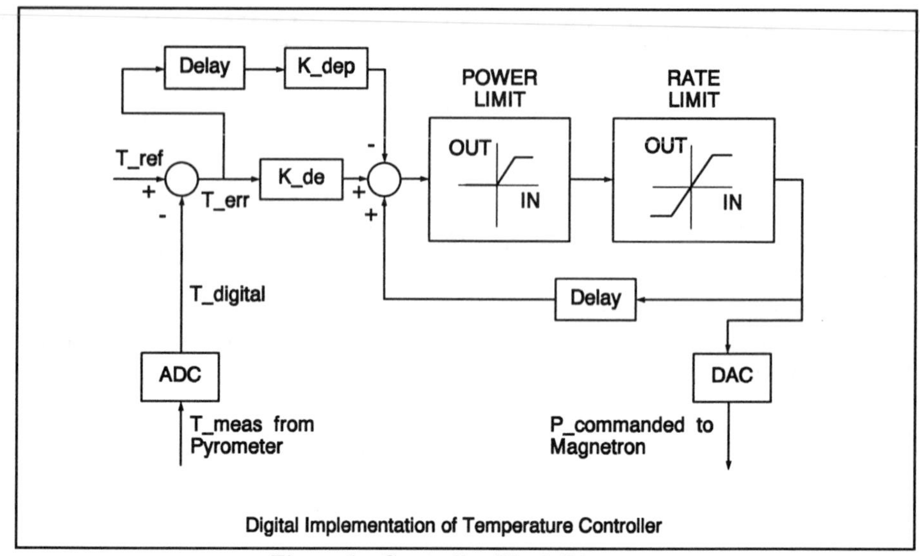

Figure 1. Controller Block Diagram.

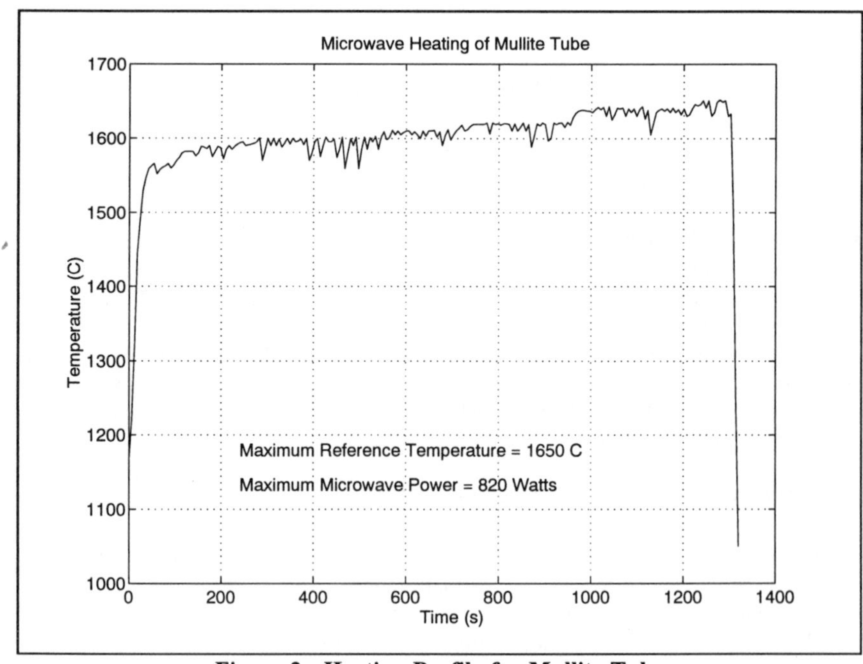

Figure 2. Heating Profile for Mullite Tube

demonstrated the ability of the control system to automatically regulate temperature.

With the mullite tubes, commanded temperatures up to 1650°C have been tested, which is above the thermal runaway temperature. Figure 2 illustrates the ability of the temperature control algorithm to prevent thermal runaway in mullite.

IRIS AND PLUNGER CONTROL

The plunger and iris are used to control the resonant frequency of the cavity and the coupling of microwave energy into the cavity. The plunger provides a movable short circuit which sets the length of the cavity and is used to control resonant frequency. The coupling of microwave energy is affected by the size of the iris. The required size of the iris and the required position of the plunger change with the temperature of the sample.

Two control strategies are used for the iris and plunger, one at low temperatures and one at high temperatures. Most of the movements of the iris and plunger take place at low and moderate temperatures, well below the temperature used for joining. The iris and plunger have to be manipulated appropriately at low power and temperature levels in order to have enough microwave energy available at the sample to heat it to the operating temperature. Therefore, the strategy is move the plunger and iris in order to minimize the reflected power. Once the initial positions of the iris and plunger are determined, a gradient search method,

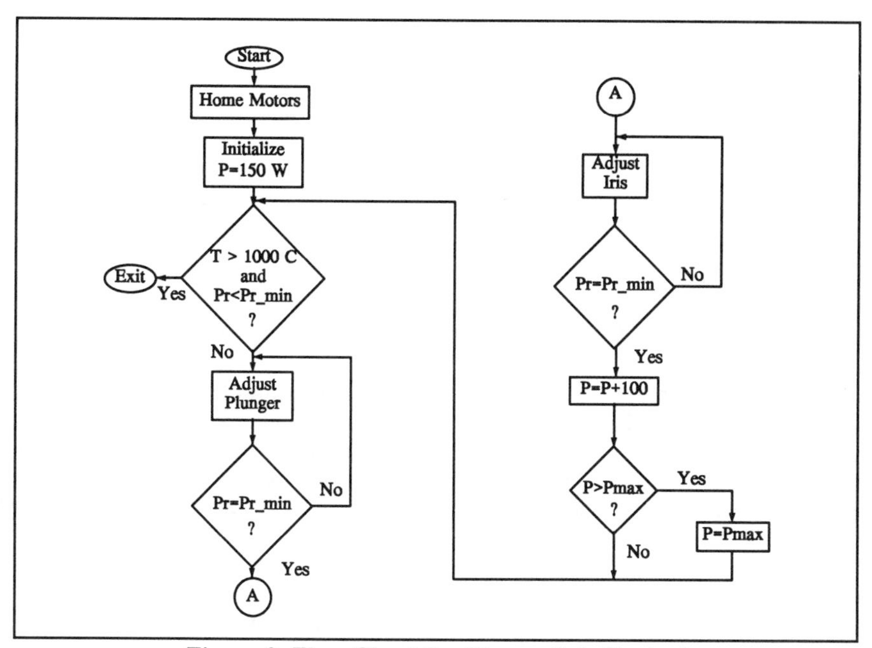

Figure 3. Flow Chart for Plunger/Iris Control

which is based on reflected power measurements, is used to control their movement. Figure 3 is the software flow chart for the control of the iris and plunger at low temperatures.

Figure 4 shows the time history of reflected power during the low temperature tuning of the cavity. The values in the plot were obtained by sampling the actual data at six second intervals. The size of iris is being automatically controlled, while the plunger was manually tuned during this experiment. The automatic control system for the iris was able to achieve a reflected power value as low as could be obtained manually.

A second control strategy is used when the temperature reaches 1000° C. Up to this point,

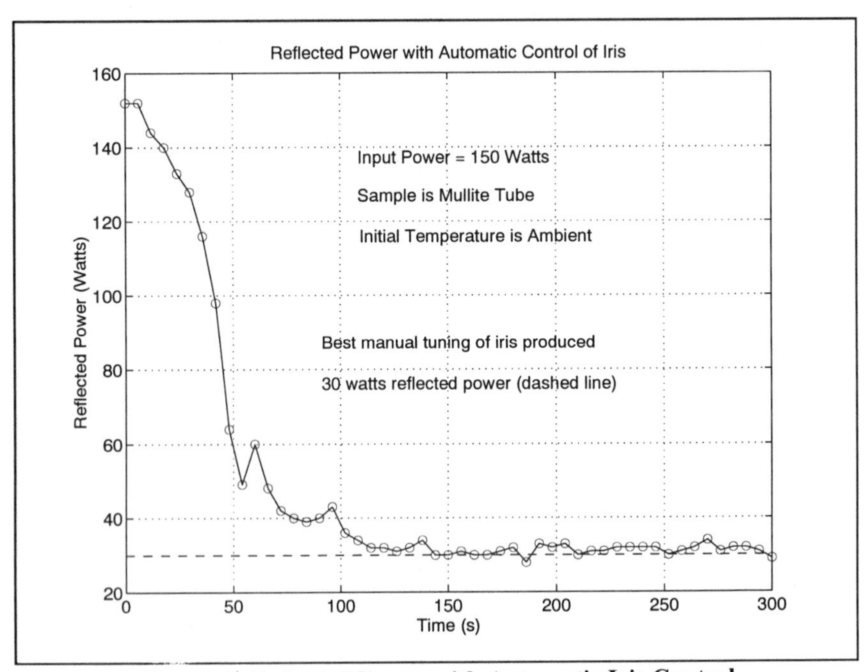

Figure 4. Reflected Power with Automatic Iris Control

there has been no control of the sample's temperature; the microwave power has been set at selected fixed values. Although the temperature has been monitored, no attempt was made to regulate it. However, above 1000°C, control of temperature becomes critically important to prevent thermal runaway and to achieve and maintain the correct temperature for joining. Although the need for plunger and iris tuning is less critical at high temperatures, there is still a need for such adjustments.

The control strategy for high temperatures is derived from an equivalent circuit model for the iris, plunger, and ceramic sample, by representing them as complex impedances located at appropriate points along an electric transmission line; the values of the impedances vary with temperature [4]. Minimizing the reflected power becomes equivalent to matching the impedances in this circuit model. The Marcuvitz model was used to represent the impedance of the ceramic rod being heated [5]. To minimize the reflected power, the expressions shown in equations (6) and (7) have been derived for the optimum iris size (X_1) and plunger position (l_2). In those equations, a is the width of the cavity, $\alpha=2\pi R/\lambda_0$, R is the sample radius, λ_0 is the free space wavelength, λ_g is the wavelength of the lowest mode in the cavity, $\beta=2\pi/\lambda_g$. The variables b_r and g_r are defined in equation (8).

$$X_1 = \left(\frac{2a}{\pi}\right)\cot^{-1}\left\{\left[\left(\frac{a}{\lambda_g}\right)\sqrt{\frac{1-g_r}{g_r}}\right]^{1/2}\right\} = \left(\frac{2a}{\pi}\right)\cot^{-1}(B) \tag{6}$$

$$l_2 = \left(\frac{1}{\beta}\right)\cot^{-1}\left[b_r \pm \sqrt{g_r(1-g_r)}\right] = \left(\frac{1}{\beta}\right)\cot^{-1}(D) \tag{7}$$

$$g_r = \left(\frac{\alpha^2\lambda_g}{a}\right)\kappa'' \quad , \quad b_r = \left(\frac{\alpha^2\lambda_g}{a}\right)(\kappa'-1) \tag{8}$$

Although equations (6) and (7) give the optimum values, they assume exact knowledge of the material properties of the sample being heated. To reduce the effects of this dependency, an incremental control strategy is developed. Derivatives of X_1 and l_2 with respect to temperature can be taken, and these new expressions can be used to incrementally vary the iris and plunger positions based on temperature measurements. The derivative expressions are

$$\frac{dX_1}{dT} = \left(\frac{a}{2\pi\sin^2(B)}\right)\left(\frac{a}{\lambda_g}\right)^{1/2}\frac{1}{g_r^2}\left(\frac{1}{g_r}-1\right)^{-3/4}\left(\frac{\alpha^2\lambda_g\ln(10)\kappa''}{700a}\right) \tag{9}$$

$$\frac{dl_2}{dT} = \pm\left(\frac{1}{2\beta}\right)\frac{1}{\sin(D)}\left[\frac{1}{g_r(1-g_r)}\right]^{1/2}(1-2g_r)\left(\frac{\alpha^2\lambda_g\ln(10)\kappa''}{700a}\right) \tag{10}$$

These derivatives are used to compute the incremental change in iris size and plunger position by multiplying them by measured changes in temperature, Δ_T:

$$\Delta_{X_1} = \frac{dX_1}{dT} \Delta_T \quad , \qquad \Delta_{l_2} = \frac{dl_2}{dT} \Delta_T \tag{11}$$

Using these expressions in equation (11), the required positions of the iris and plunger can be periodically updated. Based on empirical evidence, it appears that values for Δ_T in the range of 25°-50° C will be adequate for controlling the coupling and resonant frequency in the high temperature regime.

CONCLUSIONS

Although the plunger control system has not been fully tested yet, the ability of the temperature control system and the iris control system have been shown to fulfill their goals. The temperature controller is able to regulate temperature accurately, and it can be used to achieve temperatures above the thermal runaway temperature without runaway. The iris controller is able to minimize the reflected power so that high temperatures can be reached.

ACKNOWLEDGMENT

The authors thank Dr. Murray Black of George Mason University and Dr. Richard Silberglitt and Dr. Iftikhar Ahmad of FM Technologies, Inc. in Fairfax, VA for their helpful advice and constant support during this research.

REFERENCES

1. H. Fukushima, T. Yamanaka, and M. Matsui, J. Material Research, Vol. 6, No. 2, p. 397 (1990).
2. G.O. Beale, F.J. Arteaga, and W.M. Black, "Design and Evaluation of a Controller for the Process of Microwave Joining of Ceramics," IEEE Transactions on Industrial Electronics, Vol. 39, No. 4, pp. 301-312 (1992).
3. G.O. Beale, and F.J. Arteaga, "Automatic Control to Prevent Thermal Runaway During Microwave Joining of Ceramics," Proc. of Spring Meeting of the Materials Research Society, pp. 265-270, San Francisco (1992).
4. M. Li, G.O. Beale, and W.M. Black, "Feedback Control of Temperature Profiles During Microwave Joining of Ceramics," Proc. of the Spring Mtg. of the Mat. Res. Soc., pp. 643-648, San Francisco (1994).
5. N. Marcuvitz. Waveguide Handbook, Vol. 10, McGraw-Hill (1951).

MICROWAVE JOINING OF ZINC SULFIDE

R.R. Di Fiore and D.E. Clark
University of Florida, Gainesville, FL

ABSTRACT

Zinc sulfide (ZnS) is useful as a transmission window due to its high transmissibility in the infrared (IR) region. A thin interlayer was used to join ZnS bars using microwave hybrid heating. The resulting joint was examined using FT-IR spectroscopy to determine the transmission of IR light and mechanically tested using a four point bend test to determine the strength of the joint.

INTRODUCTION

Zinc sulfide has a very low vibrational frequency, and exhibits excellent transmission in the infrared. Typically, ZnS has a transmission range of 0.5 - 12 μm, with a refractive index of 2.2 at 3 μm [1]. The ZnS substrate used in this study was grown by chemical vapor deposition (CVD). The yellow color of the samples indicated the presence of sulfur and zinc hydride impurities. The zinc hydride impurity was characterized by an absorption peak at 6.1 μm (1639.3 cm^{-1}) [2,3]. All samples examined using Fourier transform infrared spectroscopy exhibited this phenomenon.

Zinc sulfide is used as an infrared window in IR detection systems. Manufacture of the radome from CVD grown ZnS can be a complex and expensive process. An alternative is the growth of simple geometry ZnS, which could then be joined to form the radome. Since zinc sulfide tends to oxidize rather than bond to itself, the use of an interlayer is necessary [4,5,6]. Several materials were tested and the silicate glasses were found to be most compatible. Since silicate glasses have distinct absorption bands in the infrared, it was necessary to limit the interlayer width to less than 50 μm. This reduced the absorption peaks due to silica. To optimize performance, the objective was to form a joint with IR transmission and mechanical strength approaching that of the as-received ZnS.

One of the advantages of using microwave energy is selective heating. This can be achieved by using a combination of microwave hybrid heating (MHH) and a microwave absorbent material [7]. The silicate glasses used as interface materials were chosen due to their dielectric properties (tan $\delta = 0.01$).

EXPERIMENTAL PROCEDURE

Zinc sulfide samples were obtained as ASTM test bars (13 mm × 6.1 mm × 2.3 mm). The surfaces to be joined were prepared with 250 grit SiC. A rough surface was desired to improve the strength of the joint. Several materials were used as interlayers: silica and alumina gels, soda-lime-silica (SLS) glass (in the form of 160 μm thick cover slips) and a borosilicate frit (<38 μm diameter). The SLS glass and borosilicate frit were chosen as interlayer materials. Zinc sulfide powder (particle size <38 μm) was added to the borosilicate frit to increase the index of refraction of the interlayer. Interlayer frits were prepared with the weight percent of ZnS varying from 10 % - 50 %. Frits of greater than 10 weight percent ZnS were found to have a large volume fraction of voids and could not be used as an interlayer.

The interlayer material was layered between two ZnS bars and placed in a zirconia sample holder to insure that the ZnS assembly was aligned during processing (Figure 1). The zirconia, which suscepts at temperatures above 600 °C, helped to maintain a uniform temperature along the length of the ZnS assembly. The refractory assembly was placed within a modified 900 watt, multimode commercial microwave oven.

A 0.5 - 1.0 kg load (36 - 72 kN) was applied along the longitudinal axis to force the interlayer into the irregular surface of the ZnS. The load reduced the width of the SLS and borosilicate frit interfaces. Microwave hybrid heating was used to reach the processing temperatures (Figure 2) [7,8]. Four different atmospheres were

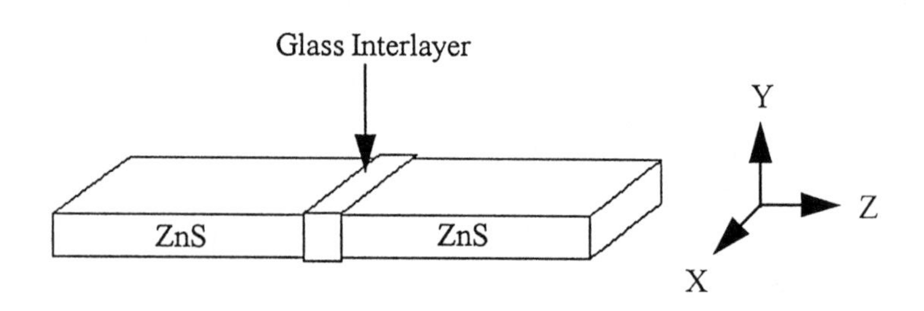

Figure 1. Orientation of the joined ZnS bars.

investigated: nitrogen, helium, air in the presence of carbon powder and silicon carbide, and air with silicon carbide alone. Silicon carbide (18 mesh ASTM) was chosen as the susceptor due to its high loss factor (tan δ). It was observed that the use of nitrogen and helium atmospheres did not result in joining. Processing in air resulted in heavily oxidized joined specimens. In the cases where carbon powder (200 mesh) was used with air, it provided an atmosphere which greatly reduced oxidation of the ZnS surface. These ZnS assemblies were processed at different temperatures to determine if there was an effect on the mechanical and optical properties of the ZnS.

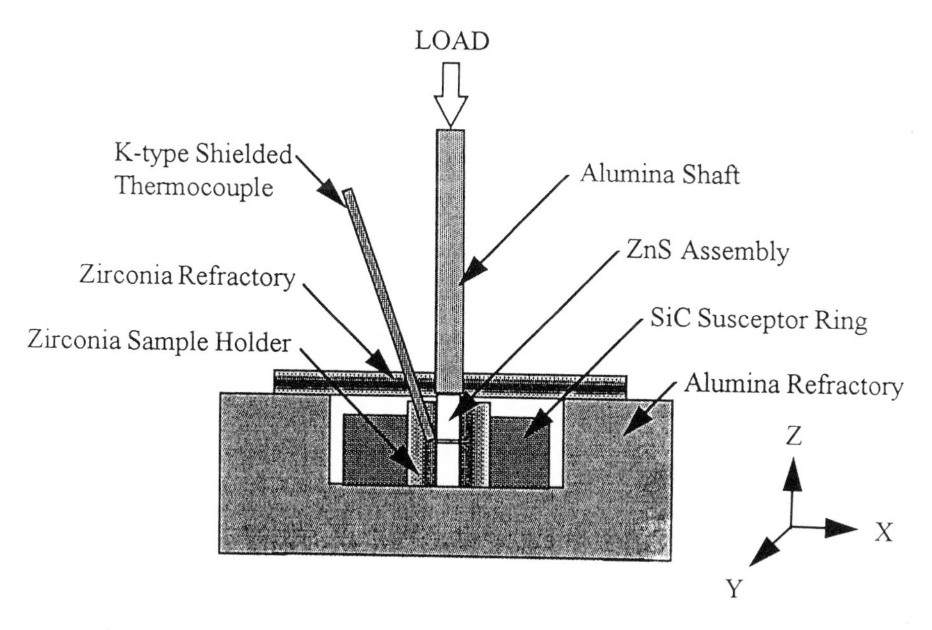

Figure 2. Microwave refractory assembly for the joining of ZnS.

RESULTS

In the present study, the joining temperature was approximately 600 °C for the borosilicate interlayer and 450 °C for the SLS interlayer. Processing times were varied between 10 to 15 minutes.

Zinc sulfide was successfully joined using two types of interlayers processed in air with carbon powder; a SLS glass and a borosilicate frit. In contrast, there was no evidence of joining in an inert atmosphere (nitrogen, helium). The borosilicate frit resulted in a thinner interlayer (30-40 μm) than the SLS glass (50-75 μm) under an applied load. The frit, mixed with ethanol, was easily applied to the surface and shows promise as an interlayer, especially when joining complex shapes. Microwave hybrid heating resulted in processing times as low as 10 minutes for ZnS joined with the SLS glass or borosilicate frit. The as-received ZnS was found to exhibit crystal growth when held at elevated temperatures (>450°C) for more than twenty minutes, based on visual observation. This was not found to have an effect on the IR transmission spectra of as-received samples.

Zinc sulfide bars of similar dimensions were analyzed using FT-IR spectroscopy in transmission [9]. Samples were aligned along the longitudinal axis so that the infrared beam passed directly through the joint (Figure 3). The transmission through joined samples compared favorably to the as-received ZnS for interlayers of less than 50 μm in thickness. The absorption peak due to zinc hydride (1640 cm^{-1}, or 6.1 μm) was eliminated in the ZnS samples processed in the microwave at temperatures above 450 °C. Response in the infrared range of interest (1 - 10 μm) was 80% - 85% of the transmittance of the as-received ZnS test bars (Figure 4). There is a mismatch in the index of refraction (η) of ZnS (2.2 at 3 μm) compared to that of SLS (1.51) and borosilicate glass (1.47). The η mismatch can lead to the reflection of the infrared light at both the ZnS-glass and glass-ZnS interfaces. If the incident light beam is perpendicular to the interface, the internal reflection at each interface is approximately 4% - 5%. This can be reduced by matching the index of refraction of both materials.

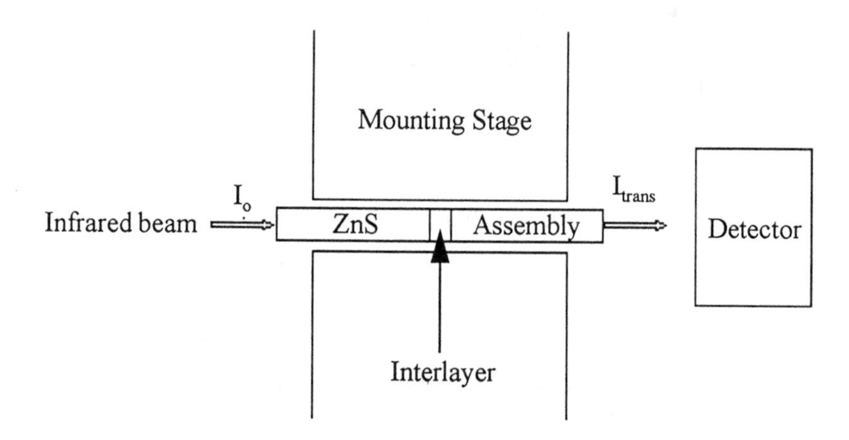

Figure 3. Alignment of joined ZnS for FT-IR analysis.

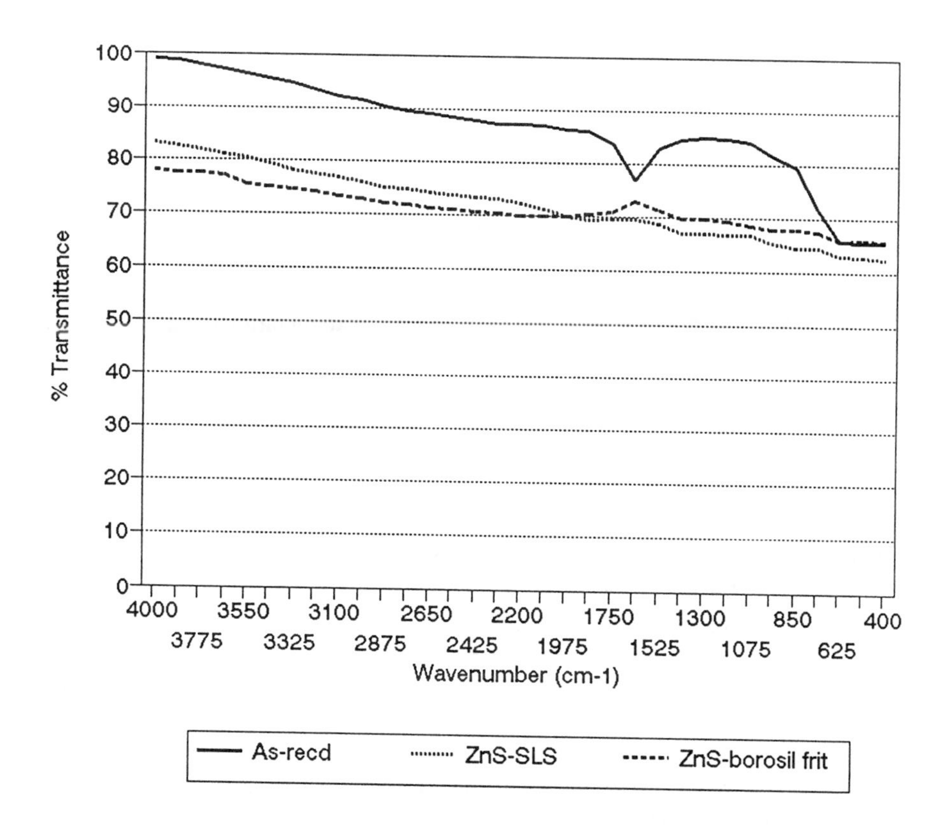

Figure 4. FT-IR spectra of ZnS (as-received), ZnS/SLS, and ZnS/borosilicate frit.

The surfaces of all samples were polished with 0.7 μm alumina powder before mechanical testing. The interlayer joint was indented with a 2 kg load and placed on a four point bend apparatus to measure the fracture strength (Figure 5). Fracture strength values were obtained for as-received ZnS bars and joined samples (3-4 samples each) for each processing schedule The four point bend test caused fracture along the ZnS-glass interface. All joined specimens tested in this manner failed at the interface.

The processing temperature had an effect on the strength of the ZnS. The fracture strength of the as-received ZnS was 32.8 MPa; ZnS at 450 °C, 38.9 MPa; and at 650 °C, 32.9 MPa (Table 1).

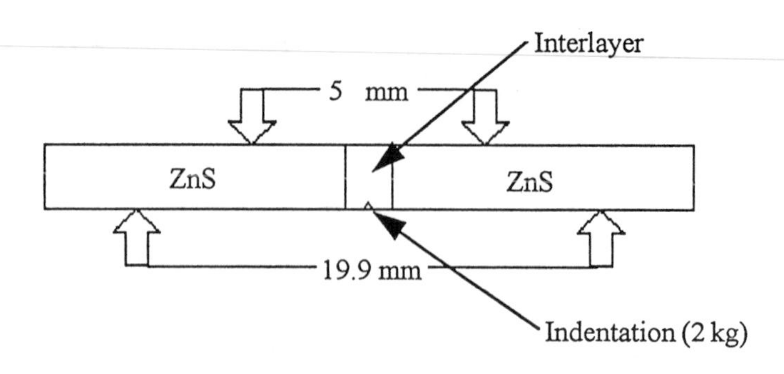

Figure 5. Four-point bend test of joined ZnS.

The ZnS test bars joined using an SLS glass interlayer had a strength which was 65% of the ZnS processed at 450 °C. The ZnS joined with borosilicate frit (at 600 °C) had a strength comparable to that of ZnS processed at the same temperature (33.7 MPa and 32.9 MPa, respectively). A higher processing temperatures was required for the borosilicate frit due to a higher softening temperature. The addition of ZnS to the borosilicate frit increased the softening point and also exhibited void formation within the interface. The void formation may have resulted in a lower fracture strength.

Table 1. Average fracture strength of ZnS test bars, ZnS/SLS and ZnS/borosilicate frit joined specimens (MPa).

Processing		Interlayer Material:		
Temperature	Zinc sulfide	SLS	Borosilicate	ZnS/borosilicate
25 °C	32.7 (as-rec'd)	-	-	-
450 °C	37.9	24.5	-	-
600 °C	32.9	15.0	33.7	26.7

SUMMARY

Zinc sulfide has been successfully joined using microwave energy. This was achieved using a combination of microwave hybrid heating and a microwave absorbent interface. Interface materials were developed using either SLS glass, borosilicate frit, or a ZnS-borosilicate frit. The infrared transmission through the joint was 80 %- 85 % of that of ZnS. The fracture strength of joined specimens was comparable to that of as-received ZnS processed at the same temperature.

ACKNOWLEDGMENTS

We would like to thank Dan Harris of China Lake and Dr. Steve Fishman of ONR for their guidance and financial support. Also, a special thank you to Guy La Torre at the Advanced Materials Research Center at the University of Florida (UF), and Dr. J.J. Mecholsky, Jr. and Joan Chan of UF.

REFERENCES

1. P. N. Kumta and S. H. Risbud, "Novel Glasses in Rare-Earth Sulfide Systems," *Am. Ceram. Soc. Bull.,* **69** [12] 1977 - 1984 (1990).
2. K. L. Lewis, G. S. Arthur and S. A. Banyard, "Hydrogen Related Defects in Vapor-Deposited Zinc Sulfide," *Journal of Crystal Growth*, **66** 125 - 136 (1984).
3. C. H. Lee and C. Y. Pueng, "Preparation and Properties of ZnS Thin Films by Low-Pressure Metalorganic Chemical Vapour Deposition," *Journal of Materials Science*, **28** 811 - 816 (1993).
4. C. H. Mathewson, Zinc, The Science and technology of the Metal, Its alloys and Compounds, Reinhold Publishing Corp., London, England (1960).
5. A. Celikkaya and M. Akinc, "Morphology of Zinc Sulfide Particles Produced from Various Zinc Salts by Homogeneous Precipitation," *J. Am. Ceram. Soc.*, **73** [2] 245 - 250 (1990).
6. W.W. Chen, J. M. Zhang, A. J. Ardell, B. Dunn, "Solid-State Phase Equilibria in the ZnS-CdS System," *Mat. Res. Bull.* **23** 1667 - 1673 (1988).
7. W. H. Sutton, "Microwave Processing of Ceramic Materials," *Am. Ceram. Soc. Bull.*, **68** [2] 376 - 385 (1989).
8. D. Palaith and R. Silberglitt, "Microwave Joining of Ceramics", *Am. Ceram. Soc. Bull.*, **68** [9] 1601 - 1606 (1989).
9. W. H. Dumbaugh, "Infrared Transmitting Glasses", *Optical Engineering*, **24** [2] 257 - 262 (1985).

APPARATUS FOR THE JOINING OF CERAMICS USING MICROWAVE HYBRID HEATING

A.D. Cozzi and D.E. Clark
Department of Materials Science and Engineering
University of Florida
Gainesville, FL 32611

M.K. Ferber and V.J. Tennery
High Temperature Materials Lab
Oak Ridge National Labs
Oak Ridge, TN 37831

ABSTRACT

An apparatus was designed and constructed to facilitate the joining of ceramics in a microwave field. The microwave unit used is a modified Goldstar MA-1172M household microwave oven. The maximum load that can be applied is 2.24 kN. Temperature can be monitored with either a shielded type R thermocouple or an optical pyrometer. Measurements of the temperature and applied load are collected remotely.

INTRODUCTION

Joining is an important step in the fabrication of engineering structures. While it is often preferable to fabricate the product in one piece, it is not always possible to achieve this. In some cases, forming a single piece is not even the prescribed method for manufacturing the desired shape. This is most apparent in the case of a final product that consists of more than one material. Other examples where joining is the preferred method of manufacture are when the piece is a complex shape that would require either a complicated mold or extensive machining.

The applications for the joining of ceramics are extensive. Ceramic/metal joints are the most common. They appear in light bulbs connecting the metal base to the glass tube, substrates for integrated circuits, and in spark plugs attaching the metal core to the ceramic sheath.

The use of ceramic/ceramic joining is popular in the traditional ceramic industries such as joining bricks with mortar or attaching a handle to a ceramic mug. It is less common but no less important in advanced ceramics. Ceramic/ceramic bonds can provide increased resistance to operating temperature over metallic alloys that require costly and/or imported metals such as chromium and cobalt.

Until recently, most of the work performed in the joining of materials has used conventional heating. That is, the specimens to be joined were heated in either a resistance or RF heated furnace. Lasers have also been employed to heat the workpiece. Lately, microwaves have been used to provide the energy necessary to heat the specimens to temperatures suitable for joining.

Loehman[1] discusses the potential for saving money and time through the use of microwaves. He also mentions promise for the improvement of microstructure and hence, mechanical properties. Using a single mode microwave, Palaith et al.[2] joined mullite rods. The temperature was as high as 1300°C and the pressure was 9 MPa. They reported an increase in strength of the joined material over the starting material and attributed it to local densification. The integrity of the joint during heating was evaluated by monitoring the response from an acoustic wave transmitted through the specimen. The authors also provide background on the mechanisms that enable microwaves to heat materials. Alumina, ranging purity from 92% to 99% were joined by Fukushima et al.[3] In the 92% and 96% purity alumina, direct joins were obtained that had strengths near the of the original material. The pressure was nominal at 0.6 MPa and the temperatures used were as high as 1850°C. The 99% purity alumina was joined using sheets of the lower purity alumina as an interface. The temperatures used were as high as 1720°C and the pressure varied up to 2.4 MPa. Strengths were reported up to 90% of the starting material. Al-Assafi[4] used a multi-mode microwave to join 94% and 96% alumina. An alumina sol-gel precursor was used as the interlayer. The temperatures used for joining ranged from 1350°C to 1650°C. Microhardness measurements across the joint area revealed little or no difference in the hardness from the bulk. Micrographs of the fracture surface of the alumina joined at 1650°C exhibit no discontinuities at the joint.

The objective of this study is to design, build and demonstrate the effectiveness of an apparatus for the microwave joining of ceramics.

EXPERIMENTAL PROCEDURE

A home model microwave oven* was modified to allow for the application of pressure and measurement of temperature, figure 1. The turntable assembly was removed from the base of the microwave oven. An aluminum fan was attached to the turntable motor which was mounted opposite the waveguide. Twenty-five millimeter holes were machined in the casing to accommodate the alumina pushrod for pressure application, sighting of the specimen with the pyrometer and for a feed through for the thermocouple. Remote control of an air cylinder is used to apply pressure.

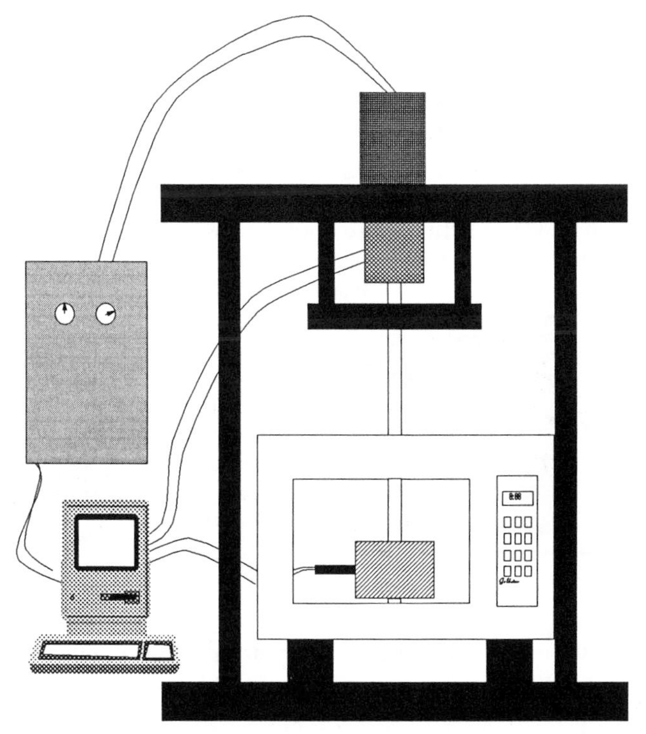

Figure 1. Apparatus used for microwave joining.

The pressure is adjusted using feedback from a load cell mounted in-line with the air cylinder. The alumina pushrod passes through a copper tube inserted in to top of the microwave to attenuate the microwaves that may pass out of the cavity. The

* Goldstar model MA-1172-M, 1.1 cu. ft., 1000 watts

thermocouple* has a platinum alloy sheath to prevent interference from the microwaves. It enters the microwave through the back wall of the cavity and is positioned approximately two millimeters from the joint. The optical pyrometer** focuses on the joint area of the bars through a hole in the front window screen. The material used for this research are 220 millimeter Coors AD995, 99.5% alumina rods. From the longer rods, bars 25 mm long by 19 mm square are cut for the joining experiments. The ends to be joined are machined with a 240 grit diamond wheel. A 0.2 - 0.3 mm interlayer of Coors AD96, 96% alumina was used between the pieces to facilitate joint formation. The surface of the interlayer was ground flat in the same manner as the bars to be joined. The two end members and the interlayer are placed in a susceptor made from 30% silicon carbide granules and 70% high-purity alumina cement***. This is encased in a low density aluminosilicate refractory and placed on a refractory brick. The whole assembly, figure 2, is placed in the microwave oven and the pieces to be joined are aligned with the pushrod. A nominal load is used to hold the assembly in place during heating. The alumina is heated to the desired temperature and the chosen load is applied. After a given time, the load is removed and the microwave oven is permitted to cool.

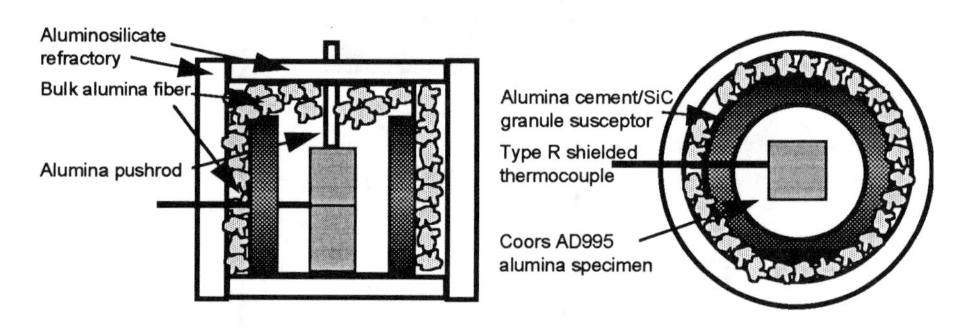

Figure 2. Microwave hybrid heating assembly used to join alumina.

For comparison purposes, the experiment was also performed in a conventional furnace. Following the same procedure, the temperature, pressure and time routine were matched as closely as possible. The only variation from the

---

*ARi Industries, Inc. model # HIT-99N-4DR9AA-36

**IRcon model # UX-20

***Alcoa AL-66 cement

microwave heating was that the conventional heating took more time to reach the test temperature.

The maximum joining temperature used for this study was 1400°C in all cases. This temperature was chosen to minimize the possibility of grain growth of the high-purity alumina. Previous work has shown that time is not an important factor[3]. Thirty minutes was chosen as the time under load, at temperature for all experiments. The pressures used ranged from the nominal pressure of 61 kPa to a maximum of 3.03 MPa.

The joined bars were then machined into 3 mm x 4 mm x 50 mm specimens for mechanical testing. Depending on the fragility of the joint, two to eight specimens were made from each experiment. The strength of the specimens was measured using a four point bend test. Strength tests were performed at room temperature and at 1000°C.

Fracture analysis of the specimens was performed on a stereo microscope. Joined sections were polished to examine the joint area.

RESULTS

During the joining process, the pressure and the temperature were both controlled. In a matter of moments, the pressure could be raised or lowered throughout the entire available range. The setpoint of 1400°C was obtained in approximately one hour.

The strength of the specimens joined using 1.2 MPa pressure in the microwave oven varied from 62 to 127 MPa. Only two of the conventionally joined specimens endured the machining process. These specimens had strengths around 37 MPa. Figure 3 is the comparison of the strengths of both the microwave and conventionally joined bars. Under similar joining conditions, the specimens joined in the microwave had an average strength three times that of the conventionally joined specimens. This is not to say that the conventionally joined bars would have similar strengths given longer exposure times at temperature. The increase in strength may be due to the heightened microwave absorption of the interlayer generated by the greater dielectric losses with respect to the high-purity alumina. This phenomenon can produce localized heating in the joint region. Figure 4 is the effect of pressure applied during joining on strength. The bars joined with only a nominal load did not join well enough to machine specimens. A fifty percent increase in pressure raised the breaking strength from an average of 111 MPa to 171 MPa. This is a significant improvement in strength. The greater load represents only half the load of which the apparatus is capable of applying. The as-received specimens had an average room temperature strength of 303 MPa. In bend tests performed at 1000°C, the average strength of the as-received specimens was 176 MPa. This is 58% of the value of the specimens tested at room temperature. The results of the specimens joined using 1.8 MPa pressure followed a similar trend. The average strength of these specimens at 1000°C was 67 MPa, 39% of the room temperature strength. Figure 5 is the strength of both the microwave joined and the as-received specimens tested at room temperature and 1000°C.

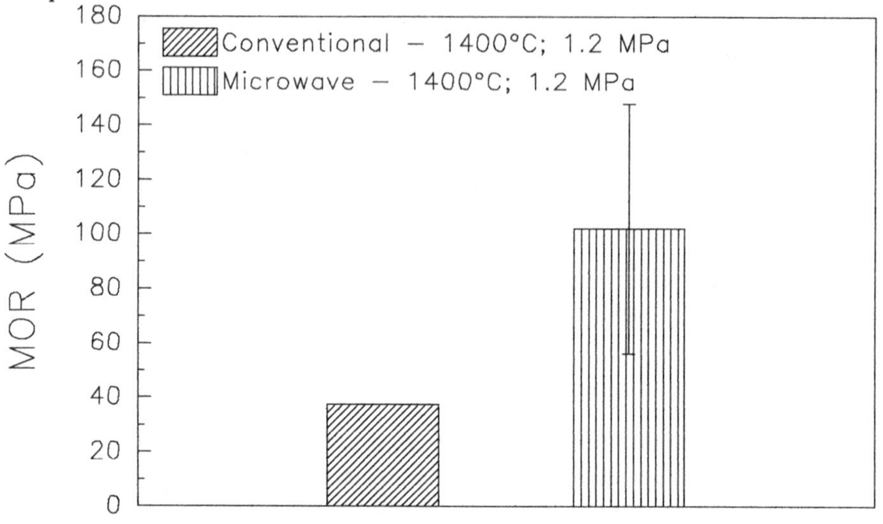

Figure 3. Strength of microwave and conventionally heated specimens joined at 1400°C using 1.2 MPa pressure for 30 minutes.

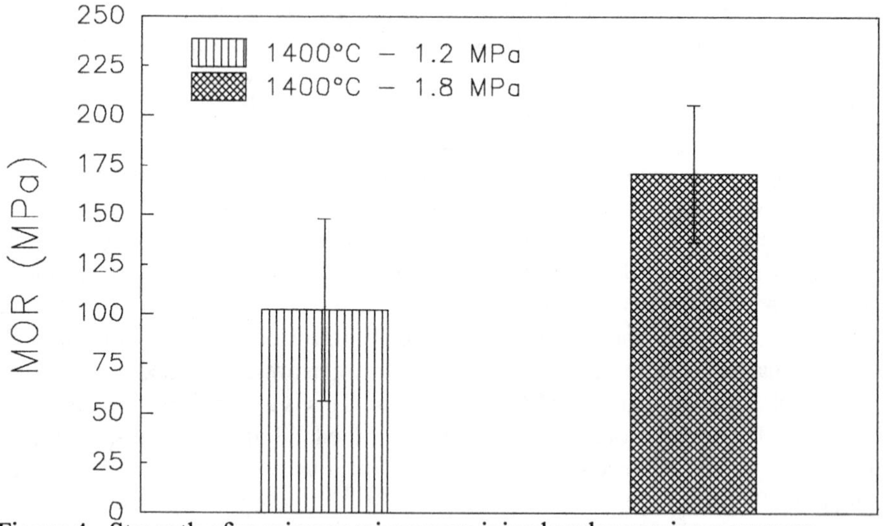

Figure 4. Strength of specimens microwave joined under varying pressures.

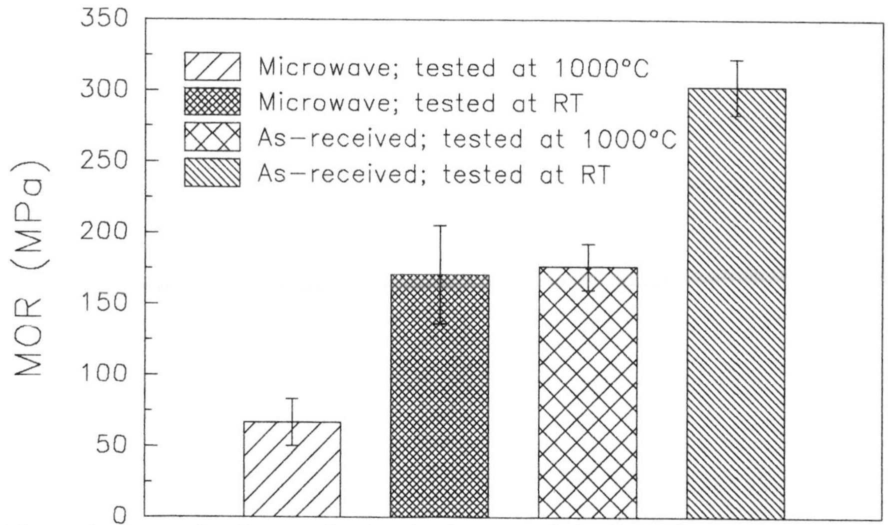

Figure 5. Strength of as-received and microwave joined specimens at room temperature and at 1000°C.

In all cases, failure occurred at the interface between the 99.5% Al_2O_3 and the 96% Al_2O_3. This usually consisted of the delamination of the two surfaces. In one instance, however, the crack left the interface and entered the 99.5% Al_2O_3. Figure 6 is a photograph of the fracture surface of the bar joined at 1400°C for 30 minutes using 1.8 MPa pressure. SEM confirmed that the interlayer did not show any porosity along the joint. The transition from the interlayer to the end members, figure 7, was well-defined.

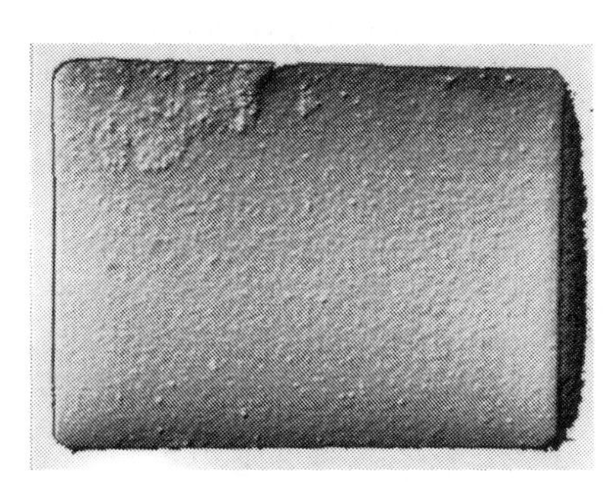

Figure 6. Fracture surface of 96% Al_2O_3 with 99.5% Al_2O_3 affixed. (35x)

Figure 7. Joint region showing the 96% Al_2O_3 interlayer between the 99.5% Al_2O_3.

CONCLUSIONS

Using hybrid heating, alumina could be heated to a temperature that permits joining. This can be done at a temperature low enough to minimize grain growth. A lower purity alumina can be used as an interlayer for joining high-purity alumina. The use of microwave heating may contribute to the increased strength due to selective heating of the interlayer. Further testing is needed to determine if this is responsible for the increased strength. The applied loads can be varied continuously during the microwave heating process. The pressure applied during the joining process has a significant effect on the strength of alumina. Although the strengths of the microwave joined specimens were only half that of the as-received alumina, these experiments were preliminary in order to determine the viability of the microwave joining apparatus.

ACKNOWLEDGEMENTS

Research sponsored by the U.S. Department of Energy, Assistant Secretary for Energy Efficiency and Renewable Energy, Office of Transportation Technologies, as part of the High Temperature Materials Laboratory Fellowship Program, under contract DE-AC05-84OR21400 with Martin Marietta Energy Systems, Inc.

REFERENCES

1. R.E. Loehman, "Problems and Opportunities in Microwave Joining of Ceramics," in Microwaves: Theory and Applications in Materials Processing II, Ceramic Transactions, Vol. 36, pp. 417-430 eds. D.E. Clark, W.R. Tinga and J.R. Laia Jr., American Ceramic Society, Westerville, OH (1993).
2. D. Palaith, R. Silberglitt, C.C.M. Wu, R. Kleiner and E.L. Libelo, "Microwave Joining of Ceramics," pp. 255-266 in Microwave Processing of Materials, Symposium Proceedings, Vol. 124. Eds. W.H. Sutton and I. J. Chabinsky. Materials Research Society, Pittsburgh, PA, (1988).
3. H. Fukushima, T. Yamanaka and M. Matsui, "Microwave Heating of Ceramics and it's Application to Joining," pp. 267-272 in Microwave Processing of Materials, Symposium Proceedings, Vol. 124. Eds. W.H. Sutton and I. J. Chabinsky. Materials Research Society, Pittsburgh, PA, (1988).
4. S. Al-Assafi ,Microwave Joining of Alumina Ceramics with Alumina Gel, Masters Thesis, University of Florida, (1992).

Processing and Thermal Effects

CHARACTERIZATION OF MICROWAVE PROCESSED ZnO

L. P. Martin[*], D. Dadon[**], D. Gershon[***], B. Levush[***], Y. Carmel[***] and M. Rosen[*]

[*]Department of Materials Science and Engineering, Johns Hopkins University, Baltimore, MD.
[**]N.R.C.N., P.O. Box 9001, Beer Sheva, Israel.
[***]Laboratory for Plasma Research, University of Maryland, College Park, MD.

ABSTRACT

ZnO samples were prepared by microwave and conventional sintering processes using identical time-temperature profiles. Comparison of post-sintering ultrasonic properties as a function of residual porosity over the range of 0-35% indicated no difference in the velocity-porosity relations for the samples prepared by the two techniques. An empirical model was fit to the ultrasonic velocity and elastic moduli data. The real and imaginary parts of the permittivity of the samples processed by the two techniques possessed similar frequency dependence. In both cases there was an increase in the measured permittivity values with decreasing porosity, however the microwave sintered samples had considerably higher values. The crystallographic features and the morphology of the samples were inferred from XRD and SEM examination, respectively. It was observed that there was significantly more densification at 850° C for microwave sintering than for conventional sintering.

INTRODUCTION

In-situ characterization of ceramic materials during sintering may yield significant insight into the mechanisms of the sintering process and the optimization of processing parameters. Many experimental and empirical ultrasonic techniques for determination of elasticity-porosity relations in polycrystalline porous materials have been reported in the literature[1-6]. In addition, complex permittivity measurements provide understanding of the microwave-material interaction during microwave processing. The complex permittivity depends on temperature, frequency, and the density of the absorbing material. The density is, in turn, a function of both time and temperature during the sintering process. The complex permittivity, combined with heat capacity and thermal conductivity, can be used to determine the temperature profile within a sintering part. This information can be used to predict densification of the part during sintering, and is essential for understanding and optimizing the process[7-11].

The purpose of the present research effort is to develop in-situ real time diagnostic capabilities for monitoring the development of ultrasonic and dielectric properties during the microwave sintering of ceramic materials. The initial thrust of this effort has been twofold. First, a laser-based technique has been developed for in-situ monitoring of ultrasonic velocity during conventional sintering. Second, extensive post-sintering characterization of conventional and microwave sintered ZnO samples has been performed and is presented here. The next phase of the project will be to adapt the ultrasonic and dielectric diagnostic capabilities for in-situ microwave sintering.

In this study ZnO samples were sintered to various densities in both a conventional furnace and an overmoded 2.45 GHz microwave applicator. Sintering schedules in the microwave applicator were designed to mimic those in the conventional furnace. This was done to enable observation of property and/or morphology differences caused by the different sintering environments. Future work will include characterization of samples sintered at much higher heating rates in the microwave environment. The elastic parameters, as determined by ultrasonic techniques, were compared for the various samples sintered by these techniques. An open ended coaxial probe coupled with a vector network analyzer was used to evaluate the dielectric properties of the samples. Scanning electron microscopy was used to compare the porosity for samples of intermediate density prepared by each of the sintering processes.

EXPERIMENTAL

A commercial zinc oxide powder (Cerac Z-1012) with a -200 mesh particle size was uniaxially pressed without binder to 37 MPa. Typical green samples were 40 g, with average porosity of 46.5%. SEM analysis determined that the individual powder particles were porous, with features of the order of 1 μm or less. The samples were cylindrical, with a diameter of 31 mm and approximate thickness of 17 mm. Sintering schedules included a slow ramp (1° C/min) to 200° C to allow time for water evolution. Most of the samples used in this study were subsequently subjected to a heating rate of 15° C/min to the maximum sintering temperature, and then allowed to cool with no dwell time. Samples were prepared for microscopic analysis by polishing to 1 μm alumina powder, and etching with a 5% acetic acid solution.

Conventionally sintered samples were treated in a tube furnace. The openings at the ends of the furnace were insulated with alumina fiber (wool), and the sample temperature was measured by a type K thermocouple placed in direct contact with the sample surface. The sintering atmosphere was air. All samples were allowed to cool down to ambient temperature in the furnace.

Microwave sintering was performed in a highly overmoded 2.45 GHz applicator (MMT Model 101). The samples were insulated in an alumina enclosure and loosely packed in powder of the same composition as the samples. For all samples sintered in the microwave applicator, a type K thermocouple inserted into a hole drilled into the core of the sample was used for temperature measurement and control. Additional measurement was made using a second type K thermocouple at the surface of the sample. The atmosphere was air, and the maximum power levels used in the heating were less than 1 kW. Cooldown of the samples was designed to mimic the cooling rate of the samples prepared in the conventional furnace.

Density measurements were performed by dimensional measurement and weighing. Thus, porosity values, as inferred from the density measurements, relate to the total porosity in the sample. Porosity P was determined from

$$P = 1 - \frac{\rho}{\rho_0}, \qquad (1)$$

where ρ represents the density as determined experimentally from mass and volume, and ρ_0 represents the theoretical (x-ray) density, 5.61 gm/cm^3, of ZnO.

Ultrasonic measurements were performed using 5 MHz longitudinal and shear wave transducers. The transducers were excited by timed pulses from a Panametrics 5052PR pulser-receiver. Time of flight measurements for the ultrasonic signals were performed using a pulse-echo overlap method on a LeCroy digital oscilloscope (Model 9450). Sound wave velocity was determined by measuring the sample thickness with a micrometer, and dividing the pathlength (twice the thickness) by the time of flight. Considerable care was taken to avoid contamination of the porous samples with couplant. Each data point presented in the ultrasonic data represents the numerical average of 3 independent measurements. Scatter in these measurements was typically less than ±0.5%.

An open ended coaxial line technique was used to measure the dielectric properties of the samples at room temperature over a frequency range from 0.2 to 20 GHz. The probe used was a HP 85070B dielectric probe. A vector network analyzer was used to evaluate the complex permittivity, including dielectric loss factor. Since surface features can play an important role in the measured dielectric properties, all samples were sanded flat with 240 grit paper before evaluation. Surface roughness and flatness were not, however, measured quantitatively.

RESULTS AND DISCUSSION

Figure 1 shows the final density versus maximum sinter temperature for several ZnO samples sintered by conventional and microwave techniques. Temperatures measured by the core thermocouple are shown for the microwave sintered samples, while conventionally sintered sample temperatures refer to the surface temperature. The density is expressed as percent of the theoretical density ρ_0. The samples were all heated at 1° C/min to 200° C, then at 15° C/min to the maximum temperature. Upon reaching the maximum temperature, the conventional furnace was shut down and the samples were allowed to cool with the furnace. Microwave heating and cooling schedules were used which duplicated those of the conventional furnace. Samples were prepared in this fashion in order to isolate any effects on the structure or properties resultant solely from the use of microwaves for the sintering process. It has been noted in the literature that heating rate may affect the densification for both microwave and conventional sintering[12], and it was desired to eliminate such effects between samples in this study. It has also been noted that there may be substantially lower porosity in ZnO samples sintered by microwaves at around 850° C than in conventionally sintered samples prepared at the same temperature[12]. This effect is present in our data and is illustrated in the micrographs shown in Figure 2.

Figure 2 shows SEM micrographs of two samples sintered to 870° C, one by conventional and one by microwave heating. Microscopic analysis of several conventionally sintered samples indicates that there is no significant density gradient through the samples, while the microwave sintered samples tend to show a somewhat more porous structure near the surfaces than in the center. This effect, however, would be to lower the apparent bulk density of the microwave

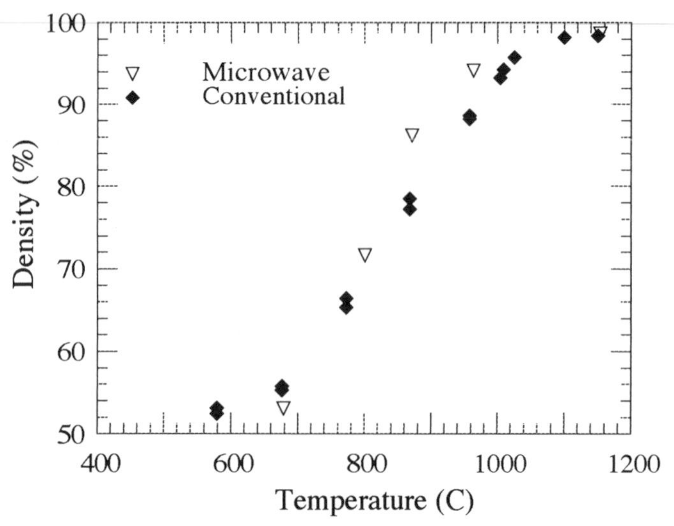

Figure 1. Density vs. maximum temperature for microwave and conventional sintered ZnO.

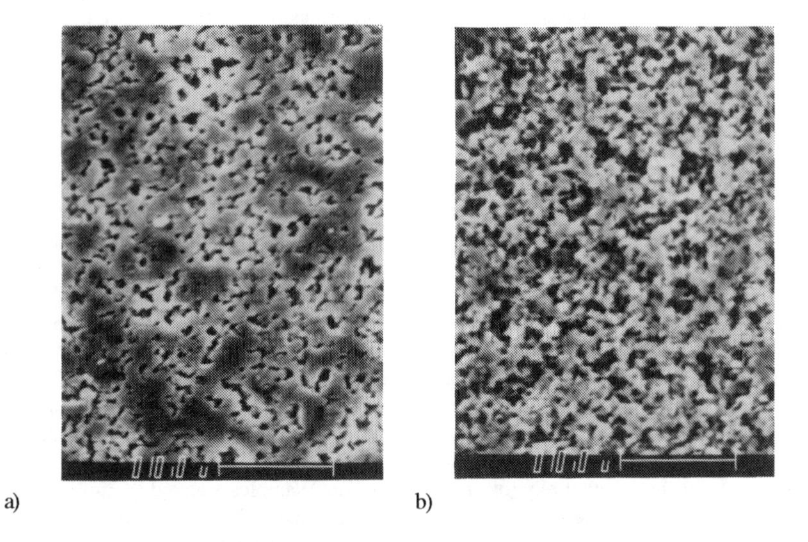

a) b)

Figure 2. Microstructure of ZnO samples sintered to 870° C. a) applicator and b) conventional furnace.

sintered samples. In addition, the micrograph of the microwave sample shown in Figure 2 was taken from a region of the sample very near the tip of the thermocouple used to measure the sample temperature. Thus, it is felt that these micrographs are representative of the porosity of this material sintered to 870° C in both conventional and microwave environments. This illustrates that for the heating rate of 15° C/min, samples of this material sintered to this temperature by microwaves tend to develop a significantly less porous microstructure.

Figure 3 shows the ultrasonic velocity versus porosity for both shear (V_s) and longitudinal (V_l) waves in samples sintered by conventional and microwave processes. The data cover a range of porosity from near 0% to over 30%. It has been proposed that a relation of the form

$$V = V_0(1 - P)^n \tag{2}$$

provides the most physically meaningful representation of the ultrasonic velocity-porosity relation[4,5]. In this expression V_0 is the ultrasonic velocity for a fully dense material and n is an empirically determined parameter dependent upon pore size distribution and pore geometry. A value of n = 1 provides a good fit for a number of materials and yields a linear relation between velocity and porosity[5]. Least squared regression analysis was used to fit equation (2) to the data, and the results can be seen in Table I. There is effectively no difference in the parameters describing this velocity-porosity relation for the microwave versus conventional sintered samples. It is also apparent that the best fit for the longitudinal velocity in this material system is n = 1.5. The predicted longitudinal velocity at 0% porosity of 5928 m/s, determined by averaging the values for the conventional and microwave sintered samples, is consistent with published values determined experimentally (6000 m/s at 3% porosity)[13] and analytically from the single crystal elastic moduli (6000 m/s at 0% porosity)[14]. Similarly, the predicted shear velocity at 0% porosity, 2852 m/s compares well with analytic predictions based upon single crystal moduli (2831 m/s)[14].

Table I: Ultrasonic Parameters Determined from Equation (2):

		V_0 (m/s)	n
V_l	Conventional	5972	1.54
V_s	Conventional	2868	0.97
V_l	Microwave	5884	1.55
V_s	Microwave	2835	0.95

The ultrasonic velocities of the shear and longitudinal waves in an isotropic, linearly elastic bulk solid are related to the density and the elastic moduli by the following relationships

$$E = V_l^2 \rho \frac{(1+\nu)(1-2\nu)}{(1-\nu)}, \tag{3}$$

$$G = V_s^2 \rho \tag{4}$$

and

$$\nu = \left(1 - \frac{V_l^2}{2V_s^2}\right)\left(1 - \frac{V_l^2}{V_s^2}\right)^{-1}, \tag{5}$$

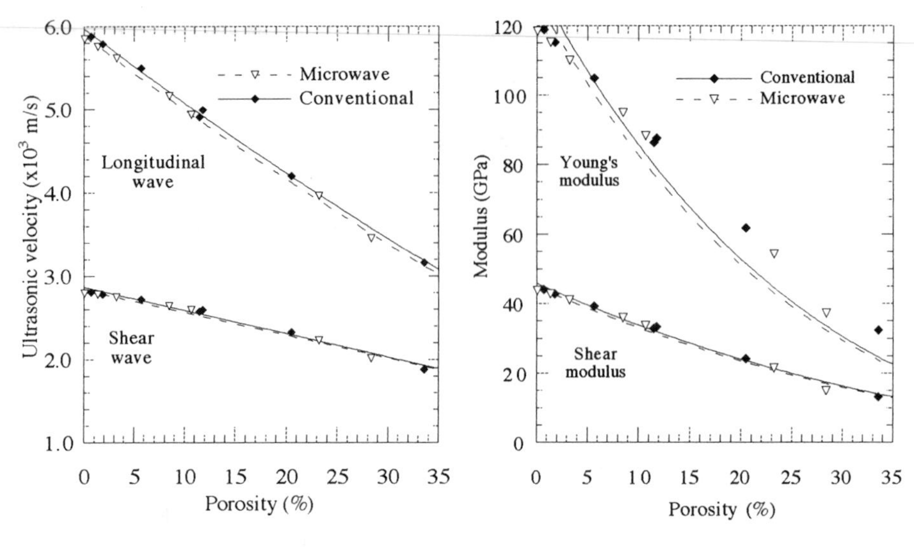

Figure 3. Ultrasonic velocity vs. porosity for microwave and conventional sintered ZnO.

Figure 4. Moduli vs. porosity for microwave and conventional sintered ZnO.

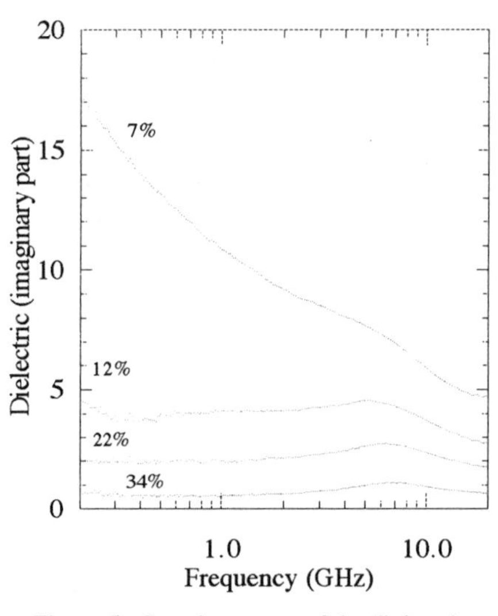

Figure 5a. Imaginary part of the dielectric constant vs. frequency for conventional sintered ZnO. Labels indicate porosity.

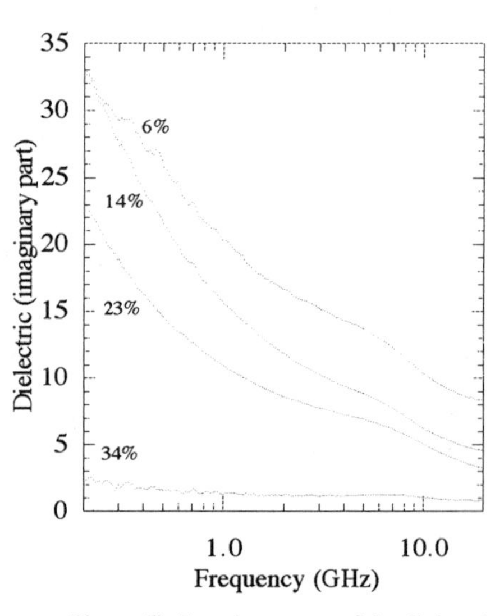

Figure 5b. Imaginary part of the dielectric constant vs. frequency for microwave sintered ZnO. Labels indicate porosity.

where E and G are the Young's and shear moduli, respectively, and ν is Poisson's ratio. Moduli values determined using equations (3), (4) and (5) are shown in Figure 4. In this figure, the Young's and shear moduli are plotted as a function of total porosity for both the conventional and microwave sintered samples. The expected value for the Young's modulus in a 0% porosity sample may be determined by the intercept at approximately 125 GPa. This correlates well with published values calculated from single crystal elastic moduli (123.6 GPa)[14]. Similarly, the shear modulus compares well with calculated values (45.5 GPa)[14].

Assuming Poisson's ratio to be constant with respect to porosity, equation (2) leads directly to an expression of a similar form for the Young's modulus

$$E = E_0(1-P)^{2n+1} \tag{6}$$

where E_0 is the corresponding Young's modulus of a completely dense sample of the material[4]. A similar expression is easily derived for the shear modulus. Least squared regression analysis was used to fit equation (6) to the experimental data using the parameters from Table I. It is evident that the curves deviate slightly from the data as the porosity increases. This is attributed to the variation of Poisson's ratio with porosity. Recall that the treatment which produced equation (6) assumes Poisson's ratio to be constant with respect to porosity.

Complex permittivity measurements were performed over a frequency range of 0.2-20 GHz for the samples sintered by the two techniques. The real part of the permittivity for the samples with lowest porosity ($P < 15\%$) possessed similar frequency dependence, but the values for the microwave sintered samples were higher than those of the conventionally sintered samples by a factor of about 2. For the higher porosity samples, there was a weak frequency dependence, and the values for both the conventional and microwave sintered samples were comparable. In both cases there was an increase in the measured value with decreasing porosity over the porosity range in question.

The imaginary part of the dielectric constant for two sets of samples are shown in Figures 5a and 5b. Labels in the figures indicate the porosity level for each sample examined. The results for both sets at the highest porosity show similar values and frequency dependence. Both sets of samples exhibit increasing values with decreasing porosity, however over most of the porosity range the values for the microwave sintered samples are significantly higher than for the conventional sintered samples. This may be due to the partial pressure of oxygen, P_{O2}, in the two sintering environments. P_{O2} is well known to have a strong effect on the stoichiometry, and thus the conductivity, of ZnO[15]. The P_{O2} was not controlled in either sintering environment. It was observed by the authors, however, that samples sintered in nitrogen had measured dielectric values as much as 10x higher than those sintered in air.

CONCLUSIONS

Samples of ZnO were sintered by both microwave and conventional. The ultrasonic and elastic properties were determined as a function of porosity, and no difference was observed between the samples prepared by the two techniques. It was noted that at a sintering temperature in the vicinity of 850° C, there was a somewhat less porous structure in the microwave sintered samples than in the conventionally sintered samples. This is consistent with observations of previous investigators[12]. Measured permittivities for samples sintered by the two techniques both exhibited increasing values with decreasing porosity over the porosity range of 5-35%.

ACKNOWLEDGMENTS

The authors wish to thank Dr. V. Chakarov and Ms. E. Salama for their technical assistance. Also, J. Rodgers, F. Pyle and D. Cohen. Thanks to Dr. E. Lindgren for many valuable discussions. This work was supported by the Division of Advanced Energy Projects, U.S. DoE under contract # DE-FG02-94ER 12140.

REFERENCES

1. A. Nagarajan, "Ultrasonic study of elasticity-porosity relationship in polycrystalline alumina," *J. Appl. Phys.*, **42**, 3693-3696 (1971).
2. R. M. Spriggs, "Expression for effect of porosity on elastic modulus of polycrystalline refractory materials, particularly aluminum oxide," *J. Am. Ceram. Soc.*, **44**, 628-629 (1961).
3. E. P. Papadakis and B. W. Peterson, "Ultrasonic velocity as a predictor of density in sintered powder metal parts," *Materials Evaluation*, **37**, 76-80 (1979).
4. K. K. Phani and S. K. Niyogi, "Porosity dependence of ultrasonic velocity and elastic modulus in sintered uranium dioxide- a discussion," *J. Mat. Sci. Lett.*, **5**, 427-430 (1986).
5. D. J. Roth, D. B. Stang, S. M. Swickard, M. R. DeGuire and L. E. Dolhert, "Review, modeling, and statistical analysis of ultrasonic velocity-pore fraction relations in polycrystalline materials," *Materials Evaluation*, **49**, 883-888 (1991).
6. D. S. Kupperman and H. B. Spriggs, " Ultrasonic wave propogation characteristics of green ceramics," *Ceram. Bull.*, **63**, 1505-1509 (1984).
7. W. Sutton, "Microwave processing of materials," *MRS Bull.*, **18**, 22-24 (1993).
8. M. Iskander, "Computer modelling of numerical simulation of micrtowave heating systems," *MRS Bull.*, **18**, 30-36 (1993).
9. W. Sutton, "Microwave processing of ceramic materials," *Ceram. Bull.*, **68**, 376-385 (1989).
10. M. Janney and H. Kimrey, "Diffusion-controlled processes in microwave-fired oxide ceramics," *Mat. Res. Symp. Proc.*, **189**, 215-226 (1990).
11. A. Birman, B. Levush, Y. Carmel and M. Rosen, "Modelling of multi-frequency microwave processing of ceramics," this conference.
12. Y. Ikuma, and K. Takahashi, "Rapid-rate sintering of ZnO by microwave heating," *J. Ceram. Soc. Japan*, **100**, 1306-1310 (1992).
13. K. L. Telschow, J. B. Walter, G. V. Garcia, "Laser ultrasonic monitoring of ceramic sintering," *J. Appl. Phys.*, **68**, 6077-6082 (1990).
14. O. L. Anderson, "Determination and some uses of isotropic elastic constants of polycrystalline aggregates using single-crystal data," in Physical Acoustics: Volume IIIB (Academic Press, New York, 1965), pp. 43-95.
15. W.D. Kingery, H. K. Bowen, and D. R. Uhlmann, Introduction to Ceramics (John Wiley & Sons, Inc., New York, 1976), pp. 157-162.

MICROWAVE SINTERING OF NANOCRYSTALLINE CERAMICS

R.W. Bruce,* A.W. Fliflet, L. K. Kurihara,+ D. Lewis, III, R. Rayne, G.-M. Chow, P.E. Schoen, B. A. Bender, and A.K. Kinkead,† Naval Research Laboratory, Washington, DC 20375-5346

ABSTRACT

A single-mode cavity microwave furnace, operating in the TE_{103} mode at 2.45 GHz, has been set up at the Naval Research Laboratory (NRL) and is currently being used to investigate sintering of nanocrystalline ceramics. This presentation will discuss the apparatus used and the results obtained to date. The high purity Al_2O_3 and TiO_2 nanocrystalline powders were prepared by the sol-gel method. These powders were first uniaxially pressed to 14 MPa, CIP'ed to various pressures ≥ 420 MPa and finally sectioned into wafers. The density of the green wafers was 30 to 38% TD. The wafers were heated in the microwave furnace for up to three hours at temperatures ≤ 1720°C. The temperature of the workpiece was monitored using an optical pyrometer. Final densities up to 80% TD have been obtained to date for Al_2O_3 and up to 52% TD for TiO_2. Work is ongoing to characterize the sintered compacts, optimize the casketing for this furnace, and lay the groundwork for new studies using a 35 GHz gyrotron and quasioptical gyrotron tunable from 85 to 120 GHz.

INTRODUCTION

The sintering of nanocrystalline ceramics using microwaves as the energy source has been studied by several groups [1-7]. Nanocrystalline ceramics are of interest for microwave processing because the dielectric loss factor increases significantly as the particle size decreases. Coupled with the fact that the power absorbed by any material is linearly dependent upon the frequency, raising this frequency, from, say, 2.45 GHz to 35 GHz or higher, will also raise the power absorbed [8]. A five year study was begun at NRL in the Fall of 1994, one purpose of which is to investigate microwave processing issues for nanocrystalline materials. The initial phase of this study is to focus on the use of 2.45 GHz microwave power for microwave sintering of ceramics. This effort will be followed by work at higher frequencies using a tunable quasioptical gyrotron (QOG) that will operate in the range 85 to 120 GHz, and a 35 GHz gyrotron.

EXPERIMENTAL

Sample Preparation

Nanocrystalline powders were prepared using a modified sol-gel technique. The resulting powders were then filtered, washed with water, and dried. The resulting precursor powders were filtered,

* EE Dept., United States Naval Academy, 105 Maryland Ave., Annapolis, MD 21402-5025
+ NRL/ASEE Post Doc
† SFA, Landover, MD

washed with distilled deionized water, and dried. The powders were calcined in air at 700°C for 2 hours. The calcined powders were then uniaxially pressed to 14 MPa and cold isostatically pressed (CIP'ed) to 420 MPa. In spite of the fairly high CIP pressure, the density of the green compacts was only 30% TD (theoretical density) for the titania and 37% TD for the alumina. These green compact densities are probably too low for achieving near 100% densification during pressureless sintering. The low densities are attributed to the fact that no binders, surfactants or sintering aids were used. In addition, the green density is affected by the presence of agglomerates. One goal of this investigation was to determine whether pure nanocrystalline materials could be sintered without any form of pressing or sintering aids. Individual samples were then cut using a diamond saw. The samples were placed in a box oven at 200°C until the sintering operation to minimize the absorption of water vapor prior to sintering. The individual samples were then placed into an insulating box ("casket") located at a rf field maximum of the microwave furnace. Runs with various heating profiles were then used to obtain preliminary data on densification rates and achievable final density. Of the two materials, the titania was more easily heated but, as will be described below, did not achieve final densities as high as the alumina.

The Microwave Cavity Furnace

The microwave cavity furnace was fabricated out of WR-284 rectangular waveguide (nominal length 27.25 cm). Various viewing and material loading ports were made in the narrow and broad walls for ease of loading and for in-process observation of the experiment. The viewing ports were constructed in the narrow walls using circular waveguide operating below cutoff to minimize the possibility of microwave energy leakage. They were terminated in glass-sealed apertures so that the environment during processing could be controlled. The input power to the furnace is controlled using a variable aperture iris on the source side of the cavity and tuning is achieved using a sliding short on the other end. The source of the microwave power is a Cober SF6 6 kW microwave generator. Various other microwave components are used to assist in the monitoring of the power and maintaining the environment. Figure 1 is the system configuration for most of the experiments reported here though the system is still evolving to meet processing needs.

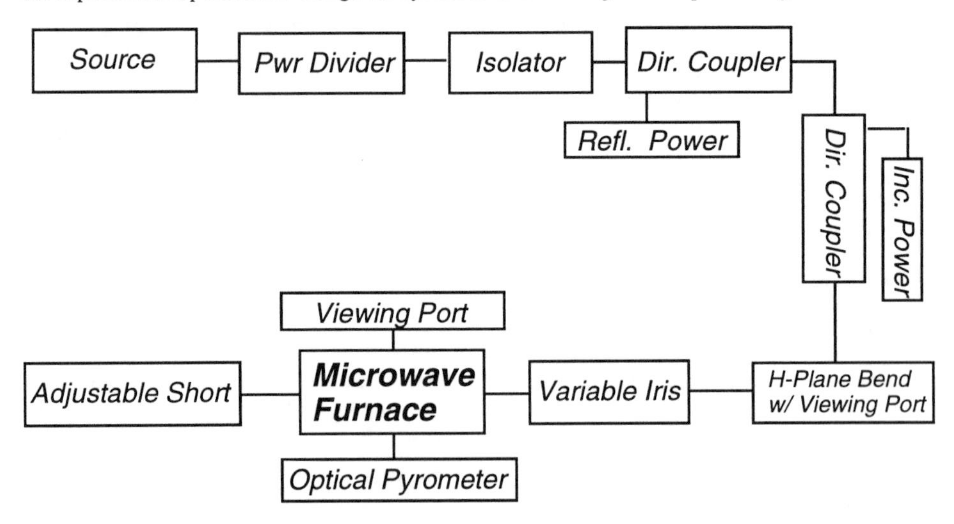

Figure 1: Microwave Furnace System

Sintering

A sufficient number of runs were made using the furnace to ensure that reliable results could be obtained. Some of these results are tabulated in Table I, below. It should be noted from this table that both the alumina and titania were only isostatically pressed to about 30% TD. This is in contrast to some work in which the starting TD's were closer to 50% TD [1,5]. Final TD's varied considerably. The first runs for the Al_2O_3 were to determine if these materials could be heated at all in the rather simple casketing system devised. The casketing usually consisted of a cavity hollowed out of Zircar SALI™ or ZYZ3™ fiberboard insulation which also had a lid to ensure that the sample was fully covered. A small hole was made in the widest side so that the surface temperature of the sample could be observed. The sample and casket were placed at the mid-point of the second maximum of the TE_{103} mode. This allowed viewing of the sample through an observation port constructed in the middle of an H-plane bend that was upstream from the furnace.

Table I. Summary of Results of Initial Sintering Experiments

Material	Sample No.	Starting Density (% TD)	Ending Density (% TD)	T_{max} (°C)	Time at T_{max} (Min.)	Reference
Al_2O_3	1 - 1			1100	30	Figures 3d, 5b
	1 - 3	36.1	51.8	1460	10	
	1 - 4	36.7	74.2	1720	25	Figure 3e
	2 - 1	37.1	81.8	1700	70	
	2 - 3	37.4	74.8	1550	120	Figure 3f
	2 - 4	37.9	68.8	1530	180	
	2 - 8	37.3	76.2	1650	71	Figures 3g, 5c
TiO_2	1 - 1	30.0	32.6	1217	2	
	1 - 2$^+$	32.6	38.7	1225	12	
	2	30.0	44.0	1230	60	Figures 4c, 6b
	3	30.0	52.0	1250	120	Figure 6c

$^+$ sample heated twice

Early in the process it was observed that the casket material could be easily heated by microwaves. This is due to the fact that in one case, silica is coated onto the fibers of the board, and in the other, the yttria performs the same role. This significantly increases the conductivity of the material and results in hybrid heating of the workpiece at low temperatures. Unfortunately, hot spots often occurred in the casket material, resulting in local melting and thermal gradients near the workpiece. Some of the cracks formed in the samples during processing can be attributed to these thermal gradients. Hot spot formation in the casket did not appear to mask the workpiece from the microwaves.

Both forward and reflected power as well as the temperature were monitored for each run. A graph of all of these values for sample 2-8 is given below. The following items should be noted from these data. Initial heating of the sample was characterized by a steady decrease in the reflected power. At approximately 35 minutes elapsed time the workpiece temperature increased rapidly before stabilizing at ~ 600°C. This temperature increase was accompanied by an increase in the reflected power. This is due to the detuning of the cavity as the sample becomes a more resistive load and the cavity less of a resonant circuit. After the cavity is significantly detuned by the changes in the material, the effect of retuning the cavity with the movable short is negligible. Optimization of power input is now achieved by increasing the opening of the iris. This accounts for the marked decrease in the reflected power at approximately 45 minutes of the elapsed time. The improved coupling is accompanied by a significant increase in the sample temperature. After this, temperature increases are achieved by increasing the forward (source) power. This is especially true at the higher temperatures at which large increases in forward power are required in order to obtain modest gains in temperature (at 120–135 minutes elapsed time). The reason for this is due to radiation losses from the sample and casketing into the relatively cold walls of the cavity (which scale as $\sim T^4$). Temperature was monitored using both a single and a two-color pyrometer. There was reasonable agreement between these two sensors especially at the higher temperatures. Forward and reflected power were monitored using S-band power sensors at the forward and reflected arms of the directional couplers [see Figure 1].

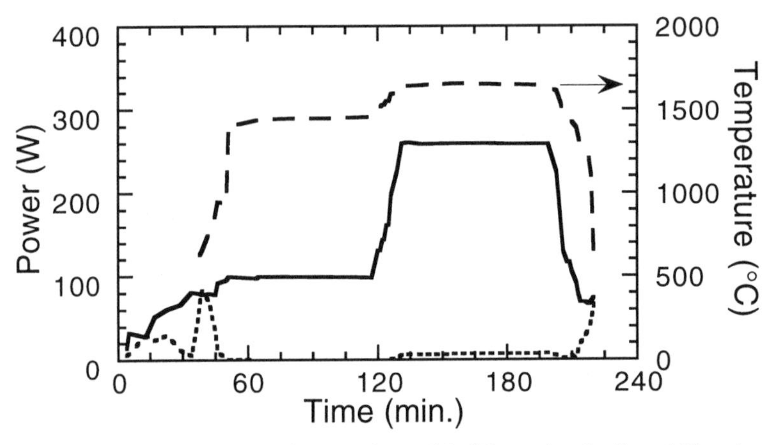

Figure 2: Temperature (dashed curve), forward (solid curve) and reflected (dotted curve) power profiles for microwave sintering of nanocrystalline Al_2O_3 ceramic sample 2-8.

RESULTS AND DISCUSSION

The physical properties of the samples were investigated before and after the microwave processing. Table II shows a summary of x-ray diffraction and scanning electron microscope (SEM) characterization of the crystalline phases formed and the average particle sizes of the oxide precursors and the final microwave treated samples. The alumina and titania x-ray diffraction data are shown in Figures 3 and 4, respectively. SEM micrographs of the alumina and titania starting materials and treated samples are shown in Figures 5 and 6, respectively. Estimates of densification were made from physical measurements.

These data show that significant densification was generally accompanied by a substantial increase in the grain size. The one exception is sample 1-4 which achieved a density of 74% at a grainsize of 11 nm. Note that the proccessing time for this sample was relatively short (25 minutes). The system of hybrid heating was usually successful though the problem with non-uniform heating (hot spots) caused a number of the samples to crack. In particular, the alumina results show a very good consistency between the methods used. It is evident from Figures 3d and 5c that the alumina processed at 1100°C retains the γ phase and does not transform to the α phase. Processing at higher temperatures (>1100°C) produces the characteristic α-alumina phase and, as expected, leads to significant grain growth (see SEM data in Figure 5d). The micrograph for titania sample 3 [Fig. 5c], which was held at 1200°C for 2 hours, shows a trapped spherical pore indicative of melting.

Table II. Summary of ceramic phases as determined by x-ray diffraction and SEM

	As Prepared	Calcined	Microwaved	Sample
Material:	TiO_2	TiO_2	TiO_2	
Phase	Rutile*	Rutile and Anatase	Rutile	
Average crystallite size (Å)		200	1300	1+
			4 μm	2
			5 μm	3
Material	AlOOH	Al_2O_3	Al_2O_3	
Phase	Boehmite	predominately γ	α above 1100°C	
Average crystallite size (Å)	< 10	< 10	20	1-1
			112	1-4
			> 1 μm	2-3
			> 1 μm	2-8

* X--Ray amorphous (see Fig. 4), however, an electron diffraction pattern observed in TEM

+ see Table I for sample processing conditions

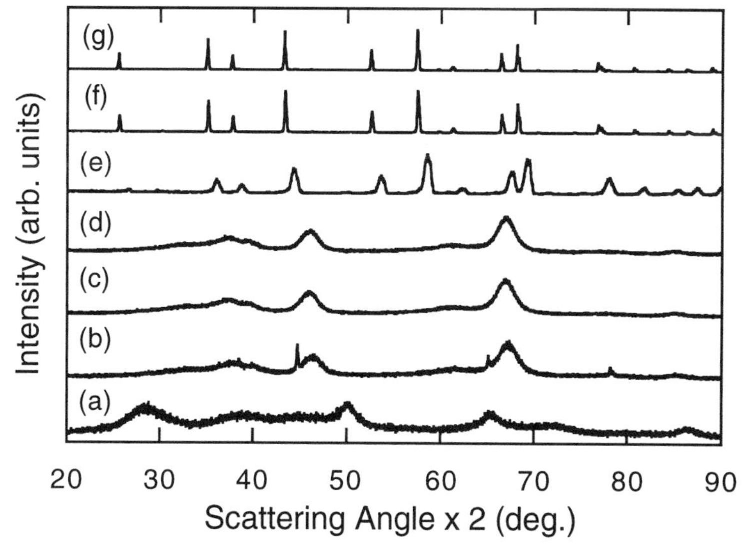

Figure 3. X-Ray Diffraction Data for Al_2O_3 samples. Curve a: sol-gel powder; b: calcined powder; c: CIP'ed green body; d: microwaved sample 1-1 (30 min.@1100°C); e: microwaved sample 1-4 (25 min. @1720°C), f: microwaved sample 2-3 (120 min. @1550°C), g: microwaved sample 2-8 (65 min.@1450°C and 71 min. @1650°C).

Notice the rather clean grain boundaries shown by the SEM data in Figures 5 and 6. This is indicative of a very pure starting material which partially accounts for the low final densities achieved. The low final density achieved is also due to the existence of agglomerates in the pre-sintered material. According to a theory concerning sintering in the presence of agglomerates, high theoretical density is not reached until grain size approaches the size of the agglomerate [9].

Approaches to achieving higher final densities currently under investigation include the use of sintering aids and application of pressure during processing (sinter forging).

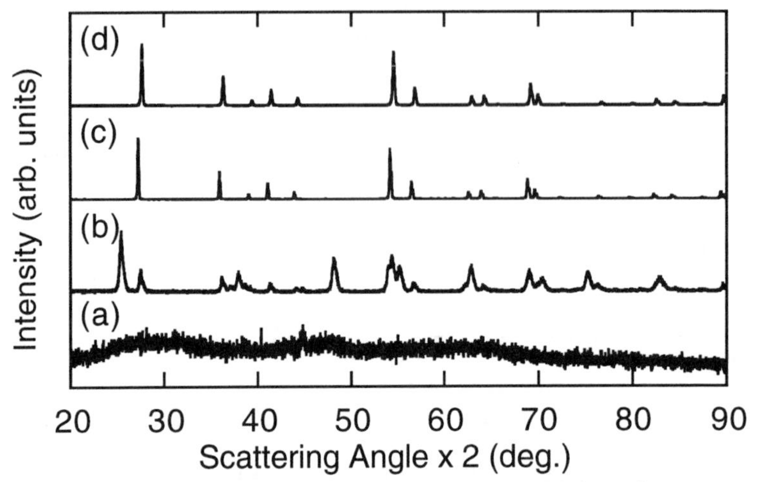

Figure 4. X-Ray Diffraction Data for TiO_2 samples. (a) sol-gel powder; (b) calcined powder; (c) microwaved sample 2 (1 hr @ 1200 °C); and (d) microwaved sample 3 (2 hr @ 1200 °C).

Methods of avoiding agglomeration during powder preparation are also being considered. The present results confirm the expectation that nanocrystalline materials couple reasonably well to 2.45 GHz microwave energy. As further evidence of this, the re-heating of partially sintered samples required more power to achieve a given temperature than the initial heating, indicating that the larger grains with their clean boundaries did not provide as much conductivity enhancement as the green compact. Similar comments can be made concerning the results for titania shown in Figures 4 and 6.

SUMMARY

A single-mode microwave cavity furnace has been constructed, instrumented and integrated into a microwave system that has been used to perform preliminary sintering experiments. Sintering of nanocrystalline materials up to 80% TD was achieved with moderate grain growth. Several methods of achieving higher final densities are under investigation.

ACKNOWLEDGMENT

This work was supported by the Office of Naval Research.

Figure 5. SEM micrographs showing the evolution of grain growth and particle size for nanocrystalline Al_2O_3. a: precursor (Boehmite) powder, b: calcined powder, c: microwaved sample 1-1, d: microwaved sample 2-8.

Figure 6. SEM micrographs showing the evolution of grain growth and particle size for nanocrystalline TiO_2. a: precursor powder, b: microwaved sample 2, c: microwaved sample 3, arrow shows trapped spherical pore indicative of melting.

REFERENCES

1. J.A. Eastman, K.E. Sickafus, J.D. Katz, S.G. Boeke, R.D. Blake, C.R. Evans, R.B. Schwarz and Y.X. Liao, "Microwave Sintering of Nanocrystalline TiO_2", MRS Symposium Proceedings, Vol. 189, pp. 273-278, Pittsburgh, PA (1990).
2. S.N. Kumar, A. Pant, R.R. Sood, J. Ng-Yelim and R.T. Holt, "Production of Ultra-Fine Silicon Carbide by Fast Firing in Microwave and Resistance Furnaces", Ceramic Transactions, Vol. 21, pp. 395-402, Westerville, OH (1991).
3. D. Vollath, R. Varma and K.E. Sickafus, "Synthesis of Nanocrystalline Powders for Oxide Ceramics by Microwave Plasma Pyrolysis", MRS Symposium Proceedings, Vol. 269, pp. 379-384, Pittsburgh, PA (1992).
4. D. Vollath, "Some Activities of Microwave Processing of Ceramics in Germany", Ceramic Transactions, Vol. 36, pp. 147-156, Westerville, OH (1993).
5. J. Freim, J. McKittrick, J. Katz and K. Sickafus, "Phase Transformation and Densification Behavior of Microwave Sintered γ-Al_2O_3", MRS Symposium Proceedings, Vol. 347, pp. 525-530, Pittsburgh, PA (1994).
6. J. Zhang, Y. Yang, L. Cao, S. Chen, X. Shong, and F. Xia, "Microwave Sintering on Nanocrystalline ZrO_2 Powders", MRS Symposium Proceedings, Vol. 347, pp. 591-596, Pittsburgh, PA (1994).
7. D. Vollath, "A Cascaded Microwave Plasma Source for Synthesis of Ceramic Nanocomposite Powders", MRS Symposium Proceedings, Vol. 347, pp. 629-634, Pittsburgh, PA (1994).
8. R.W. Bruce, "New Frontiers in the Use of Microwave Energy: Power and Metrology", MRS Symposium Proceedings, Vol. 124, pp. 3-15, Pittsburgh, PA (1988).
9. M.J. Mayo, D.C. Hague and D.-J. Chen, "Processing Nanocrystalline Ceramics for Applications in Superplasticity", Materials Science and Engineering, A166, pp. 145-159 (1993).

EFFECT OF MICROWAVE HEATING ON THE SINTERING OF SILICON NITRIDE–DOPED WITH Al_2O_3 AND Y_2O_3

Yoon Chang Kim, Sang Cheol Koh, Do Kyung Kim, and Chong Hee Kim
Department of Materials Science and Engineering, Korea Advanced Institute of Science & Technology, 373-1 Kusong-dong, Yusong-gu, Taejon 305-701 South Korea

ABSTRACT

Dense silicon nitride with relative densities of >99% was obtained by microwave heating at 2.45 GHz. The densification and the $\alpha \rightarrow \beta$ phase transformation behavior of the microwave-sintered silicon nitride were investigated and compared with those conventionally sintered which used the same sintering schedule. Microwave heating enhanced the densification and the $\alpha \rightarrow \beta$ phase transformation and the effect of microwave heating was more remarkable in silicon nitride containing larger amounts of additives.

INTRODUCTION

Silicon nitride is one of the most promising materials for use in structural applications at high temperature. However, it is difficult to densify Si_3N_4 because of its low self-diffusivity. Manufacturing a dense Si_3N_4 requires sintering additives for liquid-phase sintering [1]. Currently, hot isostatic pressing and gas pressure sintering in the presence of additives are common fabrication methods for producing a full dense Si_3N_4, but these are expensive. Since high manufacturing cost limits the introduction of silicon nitride into the market place, new processing techniques capable of lowering the cost are required.

In recent years, microwave processing has received attention in the preparation of ceramic materials. The microwave processing has several advantages over conventional heating techniques, including the reduction in manufacturing cost and sintering temperature, and the improvement in material properties [2]. Many efforts have been made over the last decade to develop the microwave sintering of various oxides and nonoxides, but less work has been reported on microwave sintering of silicon nitride.

In this work, microwave sintering of Si_3N_4 with Y_2O_3 and Al_2O_3 as additives was carried out. The densification and the $\alpha \rightarrow \beta$ phase transformation behavior of the microwave-sintered silicon nitride were compared with conventionally sintered one and the effect of microwave heating on the densification and the $\alpha \rightarrow \beta$ phase transformation was investigated.

EXPERIMENTAL PROCEDURE

High purity α–Si_3N_4 (E-10, UBE Industrial Ltd., Tokyo, Japan) powder was used as a starting material. As sintering additives, Al_2O_3 (AKP-50, Sumitomo Chemical Co., Ltd., Tokyo, Japan) and Y_2O_3 (Grade Fine, H. C. Stark GmbH, Goslar, Germany) powders were used. Alpha–Si_3N_4 and 8, 12 wt% additives (1:3 mixture of Al_2O_3 and Y_2O_3) were mixed by ball milling in ethanol using Al_2O_3 balls and a polypropylene container for 24 h. For convenience, the silicon nitride specimens with 8 wt% and 12 wt% additives are designated as 2A6Y and 3A9Y, respectively. After mixing the powder mixture was dried and granulated. The powder was die-pressed at 30 MPa and isostatically pressed at 200 MPa. The size of the resultant pressed specimens was about $\phi 24 \times 12$ mm for microwave sintering and $\phi 10 \times 5$ mm for conventional sintering.

Microwave sintering was carried out in a 2.45 GHz, 6 kW microwave furnace using an insulation box designed for this study. Fig. 1 and Fig. 2 show the microwave furnace and the insulation box, respectively. The microwave

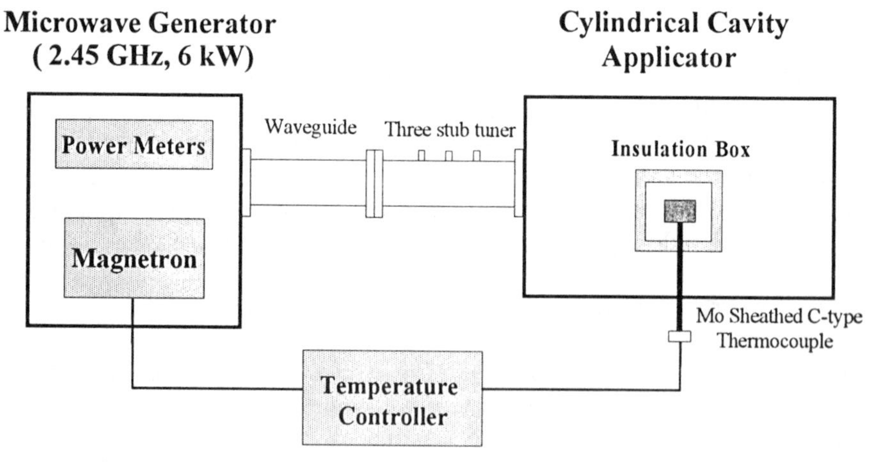

Fig. 1. A schematic of the microwave furnace used in this study.

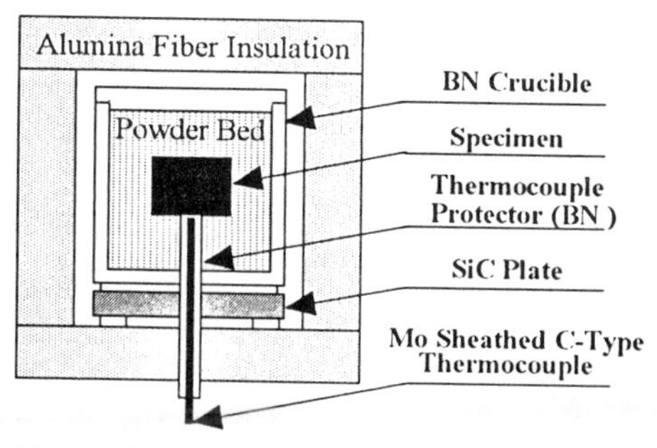

Fig. 2. A schematic of the insulation box.

furnace was built in-house and the insulation box was composed of BN crucible, alumina fiber boards, and SiC plate. The SiC plate was used as a preheater. Specimen was placed in the insulation box and then heated to sintering temperature with the heating rate of 25°C/min in nitrogen atmosphere. During the heating cycle, the temperature was monitored using a molybdenum sheathed C-type thermocouple in contact with the specimen and controlled automatically through feedback control by a programmable controller. Conventional pressureless sintering was performed in a graphite resistance furnace using the same sintering schedule. Pure Si_3N_4 powder was used as a powder bed to suppress the decomposition of specimen in both cases.

The bulk densities of the sintered specimens were measured in distilled water by the Archimedes method. The polished and etched surfaces of the sintered specimens were observed by scanning electron microscopy (SEM). The crystalline phases were identified using X–ray diffractometry (XRD) and their contents were calculated by the method proposed by Gazzara *et al* [3].

RESULTS AND DISCUSSION

Figure 3 shows the microstructures of the parts of the center and the near surface of the 3A9Y specimen that was microwave-sintered at 1750°C for 30 min. Microstructural differences were not observed. Microwave heating can generate cracks and microstructural differences within the sintered specimen because of localized thermal runaway and inverse thermal gradients. In this work, however, a crack-free specimen with homogeneous microstructure was obtained, even though a high heating rate (25°C/min) was used. This seems to be mainly due to the use of SiC plate as a preheater in the insulation box. The SiC plate heated uniformly the Si_3N_4 specimen until microwave energy was efficiently absorbed by specimen. As a result, it was possible that thermal runaway was avoided and the thermal uniformity of specimen was increased.

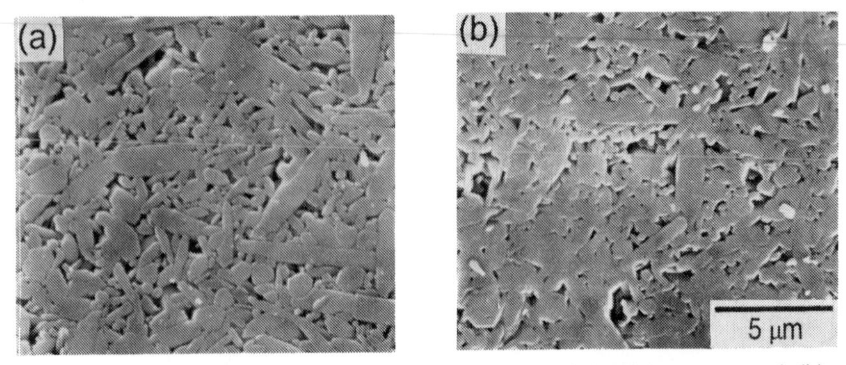

Fig. 3. SEM micrographs showing the parts of the (a) center and (b) near surface of 3A9Y microwave-sintered at 1750°C for 30 min.

Figure 4 shows the variation of relative density with sintering temperature for 2A6Y and 3A9Y. The microwave-sintered specimens exhibited higher relative density than the conventionally sintered specimens at all sintering temperatures. This indicates that the densification was enhanced by microwave heating. Especially, the 3A9Y experienced more enhanced densification by microwave heating as compared with 2A6Y. The microwave-sintered 3A9Y reached more than 99% of theoretical density at the sintering temperature of

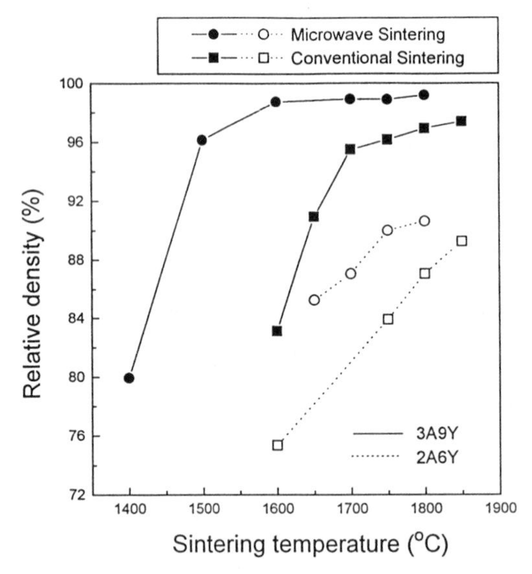

Fig. 4. Variation of relative density with sintering temperature for 2A6Y and 3A9Y. The isothermal holding time was 30 min.

1600°C. In the case of conventionally sintered 3A9Y, such a high density could not be obtained even at the sintering temperature of 1850°C.

The microstructures of the microwave-sintered and the conventionally sintered 3A9Y were compared in Fig. 5. In the case of microwave-sintered specimen, inter-locked microstructure by large elongated grains was already formed at 1700°C and little morphological changes can be seen with the sintering temperature. On the other hand, in the case of conventionally sintered specimen, most of the grains were equiaxed at 1700°C and these equiaxed grains tranformed into elongated grains as the sintering temperature increased.

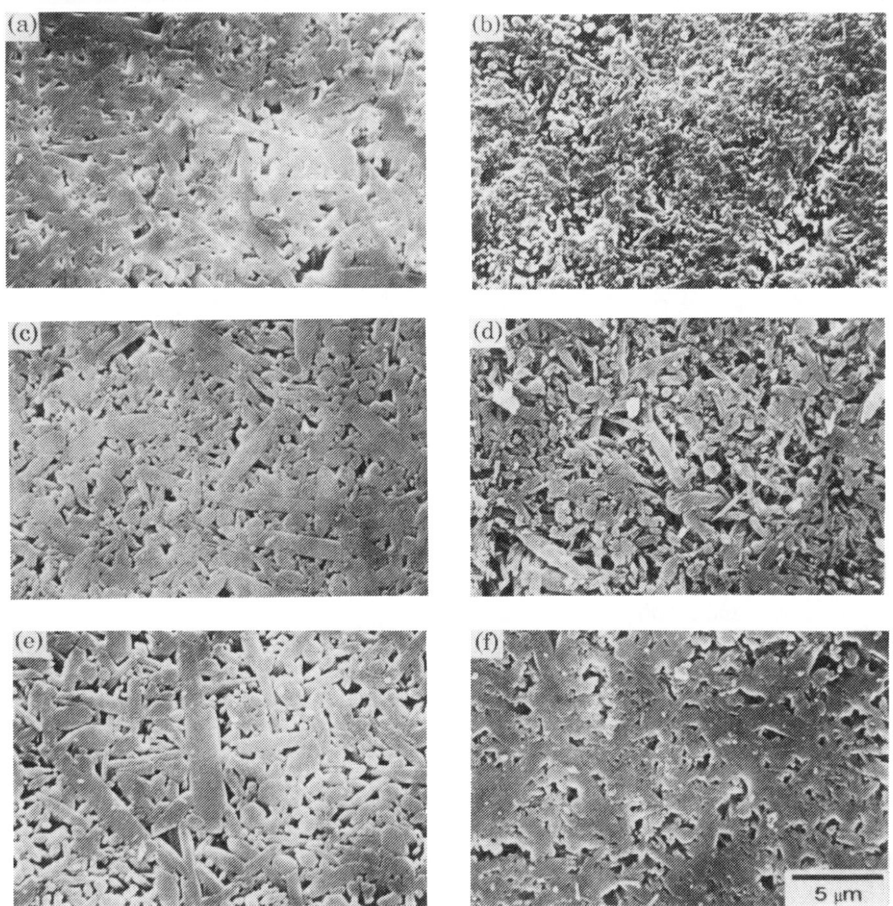

Fig. 5. SEM micrographs of 3A9Y sintered by (a, c, e) microwave and (b, d, f) conventional heating. The sintering temperatures were (a, b) 1700°C, (c, d) 1750°C, and (e, f) 1800°C.

The development of grain morphology depends directly on the transformation process of α to β phase during sintering [4]. Therefore, different behavior of α→β phase transformation is expected in the two heating methods.

Variation in the β–phase fraction for the sintered specimens is shown in Fig. 6. This figure clearly demonstrates the presence of the effect of microwave heating on the α→β phase transformation. The phase transformation was initiated and completed at the lower temperature for the microwave-sintered specimens and 3A9Y exhibited more enhanced α→β phase transformation than 2A6Y.

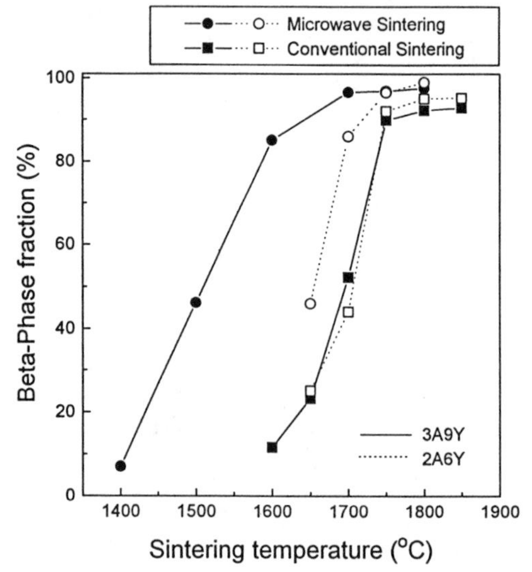

Fig. 6. Variation of β–phase fraction with sintering temperature for 2A6Y and 3A9Y.

From the above results, it was shown that the densification and the α→β phase transformation were enhanced by microwave heating and the magnitude of the enhancement was higher in 3A9Y than in 2A6Y. The reason why the densification and the α→β phase transformation were enhanced by microwave heating is not obvious yet. Two opinions, however, have been suggested about the effect of microwave heating. One is that the diffusion in liquid phase is enhanced [5] and the other is that the viscosity of liquid phase is decreased [6]. However, both the suggestions can not explain the higher magnitude of the enhancement in silicon nitride containing larger amount of additives.

In the present study, the following experiment was performed to investigate the effect of the amount of additives on the microwave sintering of Si_3N_4.

Under constant microwave power of 1.1 kW, each of specimens with different amount of additives (0, 4, 8, and 12 wt%) was heated and the variation of temperature with time was measured. The temperature profiles are shown in Fig. 7. In spite of the same microwave power, specimen containing larger amounts of additives exhibited higher temperature. This is a clear evidence that microwave heating of silicon nitride occurs predominantly by coupling between the microwave and the sintering additives. The heating rate of specimen (dT/dt)

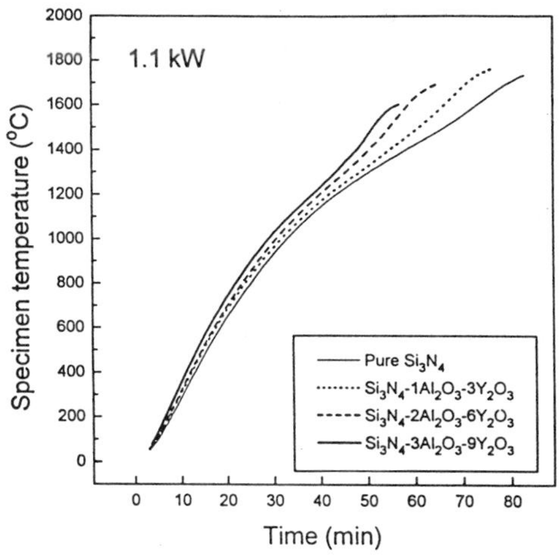

Fig. 7. Heating profiles of the specimens containing different amount of additives (0, 4, 8, and 12 wt%). The specimens were heated under the constant microwave power of 1.1 kW.

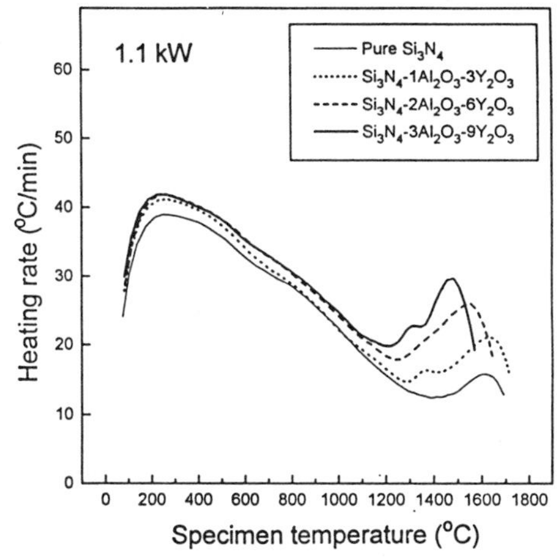

Fig. 8. Heating rate curves derived from the heating profiles in Fig. 6.

was derived as a function of temperature as shown in Fig. 8. Two peaks appeared in the figure. The first peak at lower temperature is due to the preheating of SiC plate as discussed earlier and the second peak is due to the coupling between microwaves and the specimen. As a consequence, the content of additives had a great influence on the second peak. The height of the second peak increased and the position shifted to lower temperature as the content of additives increased. This means that the degree of coupling between microwaves and the specimen was increased as the content of additives increased. Therefore, it can be concluded that the higher degree of coupling to microwaves is responsible for the more remarkable effect of microwave heating in 3A9Y.

CONCLUSIONS

Silicon nitride with near theoretical density was obtained by microwave sintering of α–Si_3N_4 with the addition of Al_2O_3 and Y_2O_3. The densification and $\alpha \rightarrow \beta$ phase transformation were enhanced by microwave heating as compared with conventional heating. The effect of microwave heating was more remarkable in 3A9Y than in 2A6Y because of the higher degree of coupling to microwaves.

REFERENCES

1. G. Ziegler, J. Heinrich, and G. Wötting, "Relationships between Processing, Microstructure and Properties of Dense and Reaction-Bonded Silicon Nitride," *J. Mater. Sci.*, **22**[9] 3041–3086 (1987).
2. W. H. Sutton, "Microwave Processing of Ceramic Materials," *Am. Ceram. Soc. Bull.*, **68**[2] 376–386 (1989).
3. C. P. Gazzara and D. R. Messier, "Determination of Content of Si_3N_4 by X–ray Diffraction Analysis," *Am. Ceram. Soc. Bull.*, **56**[9] 777–780 (1977).
4. H. Knoch, K. A. Schwetz and A. Lipp, "Microstructure Development in Silicon Nitride," ; pp. 381–392 in *Progress in Nitrogen Ceramics*, ed. F. L. Riley, Martinus Nijhoff Pub., Netherlands, (1983).
5. T. N. Tieg, J. O. Kiggans, and H. D. Kimney, "Microwave Processing of Silicon Nitride." ; pp. 267–272 in Mat. Res. Soc. Symp. Pro., Vol. 189, *Microwave Processing of Materials II*, eds. W. B. Snyder, W. H. Sutton, D. L. Johnson, and M. F. Iskander, Materials Research Soc., Pittsburgh, PA, (1991).
6. J. Zhang, L. Cao, and F. Xia, "Microwave Sintering of Si_3N_4 Ceramics," ; pp. 329–334 in Mat. Res. Soc. Symp. Pro., Vol. 269, *Microwave Processing of Materials III*, eds. R. L. Beaty, W. H. Sutton, and M. F. Iskander, Materials Research Soc., Pittsburgh, PA, (1992).

GUIDELINES FOR LARGE SCALE MW-PROCESSING OF HARDMETALS

T. GERDES*, M. WILLERT-PORADA*, K. RÖDIGER**, H. KOLASKA***

* University of Dortmund, Dept. Chem. Eng., Dortmund, FRG
** WIDIA GmbH, Essen, FRG
*** Fachverband Pulvermetallurgie, Hagen, FRG

ABSTRACT

MW-sintering of hardmetals has the potential of reducing the processing time and improving mechanical properties. However for a large scale MW-processing the influence of penetration depth of MW-radiation and EM-field pattern distortion has to be known. Because information on penetration depth calculated from dielectric property measurement or DC-conductivity is not sufficient, a method to calculate the penetration depth from calorimetric measurement is discussed.

For a homogeneous MW-sintering process the field can be adapted to the hardmetal batch by using a variable size multi-mode cavity.

INTRODUCTION

Composite materials made from refractory carbides and metal binders, known as hardmetals for nearly a century, are still occupying a large share of the cutting tools market, due to their unique combination of hardness, toughness and strength. Within the industrial countries, 80 % of the cuttings, facings and borings were made by hardmetal-tools, representing a worldwide turnover of about 8 billion US $. [1]. The growing demand for machining time reduction requires a continuous improvement of tool bit properties and overall quality within this class of materials, ideally combined with decreased costs of production.

The required microstructural features, like extremely small grain size and optimized carbon content can be achieved by vacuum sintering in large graphite furnaces. A semi batch or batch process, in which removal of pressing aids is combined with sintering is applied. Therefore, the processing time which varies between hours and days is dominated by the low heating rate required in conventional heating of large batches of materials containing processing aids.

It has been shown that microwave sintering of hardmetals has the potential of reducing processing time while preserving or even improving mechanical properties of conventional hardmetals [2]. However, penetration depth of microwave radiation and thermal conductivity are the materials parameters limiting the practicaly achievable acceleration of heating. Promising preliminary results on scale up using microwave assisted heating with 4-8 fold increase of the heating rate as compared to conventional processing are known for low loss materials [3]. For this case, the total sintering cycle is dominated by the cooling step.

For high dielectric loss materials shielding and serious distortion of the EM-field pattern is expected to occur. Therefore, homogeneous heating with increased heating rates will depend upon the particular EM-field distribution over the batch and the individual sample.

To define guidelines for large scale microwave processing of such materials the development of experimental procedures which provide that information as a function of batch size, geometry, microwave power level and microstructural properties of the green parts is required.

For a rough estimate of the heating behavior of a sample knowledge of the power distribution within the sample is necessary. Assuming that the power deposited at a distance, x, from the surface of a sample is described by an attenuation constant, α, at which the dielectric field strength decays to 1/e of its value on the sample surface, as shown in Eq.1:

$$P(x) = P_0 \cdot e^{-2 \cdot \alpha \cdot x} \qquad \text{[Eq. 1]}$$

P_o : Power density at sample surface

such an estimate can be provided by the complex dielectric constant ε^*_r of the material. As visible from Eq. 2 penetration depth δ', defined as the reciprocal value of the attenuation constant α, can be calculated from the real and imaginary part of the dielectric constant at the frequency of the EM-field used for heating:

$$\delta' = \frac{1}{\alpha} = \frac{\lambda_0}{2\pi} \cdot \left[\frac{\varepsilon'_r}{2} \left(\sqrt{1 + \left(\varepsilon''_r / \varepsilon'_r \right)^2} - 1 \right) \right]^{-1/2} \qquad \text{[Eq. 2]}$$

λ_o : Wavelength in air $\qquad$ ε'_r : Permittivität $\qquad$ ε'': Loss factor

For materials with a high electronic conductivity besides the dielectric properties DC-conductivity is often used as a measure for the penetration depth, as indicated by Eq. 3:

$$\delta'' = 0.029 \cdot \sqrt{\sigma \cdot \lambda_0} \qquad \text{[Eq. 3]}$$

σ : Resistivity [Ωm] $\qquad$ λ_o : Wavelength in air [m]

However, these calculated values are based on material parameters estimated under well defined experimental conditions, e.g., at low power levels, using a clearly defined waveform or a low level of DC-current. Under such conditions heating of the sample is avoided. Opposite to this, when a microwave heating experiment is performed, high power levels and ac-currents are applied in order to achieve reasonable heating rates. Additional effects such as microplasma ignition within pores of a green part or non linear voltage-current behavior within a green body containing a mixture of semiconducting and conducting phases have to be taken into account.

Therefore, an experimental method for penetration depth estimation has been developed, based on the measurement of heating and cooling rates under pulsed microwave exposure at power levels typical for sintering experiments. Using this method, the penetration behavior can be investigated as a function of the power level, the amount of plastifier, the sample geometry and the sintering stage. Based on this information, materials parameters for successful microwave processing can be obtained.

Optimized processing conditions can only be achieved when the particular EM-field distribution within an applicator can be adapted in accordance with materials parameters. With every variation of the batch size and geometry the field pattern can significantly change due to absorption and reflection of the microwave radiation by the high loss materials. Such changes can not be accommodated for by changing materials parameters only, but require additional measures. Therefore, a multi-mode resonator with an optional adaptation of the cavity size and geometry has been used for sintering different load sizes of hardmetals under optimized EM-field distribution conditions.

EXPERIMENTAL

Green processing: Attritor milled commercial tungsten carbide-6% cobalt powder mixtures from *Widia* (THM) were isostatically or axially pressed at 300 MPa. The content of paraffin binder varied from 2 wt.%. for pressed bodies to 10 wt.% (64 vol.% !) for extruded hardmetal parts.

Dielectric properties: The dielectric properties at 2.45 GHz for WC-Co6 at room temperature were measured by the method of *Roberts* and *von Hippel* [5] using a TE_{01} resonant cavity.

Calorimetric measurements: Data for the calorimetric calculations were obtained from the following measurements: the WC-Co green bodies were heated with 700W-pulses of 2.5 s or 5 s duration with a 30 s time interval between those pulses for 3 to 5 subsequent pulses. Further details on that method are given in [3].

Sintering: The sintering experiments were performed in a 3 kW, 2.45 GHz inertgas-cavity using an alumina fiber insulation box for casketing the material. The low temperature range was controlled by an thermocouple, the high temperature range by an optical Pyrometer.

Field distribution measurements: To characterise the field distribution upon varying the load size and geometry a thermo-sensitive paper between two glass plates is used, as shown in Figure 2.

One side of the paper is covered by a thin layer of a metallic spray, yielding a dense regular pattern of squares, each approximately 3 mm accross. The metallic layer was needed, because the paper itself did not couple very well with MW-radiation. The metal particle layer heated easily by MW-radiation and shaded the paper within seconds depending upon the local field strength. To avoid the spread of shading by thermal conductivity, a discontinuous metal layer had to be used. This was achieved by patterning the metal layer.

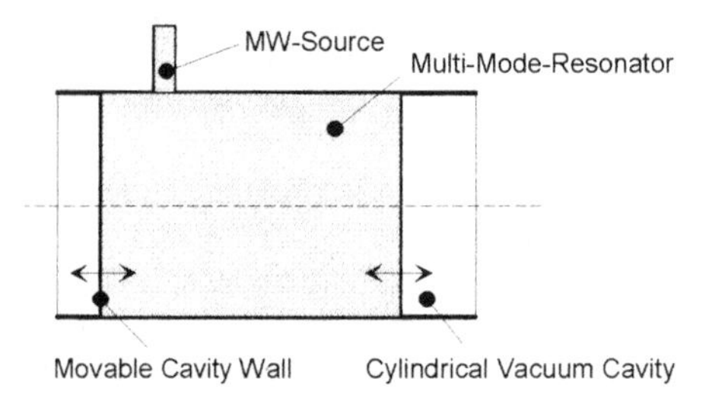

Figure 1. Schematic set-up of variable size microwave-cavity

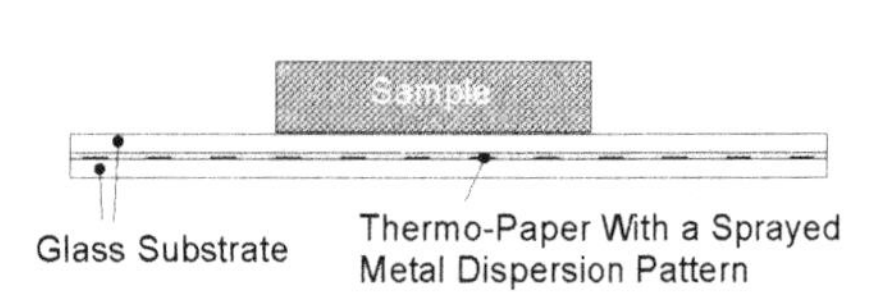

Figure 2. Arrangement for illustrating the field pattern while MW-heating

The cylindrical cavity has two moveable side walls to vary the size and position of the resonator in relation to the MW-source (Figure 1) and to optimize the field distribution at the location of the batch.

RESULTS AND DISCUSSION

The results of dielectric property measurements by the method of Roberts and von Hippel are shown in Table 1. Based on these results a volumetric heating of WC-Co6 green bodies is expected.

Table 1. Dielectric Properties of WC-Co6 Greenbodies at Room Temperature.

rel. permittivity	55.6 ± 3
loss factor	3.8 ± 0.1
penetration depth	7.5 cm

In contrast to this the penetration depth calculated from the DC-resistivity measurements of a green body yields a penetration depth in the order of 7 µm.Even if a large margin of error is considered, the difference of these two calculations can not be fully understood.

However, neither the penetration depth calculated from the dielectric properties, measured at an extremely low power level, nor the penetration depth calculated from the DC-resistivity can describe the heating behavior satisfactory. Therefore, the penetration depth has been estimated by a further method, based on heat balances. For that purpose the energy and heat balance are analyzed just before the end of the MW-pulse, and directly after turning off the MW-power.

The heat-balance of the MW-heated volume ratio V_{sk} (Figure 3.) in a small temperature interval with $V_{sk} \neq f(T)$ and the heat capacity $C_p \neq f(T)$ can be described as:

$$\frac{\partial \dot{Q}_{st}}{\partial t} = m_{sk} \cdot C_p \cdot \frac{\partial T}{\partial t} = \dot{Q}_a + \dot{Q}_i + P \qquad \text{[Eq. 4]}$$

m_{sk} : mass of the skin volume $\quad \dot{Q}$: heat flow $\quad t$: time $\quad T$: temperature

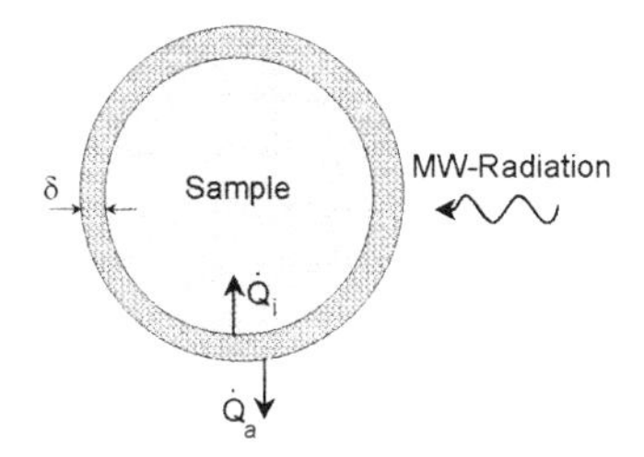

Figure 3. Heat flow in a MW-heated sample with low penetration depth.

The MW-power absorbed by the skin-volume V_{sk} is calculated as P_{spec}:

$$P_{spec} = \frac{P}{V_{sk}} = \cdot \rho \cdot C_p \cdot \left(\frac{T_3 - T_2}{t_3 - t_2} - \frac{T_4 - T_3}{t_4 - t_3} \right) \qquad \text{[Eq. 5]}$$

The meaning of the indices is illustrated by Figure 5. For the absorbed MW-power we obtain:

$$P = m_{tot} \cdot C_p \cdot \left(\frac{t_6}{t_3} \cdot \frac{T_6 - T_5}{t_6 - t_5} + \frac{(T_6 - T_0)}{t_3} \right) \qquad \text{[Eq. 6]}$$

m_{tot} : total sample mass

If the penetration depth δ is small as compared to the overall sample dimensions, it can be calculated using Eq. 5 and Eq. 6:

$$\delta = \frac{V_{sk}}{A} = \frac{m_{tot} \cdot \left(\frac{t_6}{t_3} \cdot \frac{T_6 - T_5}{t_6 - t_5} + \frac{(T_6 - T_0)}{t_3} \right)}{A \cdot \rho \cdot \left(\frac{T_3 - T_2}{t_3 - t_2} - \frac{T_4 - T_3}{t_4 - t_3} \right)} \qquad \text{[Eq. 7]}$$

A: sample surface

Figure 4. shows that the temperature increases very fast for a debindered green body during the MW-pulse but also decreases rapidly after turning off the MW-power. Such a temperature peak is typical of skin heating, with heat flowing into the samples volume and from the surface to the surrounding. A skin depth of approximately 0.5 mm results from a calculation using Eq. 6. With increasing prefiring temperature of the material the skin depth decreases. The origin for the decreasing penetration depth is the increasing connectivity of the conductive phase due to neck formation and surface area reduction.

Also the paraffin content influences the heating behavior. For low paraffin contents the typical peak of a skin heating is observed whereas for paraffin contents > 5 wt.% low cooling rates are observed, which are typical for volumetric heating, as indicated by Figure 5.

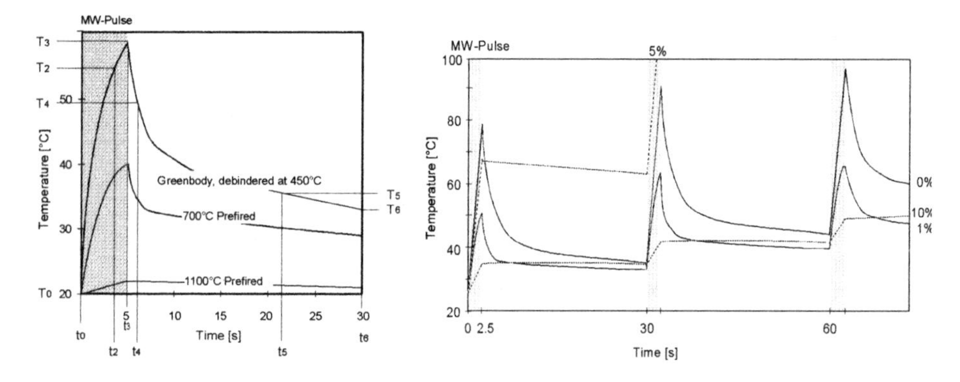

Figure 4. Pulse heating (700W/5s) of hardmetals in different sintering stages.

Figure 5. Pulse heating (700W/2.5s) for different paraffin binder contents.

The transition from skin heating to volumetric heating takes place near the percolation threshold, as visible from conductivity measurements shown in Figure 6.

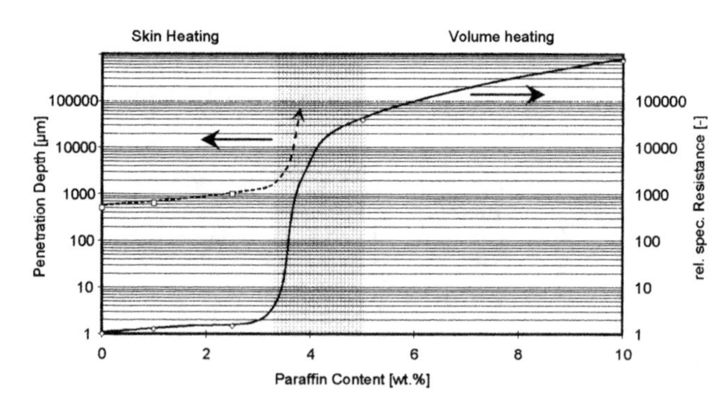

Figure 6. Penetration depth and relative resistance as a function of paraffin content.

The penetration depth calculated from the heating behavior is intermediate between the value calculated from the dielectric properties and the value obtained from DC-resistivity. When the porous body is seen as a complex network of metallized dielectrics, it shows a close similarity to a complex arrangement of strong attenuated "microstrips". Therefore a slight increase in penetration could occur, when in addition to the "bulk" particle penetration a contribution from "guided" structures within the μm-metal-semiconductor arrangements is assumed.

On the other hand the power density in a real heating experiment could cause effects like microplasma-ignition in the pores and arcing between adjacent grains. Such effects cause additional shielding of the EM-field. In this case high heating rates combined with a decrease in penetration depth would result. Such a phenomenon would strongly depend upon the power level. Because power level in direct dielectric property measurements is limited by the sensitivity of the diagnostic instruments, the calorimetric method is used to investigate the influence of power level on penetration depth.

There is no direct method to estimate the average power density at the sample surface. Therefore the penetration depth obtained from the calculation according to Eq. 7 is used to calculate P_0. form Eq.1. The results are shown in Table 2.

Table 2. Penetration Depth for Different Surface Power Densities.

Power Density [W/cm^3]	Penetration Depth [mm]
9.5	1.1
10.2	1.2
27	0.9
95	0.5

The reduction of penetration depth with increasing power density, visible from the data in Table 2 can be understood by the more appropriate plasma conditions at the higher power levels. Nevertheless, the distinction between the penetration depth from the dielectric properties (75 mm) and heat balances (0.5 mm) can not only be explained by plasma forming. The method used for measuring the dielectric properties of WC-6Co green parts, which despite their low density display a high electronical conductivity, probably reaches its natural limits regarding the requirement to exactly fit the resonator cavity.

Taking into account the highly reflecting behavior of hardmetal green parts and the high power absorption within the surface of a sample, the necessity of careful power distribution adjustment over a load is evident. In order to estimate the influence a load has on the EM-field pattern, the particular field pattern of a loaded and unloaded cavity has been recorded using the method described in the experimental part. The results are shown in Figure 8.

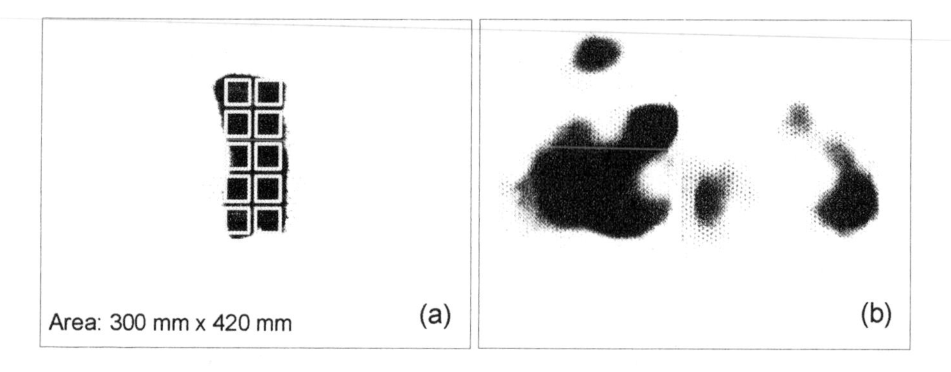

Figure 8. Field pattern of a cavity, loaded with hardmetal green bodies (a) and an empty cavity (b).

By adapting the movable cavity walls to the size and geometry of the load the field can be concentrated homogeneously over the load. The same cavity without a load shows a totally inhomogeneous field distribution. Sintering experiments can be successfully performed in such a “size adapted“ applicator.

However, using a typical sintering cycle which includes a holding time, homogeneity of microwave sintering can not be sufficiently characterized. For such a purpose, sintering is performed at a high heating rate, with no holding time and is interrupted before theoretical density is reached. In such an experiment, field distribution inhomogeneity can be judged from density differences over a batch. The results are shown in Figure 9.

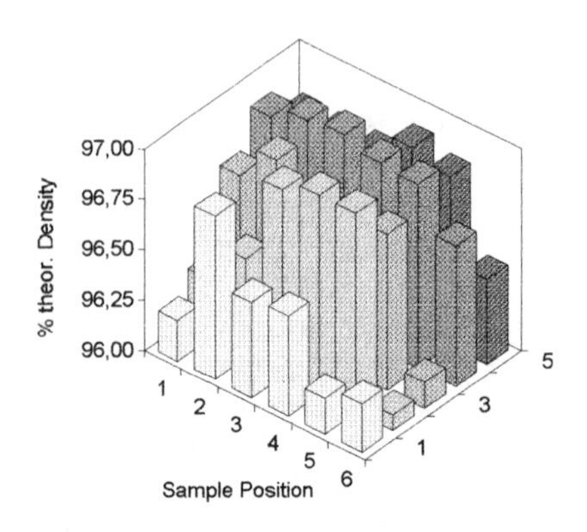

Figure 9. Distribution of relative density for a prefired batch of hardmetal tools.

Figure 10. Casketing for MW- sintering hardmetal cutting tools.

The plot shows the typical situation for a MW-heated batch with higher temperatures in the center and lower temperatures at the surface of the sample and not as expected a strong shielding effect with an inversion of the temperature gradient. The reasons for the volume heating of the batch are the adapted field and the change of penetration behavior at high temperatures

CONCLUSIONS

Comparison of penetration depths calculated from dielectric properties or from DC-conductivity with values obtained from a calorimetric measurement at power levels typical for a sintering experiment reveal large differences between values obtained from this methods. Guidelines for scale up of microwave sintering can therefore be only developed using measurements under conditions close to those of sintering experiments. Using such conditions, the interaction between materials parameters (e.g., binder content and power level) can be analyzed in order to establish optimized heating rates.

On the other hand, heating of larger batches requires knowledge of the EM-field distribution at the batch position. Field distribution measurements using a cavity with variable size and geometry provide the guideline for adjustment of the cavity size and microwave source position to the batch size.

ACKNOWLEDGMENTS

We are indebted to Dipl.-Ing. V. Pohl, Institut für Elektrowärme, Universität Hannover, for support with the dielectric measurements.

REFERENCES

1 H. Kolaska, Pulvermetallurgie der Hartmetalle, Fachverband Pulvermetallurgie, Hagen (1992).

2 T. Gerdes, M.Willert Porada,Microwave Sintering of Metal-Ceramic and Ceramic-Ceramic, Composites, Mat. Res. Soc. Symp. Proc., Vol.347, 531-537 (1994).

3 R.Wroe, Seminar lecture, Uni-Dortmund (1995).

4 T.Gerdes, M.Willert Porada, Microwave Processing of Metal-Ceramic-Composite-Material, Friedrichshafen (1994).

5 A. Roberts, A. von Hippel, J. Appl. Phys. 17, 610 (1946).

MICROWAVE PROCESSING OF POLYIMIDE THIN FILMS FOR ELECTRONICS

David A. Lewis, Susan J. LaMaire and Alfred Viehbeck
IBM T.J. Watson Research Center
PO Box 218
Yorktown Heights, NY 10598

ABSTRACT

Microwave energy was utilized to quickly and efficiently cure polyimide thin films as interlayer dielectrics in high performance multi-chip modules. The process results in a 33% reduction in raw process time for a 4 level electrical structure (8 levels of dielectric) and an even greater reduction in the manufacturing cycle time, since single part processing reduces the effect of batching for curing cycles. As a part of the feasibility study, a test vehicle was successfully completed using microwave processing for all curing steps. Materials retained the necessary mechanical properties after microwave processing and there were no problems due to the effects of the metal wiring, either in the glass ceramic substrate or in the thin film structure.

INTRODUCTION

The advantages of microwave energy in materials processing are well documented. However, greater penetration depth, volumetric heating, etc. will not result in the greater commercial exploitation of the technology, which has been limited up to this time. For any new technology to be successfully implemented, it must be a technical success in addition to providing a good business case. Although under certain circumstances microwave processing can provide significant utilities cost savings, this alone rarely justifies the additional costs of developing and installing a new technology.

One of the most important factors in the cost of a product, especially high value add products, is the manufacturing time required to complete the product.

The raw process time and the time for "batching" to accumulate sufficient parts at a particular station to initiate the next process operation can add significantly to the cost, with batching often being the most time limiting step. Even if significant time savings can be generated at one process, if there is not an overall process time improvement, the individual time savings are irrelevant.

The manufacture of multi-chip modules (MCM's) for applications in high performance computers meets the general requirements by being a high value add product and manufacturing time is an important cost factor. These components are utilizing a greater amount of polymers as dielectric materials, due to the higher electrical performance than tradiational ceramics. Although the polymeric dielectric films are thin, the substrate on which they are cast is a large thermal mass which is susceptible to cracking if large thermal gradients exist, for example if a predominantly surface heating technology is used. Hence, provided acceptable uniformity can be achieved, microwave processing is an excellent technology for this application.

The class of polymer best suited to this application are polyimides [1], due to excellent thermal and dimensional stability and low coefficient of thermal expansion in the x-y plane, which reduces the level of thermal stress in the final structure. The curing of polyimide thin films is a very complex process with solvent evaporation, imidization, large scale ordering and in some cases the breakup and or formation of complexes occurring during the curing cycle. Hence, the mechanical properties of the fully cured films are dependent on the curing cycle, including the heating rate and the maximum temperature reached. The time required in curing ovens is a significant portion of the total manufacturing time and hence a significant reduction in the curing time can impact the total manufacturing time.

To demonstrate the utility of this technology, a test structure consisting of two different types of polyimides - one requiring a full curing cycle and the other (photosensitive) a drying operation - on a ceramic multi-chip module, complete with copper wiring was fabricated. It was important that the processing not compromise the integrity of the completed package or impact other process sectors. This paper describes the results of the technology feasability studies.

EXPERIMENTAL

The microwave equipment used in this work can be broken into three parts: (a) the microwave generator, operating at 2.45 GHz, variable power to 750W; (b) power measuring equipment to maintain critical coupling in the microwave applicator and monitor the power density within the applicator; and (c) a 10-inch tunable resonant cavity based on the design of Asmussen et al.[2]. The applicator was extensively modified from the original designs to allow for solvent evolution, cleaning and increased performance with an unloaded Q'factor of greater than 10,000. The experimental layout also facilitated the monitoring of the state of cure by measuring the cavity parameters, which change as the dielectric constant of the curing material changes, thus providing a useful method for monitoring and controlling the quality control of parts during processing. The TE_{111} mode was used in this work.

The temperature of the curing film was measured using an infra-red pyrometer, mounted on the top of the cavity, focussed through a hole in the sliding short (the top of the cavity) and onto the film below. The pyrometer was selected to be sensitive at 7.9 um, which corresponds to a strong absorption in polyimide films (to prevent radiation from the substrate from affecting the temperature measurement), avoids CO_2 and solvent interference absorptions and has a high emmissivity (0.95) to enable accurate measurements.

The glass ceramic substrate was a very large load in the applicator, resulting in substantial perturbation of the electric field patterns and the order of modes. To minimize the effect of the substrate on the applicator parameters, the part was placed 0.125 mm above the applicator base on quartz standoffs with a small air gap to minimize thermal conduction from the part to the base of the applicator.

As described above, two different polyimides were utilized in this work. Polyimide I is a fully imidized photosensitive polyimide, Probimide 412 from Ciba Geigy, which only required a drying step to remove solvent after the film had been exposed and developed. BPDA-PDA, Polyimide 11 from Du Pont is in the poly(amic acid) form and must be cured in addition to solvent removal.

The structures for these materials are shown in Figure 1. The complex relationship between mechanical properties and processing cycle during microwave curing for Polyimide ll have been described elsewhere [3,4], while there is essentially no dependence of the curing cycle on the properties of Polyimide 1. The properties developed by the optomized microwave curing cycle are comparable to those obtained utilizing conventional processing techniques, but with a substantially reduced process time. Typical temperature - time processing cycles used for microwave processing for polyimides I and ll are shown in Figure 2. It is important to note that with the in-situ temperature measurements, it is possible to obtain a very high repeatability for temperature-toime profiles from part to part and prevent temperature over-runs and detect problems during the processing by unexpected deviations from the normal situation.

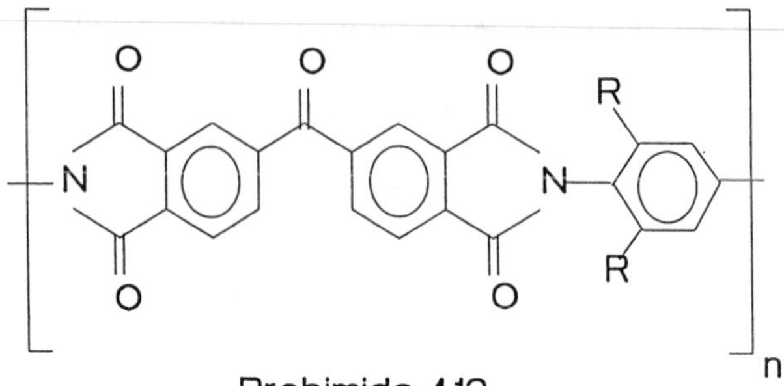

Probimide 412

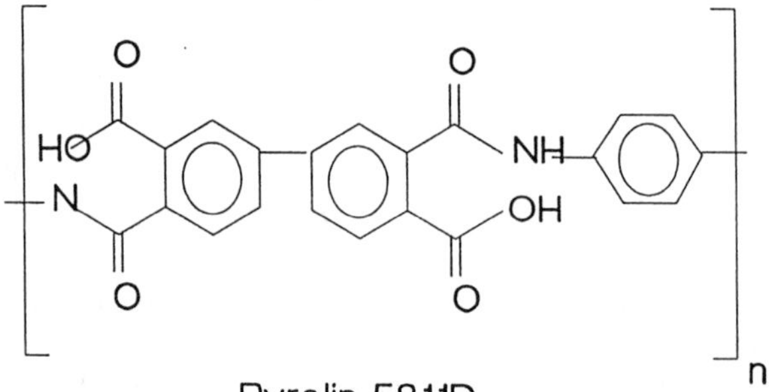

Pyralin 5811D

Figure 1. Chemical structures of polyimides I and II.

Although it is possible to process films directly after the polyamic acid was applied, to simplify sample handling and reduce contamination, films were softbaked at 100 °C for 30 minutes prior to microwave processing.

All curing was performed under nitrogen to prevent oxidation of the polymer or copper wiring in the structure.

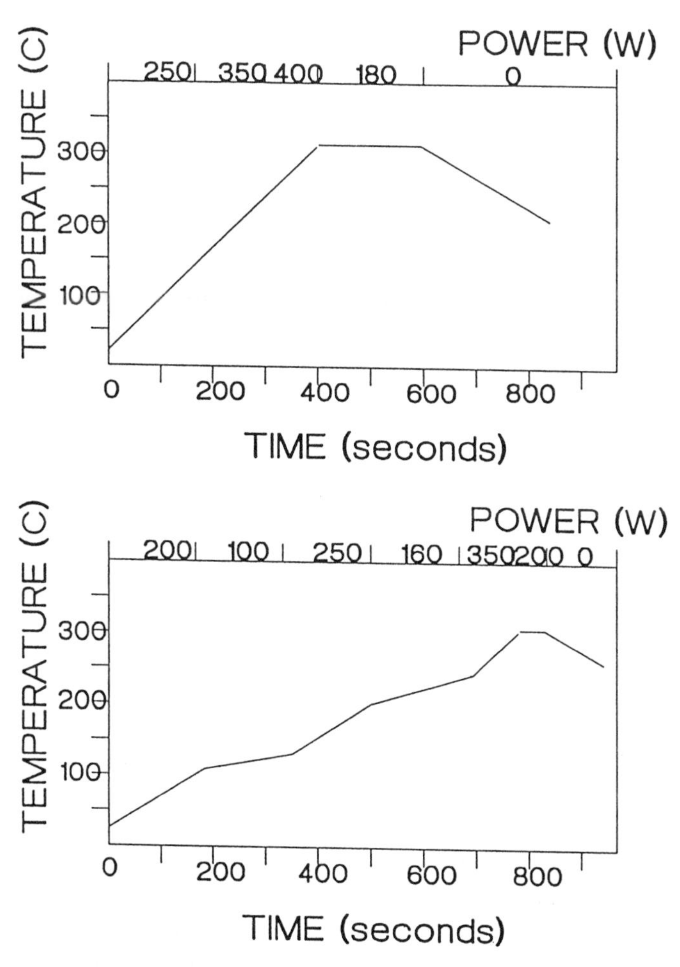

Figure 2. Temperature-time profiles for polyimides I and II during microwave processing.

RESULTS AND DISCUSSION

It is important to note that the polyimide films on the substrate (both uncured and fully cured) heated substantially faster then the substrate alone. The effect of heating the substrate more than the film produces a situation analogous to a fast thermal curing process which does not produce acceptable properties in the polyimides.

Since the part being processed is square and the applicator and mode circular, it was not surprising to observe excessive heating at the corners of the part. Polyimide I is prone to thermal degradation above 350°C in nitrogen and at about 270°C in the presence of oxygen. This sensitivity was used to detect non-uniformities in the electric field across the part during processing during early development. In order to improve the heating uniformity of the substrate, a collar [5] was designed and fabricated out of Macor, a machinable ceramic or quartz. This device had the effect of moving the electric field concentrations away from the edge of the part and into the collar, away from critical curing regions. While the collar can be made from a wide range of materials. it is preferable that it is made of a high dielectric constant material with a small loss factor, to prevent excessive heating. Additionally, the size of the collar was reduced to minimize the load in the applicator.

The purpose of building test vehicles was to resolve concerns about curing over the dense metallurgy of the thin film structure. Despite initial concerns, no problems were observed over the thin film metallurgy. Electrical testing of the completed test vehicles showed that the thin film wiring and the ceramic substrate are not damaged by the microwave curing process.

Parts were cycled to evaluate the reliability of the microwave cured portions and to further investigate potential damage to other wiring layers. No fails were observed over a number of parts.

Additionally, there is no impact of the microwave processing of the dielectric layers on any other step in the manufacturing process of these complex multichip modules, which is an important result in allowing the implementation of this technology.

Using conventional processing, the curing cycle necessary to optimal properties for these materials required heating cycles of up to 10 hours in convection ovens. The microwave process was less than 15 minutes, including cool-down. Based on a model for the manufacturing process, there is approximately a 33 % reduction in the raw process time, with up to a 44% reduction in the manufacturing cycle time for the production of Logan thin films. The larger reduction in the manufacturing cycle time is primarily due to the single part processing nature of this technology and the elimination of batching for the curing ovens. These reductions translate directly into cost savings, in both manufacturing and development, in which a faster build cycle can reduce the overall development time.

In addition to cycle time savings, this microwave system was estimated to save approximately $90,000 per year of nitrogen and electric utilities for the same throughput, compared with a conventional oven. Since the utilization of nitrogen was so small, special ;recycling filters were not necessary, resulting in additional savings over conventional technology.

CONCLUSIONS

It has been demonstrated that microwave curing is a viable process to accelerate the curing cycle of polyimides for electronics. The test structure incorporated a ceramic substrate containing electrical wiring and the copper/ polyimide thin film structure. This structure showed good electrical characteristics with no impact of the microwave processing on yield or on the reliability of parts. A very high part to part reproducibility of the processing conditions could be maintained through sophisticated process control, resulting in a robust process with an apparently wide process window. The implementation of this technology would result in a quick payback through substantial reductions in the manufacturing process time

REFERENCES

1. C. Feger and C. Feger, "Selection Criteria for Multichip Module Dielectrics," in *Multichip Module Technologies and Alternatives: The Basics* (D. A. Doane and P. D. Franzon, eds), Van Nostrand Reinhold, pp. 311-348 (1993).
2. J. Asmussen, H. H. Lin, B. Manring, and R. Fritz, "Single-Mode or Controlled Multimode Cavity Applicators for Precision Materials Processing," *Rev. Sci. Instrum.,* vol. 58, pp. 1477-1486 (1987).
3. D. A. Lewis, S. J. Lamaire, and T. L. Nunes, "Microwave Processing of Polyimide Thin Films," in *Microwave Processing of Materials IV,* Vol. 347 (M.F. Iskander, R.J. Lauf, and W.H. Sutton, eds), Pittsburgh: Materials Research Society, pp. 681-689 (1994).
4. D. A. Lewis, T. L. Nunes, and E. Simonyi, "Processing of Polyimide Thin Films Utilizing Microwave Energy," *PMSE Prepr.,* vol. 72, pp. 40-41 (1995).
5. S. J. LaMaire and D. A. Lewis, "Uniform Microwave Heating," to IBM Corporation, USA Patent 5,220,142 (1993).

PROCESSING AND CHARACTERIZATION OF A POLYMER MATRIX COMPOSITE USING VARIABLE FREQUENCY MICROWAVE HEATING

Zak Fathi and Richard S. Garard
Lambda Technologies
Raleigh, NC

John Clemons and Craig Saltiel -University of Florida-
Center for Advanced Energy Studies in Engineering
Palm Beach Gardens, FL

R. M. Hutcheon
AECL Research, Chalk River Laboratories
Chalk River, Ontario, Canada

Mark. T. DeMeuse
EniChem America Inc.,
Monmouth Junction, NJ

ABSTRACT

Variable frequency microwave energy was successfully applied to uniformly heat thermoset polymer matrix composite (PMC) materials consisting of glass-reinforced isocyanate/epoxy mixtures as well as graphite fiber-reinforced epoxy. Results of a series of materials processed via variable frequency microwave energy are compared with computed results from numerical modeling techniques. A finite difference time domain (FDTD) technique provides 2-D models of the electric and thermal field distributions inside the microwave cavity as well as inside the processed materials. The materials' physical and chemical characteristics from the empirical trials are evaluated for both variable and fixed frequency irradiation.

INTRODUCTION

Microwave heating is fundamentally different from conventional heating since microwave penetrate many polymeric materials, heating volumetrically. Temperature gradients are unavoidable during microwave processing since the energy is absorbed by the bulk and dissipated at the surface of the specimens [1,2]. In the case of low temperature microwave processes, as is the case of polymers processing, these temperature gradients are easily corrected by using microwave-transparent insulation around the samples or by controlling the temperature inside the microwave cavity using hot-air jets.

Variable Frequency Microwave (VFM)-based technology was successfully applied to uniformly heat thermoset polymer matrix composite (PMC) materials consisting of isocyanate/epoxy mixtures [2]. This paper reports on the post-curing characteristics of several isocyanate-based PMC plates. Thermal profiles as a function of processing conditions (central frequency, bandwidth and sweep rate) have been derived. The glass transition (Tg), flex modulus, flex strength and the dielectric properties of the samples processed via variable frequency irradiation have been measured. The evaluation of microwave heating characteristics has been performed on a variety of sample sizes and configurations. Comparison between microwave and conventional processing and between variable frequency and fixed frequency microwave processing were performed. 2-D numerical modeling was performed on a variety of sample materials and sample dimensions. Numerical simulations for both fixed and variable frequency irradiation were performed.

There are four adjustable processing parameters that distinguish the VFM furnace from conventional, single frequency microwave-based technologies. These controllable processing parameters include; incident frequency, bandwidth, sweep-rate and power. During processing, the incident frequency irradiated inside the microwave cavity can be tuned to increase the coupling efficiency with the material to be processed. The combination of bandwidth and sweep rate around the selected central frequency, utilized during processing, provides the necessary distribution of microwave energy to carry out uniform heating throughout the workload. The sweep rate is electronically controlled and can be varied from minutes to milliseconds. The microwave incident power can be pulsed or continuously varied to provide some control over the heating profile of the workload.

EXPERIMENTAL PROCEDURE AND RESULTS

The isocyanate/epoxy systems, developed using proprietary catalysts provided by EniChem America, Inc., have short cure --3 minutes--and very long post-cure cycles. Typically, post-curing is performed in conventional furnaces by slowly increasing the temperature inside the furnace in a step-like manner to avoid thermal gradients at the samples' surfaces that may lead to dimensional changes, i.e., warping. The recommended conventional post curing takes about 8 hours at temperatures of up to 240 °C.

A variable frequency microwave furnace has been instrumented with Luxtron microwave-transparent temperature probes. In addition to the temperature monitoring system, the variable frequency microwave prototype has a hot-air blowing unit, a vacuum pump, for evacuation of volatiles, and is completely furnished with microwave-transparent ceramic insulation.

The processing time for the postcure of EniChem isocyanate/epoxy composites was reduced from 8 hours with the conventional thermal method to 50 minutes with the variable frequency microwave approach. The thermal profiles of 10 cm x 10 cm x 0.25 cm PMC plates have been derived for various processing conditions. Figure 1a illustrates the effect of central frequency, with a 1 Ghz frequency bandwidth, on the heating profiles of the PMC plates. Figure 1b illustrates the effect of sweep rate on the heating profile generated using 4.5 Ghz central frequency. Several experiments were carried out in order to investigate the importance of soak temperatures on the curing behavior. The measured Tg values increase with increasing soak temperatures. Full curing is achieved only when the cure temperature exceeds ≈230°C. Having reached that temperature the cure seems to proceed rapidly. The heating rate utilized to reach a given cure temperature might be a factor but only apparently as a secondary effect. Two samples

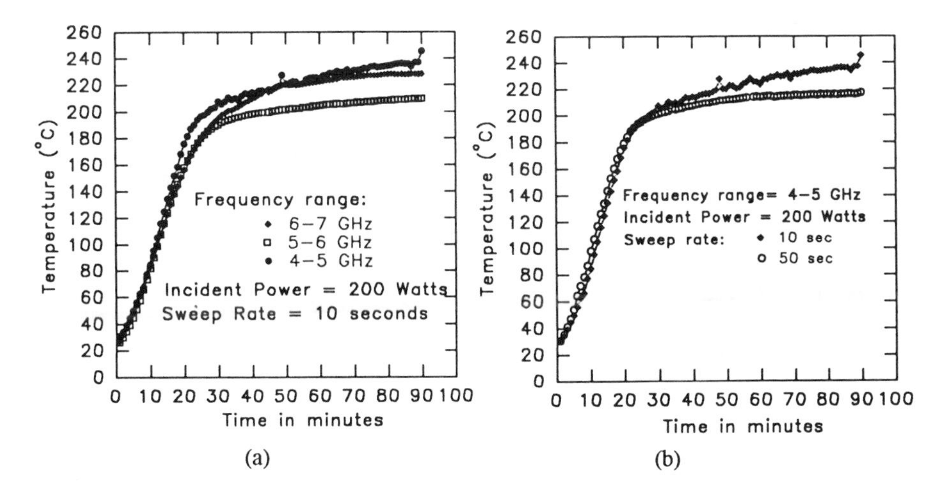

Figure 1. The thermal profile generated during variable frequency irradiation: a- the effect of frequency, and b- the effect of sweep rate.

were heated to a cure temperature of 210°C using different heating rates. Both samples exhibited comparable Tg values--220°C.

Since microwave heating is volumetric, the attainment of high heating rates does not lead to samples warping. Several samples were heat treated using high heating rates under an average microwave incident power of 280 Watts. The samples heat treated under these conditions were fully cured in 50 minutes. The glass transition temperature is plotted as a function of curing temperature (see Figure 2). As illustrated, full curing is achieved for curing temperatures above ≈230°C.

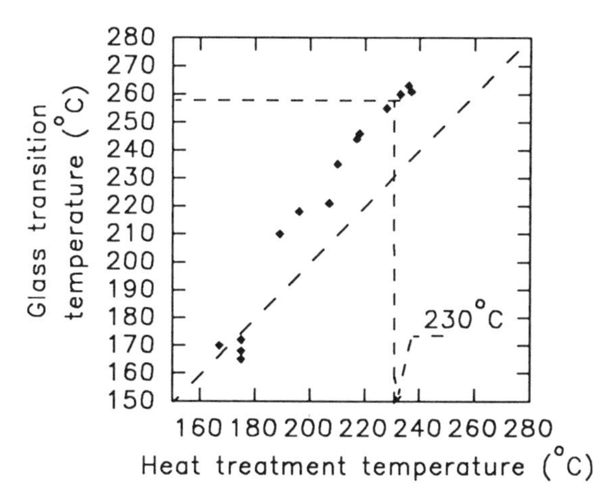

Figure 2. Glass transition versus curing temperature.

The dielectric properties of as-cured and post-cured samples have been measured. Figures 3 illustrates the relative dielectric constant, ε'_r, and the dielectric loss factor, ε''_r, as a function of temperature for the as-cured samples. As illustrated in these figures, the general trend appears to be an increase in the dielectric parameters with increasing temperatures, followed with a decrease past the 170-180 °C temperature range. A change in the charge storage capacity of the material is generally correlated to structural changes occurring at the molecular level. The structural changes occurring at the 170-180°C temperature range can be reasonably correlated to the onset of further curing--the formation of a 3-d network. Thus, the dielectric property measurements might be a good indication of the onset of post-curing. This interpretation is further supported by the results illustrated in Figure 2; the measured Tg values for the samples heat treated below 180 °C are comparable to the curing temperature, while those measured for samples heat treated above 180 °C are different.

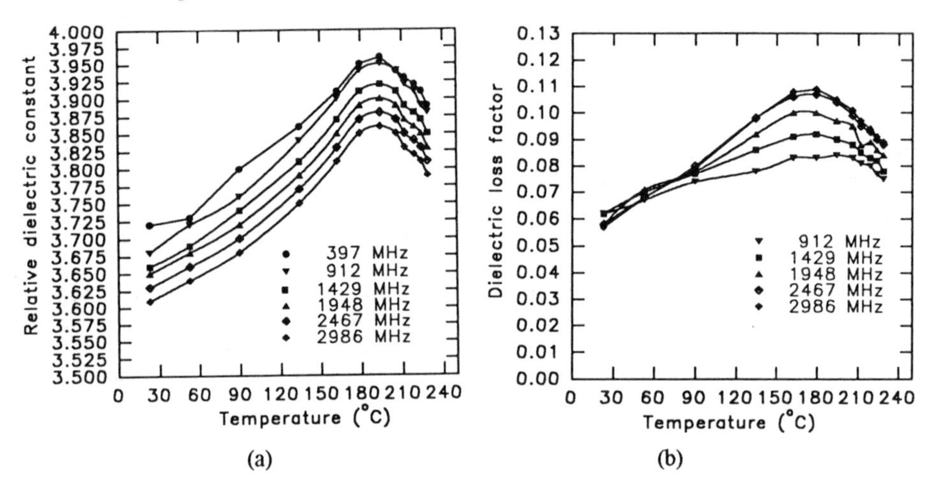

Figure 3. a- The relative dielectric constant, and b- the relative dielectric loss factor as a function of temperature for various frequencies.

The loss tangent, tanδ, of as-cured and post-cured samples are compared in Figure 4. The tanδ of the post-cured samples appear to be frequency independent and increases slightly with increasing temperature. There are no pronounced changes over the measured temperature range. This is consistent with the earlier observation since no structural rearrangements are expected in the fully cured samples.

Ten square samples, 5 cm x 5 cm x 0.25 cm, have been mounted onto a microwave-transparent ceramic tray. The assembly was then inserted in to the VFMF to simulate a small scale batch process. The samples were separated from each other by approximately 1.2 cm. The VFMF batch process was carried out at 4.5 ± 0.5 GHz at a sweep rate of 10 seconds under an incident variable frequency microwave power of 240 Watts. The cure/soak temperature was in the ≈ 230°C range. The glass transition temperatures were measured for each of the samples. The Tg measurements revealed full curing for each of the samples processed in the batch.

Heating of two (20 cm x 20 cm x 0.25 cm) PMC stacked plates was performed. Figure 5 a illustrates the experimental set up. The temperature was monitored in four different locations

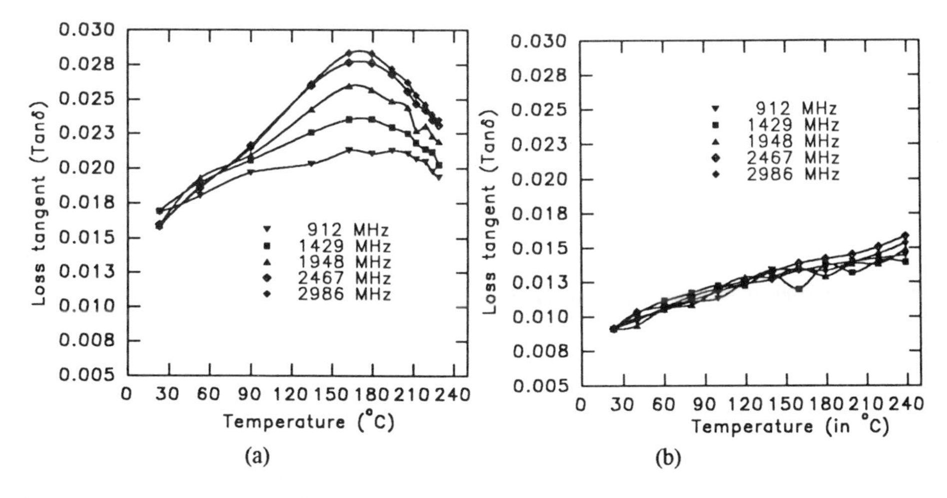

Figure 4. The loss tangent as a function of temperature for glass reinforced isocyanate epoxy mixture a- before post-cure, and b- after post-cure.

(two temperature probes for each plate). The environmental temperature was controlled to minimize heat loss from the samples' surfaces. The heating rate was rapid and the thermal profile was uniform throughout the samples (see Figure 5b).

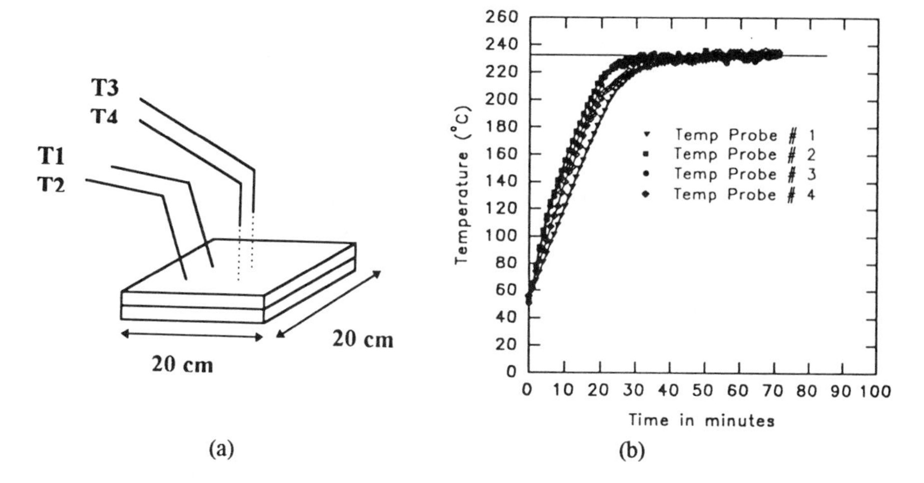

Figure 5. a- The 20 cm x 20 cm x 0.25 cm sample configuration and temperature probe placement. T1 and T2 are monitoring the temperature of the Top plate while T3 and T4 are monitoring the temperature of the bottom plate, and b- temperature profiles obtained during variable frequency irradiation (6.5 ± 0.5 GHz).

The scale-up to larger samples was successfully achieved. As indicated by the pictures (see Figure 6), there is a color change associated with curing. The color change is uniform throughout the sample processed by variable frequency irradiation (Figure 6.a).

Fixed frequency (2.45 GHz) processing was investigated to contrast and point out advantages and/or limitations, if any, when compared to variable frequency microwave irradiation. A home kitchen model, fixed-frequency microwave oven, with output powers up to 800 Watts, equipped with a turn table, was used for this comparative study. As illustrated in Figure 6b, the color change is localized to one area of the sample processed at fixed frequency. The localized heating of the sample is the physical manifestation of a non-uniform electric field inside the fixed-frequency microwave applicator.

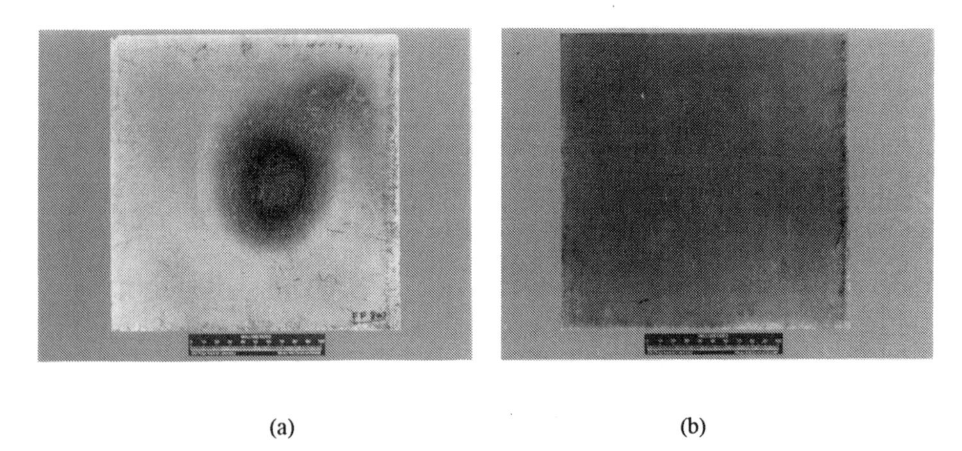

(a) (b)

Figure 6. Fixed frequency versus variable frequency illustration of 20 cm x 20 cm glass reinforced isocyanate/epoxy composites: a- fixed frequency processing, and b-variable frequency processing.

Flex modulus, strength and impact strength of fully microwave post-cured composites were measured and properties were found to be equivalent when compared to those of conventionally-cured samples. The impact strength is 5.7 KJ/m^2 for both microwave and thermally postcured composites. The flex modulus and strength for VFMF postcured composites are 5.77 ± 0.15 (GPa) and 197 ± 15.5 (MPa), respectively, while the flex modulus and strength for thermally postcured composites are 5.75 ± 0.14 (GPa) and 195 ± 11.5 (MPa), respectively.

A Finite Difference Time Domain (FDTD) technique, described elsewhere [3], was used to model the heating profiles generated during fixed frequency irradiation as well as variable frequency irradiation. Figure 7 illustrates the modeled heating profile generated within a 20 cm x 20 cm x 0.25 cm PMC during fixed frequency heating. Figure 8 illustrates the modeled thermal profile of a similar sample during variable frequency irradiation. This simulation was performed by investigating up to 41 frequencies within a 1 GHz bandwidth. The results illustrates the need to have various frequencies in order to achieve heating uniformity. The variable frequency simulation results in the establishment of a thermal gradient across the sample of about 11°C, while the thermal gradient in the fixed frequency has been evaluated to be on the order of 90°C. These numerically calculated thermal gradients are very close to those observed empirically.

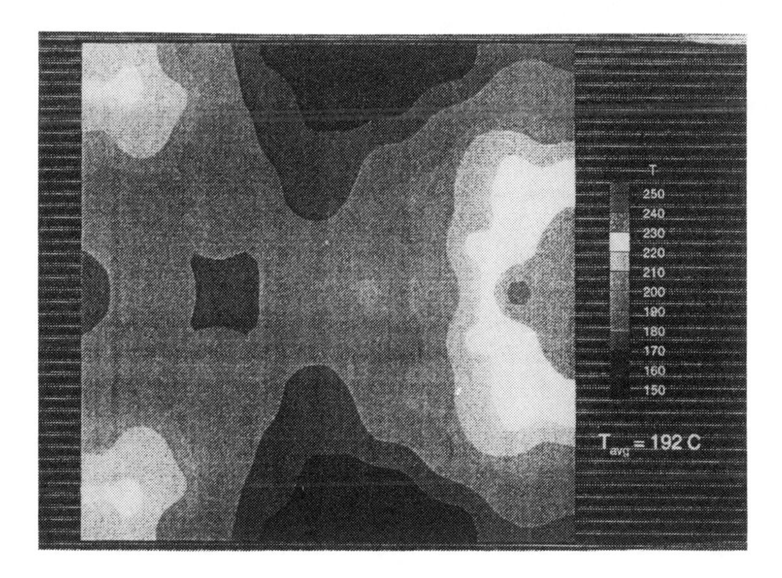

Figure 7. The simulated thermal profile of 20 cm x 20 cm PMC plates as result of fixed-frequency irradiation.

CONCLUSIONS

Variable frequency microwave energy has been used to rapidly and uniformly post-cure glass reinforced isocyanate/epoxy plates. Equivalent mechanical properties have been obtained for the microwave processed PMC samples when compared to conventionally processed samples. The variable Frequency Microwave Furnace eliminates the localized heating problems associated with fixed-frequency heating. Electronic tuning makes the VFMF a versatile processing tool applicable to an array of materials. Numerical modeling was developed and used to accurately simulate the electric field and the thermal patterns generated during fixed and variable frequency irradiation.

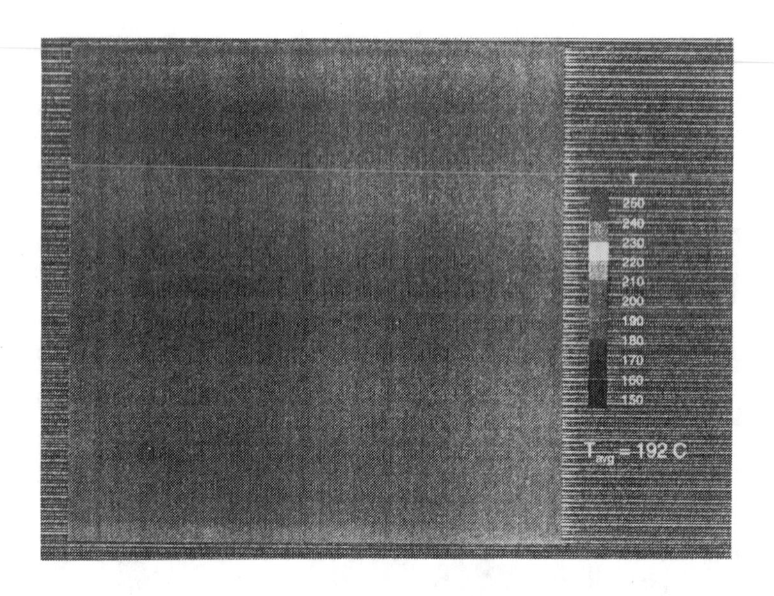

Figure 8. The simulated thermal profile of 20 cm x 20 cm PMC plates as result of variable frequency irradiation.

ACKNOWLEDGMENTS

The authors gratefully acknowledge the SBIR funding received from the Air Force of Scientific Research, Bolling AFB DC.

REFERENCES

1 D.L. Johnson, Daniel J. Skamer, and Mark S. Spotz; "Temperature Gradients in Microwave Processing," "Microwaves: Theory and Application in Materials Processing II," eds. D.E. Clark, J.R.Laia, and W.R. Tinga, Ceramics Transactions, 36, 133-146 (1993).
2 M.T. DeMeuse, and A.C. Johnson "Variable Frequency Microwave Processing of PMCs," MRS: Microwave Processing of Materials IV, 723-728 (1994)
3 J. Clemons, Masters thesis, University of Florida, 1995.

MICROWAVE ENERGY VERSUS CONVECTED HOT AIR FOR RAPIDLY DRYING CERAMIC TILE

David A. Earl
Huntington/Pacific Ceramics, Inc.
Mt. Vernon, TX 75457

D.E. Clark and R.L. Schulz
University of Florida
Gainesville, FL 32611

ABSTRACT

The purpose of this study was to determine if microwave energy could provide advantages over the conventional hot air method currently used for rapidly drying ceramic tile. Tiles consisting of a typical fast-fire body formula were dried to 0.5% moisture using a 2.45 GHz, 950W microwave oven and a natural gas-fired roller dryer. Statistical methods were employed to develop equations for predicting microwave energy consumption, tile % moisture and surface temperature given drying time, tile volume and % relative humidity. Microwave drying was found to require 36% less energy than hot air drying. Moisture was removed and surface temperature elevated at faster rates using microwave energy.

INTRODUCTION

Ceramic tiles are dried to less than 1.0% moisture before firing to increase green strength (to reduce breakage) and to avoid tile explosions due to rapid body expansion during the initial stages of firing. Current drying methods involve natural gas combustion and hot air convection.

Only about 44% of the energy produced in a conventional tile dryer goes towards heating tiles and evaporating water. The remaining energy is consumed by the dryer structure, refractories and exhausted air [1]. There are also drying speed limitations with prevailing systems. Conduction heating produces thermal gradients which can cause differential shrinkage cracks if tiles are heated too rapidly.

Microwave energy is not currently used to dry tile, but has been applied previously to drying and firing many other types of ceramics [2-4]. The objective of this study was to determine if microwave energy would provide any economic or technical advantages over the current methods used to dry ceramic tiles.

EXPERIMENTAL

The tile body selected for the drying tests consisted of a typical fast-fire wall tile formula, as shown in Table 1.

Table 1. Tile Body Formula Used in Drying Experiments.

Weight %	Material	Composition [5]
35	Illitic Clay	$Al_{2-x}Mg_xK_{1-x-y}(Si_{1.5-y}Al_{0.5+y}O_5)_2(OH)_2$
20	Wallostonite	$CaSiO_3$
25	New York Talc	$3MgO \bullet 4SiO_2 \bullet H_2O$
20	Prophyllite	$Al_2O_3 \bullet 4SiO_2 \bullet H_2O$

These materials were wet milled to produce a 7.0% moisture mix, then pressed into 5.75 cm diameter by 0.70 cm thick tile discs using a Carver lab press. Tiles were dried using a conventional natural gas-fired horizontal roller dryer or a 2.45GHz multimode oven with a glass turntable. The microwave was rate at 950W and the unit delivered approximately 500W heating power to the sample (~50%efficient).

A Scotchtrak IR-1600 portable infrared heat tracer, a Hanna portable HI-8064 thermohygrometer and a Metler digital balance were used to measure tile surface temperature, microwave cavity % relative humidity and tile weight loss, respectively, following each drying trial. To avoid water vapor condensation between the tiles and glass turntable, tiles were elevated using strips of Ultra-Poly G-24MMW plastic.

Initially, tiles exploded during microwave heating due to the rapid heating rate. To control the tile heating rate, $500cm^3$ of water was introduced into the microwave cavity as a susceptor material. Water was selected because its dielectric loss factor tends to decrease non-linearly with temperature [6]. This phenomenon results in a decrease in the amount of input power absorbed by the water during a drying cycle. Thus, power for heating the tiles gradually increases as the water temperature increases. This condition better simulates typical time/temperature curves used to dry tiles in natural gas dryers and eliminated the tile explosions.

Figure 1 shows the microwave drying set-up. A 3 ft diameter, 1725 rpm Dayton fan was used to blow air into a series of 2 mm diameter holes located on the inside wall opposite a side port. The fan was operated during some experiments to separate the effects of humidity on the dependent variables.

The natural gas roller dryer was set at a typical production peak temperature of 175°C for the drying experiments. Higher temperatures caused the tile discs to crack.

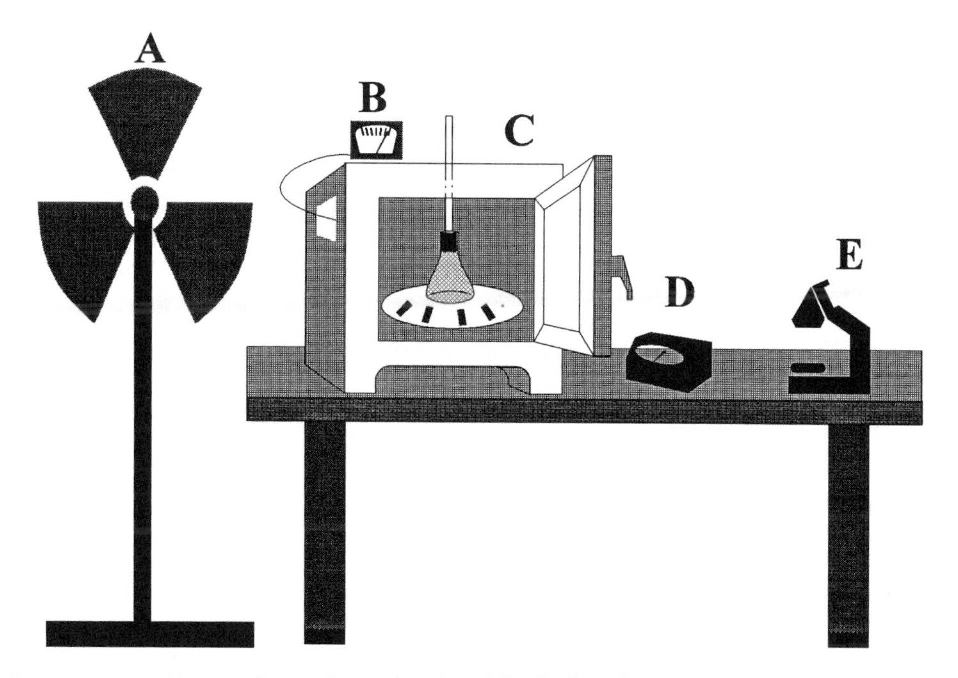

A	=	Dayton fan used to reduce humidity in the microwave
B	=	Humidity gauge
C	=	Microwave cavity including beaker with 500cc of water, rubber stopper, glass tube and plastic strips
D	=	Digital balance
E	=	Infrared heat sensor for temperature measurement

Figure 1. Microwave drying setup.

Twenty-seven microwave drying tests were performed. Drying time, tile volume in the microwave and % relative humidity were varied while tile % moisture and surface temperature were measured following each experiment. Repeat tests were also conductedto confirm test method validity. Statistical methods were used to quantify variable relationships.

Also, modulus of rupture, thermogravimetric analysis (TGA), dilatometry and average fluorine (F) evolved were compared for tiles dried with each process to less than 1.0% moisture.

In order to obtain a useful comparison between microwave and hot air drying methods, the energy consumed during both processes was estimated. The most energy efficient natural gas-fired roller dryer found during this study consumed 2411 J of energy per cm^3 of tile. Determination of energy consumed during each microwave drying test was accomplished by:

1. Energy input (E_i) = 950W microwave power input x heating time interval.

2. Water calorimetric tests were conducted to determine the relationship between average power absorbed (P_{av}) by the 500cm^3 of water during each microwave trial and heating time. The P_{av} was calculated using [7]:

$$P_{av} = \rho C_p \, dT/dt \, V \tag{1}$$

where:

ρ and C_p	=	density and specific heat of water, respectively
T	=	water temperature (°C)
t	=	microwave heating time (s)
V	=	volume of water = 500cm^3

3. A least squares regression method [9] was used to fit an equation to describe P_{av} in terms of heating time.

4. The energy consumed by the 500cm^3 of water (E_2) during each trial was estimated by

$$E_2 = \int P_{av} dt \tag{2}$$

5. The energy used for drying tiles per unit tile volume is

$$(E_1 - E_2)/(\text{tile volume, cm}^3) \qquad = \text{J/cm}^3 \tag{3}$$

This method is valid because 500cm^3 of water was used only to create variable power input over each drying time interval and would not be necessary in a microwave oven where power input could be programmed to produce specific heating versus time data.

RESULTS AND DISCUSSION

Microwave calorimetric tests revealed an inverse relationship between P_{av} absorbed by water and water temperature. This correlates to the variation in dielectric loss factor of water versus temperature observed by Datta [9]. The relationship between P_{av} absorbed by the water (watts) and microwave heating time (seconds) was found to be

$$P_{av} = 346.71 + 3108.83/t - 12.45\sqrt{t} \tag{4}$$

The regression fit was very good with an R^2 (coefficient of determination) value of 0.996 and an estimated accuracy of ± 7 watts.

By varying the amount of water and microwave heating time, it was determined that the calculated P_{av} leveled off at a maximum of 416 W. This would make the microwave oven only 44% efficient (water vapor and wall losses not considered). Generally, home microwave ovens are ≈ 50% efficient [1]. The below average efficiency may be due to the age of the unit (7 years).

The microwave experimental relationships quantified using statistical methods were:

$$E/V = 594.78 + 6.43t^2 - 22970.40/V - 42.25\sqrt{V} + 42{,}147.53t/V \quad (5)$$

$$\%M = 2.19 + 0.011V - 0.043t^2 + 3.6/t - 96.48/H \quad (6)$$

$$T = 58.61 + 26.26\sqrt{t} - 3.6 \times 10^{-6} V^3 + 6.6 \times 10^{-6} H^3 - 0.206V/t \quad (7)$$

$$d\%M/t = -0.086t - 3.601/t \quad (8)$$

$$dT/dt = 0.205V/t^2 + 13.131\sqrt{t} \quad (9)$$

where:

E/V	=	energy consumed to dry tiles per unit tile volume (energy density, J/cm^3)
V	=	tile volume (cm3)
t	=	drying time (minutes)
H	=	% relative humidity in microwave cavity
%M	=	tile % moisture
T	=	tile surface temperature (°C)
d%M/dt	=	rate of moisture removal
dT/dt	=	rate of tile surface temperature increase.

Equations 5 through 9 were used to adjust microwave drying dependent variables to conditions of 65% relative humidity which was the level measured in the hot air roller dryer. It was not possible to obtain this specific humidity in the microwave due to the limited control of air flow. Therefore the equations were used to extrapolate data in order to make parallel drying method comparisons. Figure 2 shows that it required 31% less energy per tile volume (1665 J/cm^3) to dry tiles to less than 1.0% moisture in 10 minutes when compared to hot air drying (2411 J/cm^3). Also, a 30% reduction in drying time was achieved by increasing the microwave energy per tile volume from 1665J/cm^3 to 2604 J/cm^3. Tile discs cracked when attempts were made to increase the hot air drying energy. It appears that microwave volumetric heating allows for more energy input and faster drying without cracking the tiles. The surface temperature of microwave dried tiles was elevated faster than hot air dried tiles with similar energy inputs as shown in Figure 3.

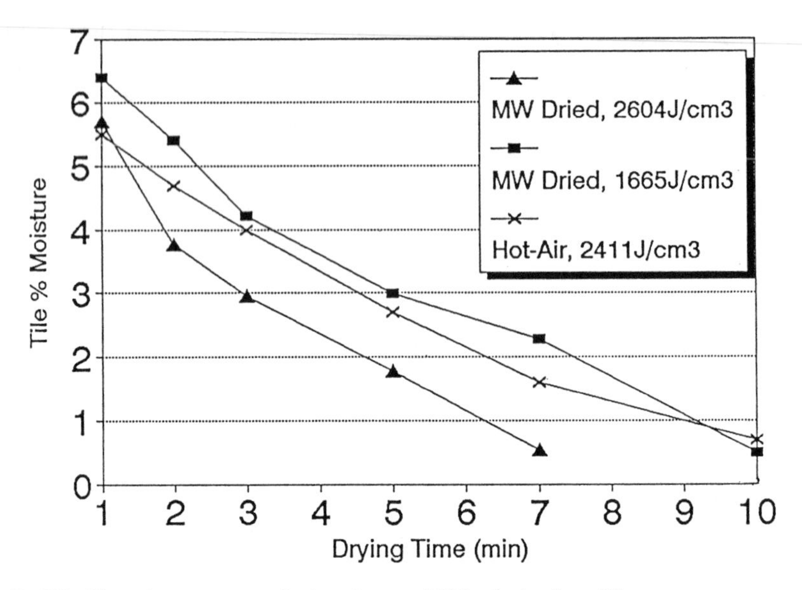

Figure 2. Tile % moisture versus drying time at 65% relative humidity.

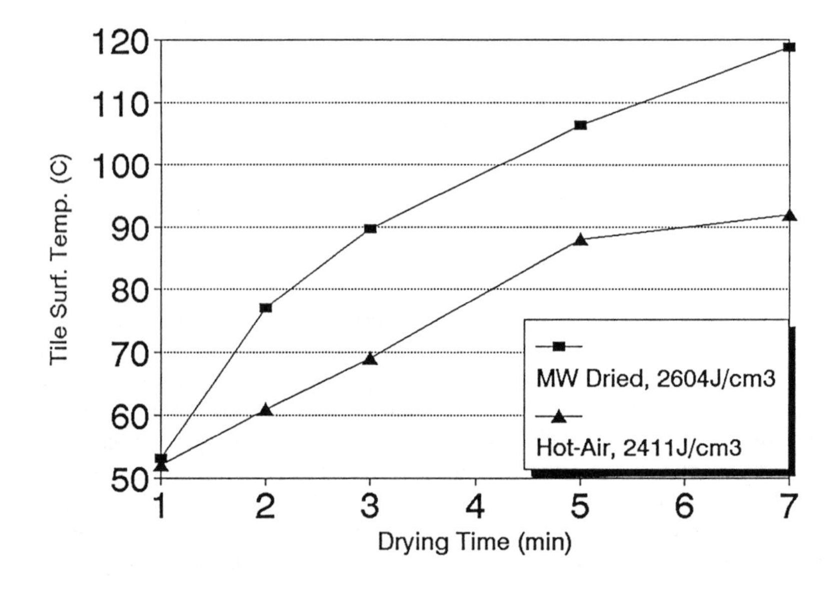

Figure 3. Tile surface temperature versus drying time at 65% relative humidity.

Figure 4 demonstrates that the rate of moisture removal in the early drying stages was much higher during microwave processing. This may be due to the fact that overall volumetric heating is a much faster process than conduction heating of a ceramic due to its low thermal conductivity [10]. Tile volume and drying time in equations 5 and 6 were iterated using a computer to find the minimum microwave energy per unit tile volume necessary to dry tiles to 0.5% moisture. Figure 5 shows that a minimum input energy per unit tile volume of 1543 J/cm^3 was required to dry 492 cm^3 of tile to 0.5% moisture. The estimated drying time was 11.8 minutes. The 492 cm^3 tile volume was very close to the actual interior microwave cavity cross-sectional area. Therefore, the microwave dried most efficiently when loaded with a single layer of tiles. Energy consumption savings for this case when compared to hot air drying was 36%.

Modulus of rupture, TGA and dilatometric analysis of dried discs yielded no significant difference between microwave and hot air processed samples.

An analysis of fluorine evolved from two tiles during both drying methods showed that the microwave processed samples evolved 33% less fluorine, on average. These results were statistically inconclusive due to the small sample size and therefore, should be re-tested.

SUMMARY

The replacement of hot air methods for drying ceramic tile with microwave energy has the potential to reduce energy costs. This study found 36% less energy was required. Tile moisture was removed and surface temperature was elevated at faster rates with microwave energy when compared to hot air. Moisture content and surface temperature of tiles dried with microwave energy were found to be non-linearly related to power input, tile volume, heating time and % relative humidity. More research should be conducted to confirm that less fluorine is evolved during microwave processing and to determine if this translates to a total reduction after drying and firing.

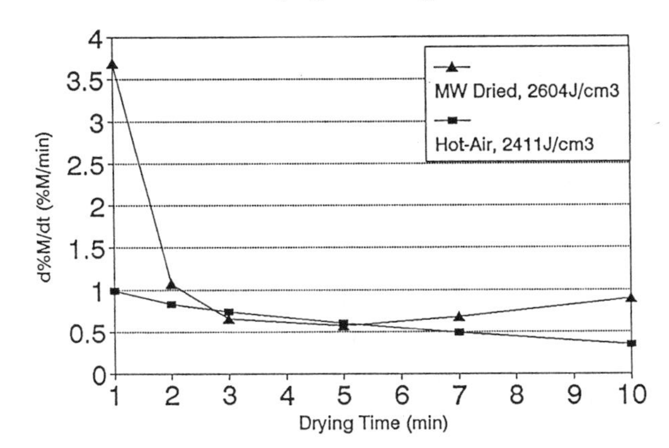

Figure 4. Rate of moisture removal versus drying time at 65% relative humidity.

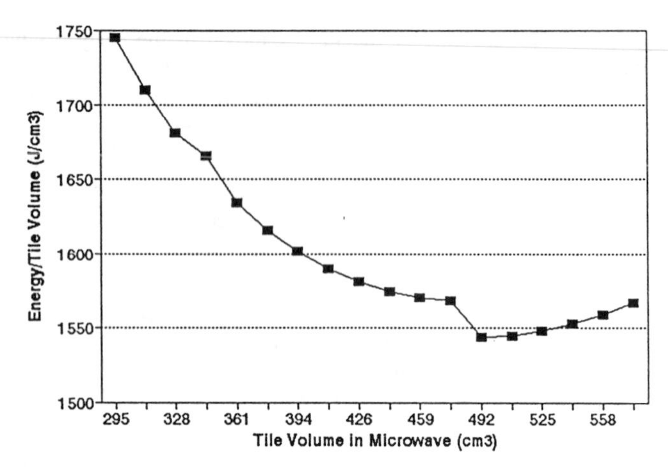

Figure 5. Energy density to dry tiles of 0.5% moisture versus tile volume during microwave drying at 65% relative humidity.

.ACKNOWLEDGMENTS

The authors thank Huntington/Pacific Ceramics, Inc. and Florida Tile Industries, Inc. for support during this work.

REFERENCES

1. G. Nassetti and G. Timellini, **Infrared Radiation and Microwaves in the Drying Process for the Production of Ceramics**, Italian Ceramic Center Int. Pub., Bologna, Italy (1990).
2. D.E. Clark, **Microwave Processing of Materials**, Workshop given for local industry at the University of Florida, Gainesville, FL (1989).
3. G. Fisher, Ceramic Industry, 121, pp. 40-42 (1983).
4. L. Sheppard, Amer.Cer.Soc. Bull., 67, pp. 1556-1561 (1988).
5. R. Grimshaw, **The Chemistry and Physics of Clays**, 4, pp. 97-124 (1971).
6. W.H. Sutton in, **Microwaves: Theory and Application in Materials Processing II**, Ceramic Transactions, Vol. 36 (D.E. Clark, W.R. Tinga and J.R. Laia, Jr., eds), American Ceramic Society, Westerville, OH, pp. 10 (1993).
7. A. Metaxas and R. Meridith, **Industrial Microwave Heating,** Peter Peregrinus, London, UK (1983).
8. R. Bowles and S. Haller, **Problem Solving with Experimental Design, Multiple Correlation, Multiple Property Optimization and Partitioning Variability**, SSI Management Consultants, Inc., Cleveland, OH (1990).
9. A. Datta, Chem. Eng., 86[6], pp. 47-51 (1990).
10. **Chemical Engineering Handbook**, 5th Ed. (R. Perry and C. Chilton, eds), 10-6, McGraw-Hill, NewYork, NY (1973).

DENSIFICATION OF LARGE SIZE ALUMINA OBJECTS BY CONTINUOUS MICROWAVE SINTERING

Jiping Cheng, Yi Fang, Dinesh K. Agrawal, and Rustum Roy
Intercollege Materials Research Laboratory, The Pennsylvania State University
University Park, PA 16802

ABSTRACT

A new apparatus designed for the continuous microwave sintering is described. This makes it possible to process ceramic rods or tubes with diameter of 10-100 mms and length of 1-2 meters. In this work, high purity alumina rods of 15 mm diameter were sintered by the continuous microwave sintering, and the effects of the temperature distribution and the feeding speed on the sample were investigated. The results indicate that both temperature gradients in the sintering cavity and the feeding speed are the key parameters during the continuous microwave sintering for making high quality ceramic products.

INTRODUCTION

Microwave processing of ceramics started several years ago with the aim at energy savings and technological benefits over conventional processing[1]. The acceleration of the sintering of ceramics by microwave heating has been experimentally established, and the accelerated sintering seems to be attributed to the enhancement of diffusion by microwave heating[2], which results in a more rapid sintering than in a conventional furnace. At present, one of the main issues is how to scale-up the microwave sintering technology and transfer it to the ceramic industry. Currently, some exploratory experiments have been done to develop laboratory scale microwave sintering to ceramic production industry[3-5].

As a new development in the microwave application, the continuous microwave sintering was introduced by the present authors last year[6]. By this technique, commercial ceramic rollers with 40 mm diameter and 2400 mm length had been successfully microwave sintered, and remarkable energy savings and economical benefits are possible when compared with the conventional processing. In the present work, high purity alumina rods were sintered by the continuous microwave sintering technique, and the densification behavior of the ceramic body during the processing was investigated.

EXPERIMENTAL

A new apparatus was designed for the continuous microwave sintering. It is equipped with a 5 KW 2.45 GHz microwave power supply, and a 64 liter rectangular multimode cavity with a sample feeding device. An alumina fiber cylinder (Zircar Co.) is used as insulator inside the cavity (Figure 1).

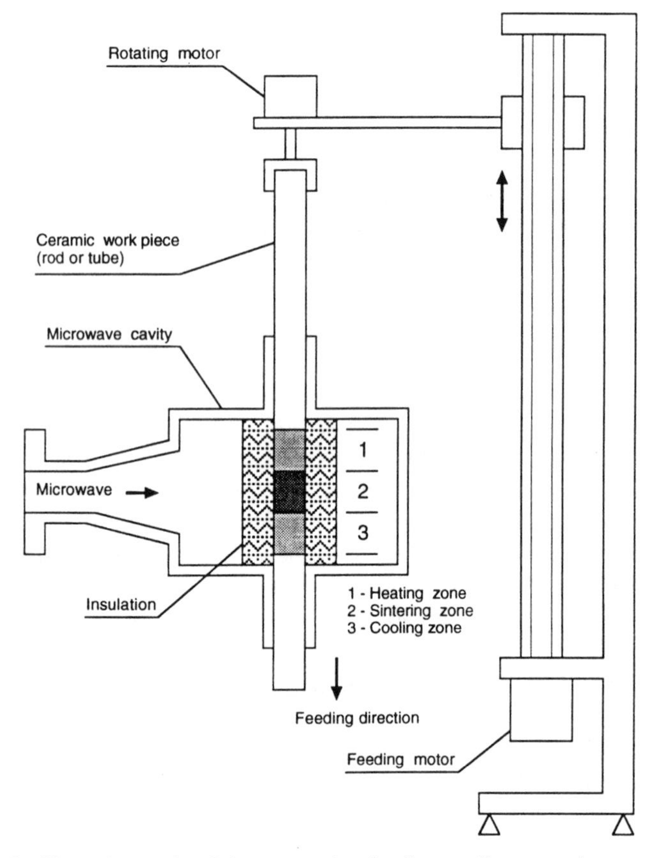

Figure 1. The schematic of the apparatus for the continuous microwave sintering

The high purity alumina sample rods were prepared as follows: the starting alumina powder (Fisher Scientific Co.) was calcined in an electric furnace at 1300 °C for 4 hours, the calcined powder added with 0.2 wt% $Mg(NO_3)_2$, 8 wt% polyacrylamide (Polyscience, Inc.) and 12 wt% water was mixed in a two-blade blender for 15 minutes. The mixture was extruded into rods of 15 mm in diameter and 600-800 mm in length in an extrusion apparatus (C.W. Brabender Instruments, Inc.). The extruded rods were dried at room temperature for 2 days.

To assemble the sample rod for the continuous microwave sintering, the rod was held by a metal clamp which was connected to the axle of the rotating motor, hung into the cavity with one end of the rod located in the center area of the cavity. At first, the sample rod was fixed in the original position while rotating only, without vertical feeding movement. Microwave power was transported into the cavity to heat the sample at a heating rate of 50 °C/min. When the sample reached the set temperature, the feeding motor was started to feed the sample at the desired speed. At this stage, the temperature was monitored by an infrared pyrometer (Accufiber Inc.), and controlled at a desired value (usually, ±10 °C of the set temperature).

To measure the temperature gradient along the sample rod during the processing, the sample rod was fed down very fast and the temperature changes read by the pyrometer were recorded with respect to the distance of the sample from its original position.

The microstructure of the sintered samples was characterized by SEM, and the densities were measured by the Archimedes method.

RESULTS AND DISCUSSION

In the beginning, in order to determine the optimum sintering procedure, the sample rod was microwave sintered in a fixed position (feeding speed = 0). Figure 2 shows the density changes along the length of the rod in this case. After 5 min. dwelling at 1400 °C, a high density of 98 % T. D. was obtained in the center part of the rod exposed to the highest temperature zone. The density decreased gradually with the distance away from the center along the length in both directions, due to the decreasing temperature. It was observed that there was no sintering effect when the temperature was lower than 1200 °C.

During the continuous sintering, the density of the sample rod changed with the feeding speed as shown in Figure 3. When the feeding speed was lower than

5 mm/min, a uniformly sintered sample was achieved. From Figure 2, we can see that the uniform high temperature zone is ~30 mm long along the length direction. A feeding speed of 5 mm/min means that each part of the sample rod dwelt at this high temperature zone (1400 °C) for 5-6 minutes. It has been proved in the previous fixed sintering experiment (Fig. 2) that the sample can be sintered very well in this case. Figure 4 shows the microstructures of the sample rod along the length direction. A uniform and dense microstructure was obtained in the well-sintered area of the sample rod.

Since the sample rod is being fed continuously during the continuous microwave

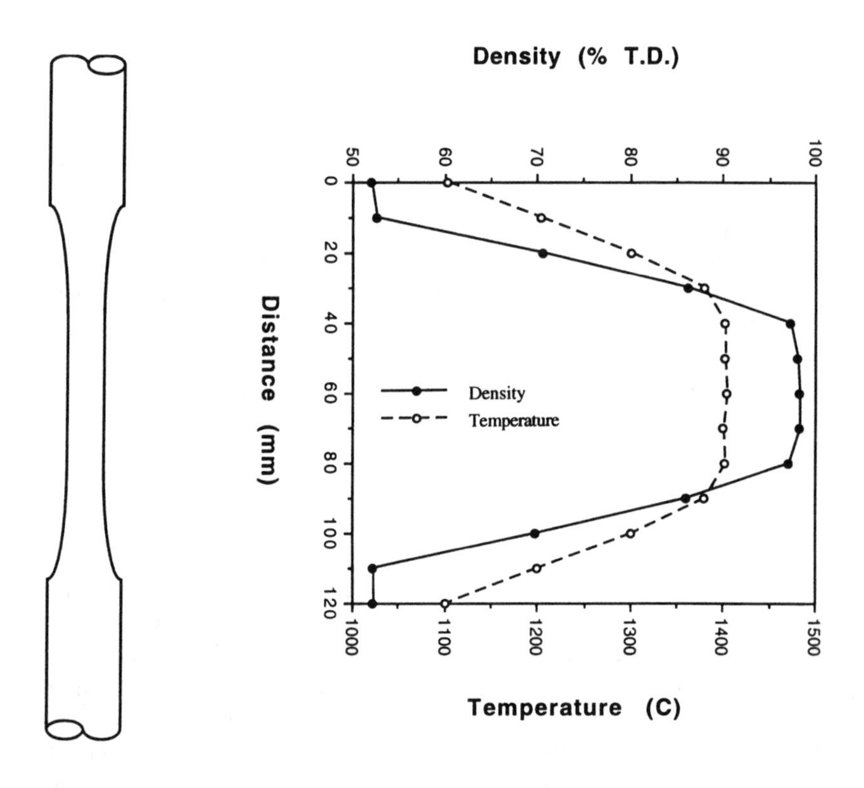

Figure 2. The longitudinal density profile of the sample rod in a non-feeding sintering process with heating rate of 50 °C/min and dwell time of 5 min. The center part of the rod was located in the center area of the cavity, the rotating rate was 6 rpm., and the feeding speed was 0.

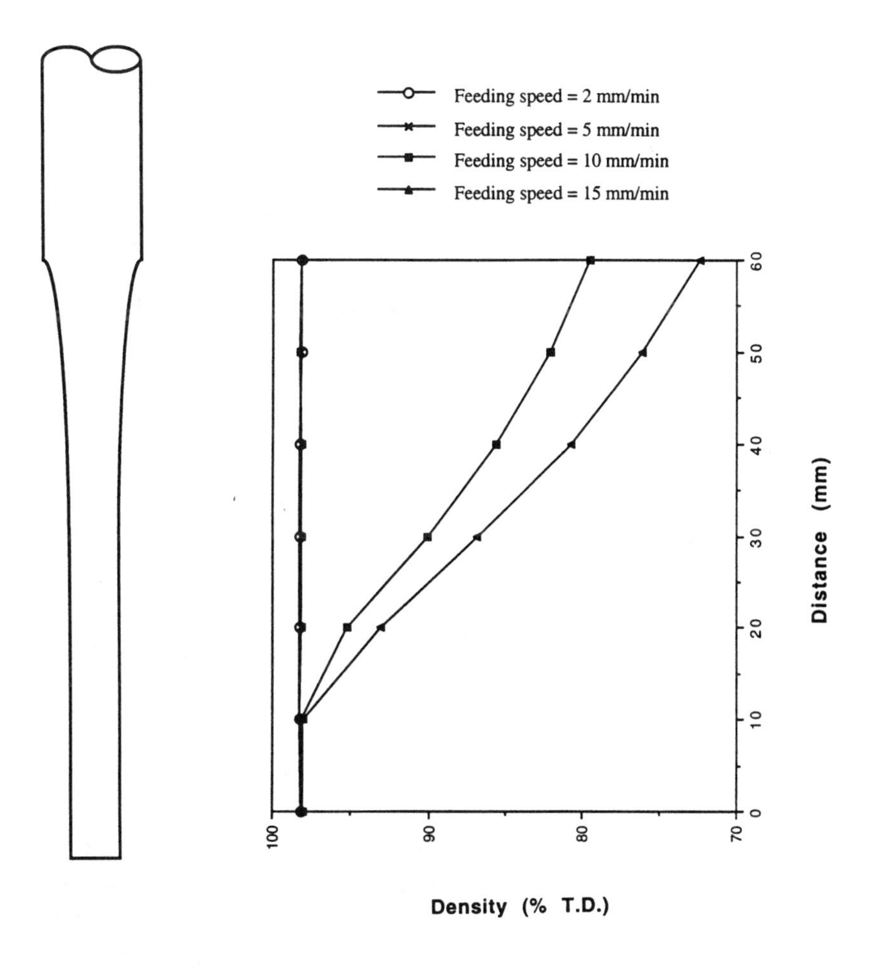

Figure 3. The density profile along the length of the sample rod processed by the continuous microwave sintering. The lower end of the rod was first put in the center of the cavity where it was heated to 1400 °C, and then the rod was fed downwards at various speeds.

sintering, there is no limitation in the length of the product that can be processed by this technique. This makes it possible to scale up the microwave sintering for certain ceramic products such as rods and tubes. The output mainly depends on the feeding speed and sintering temperature during the processing. Higher feeding produce higher yield. There are two ways to increase the feeding speed. The first is to enlarge the high temperature zone, which requires a design modification and adjustment of the sintering cavity. The second is to reduce the sintering time by

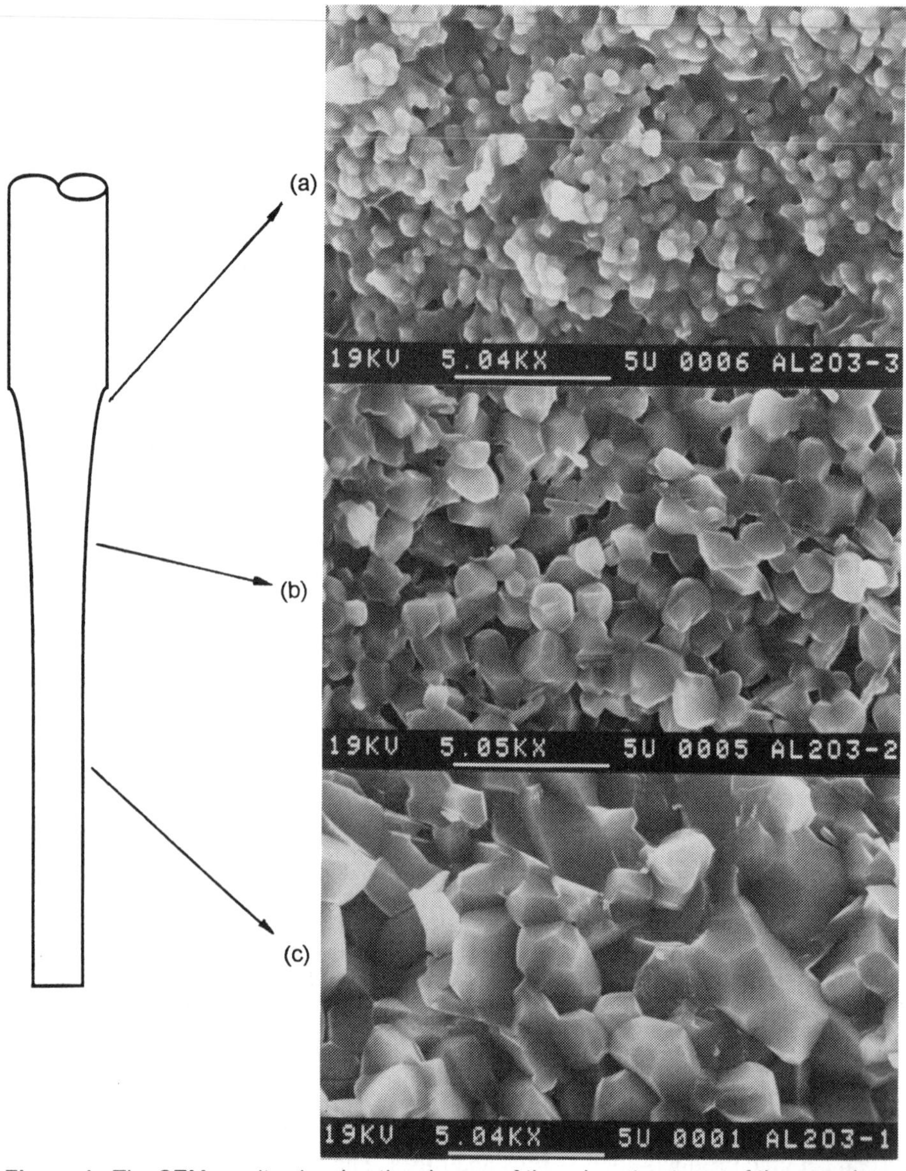

Figure 4. The SEM results showing the change of the microstructures of the sample rod during the continuous microwave sintering at 1400 °C with a feeding speed at 5 mm/min.
(a) In the partly-sintered body area of the rod with a density of 72.0 % T.D.
(b) In the partly-sintered body area of the rod with a density of 81.5 % T. D.
(c) In the well-sintered body area of the rod with a density of 98.2 % T. D.

increasing the sintering temperature, which should meet the demands required for sintering ceramic materials. In some case, an exceedingly high feeding speed may cause non-uniform heating which results in cracks or bending of the product. Efforts are being made by researchers in this laboratory to improve continuous microwave sintering techniques for their application to the ceramic industry.

CONCLUSIONS

High purity alumina rods 15 mm in diameter were sintered using a continuous microwave sintering process. A uniform and high density product was obtained when the sample rod was sintered at 1400°C with a feed speed of 5 mm/min. Higher feed rates can produce a higher yield, however it must be controlled within a certain range according to the temperature distribution in the sintering cavity to meet the processing requirements of a specific ceramic product.

REFERENCES

1. W.H. Sutton, *Am. Cer. Soc. Bull.*, 68[2], pp. 376 (1989).
2. J.D. Katz, R.D. Blake and V.M. Kenkre in, **Microwaves: Theory and Application in Materials Processing**, Ceramic Transactions, Vol. 21 (D.E. Clark, F.D. Gac and W.H. Sutton, eds), American Ceramic Society, Westerville, OH (1991).
3. M.C.L. Patterson, P.S. Apte, R.M. Kimber and R. Roy in, **Microwave Processing of Materials III** (R.L. Beaty, W.H. Sutton and M.F. Iskander, eds), Proceedings of the Materials Research Society, Vol. 269, pp. 291-300 (1992).
4. F.C. R. Wroe in, **Microwave Processing of Materials IV** (M.F. Iskander, R.J. Lauf and W.H. Sutton, eds), Proceedings of the Materials Research Society, Vol. 347, pp. 47-56 (1994).
5. J.O. Kiggans, T.N. Tiegs, H.D. Kimrey and J-P Maria in, **Microwave Processing of Materials IV** (M.F. Iskander, R.J. Lauf and W.H. Sutton, eds), Proceedings of the Materials Research Society, Vol. 347, pp. 71-76 (1994).
6. J. Cheng, Y. Fang, D.K. Agrawal, Z. Wan, L. Chen, Y. Zhang, J. Zhou, X. Dong and J. Qui in, **Microwave Processing of Materials IV** (M.F. Iskander, R.J. Lauf and W.H. Sutton, eds), Proceedings of the Materials Research Society, Vol. 347, pp. 557-562 (1994).

Processing Parameters Influencing the Sintering of Alumina in a Conventional and 14 GHz Microwave Furnace

D.J. Grellinger[1], J.H. Booske[1], S.A. Freeman[2], and R.F. Cooper[3]
University of Wisconsin—Madison
[1]Department of Electrical and Computer Engineering
[2]Material Science Program
[3]Department of Material Science and Engineering
Madison, WI 53706

ABSTRACT

Experimental design and measurement results are presented for processing conditions that affect the final density of conventionally-sintered and microwave-sintered alumina. The microwave-sintered specimens were processed in a 14 GHz multimoded cavity. Statistical experimental design methods were used to examine the roles of process temperature, soak time, and heating rate in both furnaces. Future experiments are outlined to ascertain if optimum processing conditions also translate to improvements in engineering properties such as fracture strength and indentation toughness.

INTRODUCTION

Microwave processing studies have demonstrated a potential for rapid, efficient heating of ceramic materials. This is a direct result of the fundamental difference between the heating mechanisms of a conventional and microwave furnace. In a microwave furnace direct heating is possible based on microwave absorption principles and the intrinsic material properties of the sample, such as conductivity and dielectric constant. The ability to volumetrically heat a ceramic sample with microwave energy implies opportunities for more efficient heating and better heating control.

Several studies have investigated microwave processing of alumina at 2.45 GHz[1,2,3]. These studies have typically employed hybrid heating techniques to heat samples through low temperature regimes where alumina does not readily couple to the microwave field. Many materials have been successfully processed using these hybrid heating techniques, and in some instances (although not ubiquitously) results such as lower processing temperature, improved microstructures, or higher densities have been reported [4,5,6].

Because microwave absorption is also frequency dependent, studies at 28 GHz have demonstrated that hybrid heating techniques are not necessary to heat a sample initially at room temperature [7]. Results at 28 GHz also claim more rapid densification, even at lower temperatures, than samples processed in a conventional furnace[8].

This study represents initial efforts to sinter Al_2O_3 at 14 GHz at the University of Wisconsin—Madison. Microwave-processed samples were heated in a multimoded cavity as part of a 2 kW microwave furnace designed and constructed by the authors. All samples were processed at atmospheric pressure without the use of any auxiliary susceptors as in the hybrid heating configurations.

EXPERIMENTAL APPROACH

To study the effects of processing temperature, soak time, and heating rate on the final sample density, in addition to comparing microwave versus conventional processing, we used a statistical approach known as a full factorial experimental design. This approach allowed us to study the effects of the three processing conditions in both the conventional and microwave furnaces. In addition, we also directly compared the differences between the two furnaces at a constant heating rate. In the conventional furnace, we replicated the experiment with two different alumina powders. Therefore we conducted four separate full factorial experiments each consisting of three variables at two levels. To sinter samples under all possible conditions required eight runs for each experiment (2^3). The eight different conditions for each experiment appear in Table 1. All runs were made in a random order and the final density of each part was measured and recorded.

Factorial #1: Micro./Conv. (alumina #1)			Factorial #2: Microwave (alumina #1)		
level / variable	low	high	level / variable	low	high
furnace	conventional	microwave	temperature	1441°C	1523°C
temperature	1441°C	1523°C	soak time	30 min.	2 hours
soak time	30 min.	2 hours	heating rate	4°C/min.	2°C/min.
Factorial #2: Conv. Furnace (alumina #1)			**Factorial #4: Conv. Furnace (alumina #2)**		
level / variable	low	high	level / variable	low	high
temperature	1441°C	1523°C	temperature	1441°C	1523°C
soak time	30 min.	2 hours	soak time	30 min.	2 hours
heating rate	8°C/min.	4°C/min.	heating rate	8°C/min.	4°C/min.

Table #1. Processing conditions for four full factorial experiments.

Conventional Furnace Approach

For the conventional furnace runs we used a mass of 12 grams and fired the samples in a regular tube furnace at atmospheric pressure. The sample compacts were formed in a 1-1/8 inch diameter die pressed axially to 5000 pounds. Both powders used in the conventional furnace were high purity alumina doped with 500 ppm MgO. Both Sylvania CR-30 (alumina #1) and Reynolds Chemicals RCHPDBM (alumina #2) alumina were used in the conventional furnace investigations. The green density of alumina #2 was approximately double that of alumina #1 after pressing. During firing, the sample was in constant contact with a type-R thermocouple used to measure sample temperature.

Microwave Furnace Approach

In the microwave runs, a 22 gram powder compact of alumina #1 was prepared under the same conditions as in the conventional furnace approach.

Inside the microwave furnace the sample was enclosed in a refractory fiberboard insulation casket to reduce heat loss from the sample. Inside the insulation casket, the sample was suspended by 2 dense alumina rods (1/8 inch diameter) to prevent thermal runaway in the casket which occurred in runs where the sample was in direct contact with the insulation material. In these runs, the contact between the sample and the insulation raised the temperature of the fiberboard until the insulation began to couple strongly to the microwave field. This led to densification of the fiberboard which eventually resulted in the destruction of the insulation system.

A type-S thermocouple, used to measure sample temperature, contacted the bottom of the sample at all times and contributed to the support of the sample. The thermocouple wires were enclosed in a platinum sheath to prevent arcing and false temperature measurements in the microwave field [9]. The platinum sheath was covered by an alumina tube for additional protection. The thermocouple passed through a port in the cavity wall and was used as the process variable in a PID controller. A diagram of the insulation system appears in Figure #1.

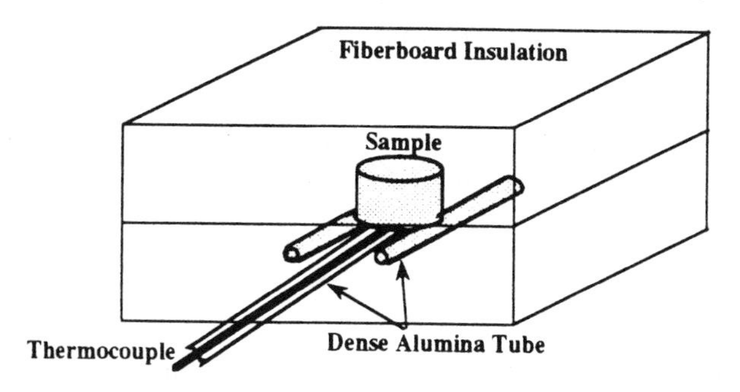

Figure #1. Insulation system and sample inside the microwave furnace.

RESULTS AND DISCUSSION

For each combination of processing conditions, an alumina sample was sintered and measured for final density. Sintering density was measured geometrically due to a present inability to make reliable Archimedes principle density measurements. Density measurements for all four full factorial experiments are presented in Figure #2 in terms of % theoretical density.

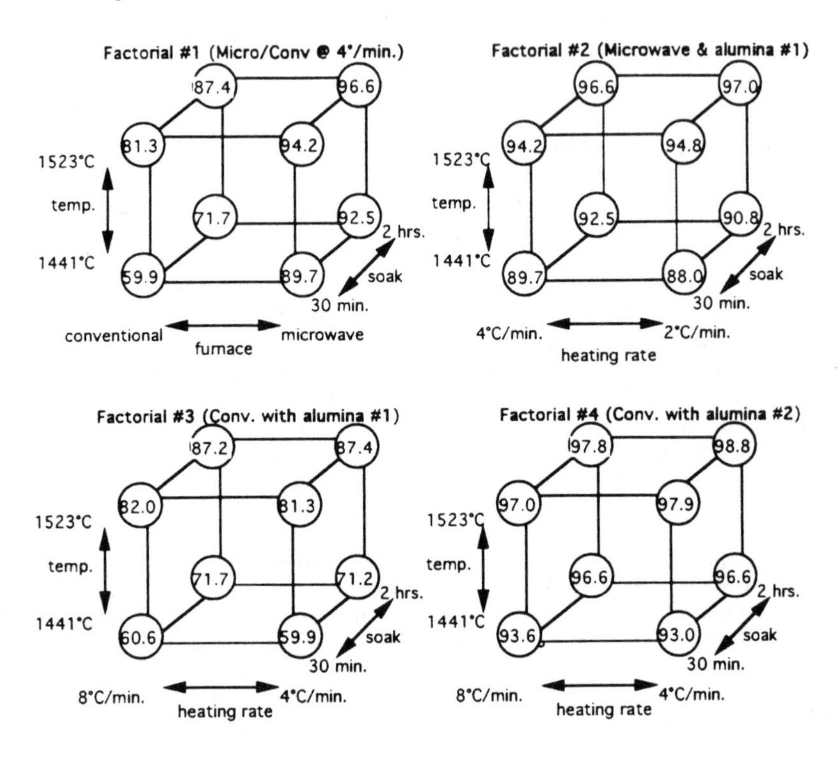

Figure #2. Measurements of % T.D. under all experimental conditions.

Visual inspection of the density measurements in Figure #2 reveals several relationships between the processing conditions and final density. Moving from any corner of any cube in Figure #2 to any of the three neighboring corners represents a change in only one of the processing conditions. Conditions specific to each cube are indicated directly above the cube in parentheses.

In factorial experiment #1, a direct comparison can be made between the conventional furnace and the microwave furnace. Moving from any location on the left face of the cube to the corresponding location on the right face of the cube represents a furnace comparison with all other processing conditions remaining constant. It is evident from Figure #2 that we observed an increase in the final density for samples heated in a microwave furnace. Comparing the other two sets of opposite cube faces in factorial #1 also shows that increased processing temperature as well as increased soak time also results in an increased final sample density. The relationships between final density and processing temperature and between final density and soak time are in accordance with accepted sintering models.

In factorial experiments #2 - #4 comparisons are made among processing conditions within each furnace. It is again evident that increased processing temperature and increased soak time result in samples of higher density in all cases studied. In addition, conventionally-sintered samples prepared with alumina #2 have higher final densities than those prepared with alumina #1. This is consistent with the samples of alumina #2 pressing to higher green densities than those of alumina #1 as previously discussed.

The relationship between final density and heating rate is not so clear in this investigation. This is the comparison between the left and right cube faces in factorial experiments #2 - #4. Final % T.D. comparisons between these two rates differ only within a few percent. Future experiments will increase the difference between the high and low levels of heating rate in order to determine the relationship with final density.

No definitive explanations are attempted at this point for the higher densities observed for microwave processing over conventional processing. However, several points and their implications are available for discussion based on the data. First, both thermocouples recorded surface temperature data. Calibration measurements ruled out systematic errors in the instrumentation. However, the thermocouple for the conventional oven was unshielded, whereas the thermocouple in the microwave oven was enclosed in both a metallic (platinum alloy) and an alumina sheath. Future measurements will investigate whether significant temperature gradients exist across these shields, thereby rendering significant underestimation of the sintered sample's surface temperature. However, prior studies comparing numerous methods of surface temperature measurement indicate that similar sheathing introduces less than 20 ^oC systematic error in the surface temperature measurement [9].

Assuming for the moment that the microwave surface temperature measurements are accurate to within 20 ^oC, we consider factorial experiment #1 which indicates a higher final density (89.7%) for the microwave sintered sample at (surface temperature) ~ 1440 ^oC than for the conventionally sintered sample (81.3%) at ~ 1520 ^oC. If any apparent enhancements associated with the microwave sintering are due solely to inverted temperature profiles (i.e., interior temperature higher than the surface temperature), then the data indicate that the microwave-sintered specimen's internal temperature must be at least ~ 100 C hotter than its surface temperature (cf. Fig. 2, Factorial experiment #1). Controlled experiments with deliberate temperature gradients in conventionally sintered specimens [10] demonstrate that temperature differences of 100 °C within the same specimen result in measurably different local densities. Hence, if the differences in the final density between the microwave- and conventionally-sintered samples are to be explained by inverted temperature profiles, then there should be a measurable difference in the local final density between the center and the surface regions of the samples. Again, this measurement will be completed in the near future to test this hypothesis.

CONCLUSIONS

It has been demonstrated in this investigation that samples processed in a 14 GHz microwave furnace have higher final densities than those processed in a conventional furnace with all other conditions remaining constant. While we offer no explanation behind this relationship, the apparent enhancements in the microwave processed case are consistent with other significant enhancements claimed over conventional processing. In future investigations we intend to further investigate this processing difference by examining both local and bulk sample properties such as fracture strength, indentation toughness, and microstructure.

In addition, we have also demonstrated that final samples densities increase with increased processing temperature and increased soak time. This is consistent with accepted sintering models.

ACKNOWLEDGMENTS

We would like to acknowledge EPRI, NSF (PYI Program), NASA, and the much-appreciated helpful advice and contributions of Dr. Mark Janney (ORNL) and Todd King (UW).

REFERENCES

1. W.H. Sutton, "Microwave Processing of Ceramic Materials," in *Ceramic Bulletin*, Vol.6, No. 2, pp. 376-386 (1989).

2. M.A. Janney, H.D. Kimery, and J.O. Kiggans, "Mirowave Processing of Ceramics: Guidelines Used at the Oak Ridge National Laboratory," in *Microwave Processing of Materials III*, MRS Symp. Proc., Vol. 269, pp. 173-185 (1992).

3. Y.-L. Tian, H.S. Dewan, M.E. Brodwin, and D.L. Johnson, "Microwave Sintering Behavior of Alumina Ceramics," in *Advances in Sintering*, pp. 391-401 (1990).

4. H.D. Kimery, J.O. Kiggans, M.A. Janney, and R.L. Beatty, "Microwave Sintering of Zirconia Toughened Alumina Composites," in *Microwave Processing of Materials II*, MRS Symp. Proc., Vol. 189, pp. 243-255 (1991).

5. Arindam De, Iftikhar Ahmad, E. Dow Whitney, and David E. Clark, "Microwave (Hybrid) Heating of Alumina at 2.45 GHz: I. Microstructural Uniformity and Homogeneity," in *Microwaves: Theory and Applications in Material Processing*, Ceramic Transactions, Vol. 21, pp. 319-328 (1991).

6. Arindam De, Iftikhar Ahmad, E. Dow Whitney, and David E. Clark, "Microwave (Hybrid) Heating of Alumina at 2.45 GHz: II. Effect of Processing Variables, Heating Rates and Particle Size," in *Microwaves: Theory and Applications in Material Processing*, Ceramic Transactions, Vol. 21, pp. 329-340 (1991).

7. M.A. Janney, and H.D. Kimery, "Microwave Sintering of Alumina at 28 GHz," *Ceramic Powder Science, II*, pp. 919-924 (1988).

8. M.A. Janney, and H.D. Kimery, " Microstructure Evolution in Microwave-Sintered Alumina," in *Advances in Sintering*, pp. 382-390 (1990).

9. D.J. Grellinger, and M.A. Janney, " Temperature Measurement in a 2.45 GHz Microwave Furnace," in *Microwaves: Theory and Applications in Material Processing II*, Ceramic Transactions, Vol. 36, pp. 529-538 (1993).

10. D. Beruto, R. Botter, and A.W. Searcy, "Influence of Temperature Gradients on Sintering: Experimenal Tests of a Theory,", *J. Am. Cer. Soc.* , Vol. 72, pp. 232-235 (1989).

SINTERING OF ALUMINA CERAMICS IN A SINGLE MODE CAVITY UNDER AUTOMATED CONTROL

Ki-Yong Lee and Eldon D. Case
Department of Materials Science & Mechanics
Jes Asmussen, Jr. and Marvin Siegel
Department of Electrical Engineering
Michigan State University, East Lansing, MI 48824

ABSTRACT

An automated processing system featuring a single-mode microwave cavity operated at 2.45 GHz has been used to sinter a series of alumina powder compacts. The automated control allows repeatable heating schedules for the processing. The resulting sintered alumina specimens were crack-free and had small, uniform grain sizes.

INTRODUCTION

Microwave sintering can be an attractive processing technique since materials can be heated directly and/or indirectly via the interaction with electromagnetic fields. Typically microwave processing yields ceramic specimens having high densities and fine grain sizes. Also, microwave processing allows high heating rates (short processing times) compared to conventional heating.

Multimode microwave cavities [1-6] require very high power due to the relatively low coupling efficiency of microwave energy with materials. Also, a multimode cavity's nonuniform electromagnetic field distribution can result in inhomogeneous heating and hot spots depending on the volume of the processed material and the location of the material within the microwave cavity [7]. Nevertheless multimode cavities are frequently used due to their low cost, ease of construction and adaptability [8].

To improve process control and heating uniformity, several investigators have designed and used single-mode microwave cavities [8-13]. Single mode cavities can maximize the electrical field strength at the location of processed material, potentially yielding higher heating rates and more uniform heating than is the case for conventional multimode cavities [7-9, 14]. Using an internally-tuned single-mode circular cylindrical cavity, Asmussen et al. [8, 14] demonstrated that microwave energy can be coupled efficiently into either low loss or lossy materials.

Most single mode cavities used for microwave processing ceramics are tuned by manually adjusting the positions of the electrical short and the probe to maximize the energy absorbed by the process material. Since the dielectric properties of material are functions of both porosity and temperature, the dielectric properties of the ceramic change during processing. This change in dielectric properties requires that the cavity be tuned continuously to maintain efficient microwave coupling and an optimum heating rate. Manual cavity tuning, however, is too slow and cumbersome to allow one to precisely tune the cavity as the material's dielectric properties can change rapidly during processing.

Recently Asmussen and Siegel et al. [15] developed a single-mode cylindrical cavity that is tuned by computer-automated adjustments of the sliding electrical short and probe positions. Two computer-driven controllers (Microstep Drive Sx Series, Compumotor, Fauver, MI) allow a rapid fine-tuning of the short and the probe positions to with an accuracy of ±0.1 mm. In this study Asmussen and Siegel's computer-controlled microwave processing system was used to obtain very similar heating schedules during processing of a series of alumina powder compacts. Such repeatability would be difficult using manual tuning techniques.

EXPERIMENTAL PROCEDURE

Experimental apparatus

Figure 1 is a schematic of the microwave sintering apparatus. The microwave power supply (Sairem, Model MWPS 2000, Wavemat Inc., Plymouth, MI,) used in this study can supply from zero to 2000 Watts of continuous wave microwave power at 2.45 GHz. Microwave power is generated by a magnetron. Waveguides then feed the microwave power into the cavity through an adjustable tuning probe. Analog power meters connected to the waveguide through attenuators indicate the forward and reflected power levels.

The single mode microwave cavity (Model CMPR-250, Wavemat Inc., Plymouth, MI.) used in the study can be internally tuned to resonate in many different microwave cavity resonant modes by adjusting the short and the power launch probe

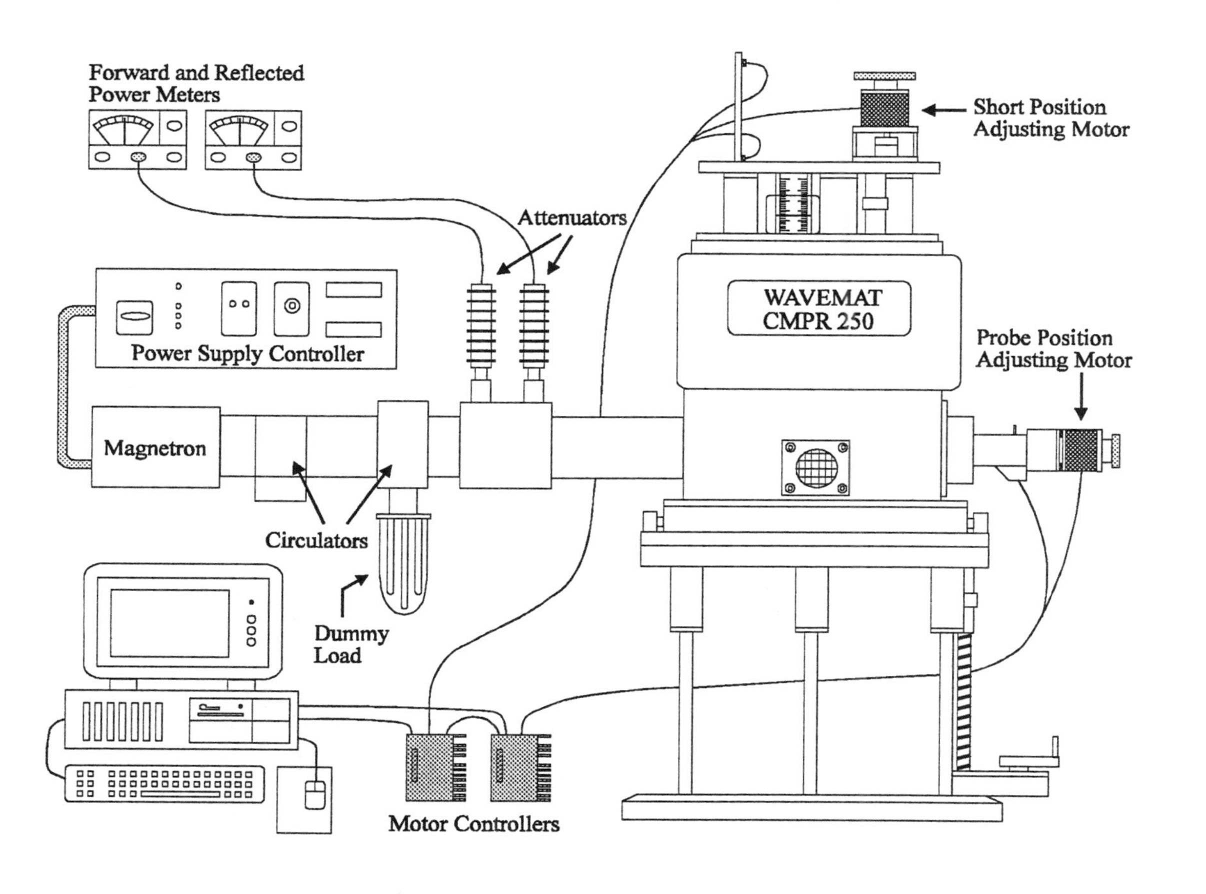

Figure 1. Schematic of microwave sintering apparatus.

(Figure 1). Cavity tuning was performed by continuous adjustments of the probe positions in order to minimize the reflected power after every change of forward power.

An optical pyrometer (Accufiber Optical Fiber Thermometer, Model 10, Luxtron Co., Beaverton, Oregon) was used to measure temperatures ranging from 500°C to 1900°C with an accuracy of ±1°C. A 5 mm diameter hole through the wall of the insulation casket allowed the pyrometer to be sited on the specimen during the experimentation.

Materials

Using the computer controlled single mode cavity, powder compacts of three different commercial alumina powders (Alcoa A-16 SG, Sumitomo AKP-30, and Sumitomo AKP-50) were processed. The average particle size for each of the powders is listed in Table 1. Each powder compact was cold pressed at about 20 MPa into a disk about 22 mm in diameter and 1.7 mm thick. The initial green density of the compacts were approximately 50 percent with respect to the theoretical density of 3.987 g/cm^3 for alumina [16].

Microwave Sintering

Each of the alumina powder compacts were heated to 1575°C using the TM_{012} microwave cavity resonant mode and then held at 1575°C for 30 minutes. Alumina, which is a low loss material, has a loss tangent ranging from 0.0003 to 0.002 [17]. The low loss tangent of alumina at or near room temperature limits direct microwave coupling to the alumina. However, the alumina specimens were placed in a casket composed of a zirconia cylinder (Type ZYC, Zircar Products Inc.) 10 cm diameter and 7 cm height. Top and bottom discs for the casket were made from alumina insulating board (SALI, Zircar Products Inc.). The typical loss tangent value of zirconia is about 0.01 at room temperature [17]. Thus the zirconia casket helped to heat the alumina by radiant heating as well as providing thermal insulation.

RESULTS AND DISCUSSION

Under computer-automated control, the heating schedule from about 800 degrees to the maximum temperature of 1575 C was very similar for specimens of each of the alumina powder types (Figures 2 and 3). Over this temperature range, the heating schedule for the alumina also was very similar to the "empty casket" (no alumina specimen present) runs. However, the power level at which significant coupling first occurred (as evidenced by an initial, rapid temperature rise) did vary in the following way. For the AKP-30 and AKP-50 coupling became significant at about 600 Watts of input power, while the A16-SG and the empty casket first coupled at a input power of about 300 Watts and 275 Watts, respectively (Figure 3).

The rapid temperature rise subsequent to the initial coupling with the microwave power may be due to hot spots in the zirconia casket cylinder, which were observed as bright spots with the unaided eye. The differences in the initial heating of the AKP grade powders and the A16 powders may be differences in the dielectric properties of the powders.

The sintered alumina specimens had densities ranging from 97.1 to 99.6 percent of theoretical (Table 1), as measured by Archimedes method. Average grain sizes (Table 1) were determined by the linear intercept method using SEM micrographs of fracture surfaces (Figures 4 and 5).

Table 1. Results on microwave sintered alumina

Material	Average particle size (μm)	Average grain size (μm)	% Densification
A16-SG	0.52	1.94 *	99.6
AKP-30	0.41	4.85 *	98.4
AKP-50	0.23	3.68 *	97.1

* Average intercept length was multiplied by a stereographic correction factor 1.5 to obtain the average grain size [18].

CONCLUSIONS

This study has demonstrated that an automated processing system featuring a single-mode microwave cavity can be controlled to provide repeatable heating schedules for a series of alumina powder compacts. The resulting alumina specimens were near theoretical density with a uniform, small-grain sized microstructure. Future studies will employ the automated processing system to sinter other ceramics and ceramic composites.

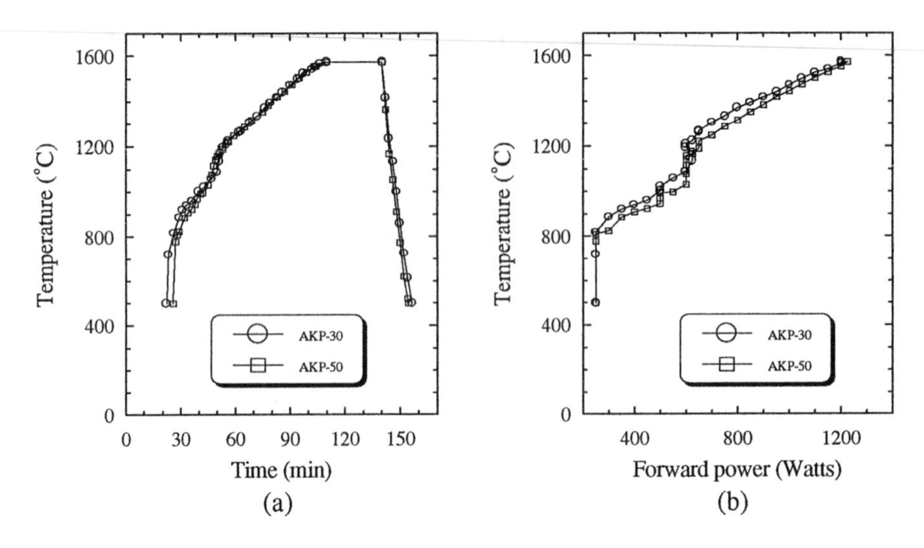

Figure 2. Heating schedule (a) and a plot of temperature vs. forward power (b) for AKP-30 and AKP-50.

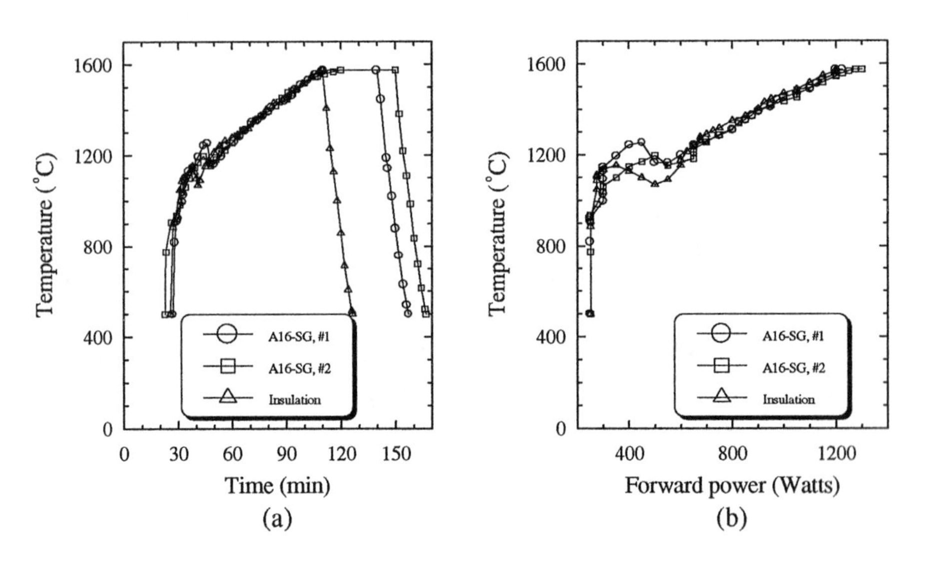

Figure 3. Heating schedule (a) and a plot of temperature vs. forward power (b) for Alcoa A16-SG.

Figure 4. Fracture surface of microwave sintered AKP-30 (Bar represents a length of one micron).

Figure 5. Fracture surface of microwave sintered A16-SG (Bar represents a length of one micron).

REFERENCES

1. A. De, I. Ahmad, E.D. Whitney, and D.E. Clark, Cer. Trans., vol. 21, pp. 319-28 (1991).
2. Y. Fang, D.K. Agrawal, D.M. Roy and R. Roy, Cer. Trans., vol. 21, pp. 349-356 (1991).
3. M.A. Janney, C.L. Calhoun, and H.D. Kimrey, Cer. Trans., vol. 21, pp. 311-318 (1991).
4. H.D. Kimrey, J.O. Kiggans, M.A. Janney, and R.L. Beatty, Mat. Res. Soc. Symp. Proc. vol. 189, pp. 243-256 (1991).
5. M.C.L. Patterson, P.S. Apte, R.M. Kimber, and R. Roy, Mat. Res. Soc. Symp. Proc. vol 269, pp. 291-300 (1992).
6. R.L. Smith, M.F. Iskander, O. Andrade, and H. Kimrey, Mat. Res. Soc. Symp. Proc. vol 269, pp. 47-52 (1992).
7. W.H. Sutton, Am. Ceram. Soc. Bull. 68[2] pp. 376-386 (1989).
8. J. Asmussen and R. Garard, Mat. Res. Soc. Proc. vol. 124, pp. 347-352 (1988).
9. Y-L. Tian, Cer. Trans. vol. 21, Amer. Cer. Soc., pp. 283-300 (1991).
10. B.Q. Tian and W.R. Tinga, Cer. Trans. vol. 21, Amer. Cer. Soc., pp. 647-654 (1991).
11. J.F. Gerling and G. Fournier, Cer. Trans. vol. 21, Amer. Cer. Soc., pp. 667-674 (1991).
12. H.S. Sa'adaldin, W.M. Black, I. Ahmad and R. Silberglitt, Mat. Res. Soc. Symp. Proc., vol. 269, pp. 91-96 (1992).
13. D.S. Patil, B.C. Mutsuddy, J. Gavulic, and M. Dahimene, Cer. Trans. vol. 21, Amer. Cer. Soc., pp. 301-309 (1991).
14. J. Asmussen, H.H. Lin, B. Manring, and R. Fritz, Rev. Sci. Instrum., 58 [8] 1477-1486 (1987).
15. J. Asmussen, Jr. and M. Siegel, to be published.
16. National Bureau of Standards (U.S.), Circ. 539, vol. 9, page 3 (1959).
17. R.C. Buchanan, ed., Ceramic Materials for Electronics, Marcel Dekker, Inc., N.Y., N.Y., page 4 (1986).
18. E. E. Underwood, A. R. Colcord, and R. C. Waugh, pages 25-52 in R. M. Fulrath and J. A. Pask, eds., Ceramic Microstructures, John Wiley and Sons, New York (1968).

SINTERING OF CERAMICS USING LOW FREQUENCY RF POWER

J.B.O. Caughman, D.J. Hoffman, F.W. Baity, M.A. Akerman, S. C. Forrester, and M.D. Kass
Oak Ridge National Laboratory
Oak Ridge, TN 37831-8071

ABSTRACT

Sintering with low frequency rf power (~50 MHz) is a new technique with unique capabilities that has been used to sinter a variety of ceramic materials, including zirconia-toughened alumina, alumina, silicon carbide, and boron carbide. Processing with low frequencies offers many advantages compared to processing with conventional microwave frequencies (915 MHz and 2.45 GHz). Because of the longer wavelength, the rf electric field penetrates materials more than microwaves. This effect allows the processing of a wider variety of materials and allows for an increase in the physical size of the material being processed. In addition, the material is heated in a single mode cavity with a uniform electric field, which reduces the occurrence of hot-spot generation and thermal runaway effects. This technique has been used to sinter large crack-free alumina samples (3" square) to >97% density. The sintering and/or annealing of a number of carbide materials has been demonstrated as well, including silicon carbide, boron carbide, tungsten carbide, and titanium carbide.

INTRODUCTION

A great deal of work has been done in the past ten years with the use of microwaves in the sintering of ceramics [1,2]. An advantage of microwave sintering over conventional sintering is related to the volumetric interaction of the electromagnetic fields with the ceramic. This volumetric coupling to the ceramic leads to a higher heating efficiency and faster processing times than those achievable in conventional furnaces. However, because of its small penetration depth, direct microwave sintering (not hybrid heating) is mostly limited to materials with relatively small volumes and/or low dielectric loss tangents.

The heating of the ceramic is obtained by the coupling of the electric field with the dielectric losses of the material. The role of the electric field and the

---

Research sponsored by the U.S. Army Tank Command under Interagency Agreement 1969-C101-A1 under Martin Marietta Energy Systems, Inc., contract DE-AC05-84OR21400 with the U.S. Department of Energy.

applied frequency in sintering can be seen by looking at the power deposition in a material. The amount of power per unit volume that is coupled into the ceramic by electric fields is given by [3]

$$P = \frac{1}{2}\,\omega\varepsilon(\tan\,\delta)E^2\,.$$

This equation shows that the losses are proportional to the frequency (ω), the loss tangent ($\tan\,\delta$), and the square of the electric field (E). In the typical microwave range, the frequency is high (10^{10} Hz), the electric field is low (<100 V/cm), and the loss tangent is generally high (10^{-2}).

The distance that the electric field can penetrate a ceramic is given by [4]

$$x = \frac{\sqrt{2(1 + \sqrt{1 + \tan^2\delta})}}{\omega\,\tan\,\delta\sqrt{\mu_o\varepsilon'}}\,,$$

where x is the e-folding penetration depth, μ_o is the material permeability (assumed free space), and ε' is the permittivity. An example of the penetration depth as a function of frequency is shown in Fig. 1 for boron carbide (B_4C) at 1100°C and alumina at 23°C. Only low frequencies can penetrate high-loss materials (such as B_4C). Sintering of relatively large samples of high-loss materials with a uniform electric field can be accomplished when samples have a characteristic length on the order of 0.2λ. These considerations led to an exploration of the application of rf energy to the sintering process.

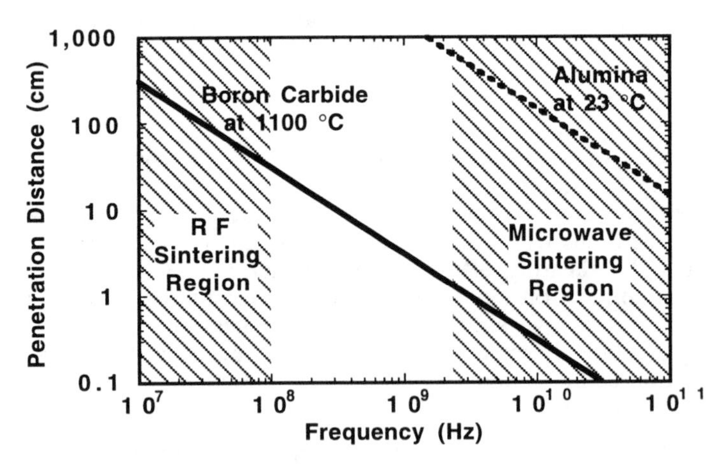

Fig. 1 Depth of electric field penetration as a function of frequency for a low-loss and a high-loss material.

RF SINTERING

The advantage of rf sintering over microwave sintering is related to the higher electric fields and greater sample penetration depth obtained with sintering at rf frequencies. For rf sintering, the frequency is lower (10^8 Hz), the loss tangent is generally lower (10^{-3}), and the electric field is higher (1000 V/cm). Thus, compared to most microwave techniques, rf absorption is reduced by three to four orders of magnitude in the product of frequency and loss tangent, but electric field effects are up to four orders of magnitude higher because the field is 10–100 times greater. Therefore, rf sintering has roughly the same total power deposition as microwave sintering but has a much higher electric field and can be used to determine whether the critical component of electromagnetic energy sintering is the strength of the electric field or the value of the excitation frequency. In contrast to microwave sintering, the wavelengths of the rf waves are considerably greater than the dimensions of the sintering cavity. This generally results in a single permissible mode of operation with a uniform electric field, requiring tuning and impedance matching circuitry not commonly found in microwave furnaces.

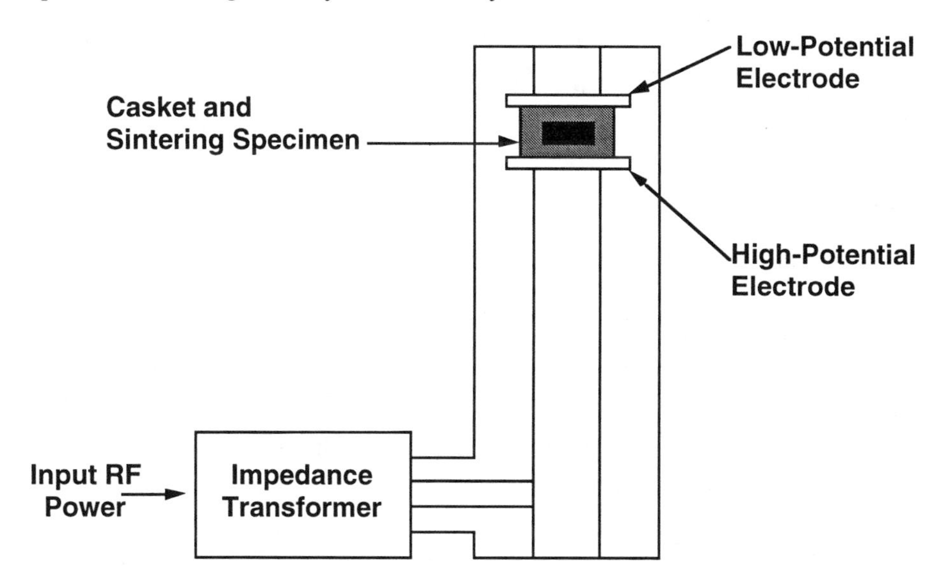

Fig. 2 Schematic of the rf sintering furnace. The casket assembly is placed in the high-field region between the electrodes.

A schematic of the rf sintering furnace is shown in Fig. 2. The cavity is composed of a quarter-wave coaxial waveguide resonator, with the center conductor grounded at one end and an electrode placed at the high voltage end. An additional electrode (near ground potential) completes the circuit. The rf power is coupled through a variable impedance transformer near the base of the cavity, which keeps the cavity tuned and eliminates reflected power during the run. The resonant structure is designed to produce a high electric field at the specimen to be sintered. The sintering specimen is typically packed in insulation inside a casket. The casket assembly is then placed in the high field region between the electrodes and is electrically equivalent to a lossy dielectric between the two plates of a capacitor. Since the cavity is coaxial, an equivalent circuit was analyzed by using a lossy transmission line model that included dielectric losses in the ceramic and resistive losses in the cavity. For a 5 cm thick lossy dielectric slab placed between the electrodes, calculations have shown that electric fields of 1-5 kV/cm can exist for a loss tangent variation of 10^{-2} - 10^{-4} and an input power of 2 kW. An electrostatic analysis of the geometry similar to that shown in Fig. 3 was performed in which a test specimen (dielectric constant, $\varepsilon = 8.9$) is surrounded by loose dielectric fiber ($\varepsilon = 2.3$). The analysis showed that the field is uniform to within a few percent and is orders of magnitude higher than achievable microwave fields.

To date, two rf sintering chambers have been made. A smaller chamber that operates between 50-55 MHz is used for small samples. The electrodes have a diameter of 13 cm and a spacing of 5-9 cm. A larger chamber that operates at 35-40 MHz is generally used for larger samples or where larger amounts of insulation are needed. The electrodes have a diameter of 36 cm and a spacing of 11 -15 cm. Several different materials have been processed and are summarized in the next section.

PROCESSING RESULTS

Zirconia-toughened Alumina

The density as a function of processing temperature has been determined for the rf sintering of zirconia-toughened alumina (ZTA). This material was chosen so that comparisons could be made to previously published results [5]. The samples were composed of 40 wt % zirconia (Tosoh TZ-2Y) and 60 wt % alumina (Sumitomo AKP 50) and prepared as previously described. The mass of the samples were typically 18-20 g. The rf processing frequency was 52 MHz, and the processing atmosphere was flowing nitrogen at slightly over 1 atm of pressure.

The samples were arranged in a casket as shown in Fig. 3. They were placed inside a 7.62 cm diameter, 4.76 cm thick boron nitride casket and surrounded by a 50 wt % mixture of alumina and zirconia fiber (Zircar). The casket was placed between the electrodes of the smaller rf sintering furnace. Either graphite or alumina fiberboard plates were placed on both sides of the casket. The temperature of the sample was measured optically by looking at the sample through a sight tube. An Inframetrics 600 Infrared camera was used for measuring temperatures below 1100 °C and an Ircon 2-color pyrometer was used for temperatures > 1100 °C. The densities were measured using the Archimedes technique in 200-proof ethanol.

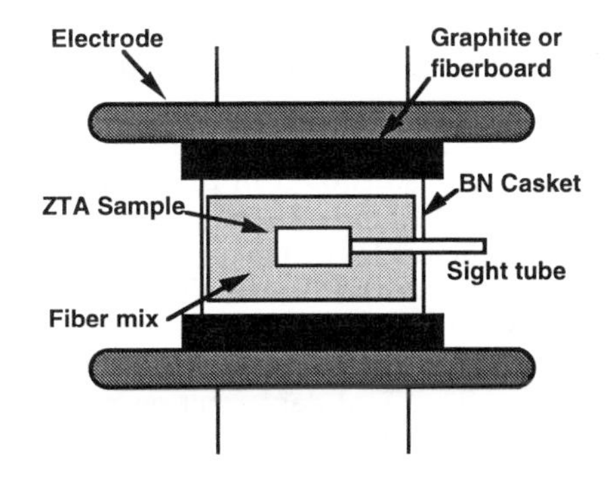

Fig. 3 Arrangement of the casket between the electrodes.

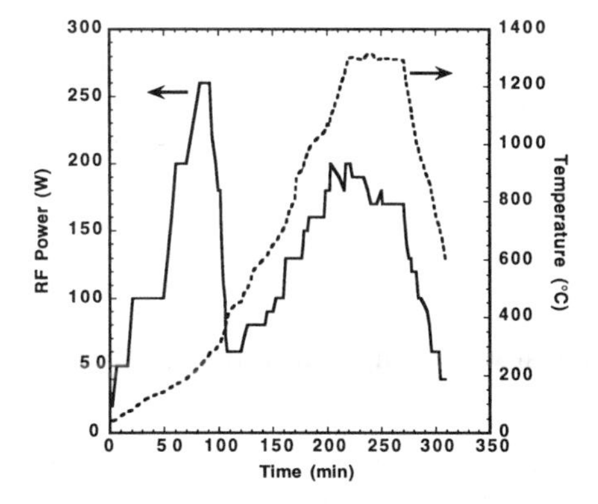

Fig. 4 RF power and sample temp during a ZTA sintering run.

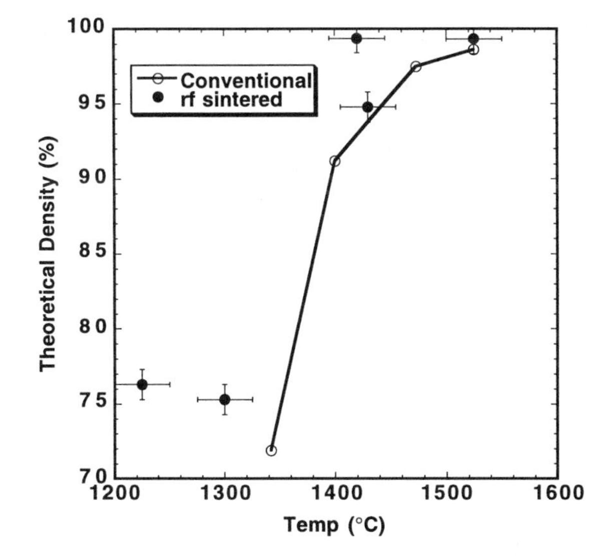

Fig. 5 Density as a function of processing temperature for rf sintered ZTA.

Heating rates and power levels were dependent on the temperature of the sample/casket assembly. A plot of rf power and temperature as a function of time for a typical run is shown in Fig. 4. Initial heating rates of the sample were roughly 2.5 °C/min until dielectric relaxation occurred, which was between 250-350 °C. After this point, the power required to maintain the temperature dropped, and the heating efficiency increased significantly. Heating rates increased to roughly 10 °C/min up to the desired temperature. The samples were held at the target temperature for 1 hour.

The density as a function of temperature is shown in Fig. 5. For comparison, the sintering curve for conventional processing [5] is also shown in the figure. The uncertainties in the temperature come from the difficulty in focusing the pyrometer through a narrow sight tube. In addition, the center of the sample will be hotter than the edge, and the temperatures shown in the figure should be interpreted as a lower limit for the actual temperature. The dependence of the density on temperature for the rf sintered samples is similar to the temperature dependence of the conventionally sintered material. The final densification was >99% for the higher temperature samples. Because of the uncertainty of the temperature, however, it is difficult to conclude that there is a significant reduction in the sintering temperature for these samples.

Alumina

Sintering of alumina has also been achieved using low frequency power at 52 MHz and 38 MHz. The samples were 99.5% pure alumina (Coors 995) and were baked to 800 °C for 8 hours in air to remove the binder before being sintered. The size and mass of the green bodies varied and ranged from 40 g (4.5 cm x 4.5 cm x 0.9 cm) to 380 g (9.0 cm x 9.0 cm x 1.8 cm). The initial density was about 55%. As with the ZTA, the alumina was processed in a flowing nitrogen atmosphere at slightly over 1 atm.

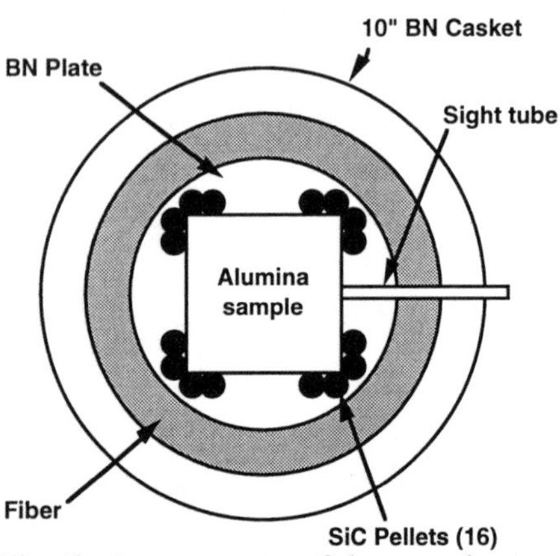

Fig. 6 Arrangement of the sample and casket for alumina runs.

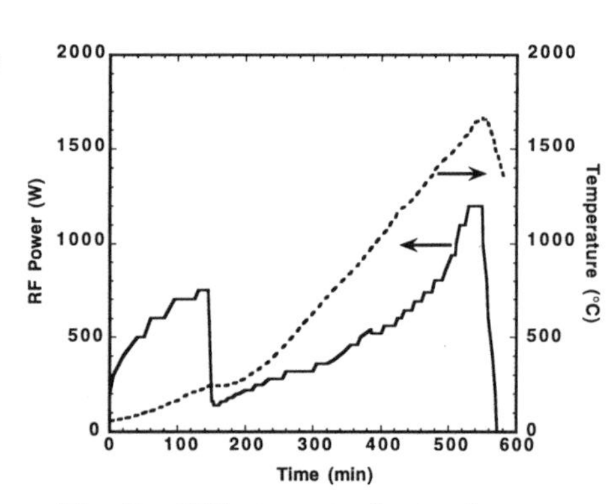

Fig. 7 RF power and sample temp during a 330 g alumina run.

The samples were arranged in a casket as shown in Fig. 6. The tiles were placed on a thin BN plate (0.32 cm) inside a 25.4 cm diam BN casket (11.4 cm high). For the larger samples, a thin layer of alumina powder was used as an interface with the BN plate to reduce surface friction. Silicon carbide pellets (1.3

cm diam, 1.3 cm thick, 5.3 g) were placed at the corners of the sample. The pellets help during the initial heating of the tile and were placed at the corners to reduce the thermal gradient in the sample. A mixture of alumina and zirconia fiber (50/50) was used as the insulation, and the temperature was measured through a sight tube as described above.

The power and temperature during a typical run are shown in Fig. 7 for an experiment with a 330 g sample. Initial heating was slow because of the low loss tangent of the alumina. The presence of the silicon carbide helps in the initial heating phase, but a different arrangement may be needed to optimize the low temperature performance. After relaxation occurs, the heating efficiency increases and the sample readily absorbs power and heats. The heating rate was kept to less than 5 ˚C/min to reduce the chance of cracking in the larger samples. Samples were typically heated to at least 1600 ˚C and held there for 30 minutes before cooling. For a 330 g sample, a power of 1200 W was needed to heat the sample to 1650 ˚C. The samples typically densified to 96-98%. Larger samples (800 g, 4.75 cm x 4.75 cm, x 2.4 cm greenbodies) are currently being attempted.

Carbides

One of the advantages of rf sintering is the ability of the electric field to penetrate high-loss materials, such as silicon carbide and boron carbide. Both of these materials have been sintered using rf power. Other carbides, such as tungsten carbide and titanium carbide have been either sintered or annealed. The results are summarized in Table I. Both silicon carbide and boron carbide have been sintered to > 83% dense. Accurate temperature measurements for these samples have not been made because of clogging of the sight tube during the experiments. However, the densities that have been obtained are consistent with heating the material to >2000 ˚C. An insulation package that can handle temperatures in excess of 2000 ˚C is also an issue. Alumina fibers typically fail when the temperature exceeds around 1700 ˚C. Recent experiments using a combination of zirconia fiber and a mixture of BN powder with glassy carbon [6] have shown promising results.

Table I. Carbide and other materials processed to date.

Material	Process	Temp (˚C)	% Density	Comments
Silicon carbide (Cercom)	sinter	≥ 2000	84	12.5 g sample (Limited by current insulation scheme)
Boron carbide (2.5 wt % C [7])	sinter	> 1700	97	Repeatability limited by current insulation scheme
Zirconia	sinter	1410	98.4	9.2 g sample
Titanium carbide	anneal	1500	-	31 g sample
Tungsten carbide (6 wt % Co)	sinter	1500	97	14 g sample

SUMMARY

The use of low frequency (< 55 MHz) rf power to sinter ceramics has been demonstrated. Sintering with lower frequencies than microwaves has the advantage of being able to penetrate large or high-loss materials with higher electric fields. Because the rf wavelengths are longer than the dimensions of the furnace, the mode of operation can be controlled to ensure nearly uniform electric fields in the region of interest. Processing to temperature up to 1700 °C is now routine, and progress is being made to extend the temperature range to > 2000 °C. The sintering of ZTA and large (> 330 g) alumina samples (without cracking) has been accomplished, along with the sintering and/or annealing of high-loss carbide materials, such as boron carbide and silicon carbide. This low frequency operating regime offers a unique opportunity to determine the effects of sintering due to electric field interactions with materials.

REFERENCES

[1] *Microwaves: Theory and Application in Materials Processing II*, edited by D. E. Clark, W. R. Tinga, and J. R. Laia, Jr., *Ceramic Transactions* **36** (1993).

[2] *Microwave Processing of Materials IV,* edited by M. F. Iskander, R. J. Lauf, and W. H. Sutton, Materials Research Society Symposium Proceedings, Vol. 347 (1994).

[3] W. H. Sutton, "Microwave Processing of Ceramic Materials," *Ceram. Bull.* **68** (2) 376–386 (1989).

[4] A. R. von Hippel, "Descriptions of Dielectrics by Various Sets of Parameters," in *Dielectric Materials and Applications,* edited by A. R. von Hippel, Technology Press of MIT/John Wiley & Sons, New York, p.12 (1954).

[5] H. D. Kimrey, J. O. Kiggans, M. A. Janney, and R. L. Beatty, "Microwave Sintering of Zirconia-Toughened Alumina Composites," *Microwave Processing of Materials II*, Materials Research Society Symposium Proceedings, Vol. 189 , p. 243 (1991).

[6] C. E. Holcombe, N.L. Dykes, and M.S. Morrow, "Thermal insulation for high-temperature microwave sintering operation and method thereof," U.S. Patent Application in Progress (1993).

[7] C.E. Holcombe and N.L. Dykes, "Ultra High-Temperature Microwave Sintering," *Ceramics Transactions* **21**, p. 375 (1991).

INFLUENCE OF MICROWAVE-ANNEALING ON THE MECHANICAL PROPERTIES OF ALUMINA TILES

M. D. Kass, M. A. Akerman, F. W. Baity, Jr., J. B. O. Caughman, S. C. Forrester, and R. A. Lowden
Oak Ridge National Laboratory
P.O. Box 2009, Oak Ridge, TN 37831-8088

E. D. Brewer
Tennessee Eastman Company
Kingsport, TN 37662

W. J. Bruchey, E. J. Rapacki, L. H. Tressler
Army Research Laboratory
Aberdeen Proving Ground, MD 21005-5066

ABSTRACT

Fully dense alumina armor tiles were annealed at 1400°C for one hour using microwave radiation at 28 GHz. Ultrasonic measurements showed that the acoustic impedances, and Young's moduli of the post-annealed tiles were higher than for unannealed tiles. The post-annealed tiles also exhibited higher hardness values and improved armor performance. Microstructure analysis indicated that the grain boundary regions of the tiles were affected by the microwave heating.

INTRODUCTION

Ceramic materials are considered leading candidates as armor facing plates because of their characterized high hardness, high compressive strength, and very low density. Previous investigations have shown that titanium diboride, zirconium diboride, tungsten carbide, silicon carbide, boron carbide, and alumina all provide excellent ballistic protection. Because of its relatively low cost, alumina is usually the material of choice for most ceramic armor applications [1-3].

Several studies have attempted to identify the relationship between material parameters and ballistic performance. The research suggests that important parameters include acoustic impedance, compressive strength, and fracture toughness, but it appears that hardness may be the outstanding characteristic [1-4].

In this study, microwave energy was applied to alumina tiles to enhance bonding between grains and to relieve residual stresses which may be present. The tiles underwent ballistic testing and the results were compared to untreated tiles to determine whether microwave heating can improve the armor efficiency.

Other investigators have observed that the grain boundary regions of certain ceramic materials are altered (recrystallized) following microwave heat treatments. Their observations suggest that microwave heating alters the morphology of the grains to provide improved mechanical performance [5,6]. The objective of this study was to use microwave energy to improve the mechanical and armor performances of alumina tiles.

EXPERIMENTAL PROCEDURE

The armor tiles used in this study were supplied by Coors Ceramics, Inc. They were composed of AD995 alumina and had densities greater than 99 percent of theoretical. The tiles measured 15.2 cm square and had thicknesses of 20 mm, 30 mm, 40 mm, and 50 mm. Four tiles of each thickness were evaluated for ballistic properties; two of these were treated with microwave energy while the other two were used for baseline comparison. Three additional tiles were provided to obtain mechanical property measurements. Of these, two (a 30-mm thick tile and a 40-mm thick tile) were microwave-annealed while the other (50-mm thick) was used for baseline comparision. A total of ten tiles were treated with microwave energy.

Each of the ten tiles was placed inside an insulating casket composed of solid alumina blocks and alumina spheres as shown in Fig. 1. This insulation package was placed inside a 28 GHz microwave furnace filled with nitrogen gas. The tiles were arranged in a verticle position with the microwave input emanating from above. The tiles were heated at a rate of 4°C/min to 1400°C, held for one hour, and then cooled to room temperature at a rate of 2.5°C/min.

Acoustic material properties were determined for each tile using ultrasonic velocity measurements at 5 MHz. Through-transmission longitudinal and shear velocities were measured at four locations on each tile. The mechanical property measurements included flexural strength, fracture toughness, and hardness. Scanning electron micrographs were also taken from representative samples to observe alterations in the microstructure that may have occured as a result of the microwave treatments.

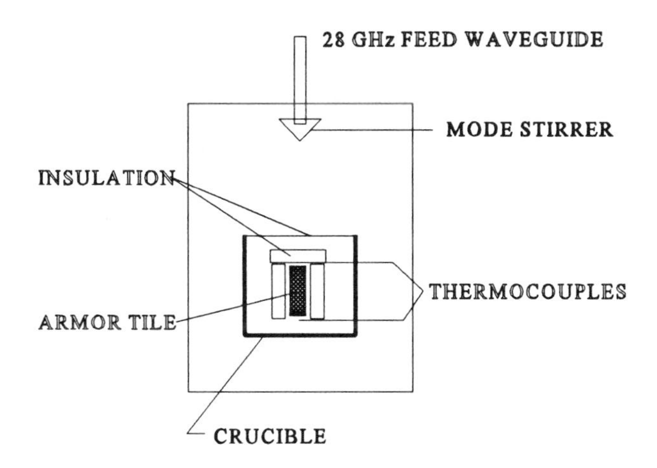

Fig. 1 Arrangement of Tile and Insulation Package

The tiles were sent to Aberdeen Proving Grounds for ballistic testing. Each tile was laterally confined in a steel frame and placed on a semi-infinite steel backup. The tiles were shot with a tungsten penetrator traveling at a nominal velocity of 1500 m/s. Following each shot, the depth of penetration into the backup was measured and used to calculate the armor efficiency of each tile. The results were evaluated for statistical validity.

RESULTS AND DISCUSSION

Acoustic Properties

The tiles which were microwave-annealed were found to have slightly higher acoustic impedances and moduli than unannealed tiles. The average ultrasonic property values for each tile thickness are shown in Table I. The bulk modulus and Poisson's ratio were also measured using ultrasonic means, but there was no discernable difference between the microwave-annealed and baseline tiles.

Table I Summary of Acoustic Properties

Tile width (mm)	Shear impedance (g/cm^2-μs)	Longitudinal impedance (g/cm^2-μs)	Young's modulus (GPa)	Shear modulus (GPa)
20	2.433	3.817	373.2	151.6
30	2.434	3.822	374.7	151.8
40	2.423	3.803	370.7	150.5
50	2.419	3.796	370.4	150.5
Baseline	2.412	3.780	367.3	149.6

There appears to be a slight decrease in the acoustic property values with increasing tile thickness.

Mechanical Properties

Mechanical properties such as flexural strength, hardness, and fracture toughness were measures from an untreated 50-mm thick tile and 2 microwave-annealed tiles (a 30-mm thick tile and a 40-mm thick tile). The results are shown in Table II.

Table II Mechanical Property Measurements from Untreated and Microwave-annealed Alumina Tiles

Tile condition	Flexural strength (MPa)	Hardness (GPa)	Fracture toughness (MPa $\sqrt{m}$)
Untreated	334.66 ± 53.91	11.25 ± 0.50	1.59 ± 0.43
Annealed:			
30 mm thick	337.62 ± 18.53	12.73 ± 0.44	1.90 ± 0.16
40 mm thick	331.53 ± 12.29	15.30 ± 0.49	2.01 ± 0.13

Flexural strength (four-point bend test) results are shown in the first column in Table II along with the standard deviation associated with an average of six tests. The results show no statistical differences in the average strengths between the annealed and untreated tiles. However, standard deviations were substantially lower for the microwave-annealed tiles than for the untreated tile.

Vickers hardness measurements (1000 gram load) are shown in the second column of Table II along with the standard deviations associated with ten tests. The results show that the hardnesses of the two annealed tiles are significantly higher than for the untreated sample. The 40-mm thick tile had hardness values which averaged approximately 36 percent higher than for the untreated tile. The standard deviations indicate that there is no statistical overlap in the data.

Fracture toughness (single-edge-notched beam) results are shown in the third column in Table II. These values represent an average of six tests. Although the average fracture toughness values of the two annealed tiles are slightly higher than for the untreated tile, they are not statistically valid because of overlap in the data. However, the standard deviation measured for two annealed tile was substantially narrower than for the untreated baseline tile. This effect was also observed for the flexural strength measurements.

These results must be carefully considered because they were measured from tiles of different thicknesses. Tile thickness can not excluded as a possible variable.

Ballistic Properties

The parameter used for evaluating the shielding performance of materials is the armor efficiency. This parameter is related to the areal densities of the armor and backup material, and also the depth of penetration into the backup.

The results of the ballistic testing are shown in Fig. 2. In this figure, the armor efficiency is plotted against tile thickness for the microwave-annealed and untreated conditions. Each data point represents an average of two specimens. The results show that the armor efficiency decreased with increasing tile thickness. The armor efficiency values for the microwave-annealed tiles were higher than for the untreated tiles at thicknesses greater than 20 mm. The efficiencies for the microwave-annealed tiles were found to average 13.4 percent higher than for the untreated tiles for all tile widths. The greatest improvement in efficiency occured at a tile thickness of 30 mm. Although these values were determined to be statistically valid, they represent an average of two data points. Further testing should be conducted to confirm the results.

A statistical analysis was performed to determine which, if any, of the acoustic properties correlated with armor performance. The evaluation found that there was a strong correlation of acoustic shear impedance with armor performance.

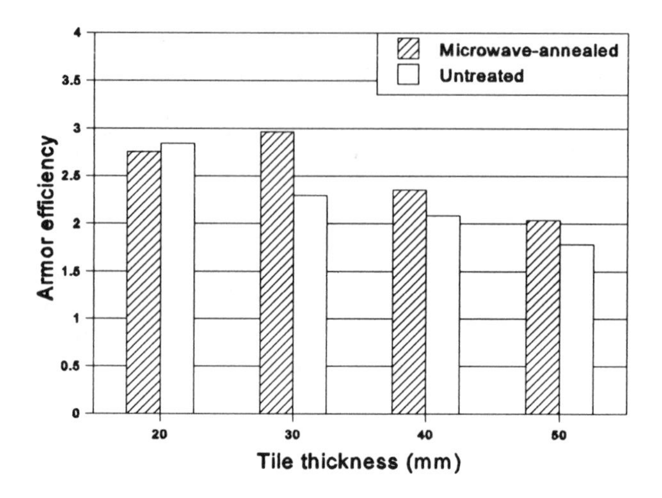

Fig. 2 Armor Efficiency Versus Tile Thickness for Microwave-annealed and Untreated Alumina Tiles

Microstructural Evaluation

Figure 3 shows SEM photomicrographs taken from the polished and etched surfaces of a microwave-annealed tile (a) and an untreated baseline tile (b). The photographs show precipitates in the grain boundaries of the untreated tile which are not present in the microwave-annealed tile. An additional comparison (not shown) performed on a conventionally-annealed tile showed a microstructure identical to the untreated sample.

Attempts to identify the composition of the grain boundary precipitate have been unsuccessful so far. An impurity analysis showed that the concentrations of Mg and Si in the microwave-annealed samples were dramatically lower than the untreated samples. This implies that these ions may have been volatized from the material as a whole.

Analysis of fractured surfaces of the alumina tiles indicates that the fracture mode is intergranular. It is reasonable to believe that the bonding between the grains of the untreated tiles was affected by the precipitates. The microwave-annealing

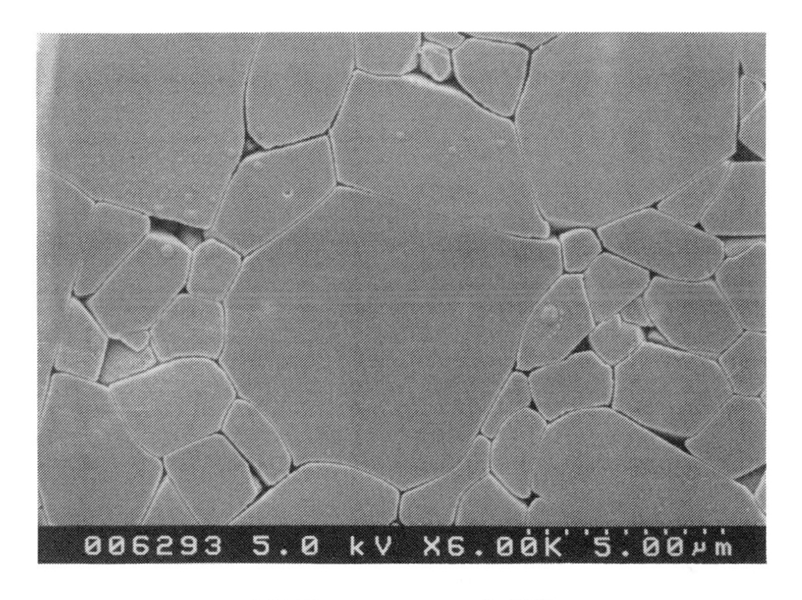

(a) Microwave-annealed Tile

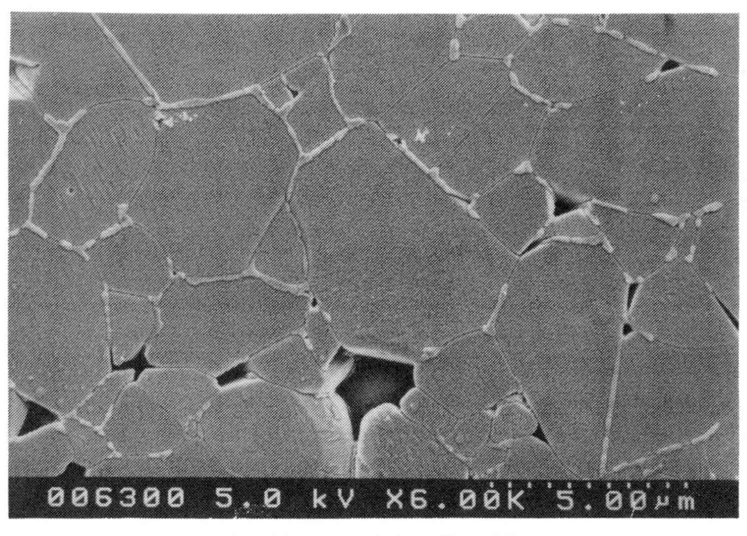

(b) Untreated Baseline Tile

Fig. 3 Microstructure of Polished Surfaces of Alumina Tiles

operation removed these boundary impurities and may have enhanced grain bonding in the process. This may explain the improved hardness and armor efficiency results obtained from the microwave-annealed alumina tiles.

CONCLUSIONS

Alumina tiles that were annealed using microwave energy had higher impedances, moduli, and hardnesses than untreated tiles. These results, however, do not exclude tile thickness as a possible variable. The annealed tiles also exhibited improved armor efficiencies which correlated with the acoustic shear impedances. Grain boundary precipitates, which were present in the unannealed tiles, were not observed for those which were microwave-annealed. These precipitates may weaken the bonding between grains. Removal of these precipitates, as a result of microwave-annealing, allows the grains to be in direct contact with each other, which may result in stronger bonding between the grains.

REFERENCES

1. G. R. Rugger, "Materials for Armor," 12th National SAMPE Symposium, 1993.

2. D. J. Viechnicki, M. J. Slavin, and M. I. Kliman, "Development and Current Status of Armor Ceramics," *Ceramic Bulletin*, 70 [6], pp. 1035-1039, 1991.

3. N. L. Rupert and R. J. Schoon, "Evaluation of High-Density Ceramics for Ballistic Application," *Dynamic Loading in Manufacturing and Service*, 2. 1993.

4. M. Akerman et al, "Microwave Processing of Silicon Carbide," *Proc. 29th Microwave Power Symposium*, Chicago, IL, pp. 144-147 (1994).

5. T. N. Tiegs et al, "Crystallization of Grain Boundary Phases in Silicon Nitride with Low Additive Contents by Microwave Annealing," *Ceramic Transactions, Vol. 36, Microwaves: Theory and Application in Materials Processing II*, Am. Ceram. Soc., Westerville, OH, pp. 259-267 1993.

6. M. K. Ferber, T. N. Tiegs, and M. G. Jenkins, "Effect of Post-Sintering Microwave Treatments on the Mechanical Performance of Silicon Nitride," *Ceram. Eng. Sci. Proc.*, 12, pp. 1993-2004, 1993.

ANALYSIS OF THE INFLUENCE OF TEMPERATURE NONUNIFORMITY ON CERAMICS SINTERING

K.I.Rybakov and V.E.Semenov
Institute of Applied Physics, Russian Academy of Sciences,
46 Ulyanov St., Nizhny Novgorod 603600 Russia

ABSTRACT

Influence of temperature nonuniformity on ceramics sintering is studied theoretically. Additional densification occuring due to the action of the thermoelastic stresses and to inhomogeneity of effective viscosity is calculated. Possibility to explain specific features observed in microwave processing of powder materials in terms of these effects is discussed.

INTRODUCTION

During sintering of ceramics, powder particles move and change shape thus providing densification of the compacts. Particle shape is modified due to plastic deformation and the driving forces for densification are the stresses that exist in the material. In isothermal processes, these stresses are caused by the tendency to minimize the total free surface and the associated excess energy. These capillary stresses are microscopic with respect to the compact, i.e., their characteristic scale of nonuniformity is comparable with the size of powder particles and they vanish upon averaging over a volume that contains many particles. The macroscopic stresses caused, for example, by an external pressure, or by nonuniform heating (the so-called thermoelastic stresses), may also act as driving forces. Specifically, thermoelastic stresses are believed to cause sintering enhancements associated with nonuniform heating [1,2].

The question of possible influence of thermoelastic stresses on densification enhancement has acquired particular significance in connection with rapid development of microwave processing. In numerous sintering experiments (see, e.g., [3]), enhancement of densification under microwave heating in comparison with conventional processing has been observed. The causes for this enhancement are not clear at this time. However, microwave heating is creates a specifically nonuniform temperature distribution over the sample during heating, and it is tempting to reduce the "microwave sintering effect" to the action of thermoelastic stresses. At the same time, there are objections to that point of view. They are based on the contradictions between the uniform nature of the observed densification and nonuniformity of the stresses, as well as the considerable densification and the relatively low magnitude of the deformation due to thermal expansion which relieves thermoelastic stresses. The role of temperature nonuniformities during densification is not obvious, and detailed analysis is necessary to clarify the situation.

In this paper, results of such analysis are presented. The process of densification of a powder material is described within the framework of a phenomenological model as the viscous flow of a continuous medium. Two possible mechanisms of the influence of thermoelastic stresses on densification are studied. First, these stresses, as stated above, can be the driving forces for viscous flow. Second, stresses with sufficient strength can overcome forces that hinder particle sliding [1] and thereby can reduce the effective viscosity of the powder. In general, thermoelastic stresses arising due to nonuniformity of temperature are inhomogeneous and can be a contributory factor in the formation of an inhomogeneous distribution of viscosity (along with the direct dependence of viscosity upon temperature). One of the obvious consequences of these effects is that densification of the material is generally inhomogeneous. Therefore, densification enhancement inhomogeneity due to thermoelastic stresses was considered.

DESCRIPTION OF THE MODEL AND THE MATHEMATICAL FORMULATION

In the phenomenological model presented, powder materials are treated as continuous media characterized by a set of parameters, such as the moduli of shear, μ, dilatation, K, the coefficient of thermal expansion, α, and others. Respectively, the dynamics of densification of such materials in nonuniform and nonstationary temperature fields are analyzed by means of the mechanics of continuous media. Assuming viscous deformations to be slow compared with sonic processes, the equations of equilibrium rather than those of motion are used:

$$\frac{\partial \sigma_{ij}}{\partial x_j} = 0. \tag{1}$$

Here σ_{ij} is the averaged (macroscopic) stress tensor that does not include capillary stresses because of their microscopic nature, and repeated subscripts imply summation. Restricting analysis to small deformations, relations of the linear theory of deformation are used, such as

$$\varepsilon_{ij} = \frac{1}{2}\left(\frac{\partial u_i}{\partial x_j} + \frac{\partial u_j}{\partial x_i}\right), \tag{2}$$

where ε_{ij} is the averaged (macroscopic) strain tensor, and u_i denotes components of the displacement vector, $\mathbf{u}$.

In the processes of plastic deformation crystalline solids exhibit viscous properties. The mechanism of viscosity is based on diffusion motion of defects in crystal [4]. To describe viscous effects in elastic media, generalized interpolation equations were used. An equation of this type was suggested by Maxwell (see, e.g., [5]) to express the relationship between the shear (antisymmetric) parts of the stress tensor, $\tilde{\sigma}_{ij} = \sigma_{ij} - \sigma_{ll}\delta_{ij}/3$, and of the strain tensor, $\tilde{\varepsilon}_{ij} = \varepsilon_{ij} - \varepsilon_{ll}\delta_{ij}/3$, in the model of a "very viscous liquid":

$$2\frac{\partial \tilde{\varepsilon}_{ij}}{\partial t} = \frac{\partial(\tilde{\sigma}_{ij}/\mu)}{\partial t} + \frac{\tilde{\sigma}_{ij}}{\eta}, \tag{3}$$

where η is the coefficient of the shear viscosity, and $\delta_{ij} = 1$ if $i = j$, $\delta_{ij} = 0$ if $i \neq j$. Distinctions between the time derivatives with respect to Eulerian and Lagrangian variables were not made.

In powder materials, as opposed to nonporous continuous media, not only the shear, but also the bulk stresses (the stresses of volumetric compression) can relax due to densification. Similar to Eq.(3), this process can be described by the following equation for the "bulk" parts of the stress and strain tensors:

$$\frac{\partial \varepsilon}{\partial t} = \frac{\partial \Theta}{\partial t} - \frac{\partial}{\partial t}\left(\frac{p}{K}\right) - \frac{p + \Gamma}{\varsigma}, \tag{4}$$

where $p \equiv -\sigma_{ll}/3$ is local pressure, $\varepsilon \equiv \varepsilon_{ll}$ is local volume strain, $\Theta \equiv \alpha T$, $T(\mathbf{r}, t)$ is the temperature of the medium, and ς is the coefficient of the bulk viscosity. It can be seen that the volume strain ε consists of additive contributions from three processes: thermal expansion (its contribution being determined by the instantaneous value of the temperature), elastic compression (its contribution being determined by the instantaneous value of the macroscopic pressure), and viscous densification which accumulates in time even if the temperature and the pressure are fixed. The process of viscous densification is of primary interest. As seen from Eq.(4), the role of the motive force for this process is played by the sum $p + \Gamma$, where Γ describes the action of the microscopic capillary (Laplacian) stresses which lead to spontaneous densification of the powder material: $\partial\varepsilon/\partial t = -\Gamma/\varsigma$ when there are no macroscopic stresses and the temperature field is stationary. For simplicity, it is assumed that Γ is spatially uniform. The above mentioned twofold effect of thermoelastic stresses on the viscous densification in terms of Eq.(4) means their contribution to p and their influence on ς.

The effective parameters of the medium $\mu, K, \alpha, \eta, \varsigma, \Gamma$ are determined by the properties of the material, as well as by the size and shape of particles, and the density of their packing. Equation (4) can be derived, and the parameters ς and Γ can be obtained by averaging the microscopic viscoelastic equations over a volume containing many powder particles (see [6] for simple examples).

To complete the mathematical formulation of the problem, it is necessary to supplement the equations (1)-(4) with the relevant boundary conditions at the surface of the macroscopic sample. Being interested primarily in processes of densification rather than of the shear deformation of the sample, it is assumed that the tangential components of the external forces at the surface are absent, and the normal components, if any, are constant and are defined by the external pressure, p_∞.

It should be noted that with these boundary conditions the following relation for the averaged pressure can be obtained from Eq.(1):

$$\langle p \rangle = p_\infty. \tag{5}$$

(Angular brackets, $\langle\rangle$, denote averaging over the entire sample volume).

When considering high temperature creep in solids, the model of viscous flow of solid matter has gained acceptance [7]. This model is valid for sufficiently slow heating or low viscosity. It is contained in the model presented as a limiting case (further referred as the "viscous limit") and can be derived from Eqs.(1)-(4) by dropping the time derivatives of stresses.

RESULTS AND DISCUSSION

1. To reveal the effects produced by thermoelastic stresses as an additional driving force for densification, consider first a medium with uniform parameters μ, K, η, ς. Averaging Eq.(4) over the volume of the sample and using

Eq.(5), we obtain in this case that

$$\frac{\partial\langle\varepsilon\rangle}{\partial t}=\frac{\partial\langle\Theta\rangle}{\partial t}-\frac{\partial}{\partial t}\left(\frac{p_\infty}{K}\right)-\frac{p_\infty+\Gamma}{\varsigma}. \tag{6}$$

It can be seen from Eq.(6) that mean densification does not depend upon the thermoelastic stresses associated with nonuniformity of the spatial distribution of temperature.

If Θ is not uniform, then the deformation is also not uniform, and Eq.(6) does not determine the densification completely. Nonuniform shear stresses and strains complicate considerably general analysis of the initial equations. However, if the deformation is sufficiently symmetrical, so that the relation $\nabla\times\nabla\times\mathbf{u}=0$ is obeyed, then drastic simplification of the problem becomes possible. Equations (1)-(4) can be reduced to local relations between the deviations $\delta p\equiv p-\langle p\rangle$, $\delta\Theta\equiv\Theta-\langle\Theta\rangle$, $\delta\varepsilon\equiv\varepsilon-\langle\varepsilon\rangle$ of the local pressure, temperature, and volume strain from their mean values:

$$\frac{\partial}{\partial t}\left[\delta p\left(\frac{3}{4\mu}+\frac{1}{K}\right)\right]+\delta p\left(\frac{3}{4\eta}+\frac{1}{\varsigma}\right)=\frac{\partial\delta\Theta}{\partial t}; \tag{7}$$

$$\left(1+\frac{4\eta}{3\varsigma}\right)\frac{\partial\delta\varepsilon}{\partial t}=\frac{\partial\delta\Theta}{\partial t}+\frac{\eta}{\varsigma}\cdot\frac{\partial}{\partial t}\left(\frac{\delta p}{\mu}\right)-\frac{\partial}{\partial t}\left(\frac{\delta p}{K}\right). \tag{8}$$

It can be seen that the deviation of the rate of the volume strain is influenced by the thermoelastic stresses. At the same time, it can be argued that this influence is not significant. Since by virtue of Eq.(7) $|\delta p|$ is limited,

$$|\delta p|\lesssim\max\frac{|\delta\Theta|}{(3/4\mu)+(1/K)}, \tag{9}$$

it can be shown that the deviation of volume strain $\delta\varepsilon$ does not exceed the maximum value of $\delta\Theta$ which is the characteristic value of that deviation in the process of nonuniform elastic thermal expansion. This is quite small, since in the experimental conditions typical of microwave processing $|\delta\Theta|\leq 10^{-3}\ldots10^{-2}$. It is interesting to note that if η/ς is constant in time, then $\delta\varepsilon$ is determined by the instantaneous values of $\delta\Theta$ and δp and hence does not change after a closed heating – cooling cycle.

2. As mentioned above, if the thermoelastic stresses exceed the capillary ones, they may lead to the reduction in viscosity (due to elimination of locks that hinder sliding) and hence to enhancement of densification (see Eq.(6)). It can be shown that this effect will be limited because

of the relaxation of thermoelastic stresses with the characteristic time $\tau = [(3/4\mu)+(1/K)]/[(3/4\eta)+(1/\varsigma)]$ (cf. Eq.(7)). Assuming that the viscosity is depressed only during the time period when $|\delta p| > p_\infty + \Gamma$, one can estimate with the help of Eqs.(7)-(8) the contribution of this effect to densification. It appears that this contribution does not exceed in order of magnitude the maximum value of $\delta\Theta$, i.e., the influence of thermoelastic stresses is insignificant (unless one imagines that the viscosity remains depressed long after the stresses have relaxed).

3. Let us now examine the effects associated with inhomogeneous viscosity. Being interested primarily in viscous densification, one may consider these effects in the "viscous limit" approximation, neglecting elastic deformations. In the case when the inhomogeneity is rather weak, $|\delta\varsigma| \ll \varsigma$, $|\delta\eta| \ll \eta$ (where $\delta\varsigma = \varsigma - \langle\varsigma\rangle$, $\delta\eta = \eta - \langle\eta\rangle$), one can linearize the starting equations of the model (rewritten for this limit). As a result, one obtains:

$$\left(1+\frac{4\langle\eta\rangle}{3\langle\varsigma\rangle}\right)\frac{\partial\delta\varepsilon}{\partial t} = \frac{\partial\delta\Theta}{\partial t} + \frac{p_\infty+\Gamma}{\langle\varsigma\rangle}\cdot\frac{\delta\varsigma}{\langle\varsigma\rangle}. \tag{10}$$

Unlike the case of homogeneous viscosity, deviations of volume strain can now accumulate with time (see the last term in Eq.(10)) and hence can exceed the quantity $max|\delta\Theta|$. However, as inhomogeneity of the viscosity is assumed to be weak, inhomogeneity of the viscous densification remains relatively small. The mean rate of densification, $\partial\langle\varepsilon\rangle/\partial t$, is not very different from the case of homogeneous viscosities.

The reasons for viscosity inhomogeneity may generally lie in nonuniform heating, dependence of viscosity upon porosity, and the action of stresses [1]. The role of each can be investigated if one assumes

$$\delta\varsigma = -\langle\varsigma\rangle(\lambda_\Theta\delta\Theta + \lambda_\varepsilon\delta\varepsilon + \lambda_p\delta p), \tag{11}$$

where $\lambda_\Theta, \lambda_\varepsilon, \lambda_p > 0$. Equation (11) makes the linearized system self-consistent. The analysis leads to the conclusion that within the approximation of weak inhomogeneity of viscosity, the deviations of the volume strain, $\delta\varepsilon$, can grow only due to the direct dependence of viscosity upon temperature.

4. If the viscosity is strongly inhomogeneous, the starting equations of the proposed model (1)-(4) can be solved analytically only for a number of special spherically-symmetrical distributions of parameters. In particular, it appears possible to solve them when viscosities and temperature have power dependences upon the radius: $\eta = \eta_R\rho^q$, $\varsigma = \varsigma_R\rho^q$, $\Theta = \Theta_0 - \Delta\Theta\rho^\beta$ (where $\rho = r/R$, $q > 0$, $\beta > 0$) at $a < r \leq R$. The densification rate obtained from

this model (provided that $a \ll R$) has the following form [6]:

$$\frac{\partial \varepsilon}{\partial t} = -\frac{\Gamma + p_\infty}{\varsigma_R + (4/3)\eta_R} \cdot \frac{q}{s}\left(\frac{r}{R}\right)^{-s} + \frac{\partial \Theta_0}{\partial t} -$$

$$-\frac{\nu}{\beta + \tilde{q}}\left[\frac{3\beta(1-\nu)}{\beta+3} \cdot \frac{q}{s}\left(\frac{r}{R}\right)^{-s} + (\beta+q)\left(\frac{r}{R}\right)^{\beta}\right]\frac{\partial \Delta\Theta}{\partial t}, \tag{12}$$

where $\nu = 3\varsigma/(3\varsigma + 4\eta)$, $s = [q + 3 - \sqrt{(q+3)^2 - 12q\nu}]/2$, and $\tilde{q} = q(\beta + 3\nu)/(\beta + 3)$.

The first term on the right of Eq.(12) expresses densification provided by the external pressure and capillary stresses, and other terms account for the thermal expansion and the contribution of thermoelastic stresses. It can be seen that the rate of densification increases towards the center, where the temperature is the highest and the viscosity is the lowest: $\partial\varepsilon/\partial t \propto r^{-s}$. The character of this increase is completely determined by the spatial distribution of viscosity (s depends only on q and ν) and appears to be the same for densification driven by thermoelastic stresses and for that provided by the external pressure and capillary stresses. It should be noted also that shear stresses arising in the sample limit its densification: even when the viscosity decreases very rapidly towards the center (so that $q \gg 3\nu$), s still cannot exceed 3ν. Therefore, strong local densification (in comparison with that at the periphery) is possible only within a relatively small volume. Unlike the previously considered cases, within this volume the densification component provided by the thermoelastic stresses can be substantially higher than the quantity $\Delta\Theta$ (which corresponds to the full difference of temperature in the sample). However, the contribution of the thermoelastic stresses into the total change in the volume of the sample remains insignificant.

The result (12) suggests that strong inhomogeneity of viscosity can lead to strongly inhomogeneous viscous densification. Strong densification can occur in a small domain around a maximum of the spatial distribution of temperature, where the effective viscosity coefficients of the material are small. It should be noted that temperature distributions having maxima are typical of processes using microwave heating. On the other hand, hot spots are frequently observed, in which the density of the initially homogeneous material increases sharply (see, e.g., [8]). Therefore, it can be surmised that formation of hot spots takes place due to a reduction in viscosity and proceeds in accordance with the described mechanism.

As for the experimentally established enhancement of densification in microwave sintering, on the basis of the above presented results we can state that it cannot be explained completely by the action of thermoelastic stresses. A similar conclusion follows from the results of recent experiments designed purposely to investigate the role of temperature nonuniformity during microwave sintering [9]. All this argues in favour of models in which the enhancement is explained in terms of a nonthermal influence of the electromagnetic field on the mass transport in the material [10].

ACKNOWLEDGMENTS

This research is supported in part by the Russian Basic Science Foundation through grant No. 95-02-05000-a. This publication was made possible by the NATO grant HTECH.LG 940364.

REFERENCES

1. Y.E.Geguzin, *Fizika Spekaniya* (Russian), 2nd ed. (Nauka, Moscow, 1984).

2. D.Beruto, R.Botter, and A.V.Searcy, *Ceram. Trans.*, **1**, 911 (1988).

3. W.H.Sutton, in *Microwave Processing of Materials III*, edited by R.L. Beatty, W.H.Sutton, and M.F.Iskander (Materials Research Society Symposium Proceedings, Vol. 269, Pittsburgh, 1992), p. 3.

4. F.R.N.Nabarro, in *Rep. of Conf. on Strength in Solids* (Physical Society, London, 1947), p. 75; C.Herring, *J. Appl. Phys.*, **21**, 437 (1950); I.M.Lifshits, *JETP* (Russian), **44**, 1349 (1963).

5. L.D.Landau and E.M.Lifshits, *Theory of Elasticity*, 3rd ed. (Pergamon, New York, 1986).

6. K.I.Rybakov and V.E.Semenov, to be published.

7. Y.Frenkel, *J. Phys. USSR*, **9**, 385 (1945).

8. J.Zhang, Y.Yang, L.Cao, S.Chen, X.Shong, and F.Xia, in *Microwave Processing of Materials IV*, edited by M.F.Iskander, R.J.Lauf, and W.H.Sutton (Materials Research Society Symposium Proceedings, Vol. 347, Pittsburgh, 1994), p. 591.

9. Y.Bykov, A.Eremeev, and V.Holoptsev, *Ibid*, p. 585.

10. K.I.Rybakov and V.E.Semenov, *Phys. Rev. B*, **49**, 64 (1994).

NUCLEATION AND CRYSTAL GROWTH OF $Li_2O \cdot 2SiO_2$ IN A MICROWAVE FIELD

A. Boonyapiwat and D. E. Clark
Department of Materials Science and Engineering
University of Florida
Gainesville, FL 32611

ABSTRACT

A comparison of nucleation and crystallization of $Li_2O \cdot 2SiO_2$ between microwave and conventional heating was investigated. Standard stereological techniques were used to evaluate the results. Nucleation and crystallization behaviors during microwave heating appeared different from conventional heating. Work is in progress to better understand the effects of microwaves on nucleation and crystallization in glass.

INTRODUCTION

The majority of glass cannot be simply converted to fine-grain glass-ceramics because the nucleation rate is very slow [1]. To make the processing of a given glass-ceramics system economically possible, a suitable nucleating agent must be used to increase the nucleation rate [2,3]. It has been shown that glass can be converted to glass-ceramics using microwave energy [4,5]. Since microwaves appear to enhance the kinetics of many process, it is possible that the use of microwave energy to process glass-ceramics may offer advantage over conventional process. Indeed, preliminary research has shown that the kinetics of glass crystallization may be accelerated in a microwave field [4,5].

The goal of this project is to investigate the effect of microwaves on nucleation and crystallization of $Li_2O \cdot 2SiO_2$ glass. Comparisons, between microwave heating and conventional heating are made on the nucleation rates and the rates of crystal growth of this system.

BACKGROUND ON GLASS-CERAMICS

Glass-ceramics are crystalline solids produced by the controlled crystallization of a specially formulated glass. Since the first invention by Stookey [6], a variety of glass-ceramics has been developed. Glass-ceramics have uniform and pore-free microstructures. Because the glass-ceramics process begins with glasses, high speed glass-forming techniques can be employed to manufacture components with a variety of complex shapes. These complex shapes can then be heat treated to yield glass-ceramics with a wide range of microstructures and properties.

The glass-ceramic manufacturing process has two steps; glass processing and glass-ceramic processing. A typical heating schedule for glass-ceramic production is shown in Figure 1. First, the glass composition is melted at the temperature, t_m, for a sufficient time to ensure homogeneity. Glass melting is completed at time, t_1, and then the glass is cooled from the melting temperature to the nucleation temperature T_n at time, t_2. During this time (t_1 to t_2), the glass is formed (casting, pressing or blowing) into its final shape. The process to convert this glass to a glass-ceramic begins at t_2. First, the glass is heat treated at T_n to create the crystal nuclei. This processing stage is called nucleation. To produce a fine grain glass-ceramic in the final product, numerous nuclei must be formed. The number of nuclei depends on nucleation time and nucleation rate which depends in turn on the nucleation temperature T_2. After the nucleation process is finished at t_3, the product is heat treated at the crystallization temperature, T_c, to grow the nuclei into larger crystals. Normally, the crystallization temperature is higher than the nucleation temperature. The percent crystallization of glass-ceramic product is controlled in this stage by controlling the crystallization time and the crystallization rate which depends on crystallization temperature T_c. After crystallization, t_4, the glass-ceramic product is cooled to room temperature.

EXPERIMENTAL

Glass Preparation and Heat Treatment

Figure 2 shows a schematic of the $Li_2O \cdot 2SiO_2$ glass-ceramics process used in this study. A batch of glass was prepared from laboratory grade lithium carbonate (Li_2CO_3) and silicon dioxide (SiO_2). The raw materials were mixed for 24 hours in a polyethylene bottle by rolling on a roller mill. The glass was melted at 1400°C for 12 hours in a covered platinum crucible, crushed and remelted to achieved good homogeneity. The glass then was cast as rods 75 mm long and 13 mm in diameter and cooled in a box furnace from

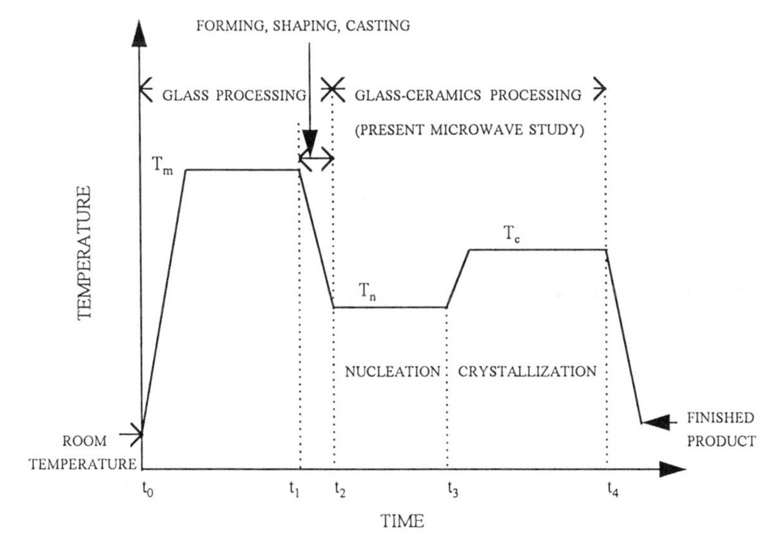

Figure 1. Heating schedule in glass-ceramics processing.

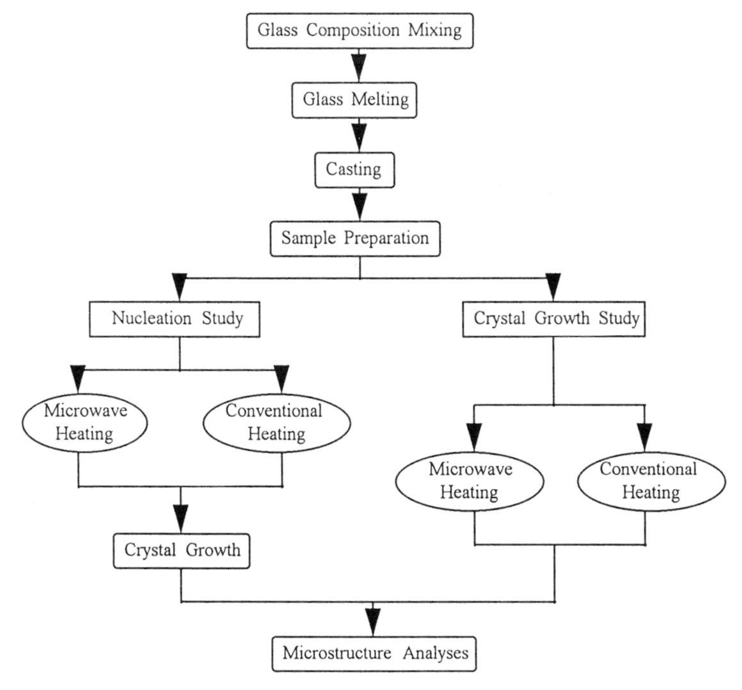

Figure 2. Experimental processing flow chart.

400°C to room temperature. Glass specimens 4 mm thick were cut from the rods and used for the nucleation/crystallization study.

Two methods of heat treatment were employed in this study; microwave hybrid heating (MHH) and conventional heating. The microwave heat treatment was carried out in a Raytheon Radarline Multimode Microwave Oven Model QMP 2101A-6 operating at 2.45 GHz. The oven has 8 magnetron tubes (800 watt each) and each magnetron has a mode stirrer. Only one magnetron tube of 800 watts was used for this study. For the conventional heating, a tube furnace was used. The temperature was controlled within ∓3°C of the equilibrium temperature for both heating methods.

MHH previously has been demonstrated by De et al [7]. For this study, a cylindrical heating cavity was made of a carved insulation brick lined with a mixture of alumina cement and a microwave susceptor (SiC). Temperature was monitored using an Inconel shielded, type-K thermocouple which was positioned so that the tip of thermocouple was about 2 mm above the sample. The temperature was controlled by adjusting the microwave duty cycle. For heating in the nucleation range (455-467°C), the microwave duty cycle was 25-30%. The microwave duty cycle was 35-40% when the heating was in the crystal growth range (567-595°C).

For the nucleation study, the traditional two-stage heat treatment was used [8]. The heat treatments for nucleation were performed in both a microwave oven and conventional tube furnace. The samples were brought to the nucleation temperature within 5 minutes. After nucleation, the samples were heat treated in a conventional tube furnace at 595°C for 15 minutes to grow the nuclei to observable sizes.

For the crystal growth study, a one-stage heat treatment (without nucleation stage) was used. As in the nucleation study, the samples were brought to the crystallization temperature within 5 minutes.

Microscopy Techniques

The samples were prepared for optical microscopy by cutting and then polishing before being etched by a solution of 2% HF for 2 minutes. The samples were then examined on a reflection microscope equipped with an eyepiece micrometer. A schematic on microscopy study is shown in Figure 3.

The nuclei densities were evaluated using standard stereological methods [9]. Since the crystals have elliptical shape, the number of nuclei per unit volume was calculated from

$$N_v = 2N_A Y/\pi k(q) \tag{1}$$

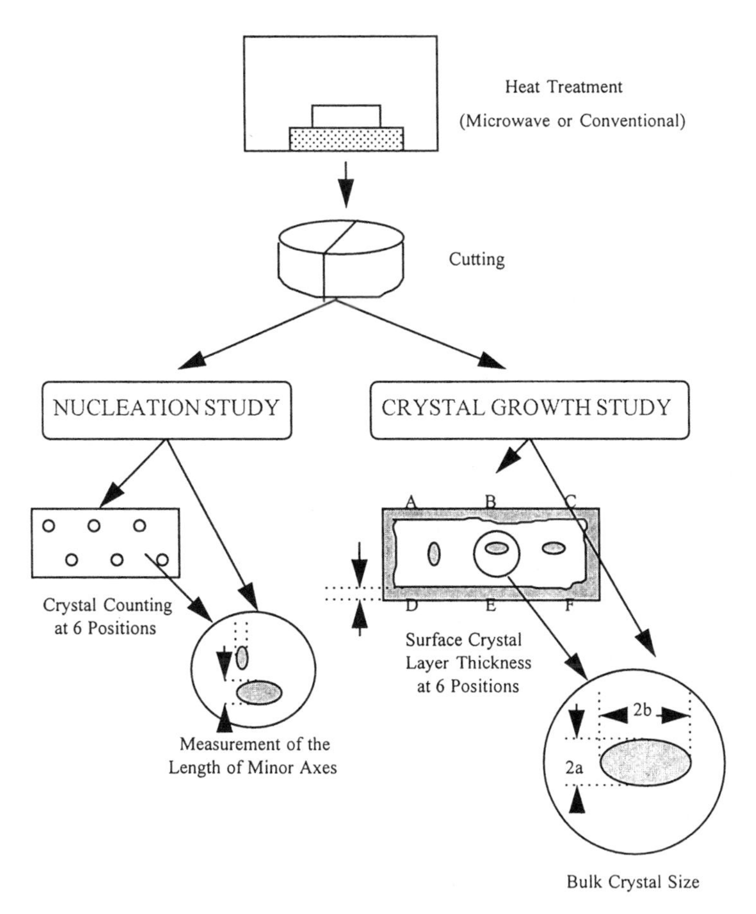

Figure 3. Schematic of microstructure study.

where: N_v = number of nuclei per unit volume,
N_A = number of crystal intersections per unit area,
Y = mean value of the reciprocals of measured minor axes,
k(q) = correction for shape factor, assumed = 1.18 [10].

For this study, N_A data were obtained by counting through the microscope at 200x magnification. Y values were obtained from measurements with the micrometer equipped on the eyepiece of the microscope. Sixty crystal intersections were measured at 1000x magnification.

For the bulk crystal size study, the crystal sizes were measured on 10 crystal intersections which exhibited large size and well defined elliptical shape. The related radii of these crystals were calculated from

$$r = (1/2)[1.5(a+b)-(ab)^{0.5}] \quad (2)$$

where: a = minor axis of elliptical crystal, and
b = major axis of elliptical crystal.

The first five of the largest radii were used to calculate the mean value of crystal size.

The advance of the crystalline layer into the interior was used for measuring the surface crystal size. Six positions on the surface of the specimens were measured.

RESULTS AND DISCUSSION

The results from the nucleation study are presented in Figure 4. The nucleation with conventional heating showed the same behavior as reported by Zanotto and James [10]. This indicated that our glass has the same nucleation behavior as normal $Li_2O \cdot 2SiO_2$. However, the nucleation in the microwave oven gave different results. Nucleation at 455°C in microwave oven was slower than in conventional heating. At a higher temperature, 464°C, nucleation in microwave field did not appear much different from the conventional heating. Since the study has been done only at two temperatures and has few data points at each temperature, we can not state a firm conclusion on the kinetics of these results. However, these results do show that glass can be nucleated in the microwave oven.

The results from the crystal growth study is shown in Table 1. Crystal growth was greater in the microwave oven than in the conventional heating. For example, the microwave heating at 567°C for 50 minutes yielded a 40 μm crystal size, whereas the conventional heating resulted in only an 11 μm crystal size. These results agree with those reported by Cozzi et al. [4] that the microwave heating yield faster crystallization than conventional heating of a pre-nucleated $Li_2O \cdot 2SiO_2$ glass.

The crystal radius as a function of time is presented in Figure 5. The data of crystal radius from conventional heating appear to agree with the results of Deubener et al. [11]. The crystal growth during the microwave heating was not only faster than in the conventional heating but also exhibited different behavior from conventional heating. Firstly, the crystal growth rate (slope of crystal radius vs time graph) in microwave heating appeared faster than in conventional heating. Moreover, the crystal growth behavior in microwave heating at 567°C appeared to be conposed of two rates, the faster crystal growth rate at the beginning and the slower rate at the later time. Both crystal growth rates in microwave heating was higher than in conventional heating.

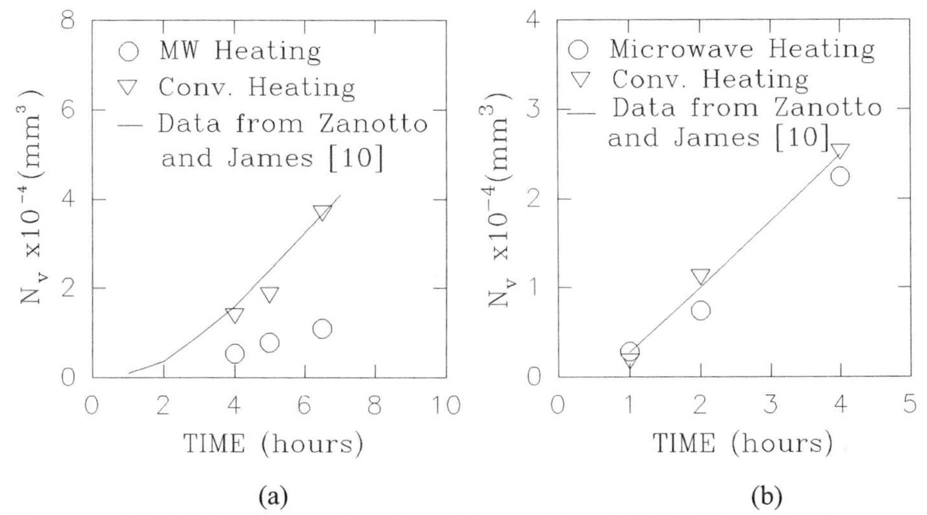

(a) (b)

Figure 4. Nuclei density as a function of time. (a) at 455°C. (b) at 467°C.

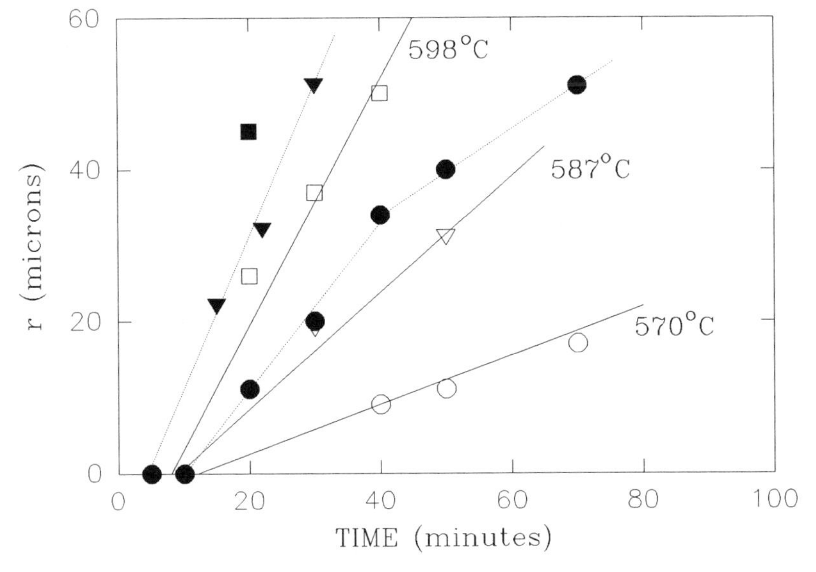

Figure 5. Crystal radius as a function of time.
● Microwave heating at 567°C. ○ Conventional heating at 567°C.
▼ Microwave heating at 583°C. ▽ Conventional heating at 583°C.
■ Microwave heating at 595°C. □ Conventional heating at 595°C.
Solid line = data from Deubener et al. [11].

Table 1. Crystal layer thickness and crystal size from microwave and conventional heating. (The value of A, B,.., F in surface crystal layer thickness were refer to the position on the samples shown in Figure 3.)

Method	Temperature (°C)	Time (min)	Surface Crystal Layer Thickness (μm)						Crystal Size (μm)
			A	B	C	D	E	F	
MW	567	20	16	20	21	17	15	20	11
		30	22	25	24	21	25	22	20
		40	39	41	50	36	35	55	34
		50	44	51	49	46	50	48	40
		70	42	55	56	55	51	54	51
	583	15	24	29	28	23	27	30	22
		22	31	33	29	30	35	32	32
		30	44	56	50	38	43	48	51
	595	20	41	39	53	34	33	55	45
		30	n	n	n	n	n	n	n
Conv.	567	40	*	*	*	*	*	*	9
		50	17	18	14	*	*	*	11
		70	23	23	21	*	*	*	17
	583	30	*	*	*	*	*	*	19
		50	31	27	29	22	22	22	30
		70	51	51	46	47	47	38	n
	595	20	28	21	23	14	15	15	26
		30	37	36	38	29	33	34	37
		40	52	53	55	44	45	47	50

n = data were not available due to too much crystal growth and crystal impingment.

* = data were not available because to the crystal layers were too thin or the thickness of crystal layer was less than 3 μm.

For the uniformity of the growth of the surface crystal layers, the conventional heating seemed to give a well defined different crystal layer thickness between the top and the bottom whereas the microwave heating did not show this phenomena. For example, the conventional heating at 567°C for 70 minutes gave 21-23 μm thickness of crystal layer at the top whereas less than 3 μm crystal layer thickness appeared at the bottom. The microwave heating seemed to yield a random variation of the thickness of the surface crystal layer. These results imply that the samples in microwave heating have a different heating pattern from those in conventional heating. In the conventional heating, samples received the heat radiated from the wall of the tube of a tube furnace. The heat reached the top of the sample before being transferred to the bottom. This resulted in more crystal growth on the top. In contrast, sample heating in microwave resulted from both microwave absorption and radiant heat emanated from the susceptor. As a result, the bottom part of sample is not much different in heating from the top.

SUMMARY

- Microwave heating can be used to produce glass-ceramics.
- The rate of nucleation in the temperature range studied appear slower in the microwave than the conventional process.
- The bulk crystal growth rate in the temperature range investigated appear higher in the microwave than in the conventional process.
- The bulk crystal growth behavior in microwave heating appear different from the conventional heating.
- The growth of the surface crystal layer in the microwave heating does not show significantly different kinetics between the top side and the bottom side as is the case in conventional heating.
- Work is in progress to better understand the effects of microwave energy on nucleation and crystallization in glass.

ACKNOWLEDGEMENTS

The authors wish to express their gratitude to G. P. LaTorre for donating equipment use and NSF and EPRI for financial support.

REFERENCES

1 P. W. McMillan, pp. 7-60, in Glass-Ceramics, Academic Press, London (1979).
2. P. F. James, J. Non-Cryst. Solids, 181, 1-15 (1995).
3. P. F. James and W. Shi, J. Mater. Sci., 28, 260-267 (1993).

4. A. D. Cozzi, Z. Fathi, R. L. Shulz, and D. E. Clark, Ceram. Eng. Sci. Proc., 14(9), 856-862 (1993).
5. T. N. Tiegs, K. L. Ploetz, J. O. Kiggans, and R. L. Yeckley, pp. 259-267, in Microwaves: Theory and Application in Materrials Processing, edited by D. E. Clark, W. R. Tinga, and J. R. Laia, The American Ceramic Society, Westerville, Ohio (1993).
6. S. D. Stookey, U. S. Patent No. 2,515,275 (1950).
7. A. De, I. Ahmad, E. D. Whitney, and D. E. Clark, pp. 329-339, in Microwaves: Theory and Application in Materrials Processing, edited by D. E. Clark, F. D. Gac, and W. H. Sutton, The American Ceramic Society, Westerville, Ohio (1991).
8. P. F. James, Phy. Chem. Glass, 15(4), 95-105 (1974).
9. R. T. DeHoff and F. N. Rhine, Trans. Met. Soc. AIME, 221, 975-982 (1961).
10. E. D. Zanotto and P. F. James, J. Non-Cryst. Solids, 74, 373-394 (1985).
11. J. Deubener, R. Bruckner, and M. Sternitzke, J. Non-Cryst. Solids, 163, 1-12 (1993).

TEMPERATURE MEASUREMENT DURING MICROWAVE PROCESSING

Greg Darby, D.E. Clark, Robert Di Fiore, Diane Folz, Rebecca Schulz, Attapon Boonyapiwat, and David Roth.
University of Florida
Department of Materials Science and Engineering
Gainesville, FL 32611

ABSTRACT

Many ceramic materials have been fabricated using sol-gel processing where the starting materials consist of a liquid organic precursor mixed with water and alcohol. The initial stages in sol-gel reactions require temperatures in the range of 100 ^{o}C or less, and therefore appear ideally suited for processing in a conventional microwave oven. In this paper we evaluate the use of several types of thermocouple geometries for measuring the temperature of liquids, including tetraethylorthosilicate (TEOS) during microwave heating. The boiling point of water is used as a reference on which to base the accuracy of our measurements.

INTRODUCTION

Temperature is one of the most important variables in the process industry. Without good temperature control, the product may be ruined. Therefore the ability to control and to monitor temperature is critical. The use of microwaves in the process industry is growing rapidly as an alternative to gas and electric resistance heating. Temperature measurement during microwave processing is not trivial, however. Microwaves may interact with the temperature sensor itself, resulting in false readings. Therefore, a reliable and accurate method of measuring temperature during microwave processing is much needed. Our objective was to evaluate methods of measuring the temperature of liquid organic precursors, specifically TEOS, heated in a microwave oven. These precursors form the basis for low-temperature fabrication of ceramics via sol-gel processing. The sol-gel process for preparing silica from TEOS involves five steps: hydrolysis, condensation, gelation, aging, and densification. The first four steps require a temperature of less than 100 ^{o}C, and thus can be conveniently carried out in a home microwave oven. In the present study, we focused our efforts on measuring/controlling the temperature during the hydrolysis and condensation steps. The boiling point of water was used as a reference temperature to compare different measurement techniques.

EXPERIMENTAL PROCEDURE

A 100 mL Erlenmeyer flask was filled with deionized water and placed onto an insulating brick located at the center of a 900 Watt, 2.45 GHz Roper Model home microwave oven (Figure 1). The tip of a partially shielded Type-K thermocouple as described in the following paragraph was positioned inside of the erlenmeyer flask at a depth of 5 5/8 inches from the mouth of the flask. The microwave duty cycle was set to 10% and the microwave power was turned on. Temperatures were recorded at 5 second intervals. The experiment was terminated when the temperature reached a constant value. This experimental procedure was repeated for duty cycles of 20%, 30%, 50%, and 100%. These experiments were also repeated for a Type-K titanium shielded thermocouple and for a Type-K alumina-shielded thermocouple. The reason for using the alumina-shielded thermocouple was to minimize the reactivity of the thermocouple with the liquid precursor, especially when higher temperatures are used. As a reference, temperatures were also recorded for boiling water using a hot plate.

The same thermocouple was used for each experiment. Only the outer shielding was different. The wire diameter of the thermocouple was 0.813 mm. The partially shielded thermocouple is defined as a Type-K chromel/alumel thermocouple with alumina fiber insulation around the thermocouple wires and inconel shielding covering the outside of the thermocouple up to the thermocouple tip, which was exposed. The titanium shielded thermocouple was exactly the same as the partially shielded thermocouple except that titanium metal was used along the length of the thermocouple and covering (or shielding) the tip of the thermocouple. The alumina shielded thermocouple was exactly the same as the partially shielded thermocouple except that alumina was used along the length of the thermocouple as well as covering the thermocouple tip. (See Figure 2.) All three thermocouples were grounded with the inconel shielding to the outer shell of the microwave during measurements. Temperature measurements were taken with an ungrounded thermocouple, and variations as large as 200 ^{o}C were observed. Grounding the thermocouple to the outer shell of the microwave appears to eliminate these large variations.

The Type-K thermocouple was selected because of its availability and temperature range (-200 to 1250 ^{o}C), which encompasses the temperature range of interest. The thermocouple was connected to a an OMEGA readout which compensated for variations in room temperature (i.e., self-compensating reference junction).

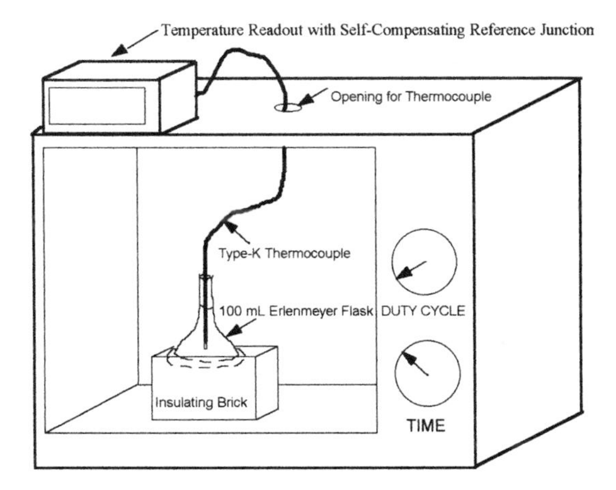

Figure 1. Microwave Apparatus for Temperature Measurements

RESULTS AND DISCUSSION

In Figure 2 are temperature versus time profiles for the partially shielded, titanium shielded, and alumina-shielded Type-K thermocouples, respectively. The scaling was cut to 150 seconds in order that all of the information could be better interpreted. Notice in all of the graphs, except when the duty cycle is 100%, that temperature versus time is not a smooth function. This may be attributed to the duty cycles of the microwave. When the power comes on, there is a sharp rise in temperature. When the power shuts off, the temperature drops. The following is a table of the average variations in heating (ie, rapid temperature drops when power shuts off during duty cycle) for duty cycles of 20%, 30%, and 50% for the various thermocouple designs.

Table I. Average Variations in Heating for Different Thermocouple Designs at 20%, 30%, and 50% Duty Cycles (in Celsius)

	20%		30%		50%	
	Increase	Decrease	Increase	Decrease	Increase	Decrease
Partially Shielded	7.82	2.85	10.29	4.64	11.00	1.25
Titanium-Shielded	7.05	2.53	6.71	1.82	14.00	2.86
Alumina-Shielded	8.13	3.07	7.14	1.67	15.43	4.29

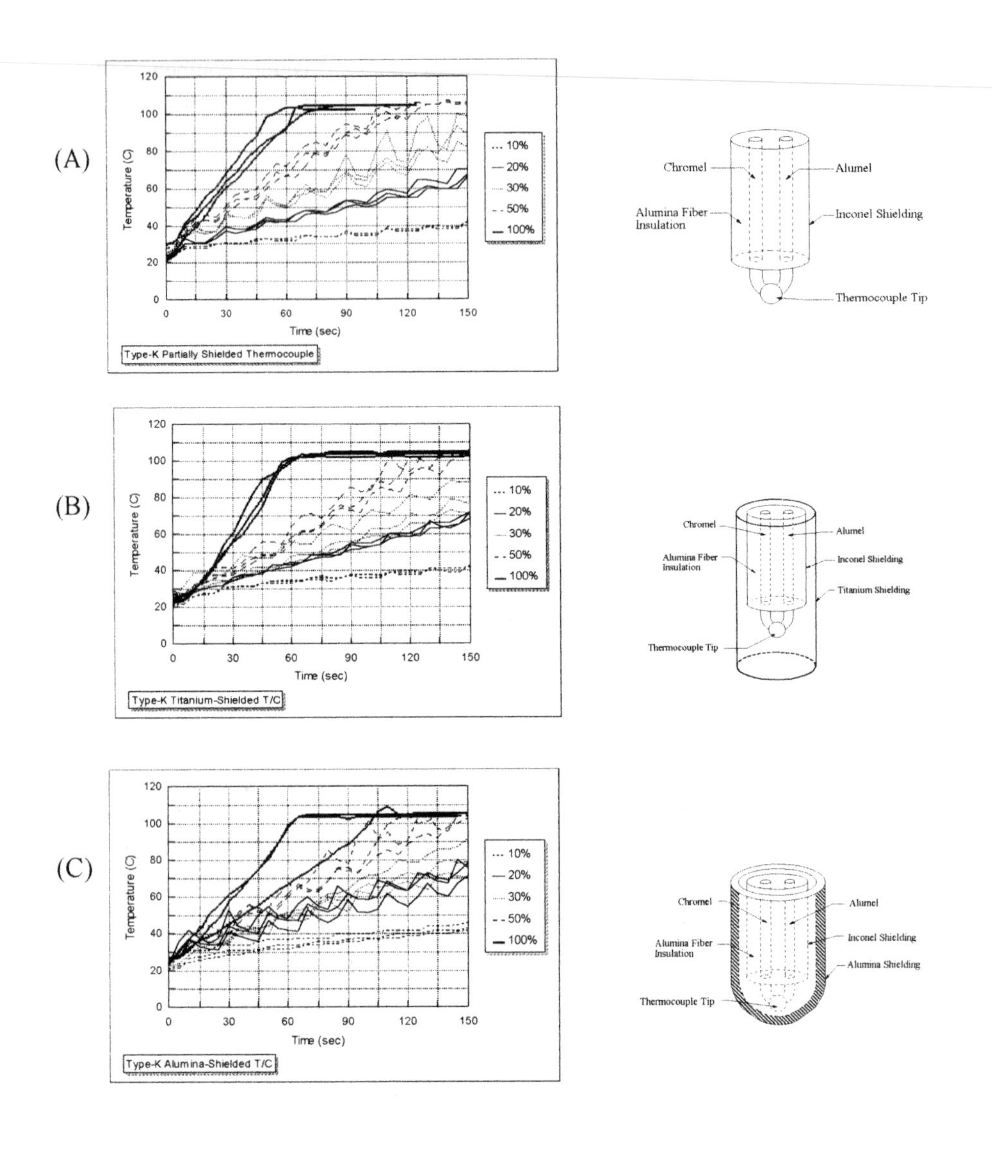

Figure 2. Temperature vs. Time Profiles During Microwave Heating.

In order to better understand these sharp increases and decreases in temperature during microwave processing, it was first necessary to examine the duty cycle profiles (ie, percent power vs. time) of the microwave oven. Cycle durations and the time between cycles were measured. Cycle durations are defined as the length of time the microwave power is on during each cycle. The time between cycles is taken from the point one cycle comes on to the point the next cycle comes on. The results are presented in Table II.

Table II. Duty Cycle for the Microwave Oven

Duty Cycle (%)	Cycle Duration (sec)	Cycle Interval (sec)
10	2.05	20.46
20	4.77	19.93
30	7.55	20.13
50	12.22	19.44

The peaks are difficult to see in the 10% duty cycle experiments because of the scaling of the axis. There are no apparent "peaks and valleys" in the 100% duty cycle experiments because the power is always on. The temperature will rapidly increase to boiling. Notice that all of the peaks in temperature, regardless of duty cycle, appear to coincide with one another; due to the time between cycles (ie, cycle interval) being constant (about 20 seconds).

It should also be mentioned that two types of boiling occured during microwave heating of the distilled water (localized boiling and bulk boiling). Bulk boiling occurred when the temperature of the water became constant (and it was visually observed that the bulk of the water was boiling) and continued until the experiment was terminated. Localized boiling occurred at temperatures below, at times well before, the measured bulk boiling point. Localized boiling was observed along the stem of the thermocouple inside the neck region of the Erlenmeyer flask as the microwave power came on (duty cycle). The localized boiling subsided as the power went off during the duty cycle.

The type of shielding on the thermocouple appears to have some effect on the temperature profiles. With the partially shielded thermocouple, the amplitude of the peaks are large, and there appears to be good repeatability of the temperature measurements; i.e. there is good matching of temperature profiles at a given duty cycle percentage. With the titanium-shielded thermocouple, the amplitude of the peaks appear to be dampened, while there still is good repeatability of the measurements. Finally, with the alumina-shielded thermocouple, the amplitude of the peaks are large, and there appears to be poor repeatability of the temperature measurements (large variances).

Due to the geometry of the thermocouple tip, it is suggested that the field gradient may be increased at the tip. If this is the case, it may contribute to the sudden increases in temperature during the microwave cycles. The titanium shielding appears to decrease this "focusing" of the microwave field at the tip of the thermocouple, leading to a dampening of the peaks. The alumina shielding appears to have no effect.

The temperature of deionized water heated by a hot plate was measured as a reference for the temperature at boiling. The purpose was to analyze the accuracy of measurement during microwave processing (microwave interactions). Temperatures at boiling for the various duty cycles are presented below:

Table III. Peak Boiling (in ^{O}C) of D.I. Water Using Either a Microwave or Hot Plate and Different Thermocouple Designs

Duty Cycle

Sensor Type	10%	20%	30%	50%	100%	Hot Plate
K-Unshielded	102.5	102.5	105.5	106.5	106.5	103.5
K-Titanium-Shielded	102.5	103.5	106.5	106.5	105.5	104.5
K-Alumina-Shielded	102.5	102.5	105.5	106.5	106.5	103.5

* Temperature measurement may vary (+/-) 0.5 ^{O}C

Referring to Table III, the type of shielding appears to have little effect on the accuracy of measurement during microwave processing. Since the temperature at boiling during microwave processing was consistently greater than that measured using the hot plate, there appears to be some heating of the thermocouple bead by the microwave field. For duty cycles of 10% and 20%, the cycle durations are apparently not long enough to contribute to the self-heating (the heat may be conducted away during the longer "cycle off" periods). For the other duty cycles the peak temperature reached is 106.5 ^{O}C, even at a duty cycles of 100%. At a duty cycle of 100%, the temperature will remain at 106.5 ^{o}C for up to 1 1/2 minutes (at which time the experiment is terminated).

We also measured temperature versus time for a silica sol during microwave processing at a duty cycle of 10% (see Figure 4). The purpose of this was to relate the above research to ceramics via low-temperature processing of sol-gels. The sample that was heated was a 50 mL solution (half of that used for the water temperature experiments) from a batch consisting of 44.5 mL TEOS, 44.5 mL Ethanol, 15 mL D.I. water, and 5 mL HCl. This batch was made using the same procedure as that for conventionally processed sols. The temperatures were measured using a Type-K partially shielded thermocouple. The experiment was terminated at 82 ^{o}C due to violent foaming. We also measured the point of foaming when heating a sol-gel using a hot plate. We measured a temperature of 82 ^{o}C for the foaming point using a Type-K partially shielded thermocouple as well as a mercury thermometer on a hot plate. We obtained agreement not only between the thermocouple and the thermometer, but between the hot plate measurement and the microwave heating measurement. Therefore, the use of shielded thermocouples appears to be a viable method for measuring the temperature of sol-gels during microwave heating.

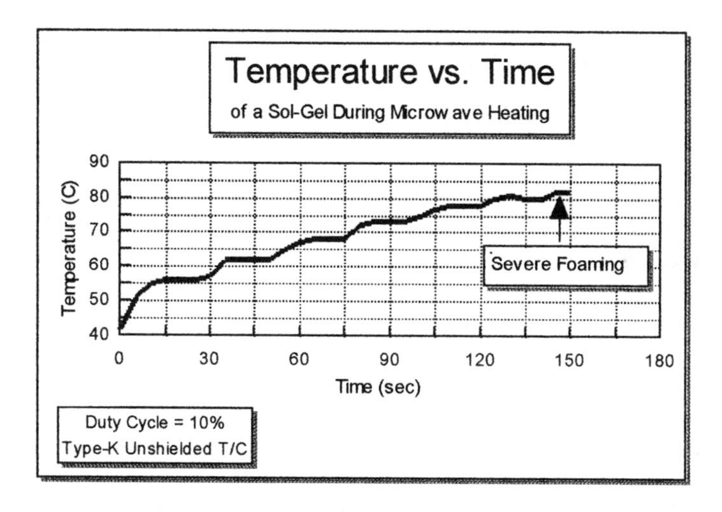

Figure 4. Temperature vs. Time of a Silica Sol During Microwave Processing

SUMMARY

Three Type-K thermocouples (partially shielded, titanium-shielded, and alumina-shielded) were evaluated as methods of measuring the temperature of a material during low-temperature microwave processing. All three thermocouples were slightly inaccurate (~3 °C) in measuring the boiling point of D.I. water during microwave heating (as compared to the boiling point of D.I. water heated by a hot plate). The temperature versus time profiles for the D.I. water were also found to be strongly dependent on the microwave duty cycle (sharp increases in temperature as the microwave power came on and decreases in temperature as the microwave power shut off). When measuring the temperature of a sol-gel during microwave heating using an unshielded Type-K thermocouple, however, we found excellent agreement between the foaming temperature during microwaving and that during heating with a hot plate. In addition, the temperature versus time profile of the sol-gel during microwave heating was not as erratic as that of the D.I. water, although the duty cycles did appear to have some effect. Therefore it appears that shielded thermocouples are a viable method for monitoring the temperature during the low-temperature processing of silica-based sol-gels.

ACKNOWLEDGEMENTS

I would also like to acknowledge the National Science Foundation and the Electric Power Research Institute for their support.

REFERENCES

1. Hogan, Paul F. and Toshiko Mori. "Development of a Method of Continuous Temperature Measurement for Microwave Denture Processing," Dental Materials Journal, Vol. 9, No. 1, 1990.

2. Pierre, Alain C. "Sol-Gel Processing of Ceramic Powders," Ceramic Bulletin, Vol. 70, No. 8, 1991.

3. Prosetya, H., and A.K. Datta, "Batch Microwave Heating of Liquids: An Experimental Study," Journal of Microwave Power, 26, 215, 1991.

4. Sutton, Willard H. "Microwave Processing of Ceramic Materials," Ceramic Bulletin, Vol. 68, No. 2, 1989.

5. Wang, T.P. "Thermocouple Selection and Maintenance - Pt. 1," InTech, April 1991.

Chemical and Reaction Synthesis

THE APPLICATION OF MICROWAVES TO RAPID CHEMICAL SYNTHESIS

R. Gedye, K. Westaway and F. Smith. Department of Chemistry and Biochemistry, Laurentian University, Sudbury, Ontario, Canada. P3E 2C6

ABSTRACT

The synthesis of a wide variety of organic and organometallic compounds can be carried out conveniently and rapidly using microwave heating. The dramatic rate enhancements which result from microwave irradiation of reactants in sealed vessels are attributed mainly to the superheating of the solvent due to the high pressures generated. However there is considerable evidence that the rates of some reactions are enhanced by microwave heating in open vessels, suggesting localized superheating or some specific microwave effect. While microwave irradiation does not usually alter the product composition of a reaction appreciably, there have been a few reports in which some product selectivity has been observed.

INTRODUCTION

Although the ability of microwaves to heat water and other polar materials has been known for over forty years, it was not until 1986 that two groups of researchers reported the use of microwave irradiation to accelerate organic reactions[1,2].

Since then, the application of microwave heating for chemical reactions has attracted considerable interest, and has been exploited in the rapid synthesis of a wide variety of compounds. Although most studies have been limited to small scale reactions the potential for scaling-up microwave assisted reactions for possible industrial applications is currently being investigated[3,4]. The so-called "microwave effect" has also received some attention. Although the acceleration of reactions by microwaves has usually been ascribed to the superheating of the solvent in sealed containers[5,6], there have been reports that some reactions occur more rapidly under microwave radiation than under conventional heating at the same temperature[7,8].

This paper gives an overview of the application of microwaves to synthesis, including its scope and limitations, product compositions, kinetics and other evidence for and against the "microwave effect".

EXPERIMENTAL

The microwave oven used in this study was a domestic Toshiba model ER-8000 BTC with 9 power settings, starting at 72W and increasing in 81W increments to 720W.

Microwave reactions under pressure were carried out in 150 mL PFA Teflon containers manufactured by the Savillex Corporation (Minnetonka, Minnesota, USA). The reaction mixture was placed in the vessel and the lid screwed on tightly. The sealed vessel was then placed in the centre of the microwave oven and heated for the specified time. After allowing to stand for 2 minutes to reduce the pressure, the vessel was removed from the oven and cooled in ice-water for 5 minutes before opening.

For the hydrolysis of N-phenylbenzamide in 20% sulfuric acid the pressure controlled system was a Floyd Inc. RMS-150 Remote Microwave Digestion System, which has a control console attached to the microwave oven and pressure transducer. The pressure vs. time operating mode was used. The reaction was carried out in a pressure control vessel which was connected to the transducer via an inlet tube. This vessel contained a pressure membrane disc designed to rupture if the pressure exceeded 200 psi. Distilled water was present above the pressure disc as well as in the inlet tube in order to monitor the pressure. Rate enhancements for comparable microwave and conventionally heated (reflux) reactions, using identical concentrations of reactants, were determined by comparing the times required to take the reaction to the same state of completion under microwave and reflux conditions respectively.

$$\text{Rate enhancement} = \frac{\text{kmicrowave}}{\text{kclassical}} = \frac{\text{reflux time}}{\text{microwave time}}$$

In the comparison of the rates of racemization of L-proline by microwave and conventional heating, the % racemization was determined by polarimetry and by high performance liquid chromatography (HPLC). The optical rotations were determined using a Rudolph Autopol II automatic polarimeter model #I1589-10 and HPLC was carried out using an apparatus consisting of a LOC Milton Roy mini pump, a Tracer 970 A variable wavelength detector and a Hewlett Packard 3380 A integrator. A 150 x 46mm chirex D-penicillamine column with Phenomenex #00F-3126-EO was used to separate the L- & D-proline in the analysis.

DISCUSSION

We first reported the use of microwave ovens for the rapid synthesis of organic compounds in 1986[1]. Some examples of these reactions included the hydrolysis of amides and esters to carboxylic acids, the esterification of carboxylic acids with alcohols, oxidation of alkyl benzenes to aromatic carboxylic acids and the conversion of alkyl halides to ethers. Rate enhancements varied from 5 to 1200 times compared to the conventionally heated reactions. Some representative reactions are given in Table 1. Shortly afterwards Giguère et al[2] reported dramatic reductions in reaction time in microwave syntheses using Diels-Alder, Claisen and ene reactions.

TABLE 1. A comparison of reaction times and yields in representative reactions using classical and microwave procedures

Compound synthesized	Procedure followed	Reaction time	Average[b] yield (%)	$\left[\frac{k_{microwave}}{k_{classical}}\right]$
Hydrolysis of benzamide to benzoic acid in water				
C_6H_5COOH	Classical	1 h	90	
C_6H_5COOH	Microwave	10 min	99	6
Oxidation of toluene to benzoic acid in water				
C_6H_5COOH	Classical	25 min	40	
C_6H_5COOH	Microwave[a]	5 min	40	5
Esterification of benzoic acid with methanol				
$C_6H_5COOCH_3$	Classical	8 h	74	
$C_6H_5COOCH_3$	Microwave	5 min	76	96
S_N2 reaction of 4-cyanophenoxide ion with benzyl chloride in methanol				
$NCC_6H_4OCH_2C_6H_5$	Classical	16 h	89	
$NCC_6H_4OCH_2C_6H_5$	Microwave	4 min[c]	93	240
$NCC_6H_4OCH_2C_6H_5$	Classical	12 h	65	
$NCC_6H_4OCH_2C_6H_5$	Microwave	35 sec[c,d]	65	1240

[a]Very high pressures developed and the study of this reaction was halted.
[b]The average yields are based on isolated yields and represent the average of at least two experiments. The S_N2 reaction was followed by titrating the chloride ion (8).
[c]Value for one run only.
[d]This reaction was done in a 50-mL Teflon bomb.

Both groups carried out their reactions in sealed vessels in domestic microwave ovens. Gedye et al[1] used Teflon vessels whereas Giguère et al[2] used glass vessels.

Since then, a large number of synthetic applications have been reported and these have been reviewed[7,9]. Of particular note are the synthesis of radiolabelled compounds containing radionuclei with short half-lives[10], the synthesis of organometallic compounds[7] and the synthesis of polymers[7,11]. The synthesis of compounds using dry-media conditions, which avoid the hazards due to the high pressures developed in sealed containers, has also received considerable attention[12,14].

We noted in our studies[15], that considerable scaling up of organic reactions was not feasible since heating larger amounts of reactants requires more energy and requires higher power-levels. The use of a continuous flow system to process larger amounts of material appears to have the greatest potential for scale-up and industrial application[3,4].

Since molecules must be polar to absorb microwave energy, only reactions in polar solvents can be accelerated appreciably. In our laboratory, it was found that dilute solutions of polar substrates in non-polar solvents are heated very slowly in a microwave oven.

Until recently, the pressures used in microwave reactions in sealed containers have been restricted to 100psi or less for safety reasons, but improvements in technology have led to the design of vessels capable of withstanding pressures of 200psi or more. We have investigated the use of a pressure-controlled system developed by Floyd Inc.[16] to study the effects of higher pressures on the rates of microwave heated reactions. Using this apparatus, reactions can be carried out both safely and very rapidly. It was shown for example that the hydrolysis of N-phenyl benzamide at a pressure of 150psi was complete in approximately five minutes whereas at 60psi, was less than 50% complete in eight minutes. Parr Teflon bombs have been developed which can withstand pressures up to 80 atm and temperatures up to 250°C[17]. However it has been reported[7] that repeated use of the vessels above 150°C can lead to distortions which reduce the safe pressure below 80 atm.

There has been considerable discussion recently on the reason for the rate enhancements observed in microwave heated reactions, and whether reactions can be accelerated in open systems as well as in sealed containers.

It was observed in our laboratory that the esterification of benzoic acid with 1-propanol occurs at approximately the same rate in an open vessel in a microwave oven as it does under conventional reflux conditions[5]. A similar result was reported by Pollington et al[6] who showed that the rates of esterification of acetic acid with 1-propanol in the presence and absence of microwave radiation at the same temperature were identical. Recent kinetic studies by Raner et al[18] on the acid-catalysed isomerization of carvone and on the Diels-Alder reaction of anthracene and diethyl maleate show that there is no significant difference between the rate of the reaction under microwave and oil bath heating at the same temperature.

On the other hand Baghurst and Mingos[7] have reported that some organometallic compounds can be synthesized more rapidly by reflux using microwave heating at atmospheric pressure than by conventional reflux. Also Berlan et al[8] have reported that two different Diels-Alder reactions occurred more rapidly under microwave heating than under conventional heating at the same temperature, and suggested that special activation by microwaves is a possibility. Kinetic studies by Lewis et al[19] showed that microwave radiation enhanced the rate of imidization of a polymer (BTDA-DDS polyamic acid) and that the apparent activation energy for the reaction was reduced from 105 to 55 KJ/mol. It was suggested that the enhancement was due to a nonuniform temperature on a molecular scale rather than a true reduction in activation energy. Baghurst and Mingos[20] have reported recently that some solvents, including water, superheat under microwave radiation even at atmospheric pressure. This would indicate that if rate enhancements do occur under microwave heating at atmospheric pressure, this is due to superheating rather than to a specific activation of molecules by microwaves.

There have also been reports that isomerization or racemization of amino acids occur more rapidly on microwave heating, at atmospheric pressure[21,22]. However our observations on the racemization of L-proline on microwave heating show that only slight increases in isomerization rates occur at atmospheric pressure compared with conventional heating. We consider that these rate enhancements are not significant enough to be evidence of any special microwave effect other than local superheating.

It is also of interest to know whether microwave heating only alters the rate of reaction or whether it can influence reaction pathways thereby leading to product compositions different from those obtained using conventional heating techniques. We have observed that using microwaves to carry out organic reactions does not change the product composition appreciably[23]. The reactions studied included the elimination vs. substitution reactions of alkyl halides with alkoxide ion and a Diels-Alder reaction in which the ratio of endo to exo products was determined. Any small changes in product composition that are observed can be attributed to the higher temperatures in the microwave reactions, rather than to specific microwave effects.

It should be noted, however, that some selectivity has been observed in the Diels-Alder reaction of 6-demethoxy-β-dihydrothebaine with an excess of methyl vinyl ketone to yield a mixture of two adducts[7]. Under conventional conditions extensive polymerization of methyl vinyl ketone occurred whereas much less polymeric material was formed under reflux conditions in a modified microwave oven at atmospheric pressure. Also Bose et al have shown recently that in some cases the steric course of β-lactam formation can be influenced using high power irradiation in a domestic microwave oven[24].

Thus, although most of the evidence up till now does not support the existance of an athermal microwave effect, the rates of some reactions are enhanced by microwave heating at atmospheric pressure, and in certain reactions some selectivity may be observed.

References

1. R. Gedye, F. Smith, K. Westaway, H. Ali, L. Baldisera, L. Laberge and J. Rousell. Tet. Lett. 27, 279 (1986).
2. R.J. Giguère, T.L. Bray, S.M. Duncan and G. Majetich. Tet. Lett. 27, 4945 (1986).
3. C. Peterson, New Scientist, Sept. 1989, 123, 44.
4. S-T Chen, S-H Chiou and K-T Wang. J. Chem. Soc. Chem. Commun. 1990, 807.
5. R. Gedye, F. Smith and K. Westaway. Can. J. Chem. 66, 17 (1988).
6. S. Pollington, G. Bond, R. Mayes, D. Whan, J. Candlin & J. Jennings. J. Org. Chem. 56, 1313 (1991).
7. D.M.P. Mingos and D.R. Baghurst, Chem. Soc. Rev. 20, 1 (1991).
8. J. Berlan, P. Giboreau, S. Lefevre & C. Marchand. Tet. Lett. 32, 2363 (1991).
9. R.A. Abramovitch, Org. Prep. Proc. Int. 23, 683 (1991).
10. S. Stone-Elander and N. Elander. Appl. Radiat. Isot. 44, 889 (1993).
11. J. Palacios, J. Sierra and M.P. Rodriguez. New Polymeric Mater., 13, 273 (1992).
12. E. Gutierrez, A. Loupy, G. Bram and E. Ruiz-Hitzky. Tet. Lett. 30, 945 (1989).
13. B. Rechsteiner, F. Texier-Bouillet and J. Hamelin. Tet. Lett. 34, 5071 (1993).
14. B. Oussaid, B. Garrigues and M. Soutiaoui. Can. J. Chem. 72, 2483 (1994).
15. R. Gedye, F. Smith and K. Westaway. J. Microwave Power and Electromag. Energy. 26, 3 (1991).
16. Floyd Inc., 5440 Hwy 55E, Lake Wylie, South Carolina, 29710.
17. Parr Instrument Co., Moline, Illinois 61265.
18. K.D. Raner, C.R. Strauss, F. Vyskoc and L. Mokbel. J. Org. Chem, 58, 950 (1993).
19. D.A. Lewis, J.D. Summers, T.C. Ward and J.E. McGrath. J. Polymer Sci. Part A (Polymer Chemistry), 30, 1647 (1992).

20. D.R. Baghurst and D.M.P. Mingos. J. Chem. Soc. Chem. Commun. 647 (1992).
21. S-T Chen, S-H We and K-T Wang. Int. J. Peptide Res. 33, 73 (1989).
22. G. Lubec, C. Wolf and B. Bartosch., Lancet. 1392 (1989).
23. R. Gedye, W. Rank and K. Westaway, Can. J. Chem. 69, 706 (1991).
24. A.K. Bose, B.K. Banik and M.S. Manhas. Tet. Lett. 36, 213 (1995).

MICROWAVE SYNTHESIS OF SIALONS

M.D. Mathis, D.K. Agrawal, and R. Roy
Materials Research Laboratory
The Pennsylvania State University
University Park, PA

R.H. Plovnick
3M Ceramic Technology Center
St. Paul, MN

ABSTRACT

Microwave synthesis of β-SiAlON has been attempted from a solid state reaction between Si_3N_4 and Al_2O_3 in a nitrogen atmosphere. The β-SiAlON reaction was complete after 15 minutes at 1500°C. No SiAlON phase was present after conventional synthesis until 1600°C under similar conditions. Due to weak initial microwave coupling of the Si_3N_4 and Al_2O_3 system as the reaction mixture heated, the clay-carbon system and the AlOOH-SiO_2-SiC-C-N system were investigated. It was found that SiAlON synthesis from those two routes yielded limited amounts of product due to difficulties encountered with reaction parameters.

INTRODUCTION

Microwave processing has become an interesting and useful tool in the sintering [1], synthesis [2-3], gel processing [4] and joining of materials [5]. Microwave processing, as opposed to conventional methods, allows a material to heat relatively uniformly and can yield unique and novel microstructures [6]. Additionally, materials can be heated rapidly during microwave processing, depending on their ability to absorb the applied energy. The mechanisms that materials use to absorb or couple with the applied field are rotation of dipoles [7] within the material, electric conduction [7] or ionic migration [8].

In this work several compositional systems were investigated for the synthesis of β-SiAlON by microwave processing. β-SiAlON is interesting because of its strength and oxidation resistance. These materials show promise as metal cutting and forming tools [9], as well as refractories [10]. Microwave processing may enhance the mechanical properties of SiAlONs as it has with other materials [6]. The silicon nitride-alumina system was studied first, due to the structural similarities between the β-silicon nitride and the β-SiAlON phases [9, 11].

In addition, carbothermal reduction reactions were investigated for β-SiAlON synthesis. This technique has been employed by several researchers [12-16] in the conventional heating synthesis of SiAlONs by nitriding various clays with carbon. Lee and Cutler [12] were the first to attempt nitridation of a kaolin-carbon mixture which gave SiAlONs and other phases.

Additionally, clays, carbon and some of the reaction intermediates, such as silicon carbide (SiC) [14], are capable of coupling with a microwave field. One possible reaction mechanism for β-SiAlON synthesis is given in equations 1-3. In a reaction attempted by Higgins and Henry [14] to form β-SiAlON from kaolin and carbon, silicon carbide is formed as an intermediate and serves with carbon as a reductant in the reaction. It is proposed that first kaolin dissociates into mullite, silica and water as shown in equation 1. Equation 2 shows that silica is reduced by carbon to give SiC and CO. Finally, mullite is reduced by SiC and C in the presence of nitrogen to give β-SiAlON and CO as shown by equation 3. This type of reaction should be valid for many clay materials, since mullite and silica are the first phases formed.

$$3(2SiO_2\text{-}Al_2O_3\text{-}2H_2O) = (3Al_2O_3\text{-}2SiO_2) + 4(SiO_2) + 6(H_2O) \quad (1)$$

$$4(SiO_2) + 12(C) = 4(SiC) + 8(CO) \quad (2)$$

$$(3Al_2O_3\text{-}2SiO_2) + 4(SiC) + 3(C) + 5(N_2) = 2(Si_3Al_3O_3N_5) + 7(CO) \quad (3)$$

The systems studied for β-SiAlON synthesis via carbothermal reduction were clay-carbon-nitrogen, and boehmite-silica-silicon carbide-carbon-nitrogen. Each system was processed by microwave and conventionally under nitrogen and the results compared.

EXPERIMENTAL

Clay-carbon-nitrogen: The clay chosen in this work was a montmorillonite (Aldrich Company (5 μm particle size)) powder. This powder was found by x-ray diffraction (XRD) to contain mica, kaolin and quartz. However, the silica-to-alumina ratio matched that of montmorillonite. The powder was mixed with graphite, giving a carbon-to-silica ratio of 3 [15].

The mixture was wet-milled overnight and dried initially by microwave and overnight by conventional oven. Samples were uniaxially pressed pellets with weights of approximately 12 grams, diameters of 1.13 inches and heights of 0.4 inches for all conventional and microwave experiments. The pellets were processed under nitrogen (Linde, Inc.).

Boehmite-silica-silicon carbide-carbon-nitrogen system (BSSCN): Boehmite-amorphous silica powders (Condea, Inc. (0.5 μm particle size)) were mixed with silicon carbide and graphite powders. Samples were uniaxially pressed pellets with weights of approximately 12 grams, diameters of 1.13 inches and heights of 0.5 inches for all experiments. Sample treatment is similar to that given in the previous section.

Silicon nitride-alumina system: Silicon nitride (H. C. Starck, Inc.) with a particle size of <0.5 μm and alumina (lot A16SG, Alcoa, Inc.) with a particle size of 0.2 μm, were first mixed in equimolar amounts, then were allowed to wet-mill overnight. The pellet heights were about 0.3 inches. Samples were treated as were the previous systems.

Microwave Equipment: The microwave apparatus used in this work was the Microwave Materials Technologies 1300 watt Model 10-1300 (Microwave Materials Technologies (MMT), Oak Ridge, TN) using an atmosphere chamber (supplied by MMT) for working in a gas environment. The features and capabilities of this unit were described previously [17].

A fibrous alumina cylinder (type-ALC, Zircar Inc) was used as outer insulation, containing the sample pellet which was surrounded by alumina bubble insulation. The top and bottom covers of the alumina cylinder were made from an alumina board (ZAL-45, Zircar, Inc.). The bottom surface had an opening drilled for thermocouple (C-type) introduction. SiC bars, with 1.0in x 0.75in x 0.25in dimensions, were used as secondary couplers [6] to aid in the heating of the pellets until the pellets coupled well with the microwave field.

Conventional Equipment: The sample mixtures were dried by a Gravity-Convected Series 200 Isotemp™ oven (Fisher Scientific Co.) at 120°C. The high temperature processing was accomplished in an atmosphere-controlled graphite resistance furnace (Astro Industries, Inc.).

Characterization: Samples were analyzed by X-ray diffraction (XRD). The unit employed was an APD Model 3600 (Phillips, Inc.) using Cu K-alpha radiation. SEM micrographs were obtained by a JEOL 6400 Scanning Electron Microscope with a Noran Voyager EDAX attachment.

RESULTS AND DISCUSSION

Clay-Carbon-Nitrogen System

Table 1 gives the XRD data for the attempted microwave synthesis of β-SiAlON from clay materials and graphite in a nitrogen atmosphere. The information in Table 1 represents the relative intensities of identified phases under a given time and temperature regime.

Table 1. XRD Results of the Microwave Clay-Carbon-Nitrogen Reaction at Various Times and Temperatures

Microwave Conditions	Phases Identified by XRD and Relative Intensities					
Temp/Time	SiC	C	β-SiAlON	J-SiAlON	mullite	other
1300°C/2min	28	100	---	---	10	SiO_2-12
1400°C/15min	100	2	3	7	---	ε–SiAlON-11
1400°C/30min	100	9	2	8	---	ε–SiAlON-6
1500°C/15min	100	3	3	8	---	---
1500°C/30min	100	---	5	---	---	AlN-20

The data show that very small amounts of SiAlON phases are synthesized beginning at 1400°C. The presence of graphite indicates that the reaction has not gone to high enough temperatures or for long enough time for completion. However, it was not possible to exceed 1500°C or remain at 1500°C for long periods of time with the present microwave and refractory equipment. The conventional synthesis results, shown in Table 2, indicate no β-SiAlON formation after 48 minutes at 1600°C and only modest amounts of the J-phase SiAlON.

Another major indication of the incompleteness of the microwave and conventional reaction was the large amounts of SiC formed. This probably results from the reaction of carbon with SiO_2, since a reduction of the amount of carbon is observed along with an increase in the amount of SiC in both microwave and conventional synthesis. As stated earlier, in the carbothermal reduction synthesis of SiAlONs from clays and other materials, SiC is an intermediate in the final stages of reaction. SiC acts as a reductant with carbon and finally reacts, with silicon entering into the SiAlON and carbon leaving as CO (see equation 3).

Table 2. XRD Results of the Conventional Clay-Carbon-Nitrogen Reaction at Various Times and temperatures

Conventional Conditions	Phases Identified by XRD and Relative Intensities					
Temp/Time	SiC	C	β-SiAlON	J-SiAlON	mullite	other
1500°C/30min	15	100	---	---	10	SiO_2-4
1600°C/30min	100	26	---	2	---	SiO_2-4
1600°C/48min	100	23	---	6	---	SiO_2-6

Boehmite-Silica-SiC-Carbon-Nitrogen System (BSSCN)

Since SiC is an intermediate product in the carbothermal reduction process, the addition of SiC as a reactant might aid in the synthesis of SiAlON. Additionally, the presence of SiC in the reactant mixture would increase the microwave absorptivity of the system. Due to the complexity of phases in the clay powder system chosen (see experimental section), boehmite (AlOOH) and amorphous silica (SiO_2) mixture was investigated for the β-SiAlON synthesis.

Tables 3 and 4 report the XRD results for this system. The information in Tables 3 and 4 can be interpreted analogously with Tables 1 and 2, with respect to the relative intensities of the identified phases.

Table 3. XRD Results of the Microwave Reaction of BSSCN at Various Times and Temperatures

Microwave Conditions	Phases Identified by XRD and Relative Intensities					
Temp/Time	SiC	C	β-SiAlON	J-SiAlON	mullite	other
1100°C/30min	100	33	---	---	---	---
1200°C/15min	100	---	1	3	81	---
1300°C/15min	100	---	13	9	18	---
1400°C/15min	100	---	18	3	2	---
1500°C/15min	100	---	30	4	---	---

The data presented in Table 3 show that more SiAlON has been produced by microwave synthesis from the BSSCN system than in the clay-graphite system mentioned in the previous section. The reaction is nucleated at about 1200°C after a 15 minute soak period. Conventionally, the J-phase can be nucleated at 1500°C after 30 minutes at temperature. The conventional synthesis method begins to nucleate the β-phase after 1600°C as shown in Table 4.

Table 4. XRD Results of the Conventional Reaction of BSSCN at Various Times and Temperatures

Conventional Conditions	Phases Identified by XRD and Relative Intensities					
Temp/Time	SiC	C	β-SiAlON	J-SiAlON	mullite	other
1500°C/30min	100	29	---	10	94	---
1600°C/30min	100	25	---	15	89	---
1600°C/48min	100	18	3	16	88	---

Both microwave and conventional synthesis attempts have led to the formation of mullite first, then SiAlON synthesis occurs as temperatures increase. Since SiC was present in the reaction mixture, and diffracts x-rays strongly, it is not possible to say if the disappearance of carbon is due to reduction of mullite to the SiAlON phase or the reduction of SiO_2 to produce more SiC. However, since SiAlON phases are formed in larger amounts than in the clay-carbon system, it may be reasonable to assume that the addition of SiC as a reactant is important to the reaction.

Currently it is difficult to directly compare the clay-carbon system to the BSSCN system from the standpoint of Al_2O_3 and SiO_2 sources. Both syntheses were attempted by different reactant routes. However, the mullite phase is initially present in both systems and SiO_2 phases are present in the clay-carbon reaction. Additionally, the initial presence of SiC in the clay-carbon system indicates the presence of additional SiO_2. The presence of mullite in both systems might indicate that the source of Al_2O_3 and SiO_2 is less important, since no SiAlON formation occurs until phases of Al_2O_3 and SiO_2 are first formed.

β-SiAlON was not the major phase in any reaction attempts using either system. This may be due to the complexity of the carbothermal reduction process [13, 16]. There are several factors that influence these reactions. Two of the influencing factors are the flow rate of nitrogen and the size of the pellet used [13, 16]. Work done by Dijen and Metselaar [16] showed that reaction rates could be enhanced by increasing the gas flow rate and reducing the pellet size. Cho and Charles [13] found that too high a flow rate leads to the formation of phases such as AlN. Particle size and pellet green density were also found to have effects on the reaction rate. Hence, more work must be done to optimize the system conditions for β-SiAlON synthesis using this route.

Silicon Nitride-Alumina System

The diphasic reaction of Si_3N_4 and Al_2O_3 to produce β-SiAlON was tried first in this investigation and remains the most successful. The reason for investigating the clay-carbon and the BSSCN systems was the fact that the reactant phases should readily couple with the applied microwave field. Clays (water-containing), graphitic carbon (electron conduction) and SiC (semiconducting nature) have absorption properties that are attractive for microwave processing. The Si_3N_4-Al_2O_3 reactants have low dielectric constants and neither is electrically conducting.

Tables 5 and 6 below report the XRD results for this system. The information in Tables 5 and 6 can be interpreted analogously with the other tables in this work, with respect to the relative intensities of the identified phases.

Table 5. XRD Results of the Microwave Reaction of Si_3N_4-Al_2O_3 at Various Times and Temperatures

Microwave Conditions	Phases Identified by XRD and Relative Intensities				
Temp/Time	α-Si_3N_4	β-Si_3N_4	Al_2O_3	β-SiAlON	x-SiAlON
1300°C/15min	65	7	100	---	---
1300°C/45min	60	7	100	---	---
1400°C/15min	50	100	55	---	---
1400°C/30min	58	6	100	---	---
1400°C/45min	64	7	100	---	---
1400°C/1hr	55	6	100	---	---
1500°C/15min	2	---	1	100	12

Microwave synthesis of β-SiAlON was nearly complete after 15 minutes at 1500°C (Table 5) Conventional synthesis gave the product phase only at higher temperatures. However, comparing the conventional methods, more β-SiAlON was produced from the Si_3N_4-Al_2O_3 system than from any other reaction route. This is likely due to the structural similarity between β-Si_3N_4 and β-SiAlON. β-SiAlON is formed when Al-O components replace Si-N in the β-Si_3N_4 structure to obey the general formula $Si_{6-z}Al_zO_zN_{8-z}$ [9, 11]. Hence it may be energetically favorable to make β-SiAlON from the Si_3N_4-Al_2O_3 route as opposed to the carbothermal reduction or other routes.

The microstructure of β-SiAlON from the Si_3N_4-Al_2O_3 route shows differences for microwave and conventional synthesis. Figures 1 and 2 are SEM micrographs of the microwave reaction after 15 minutes at 1500°C and of the conventional reaction at 1648°C for 48 minutes. The submicron-sized grains of the microwave-reacted sample appear uniform. The conventionally reacted sample appears multiphasic, which agrees with the XRD results. The presence of long column-like structures in the center of the sample is evident. These structures have sizes in the 30-40μm range. The outer edges of the conventional sample also show multiphasic regions but the sizes are in the 10-20μm range. EDAX analysis was inconclusive in the identification of components.

Table 6. XRD Results of the Conventional Reaction of Si_3N_4-Al_2O_3 at Various Times and Temperatures

Conventional Conditions	Phases Identified by XRD and Relative Intensities				
Temp/Time	α-Si_3N_4	β-Si_3N_4	Al_2O_3	β-SiAlON	J-SiAlON
1500°C/15min	54	6	100	---	---
1500°C/30min	58	7	100	---	---
1600°C/18min	57	---	100	39	1
1600°C/30min	59	---	100	77	8
1648°C/48min	54	---	100	84	12

Figure 1. SEM Micrograph of β-SiAlON Microwave Processed at 1500°C for 15 Minutes

Figure 2. SEM Micrograph of β-SiAlON Conventionally Processed at 1600°C for 48 Minutes

CONCLUSIONS

It was difficult to make any SiAlON from the clay-carbon reaction or the BSSCN system. From the XRD data it appears that higher temperatures and/or longer times are needed to completely synthesize SiAlON by microwave or conventional methods. This is rationalized by the fact that SiC, a reaction intermediate, still remains in large quantities before β-SiAlON is formed. Reaction parameters must also be controlled more precisely with respect to pellet size, nitrogen flow rates and reactant quality, to make the β-SiAlON phase.

β-SiAlON was microwave synthesized at 1500°C after 15 minutes at temperature. Conventional β-SiAlON reaction occurs at higher temperatures and longer times. SEM analysis shows submicron-sized uniform grains for the microwave-processed sample. The conventionally processed sample has a diphasic microstructure with one of the phases having considerably larger grain size. The mechanism for this synthesis likely involves β-SiAlON being formed by Al-O substitution for Si-N into the β-Si_3N_4 structure. Further work is needed to synthesize β-SiAlON with boehmite or clays and Si_3N_4 to provide initial microwave absorption for the system and a structure that is energetically favorable for β-SiAlON production.

ACKNOWLEDGEMENTS

We thank Myles Brostrom (XRD), Chris Goodbrake (SEM) and Brian Lynch (XRD) for their technical support in this project.

REFERENCES

(1) Sutton, W. H., "Microwave Processing of Ceramic Materials", *Ceram. Bull.*, 68[2], pp 376-86 (1989)
(2) Ahmad, I. and Clark, D. E., "Effect of Microwave Heating on Solid State Reactions of Ceramics", **Microwaves: Theory and Application in Materials Processing**, *Ceram. Trans.*, 21, pp 605-12 (1991)
(3) Piluso, P., Lequeux, N. and Boch, P., *Ann Chim. Fr.,* 18, pp 353-60 (1993)
(4) Roy, R., Yang, L. J. and Kormarneni, S., "Controlled Microwave Heating and Melting of Gels", *Ceram. Bull.*, 63[3], p 459 (1984)
(5) Palaith, D. and Silberglitt, R., "Microwave Joining of Ceramics", *Ceram. Bull.*, 68[9], pp 1601-06 (1989)
(6) Janney, M. A., Kimrey, H. D., " Diffusion-Controlled Processes in Microwave-Fired Oxide Ceramics" Mat. Res. Soc. Symp. Proc., 189, pp 215-27 (1991)
(7) Von Hipple, A. R., **Dielectrics and Waves**, M. I. T. Press., Cambridge, MA (1954)
(8) Neas, E. D. and Collins, M. J., "Microwave Heating", **Introduction to Microwave Sample Preparation: Theory and Practice**, (H. M. Kingston and L. B. Jassie eds), Washington DC, pp 7-14, (1988)
(9) Ekstrom, T. and Nygren, M., "SiAlON Ceramics", *J. Am. Cer. Soc.*, 75[2], 259-76 (1992)
(10) Morrison, F. C. R., Maher, P. P. and Hendry, A., *Br. Ceram. Trans. J.*, 88[5], pp 157-161 (1989)
(11) Jack, K. H., "SiAlONs and Related Nitrogen Ceramics", *J. Mat. Sci.*, 11, pp 1135-58 (1976)
(12) Lee, J. G. and Cutler, I. B., *Ceram. Bull.*, 58[9], pp 869-71 (1979)
(13) Cho, Y. W. and Charles, J. A., "Synthesis of Nitrogen Ceramic Powders by Carbothermal Reduction and Nitridation. Part 2: Silicon Aluminum Oxynitride (SiAlON)", *Materials Sci. Tech.*, 7, pp 399-406 (1991)
(14) Higgins, I. and Hendry, A., *Br. Ceram. Trans. J.*, 85, pp 161-66 (1986)
(15) Sugahara, Y., Kuroda, K. and Kato, C., "The Carbothermal Reduction Process of a Montmorillonite-Polyacrylonitrile Intercalation Compound", *J. Mat. Sci.*, 23, pp 3572-77 (1988)
(16) Dijen, F. K.and Metselaar, R., "Reaction-Rate-Limiting Steps in Carbothermal Reduction Processes", *J. Am. Cer. Soc.*, 68[1], pp 16-19 (1985)
(17) Mathis, M. D., Dewan, H. D., Agrawal, D. K., Roy, R. and Plovnick, R. H., "Microwave Processing of Ni-Al_2O_3 Composites", **Microwaves: Theory and Application in Materials Processing II**, *Ceram. Trans.*, 36, pp 431-438 (1993)

MICROWAVE INDUCED COMBUSTION SYNTHESIS OF CERAMIC AND CERAMIC-METAL COMPOSITES

T. Yiin and M. Barmatz, Jet Propulsion Laboratory, California Institute of Technology, Pasadena, CA 91109; and H. Feng and J. Moore, Colorado School of Mines, Golden, CO 80401.

ABSTRACT

Microwaves were used to induce self-propagating high temperature synthesis (SHS) in the reaction $3TiO_2+3C+(4+x)Al \rightarrow 3TiC+2Al_2O_3+xAl$. The SHS process was studied for x = 0 and 4, using slow and fast heating rates and with and without the application of uniaxial pressure. Less than 50 watts was required to internally ignite the samples using a TE_{102} microwave cavity mode. Uniaxial stresses in the range of 200 - 1400 psi were continuously applied along the vertical axis of some of the processed samples as the SHS reactions were initiated. SEM photomicrographs clearly indicate differences in the formation of whiskers and other microstructural features in the products obtained from microwave and conventional hot wire techniques. The x = 0 samples processed with fast and slow microwave heating rates reached higher densities than the conventionally processed samples.

INTRODUCTION

In recent years, there has been a renewed interest in combustion synthesis reactions, commonly called self-propagating high-temperature synthesis (SHS).[1] These SHS reactions are an attractive, energy efficient approach to the synthesis of high temperature composite materials and metastable phases.[1-3] Generally, the SHS process is ignited at the surface of the material by thermal radiation. This thermal energy can come from a laser beam[4] or a heating coil situated close to the sample surface.[5] Alternatively, the entire sample can be heated to the ignition temperature in an isothermal furnace.[6] In these types of combustion synthesis methods, a temperature gradient is produced in the material with the surface always hotter than the center.

A completely different approach for igniting the SHS process occurs when microwaves are used as the heating source. Microwaves will volumetrically heat non-conducting materials leading to an inverted temperature gradient where the center of the material is hotter than the surface. In this approach, a combustion wavefront propagates radially outward from the center of the sample. This process can produce a completely different product morphology and lead to a more complete conversion of reactants.[7]

In this study, conventional and microwave induced combustion synthesis of ceramic and ceramic-metal composites were investigated. The model reaction system used in this research was $3TiO_2+3C+(4+x)Al \rightarrow 3TiC+2Al_2O_3+xAl$, where x = 0 and x = 4. For x = 0, the combustion synthesis reaction product, TiC-Al_2O_3, does not contain any excess liquid aluminum. However, for the case x = 4, an excess amount of liquid aluminum created by the combustion synthesis reaction will infiltrate into the porous SHS ceramic composite leading to a denser product.[2] The incorporation of a ductile metal into a brittle ceramic composite matrix has considerable potential for substantial improvement in fracture toughness for both low and high temperature applications.[3]

EXPERIMENTAL PROCEDURE

Reactant powders for this study were supplied by Alfa, Inc.*, Cerac, Inc.** and Fisher Scientific, Inc.*** These powders were used as ingredients for the reaction mixture without further particle separation, purification or other processing. The reactant powders were proportionally weighed and thoroughly mixed using porcelain ball milling. The reactant mix was then uniaxially cold pressed to ~ 50% green density and dried in an oven at 150°C for 30 minutes prior to ignition. A magnetron source was used to excite a rectangular waveguide cavity in the TE_{102} mode at 2.45 GHz as shown in Fig. 1. This mode has electric field maxima along the lines defined by $z/L_z = 0.25$ and 0.75 and $x/L_x = 0.5$. There is no field dependence along the y axis. A cold-pressed cylindrical sample, 0.13 cm in diameter and 0.13 cm high, was placed on a quartz holder for one atmosphere isotropic ignition or inside a quartz tube with a piston on top to provide a vertical

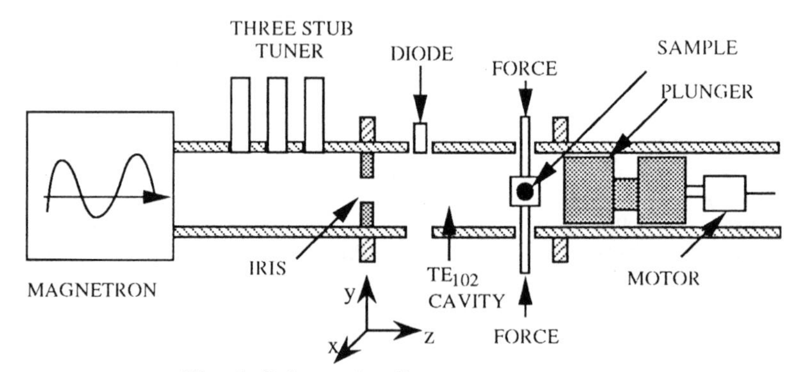

Fig. 1. Schematic of magnetron system.

* Alfa, Inc., Ward Hill, MA.

** Cerac, Inc., Milwaukee, WI.

*** Fisher Scientific, Inc., Pittsburgh, PA.

force for uniaxial pressure ignition. Samples were placed approximately half way down the y axis ($y/L_y = 0.5$) along one of the maximum E field lines ($z/L_z = 0.75$ and $x/L_x = 0.5$).

The surface temperature of the sample was measured using a non-contact Everest Intercience IR thermometer**** (range -30 - 1100°C) and/or an Accufiber non-contact pyrometer***** (range 500 - 3000°C). Densities of conventional and microwave processed samples were measured using Archimedi's principle. The microstructure of the synthesized product was then examined using a scanning electron microscope (SEM) interfaced with an energy-dispersive x-ray spectroscopy facility (EDS).

RESULTS AND DISCUSSION

Cold-pressed green samples, weighing ≈ 3 grams with 50% theoretical density, were studied using both microwave and conventional processing methods. The microwave power was controlled to ignite the samples with x = 0 using both fast and slow heating profiles. The time-temperature profiles for the fast and slow cases are shown in Fig. 2(a) and Fig. 2(b). The sample temperature was measured using both an Everest Interscience IR thermometer and an Accufiber non-contact pyrometer from the side holes of the cavity. Data from these two instruments were linked together to obtain a smooth curve. A higher starting power of ~ 75 watts was required to ignite the sample for the fast heating rate case, Fig. 2(a), as compared to ~ 50 watts starting power needed for the slow heating rate case, Fig. 2(b). The ignition temperature, as measured at the side of the sample surface, is lower (T_i ~ 300°C) for fast heating compared to slower heating (T_i ~ 650°C). The

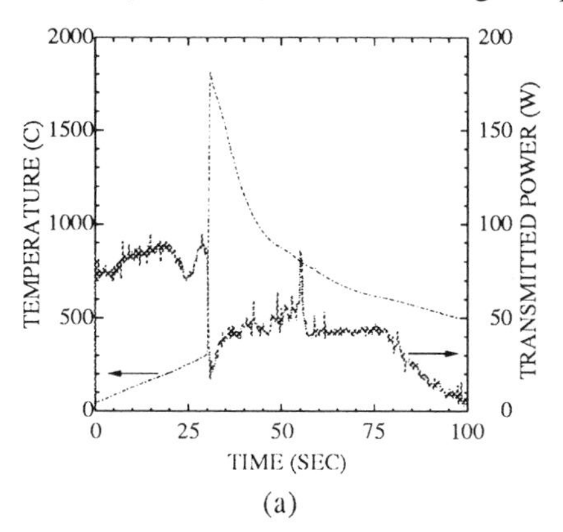

(a)

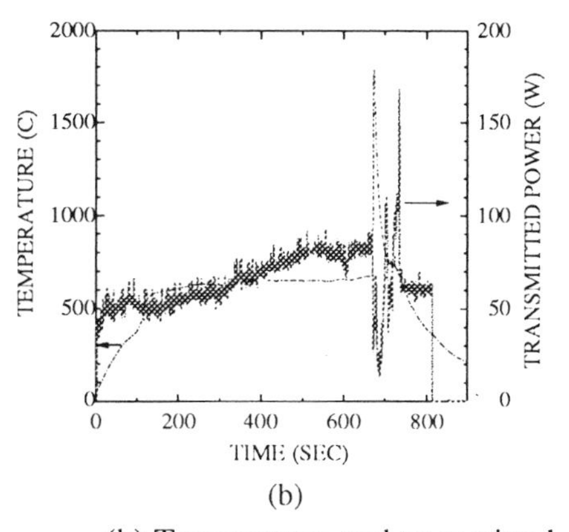

(b)

Fig. 2: (a) Temperature and transmitted power for fast heating rate (x = 0).

(b) Temperature and transmitted power for slow heating rate (x = 0).

**** Everest Intercience, Inc., Fulletron, CA.

***** Luxtron Co., Beaverton, OR.

ignition temperature, T_i, for x = 0 samples measured by the conventional technique was ~ 1200°C.[3] Thus, very high thermal gradients were produced during the microwave processing.

Figures 3(a) and 3(b) show photographs of a sample cross section cut through the center of the ignited product. The fast heating case was ignited near the center and produced equally distributed porosity except for a crack line from the center to the surface. This crack line is probably due to the high thermal gradient between the center and surface of the sample when it was ignited. The slower heating case, Fig. 3(b), has a milky outside surface and two different circular layers separated by a crack rim. These two layers are probably due to the initially melted aluminum (T_m = 660°C) being pushed toward the sample surface leading to a non-uniform distribution of the powders. When this sample was ignited, two

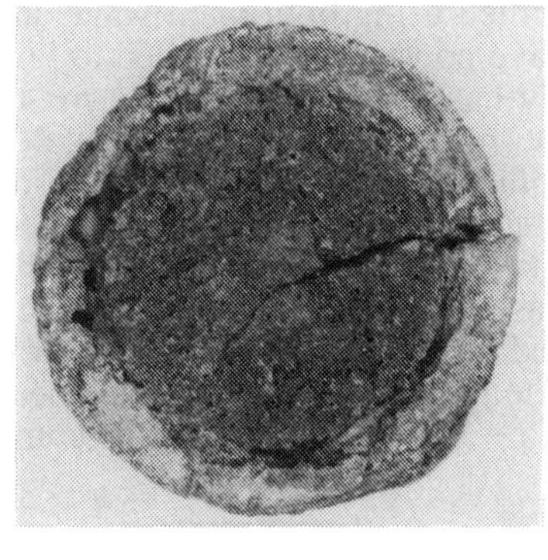

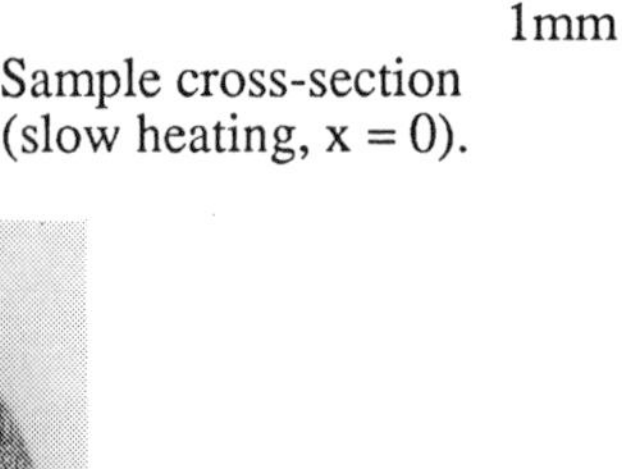

Fig. 3. (a) Sample cross-section (fast heating, x = 0).

(b) Sample cross-section (slow heating, x = 0).

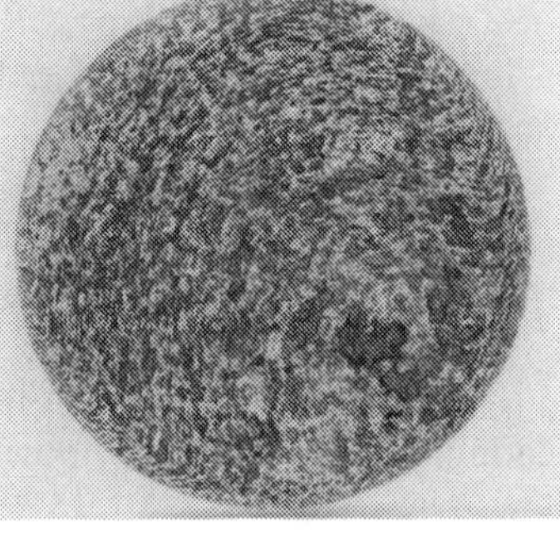

Fig. 4. Sample cross-section (fast heating, x = 4).

different phases were formed and a crack rim developed due to the thermal expansion mismatch between the two adjacent circular layers. Also, the aluminum extruded toward the sample surface became a barrier for microwave penetration and thus inhibited the further ignition of powders. These results suggest that a fast microwave heating rate can insure a more complete chemical reaction of the mixed powders leading to a more uniform microstructure. A thorough understanding of the thermal gradient problem is needed to validate this conjecture.

For the x = 4 case, slow microwave ignition was difficult due to arcing initiated by the excess aluminum at the sample surface. On the other hand, ignition using a fast microwave heating rate was successful and a uniform microstructure was observed as shown in Fig. 4. This photomicrograph clearly suggests that ignition was initiated in the lower right quadrant of the sample and the combustion wavefront then propagated radially outward.

A comparison of the densities of the processed samples is shown in Table I. For x = 0 microwave processed samples without using uniaxial stress, the fast heating rate led to a higher density as compared to slow microwave heating and conventional hot-wire techniques. For the x = 4 case, conventional and fast microwave processed SHS samples had the same density. When a uniaxial force of 1000 psi was applied during fast microwave ignition (x = 0, and 4), SHS samples with a density ~ 85% were obtained which are similar to the conventional SHS product. From these experiments, the only difference between microwave and conventional SHS ignition occurs for x = 0. When x = 4, the excess amount of aluminum uniformly infiltrated the pores for both cases.

Table I. Densities of processed samples.

Excess aluminum	Uniaxial force	Microwave ignition (%) theoretical density		Conventional ignition (%) theoretical density
x	psi	fast	slow	hot-wire technique
0	0	75	60	50
4	0	75	—	75
0	1000	85	60	85
4	1000	85	—	85

For the microstructure study, the x = 0 and x= 4 cases were considered and the conventional hot-wire SHS samples were compared to the samples ignited by fast microwave heating. For the x = 0 case, microwave SHS samples that had a higher density also had a more uniform microstructure as compared to the conventional case. The morphology of the microwave samples looked similar and contained the same complete chemical reaction composition ($TiC+Al_2O_3$) as the conventional SHS samples. For all the x = 4 cases, whiskers were observed in SEM photomicrographs for microwave and conventional processed samples.

Figures 5(a) and 5(b) show strikingly different whisker morphology between the two techniques. In the conventional case, Fig. 5(a), the excess amount of liquid Al and gases produced during SHS reaction resulted in the formation of Al_2O_3-Al hollow whiskers with a bulbous head at the top of each whisker. This behavior suggests a vapor-liquid-solid (V-L-S) mechanism of whisker formation.[3] The microwave ignited sample possessed whiskers with completely different features as seen in Fig. 5(b). These whiskers are solid, much smaller in size, and do not have the bulbous head feature. One possible explanation is that microwaves ignite the SHS sample internally and the process quickly reaches a much higher combustion temperature. The aluminum metal and even alumina may vaporize due to this higher temperature combustion synthesis process. Such a vapor-solid (V-S) reaction may result in the formation of solid Al_2O_3 whiskers, possibly related to the crystal orientation preference of Al_2O_3. In another microwave study,[8] whiskers were also formed using different carbon sources. Further research is needed to accurately define the mechanism responsible for these different whisker formations. The formation of whiskers in ceramic-metal composites using microwave SHS could lead to improve mechanical properties in the combustion synthesis products.

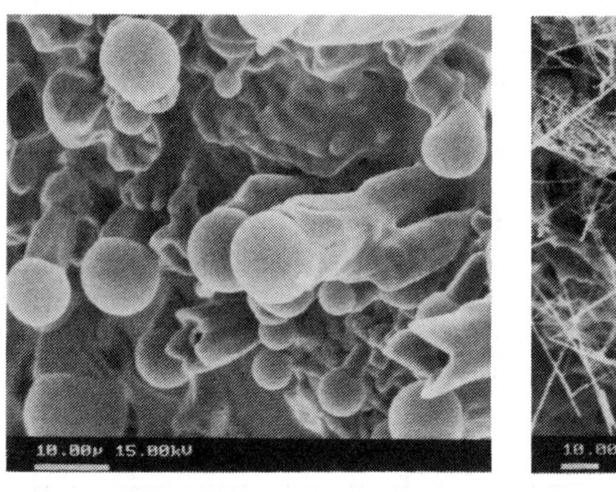

(a) (b)

Fig. 5. SEM micrograph of x = 4 (a) conventional uniaxial pressed SHS sample, (b) microwave uniaxial pressed SHS sample.

CONCLUSION

This study has demonstrated the ability to ignite the combustion synthesis process $3TiO_2+3C+(4+x)Al \rightarrow 3TiC+2Al_2O_3+xAl$ with microwave energy using slow and fast heating rates. All the slow heating rate cases produced an incomplete reaction. A uniform microstructure with 75% theoretical density was achieved for x = 0 with a fast heating rate. For x = 4, solid Al_2O_3 whiskers were observed in the microwave SHS product as compared to the hollow whiskers with bulbous heads observed in the conventional SHS product. The application of a uniaxial force during the SHS process produced 85% theoretically dense homogeneous products for both the microwave (fast heating) and conventional techniques. This investigation suggests that further studies should emphasize SHS reactions using uniaxial or isostatic pressure.

ACKNOWLEDGMENTS

The authors wish to acknowledge Mr. R. Torres from Colorado School of Mines for the powder preparation and Mr. O. Iny and Mr. R. Zanteson from the Jet Propulsion Laboratory for technical support. The research described in this article was carried out at the Jet Propulsion Laboratory, California Institute of Technology, under contract with the National Aeronautics and Space Administration.

REFERENCES

1. Z.A. Munir, and U. Anselmi-Tamburini, "Self-Propagating Exothermic Reactions: The Synthesis of High-Temperature Materials by Combustion," Materials Science Reports, pp 277-365, May (1989).
2. J.J. Moore,"Combustion Synthesis of Advanced Composite Materials," 31th Aerospace Sciences Meeting and Exhibit, AIAA Paper #93-0830 (1993).
3. H. Feng, Ph.D. Thesis, "Combustion Synthesis of Advanced Ceramic and Ceramic-Metal Composites," Colorado School of Mines, August (1994).
4. P. Dimintrou, H. Halvacek, S.M. Valone, P.G. Behrens, G.P. Hansen and J.L. Margrave, "Laser-Induced Ignition in Solid-State Combustion," AIchE J., **35**, pp 1085-1096 (1989).
5. A.G. Merzhanov, V.M. Shkiro and I.P. Borovinskaya, U.S. Pat. No. 37226643 (1973).
6. L.L. Wang, Z.A. Munir and J.B. Holt, "The Combustion Synthesis of Copper Aluminides," Metall. Trans. **21B**, pp 567-577(1990).
7. I. Ahmad, R. Dalton, and D. Clark, "Unique Application of Microwave Energy to the Processing of Ceramic Materials," J. Microwave Power, **26**, pp 128 - 138 (1991).
8. M. Willert-Porada, B. Fisher and T. Gerdes, "Application of Microwaves to Combustion Processing of Al_2O_3-TiC Ceramics," Ceram. Trans. **36**, pp 365-375, (1993).

VIBRATIONAL SPECTROSCOPY OF SILICA SOL-GEL COMPONENTS DURING MICROWAVE IRRADIATION

M.D.Roth, R.R. Di Fiore, D.C. Folz and D.E. Clark.
University of Florida
Department of Materials Science & Engineering
Gainesville, FL

C.Wehlburg, J. Szczepanski and M.T. Vala.
University of Florida
Dept. of Chemistry.
Gainesville, FL

ABSTRACT

In order to study the effects of microwaves on chemical reactions equipment was designed to acquire in-situ vibrational spectra of sol-gel components as they are irradiated with microwaves. Fourier Transform Infrared (FTIR) and Raman spectroscopy were used. A low temperature (10K) FTIR cell was used to trap samples in an argon matrix at 10^{-7} Torr. For the liquid samples no differences were seen in spectra of irradiated and nonirradiated samples, but the argon matrix isolation technique showed dramatic differences.

INTRODUCTION

Microwave processing of ceramic materials is quickly becoming popular due to its efficiency and the enhanced or different properties of the final products [1]. It has also been noticed that when silica sol is irradiated with microwaves the sol has a decreased gelation time and the gel has smaller pore radii [2]. Because materials behave differently in a microwave field the goal of this research project is to find any unusual or enhanced chemical reactions that take place in a microwave field which do not occur with conventional heating methods. To accomplish this, methods have been devised for obtaining FTIR and Raman spectra from samples as they are irradiated with microwaves. Emphasis is placed on the fact that these spectra are taken during, not after, application of the microwave field. It is hoped that from these spectra it can be determined which types of chemical reactions are enhanced by microwaves and that this information can be utilized for industrial processes.

EXPERIMENTAL PARAMETERS

A. ***Microwave***: The microwave used for these experiments is intended for creating radical species in the gas phase and can heat samples in the liquid phase. It is a Raytheon PGM-10X2 which operates at 2.45 GHz (the frequency of most industrial or home ovens) with a coaxial waveguide. It can be operated at a full range of powers with an 85 W maximum. In this study we investigated power settings of 0, 35, and 75% maximum power, corresponding to 0, 30, and 64 W, respectively.

B. ***FTIR***: The FTIR is a MIDAC model M2000. Spectra were taken at 1 cm^{-1} resolution. A polytetrafluoroethylene (PTFE or Teflon®) sample holder with AgCl windows was designed to fit inside the microwave cavity and hold two AgCl disks together (Figure 1). The sample is held between the two disks by its own capillary force.

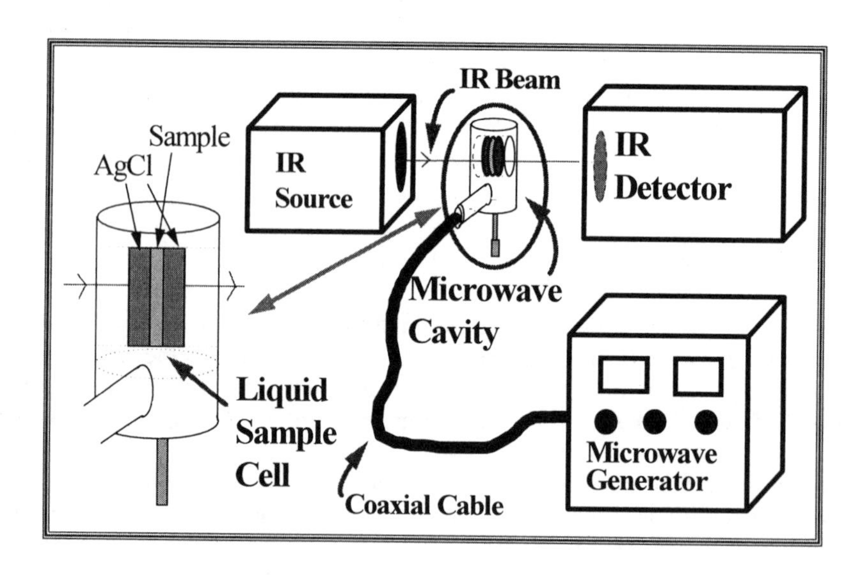

Figure 1. The FTIR microwave diagram.

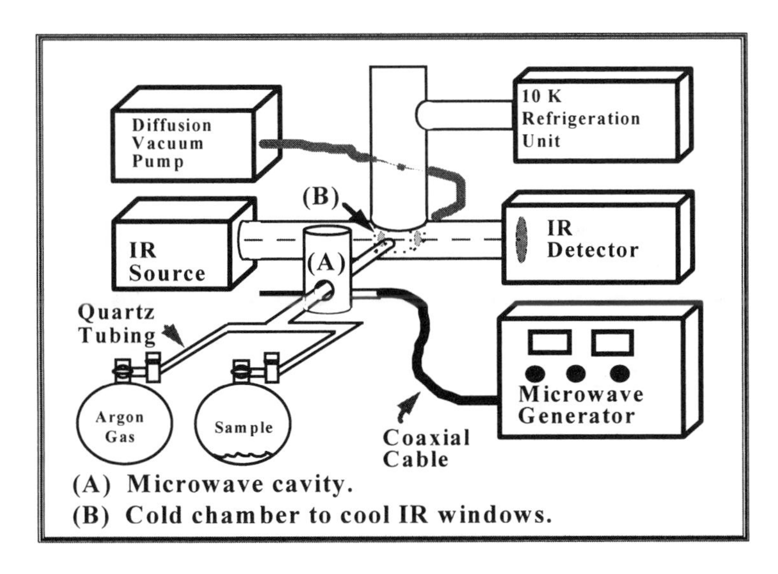

Figure 2. Matrix isolation diagram.

® du Pont, E. I. de Nemours & Co., Inc.

C. ***Matrix isolation FTIR technique*:** This is a novel method that can trap species in an argon matrix on an FTIR window at 10K or less. The molecules are isolated from each other by the argon which makes the spectra peaks much sharper. This sample is at roughly 10^{-7} Torr when it passes through the microwave field then onto the window and isolated in the argon matrix (Figure 2). Due to the low pressure any reactive species produced in the microwave field should not react together before being trapped and isolated in the argon matrix. If any reactive species (ions or radicals) are produced in the liquid phase experiments (FTIR and Raman) their lifetimes would be too short for them to be detected by FTIR or Raman spectroscopy, while matrix isolation stands a better chance of trapping these species. It should be pointed out that the matrix isolation technique is not truly in-situ, spectra are acquired after microwave irradiation.

D. ***Raman System.*** The Raman system uses a SpectraPhysics argon laser and a PC controlled Spex monochromator. The laser line was set at 488 nm with a 300 mW output and the resolution was set at 1 cm^{-1}. A motor spins the sample to prevent laser burn. See the diagram below (Figure 3) for the schematic.

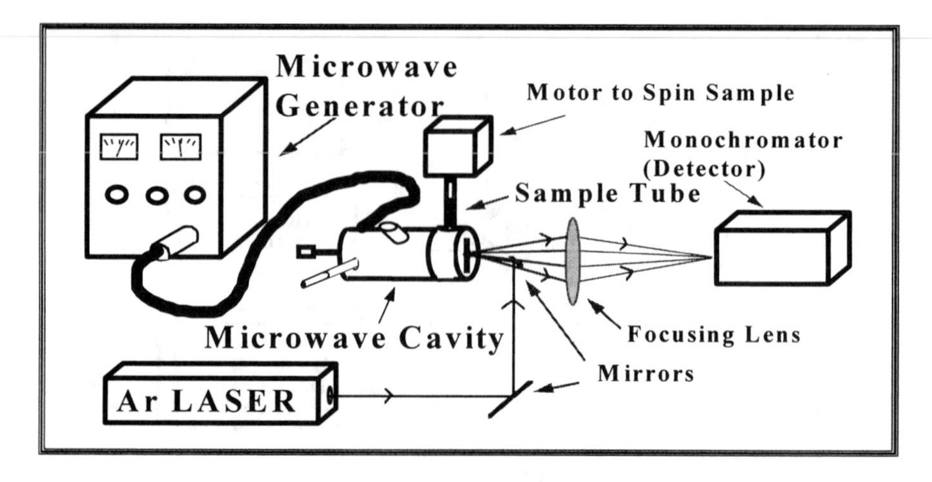

Figure 3. Raman-microwave diagram.

SOL-GEL PROCESSING

Silica sol-gel is made by reacting silicon alkoxides with water. Advantages of sol gel processing are molecular level mixing, decreased densification temperatures, porosity control, ease of dopant addition, and creation of products with many unusual properties not otherwise available [3,4].

Silica sol-gel is chosen because of the large amount of data on the system and because it has several phase changes throughout the entire process. In this study tetraethyl orthosilicate (TEOS) is chosen because of its relatively lower toxicity (when compared to tetramethyl orthosilicate, TMOS). Six stages are required to produce a solid silica sample using the sol-gel method: 1) hydrolysis, 2) condensation, 3) gelation, 4) aging, 5) drying and 6) densification. The initial reaction is the hydrolysis (1) of the TEOS:

$$Si(OC_2H_5)_4 + 4H_2O \rightarrow Si(OH)_4 + 4C_2H_5OH \quad (1)$$

And secondly is the condensation (2):

$$2Si(OH)_4 \rightarrow (OH)_3SiOSi(OH)_3 + H_2O \quad (2)$$

Further condensation leads to a polymer network and gelation (3). These reactions are by no means independent, all occur simultaneously, and the substitutions are not always complete (i.e., there is usually a small amount of ethyl groups remaining in the gel network).

The last steps, aging (4), drying (5), and densification (6) are not considered in this study, but there are some excellent reviews of the entire sol-gel process [3,4]. We are considering methods to study these last steps by spectroscopy in our future work.

The sol synthesis is outlined by Chu and Clark [5]: The molar ratio of water to TEOS is 20 and the amounts (in ml) are TEOS:44.5, ethanol:44.5, water:72.0. First half of the ethanol and the TEOS are mixed, then the remaining ethanol is added to the water and the pH adjusted to 2.0 with HCl. In the conventional process the two solutions are then mixed and heated to 60°C. In these initial experiments we have investigated only the sol-gel component TEOS and not the actual sol. This will be done in the near future.

RESULTS

The FTIR and Raman spectra (Figures 4 and 5) obtained from the liquid samples do not show any differences between the samples exposed and unexposed to microwaves. The FTIR spectra (Figure 4) display large background signals from CO_2 and water in the air, even though the system was purged with nitrogen gas. We believe this is mainly caused by the sample leaking out of the cell, yet leaving enough to be detected, and with such a weak signal the CO_2 and water cause large interferences. The negative peaks in the spectrum with the microwave set at 75% power are due to smaller amounts of CO_2 and water present during the sample scan than during background scan. The CO_2 and water peaks are easy to recognize (peaks between 1400-2000 cm^{-1} and 3500-4000 cm^{-1} are from water and the two at about 2350 cm^{-1} are from CO_2). It is obvious that the microwave field does not affect the TEOS bands under these experimental parameters.

In the Raman spectra (Figure 5), the peaks are smaller in the spectrum without microwave irradiation. This is due to the difficulty in aligning the sample. However, the relative peak heights and positions are equal, indicating no reaction.

In the matrix isolation spectra there are obviously large differences in the spectra with and without microwaves (Figure 6). Three power settings were used, 0% (off), 35%, and 75%. Except for the microwave field, all experimental parameters (deposition rate of the sample, temperature of the FTIR cell, etc.) were the same for all spectra.

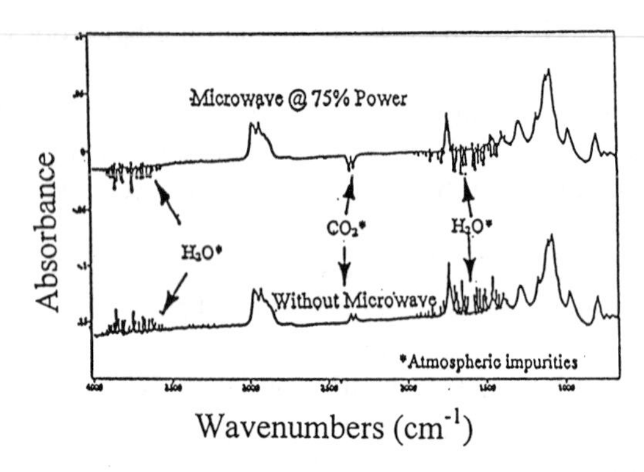

Figure 4. FTIR spectra of TEOS in liquid cell at room temp.

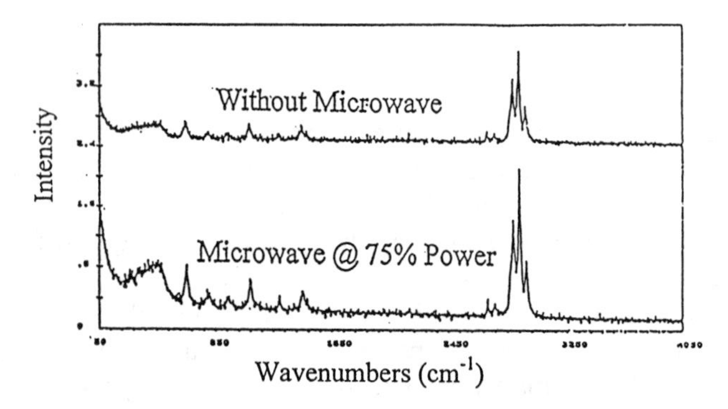

Figure 5. Raman spectra of TEOS in capillary tube at room temp.

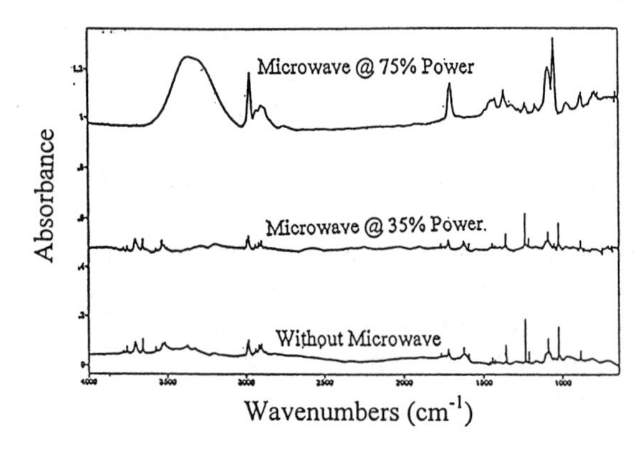

Figure 6. FTIR matrix isolation spectra of TEOS at 10 K and 10^{-7} Torr.

DISCUSSION

Microwaves, in general, heat liquids by making the molecules with a dipole rotate with the oscillating electric field. If this is the only mechanism and there are no chemical reactions or phase changes taking place, then the only effects on vibrational spectroscopy might be thermal broadening of the peaks. TEOS is not a polar molecule and it is not too surprising that no changes had taken place in the liquid phase when it was irradiated with microwaves.

In the gaseous state, however, the molecule is not be able to relax by transferring energy to neighbor molecules as rapidly as in the liquid phase. It might therefore rearrange or react with other molecules to release the energy absorbed from the microwave field. Also, when the microwave is set at 35% power, the spectrum is nearly the same as that without microwaves. This seems to indicate a minimum microwave field strength at which the reaction takes place. The new species being formed have yet to be identified. In the spectra with the microwave at 75% power there is an O-H peak formed (3100-3600 cm^{-1}) [6]. Also the other peaks seem to be rather broad for an isolation technique (which usually shows very sharp peaks). This indicates that the molecules being formed are rather large in comparison to TEOS. This is somewhat odd because of the low pressure of the system, i.e., molecules are unlikely to react together as will be shown:

The average time of exposure to the microwaves, t_{mw}, in the matrix isolation technique, can be estimated by the equation [7]:

$$t_{mw} = d_{mw}/v = d_{mw}(\pi m/8k_BT)^{1/2}=1x10^{-4}sec \qquad 1$$

where d is the width of the microwave cavity (0.020m), v is the velocity of the molecule, k_B is Boltzman's constant ($1.38x10^{-23}$ J/K), m is the mass of the molecule ($3.45x10^{-25}$ kg) and T is temperature (298 K, assumed not effected by microwaves). Equation (1) assumes that the molecules pass through the microwave cavity only once and also that the molecular weight is constant, i.e. no reaction is taking place, which is obviously not be the case.

The average mean free path, λ, of the molecule can be estimated by [7]:

$$\lambda= (2^{1/2}\pi D^2N)^{-1}=80 \text{ m} \qquad 2$$

where D is the molecule's diameter (9.3 Å [8]), and N is the number of molcules per volume and is calculated by a variation of the ideal gas equation:

$$N=P/k_BT=3.2x10^{15} \text{ molecules/m}^3 \qquad 3$$

The values are $P=1x10^{-7}$ Torr, T=298 K. With such a large mean free path it is unlikely that molecules will react together before being frozen into the argon matrix. However it is possible that they could be reacting in the matrix or on the walls of the quartz tubing.

Some possible reasons why these reactions are not seen in the liquid phase are: 1) in the liquid phase there are mechanisms by which the molecules can relax nondestructively after absorbing energy from the microwave field, 2) the energy available per molecule in the gas phase is so much higher, and 3) in the liquid phase the concentration of the reaction products may be too small to detect.

SUMMARY

Equipment was designed to acquire FTIR and Raman spectra of liquid samples in-situ with microwave irradiation. Thus far no differences have been seen in spectra with and without microwave irradiation. However, matrix isolation FTIR spectra do show differences when the sample vapor is passed through a microwave field and when it is not. The reactions taking place remain to be determined.

ACKNOWLEDGMENTS

The authors would like to thank the National Science Foundation and the Electric Power Research Institute for their generous funding.

REFERENCES

1. E.J. Pope, *Am. Cer. Soc. Bull.*, Vol. 70, no. 11, p. 1777-8 (1991).
2. R.R. Di Fiore and D.E. Clark, presented at the Annual Conference on Composites and Advanced Ceramic Materials, Cocoa Beach, FL (1995), to be published in, **Ceramic Engineering and Science Proceedings**.
3. L.L. Hench, J.K. West, *Chem. Rev.*, Vol. 90, p. 33-72 (1990).
4. C.W. Turner, *Ceramic Bulletin*, Vol. 70, no. 1, p. 1487-90 (1991).
5. P-YChu and D.E. Clark, *Spectroscopy Letters*, Vol. 25, no. 2, p. 201-20 (1992).
6. R.M. Silverstein, G.C. Bassler and T.C. Morril, **Spectroscopic Identification of Organic Compounds,** John Wiley & Sons, Inc., p. 109 (1991).
7. J.H.Noggle, **Physical Chemistry, 2nd Ed.** p. 40 & 463 (1989).
8. P. Ho and Melius, *J. Physical Chemistry*, Vol. 99, no. 19, p. 2166-76 (1995).

MICROWAVE SYNTHESIS OF ALUMINUM TITANATE IN AIR AND NITROGEN

M.D. Mathis, D.K. Agrawal and R. Roy
Materials Research Laboratory
The Pennsylvania State University
University Park, PA

R.H. Plovnick
3M Ceramic Technology Center
St. Paul, MN

R.M. Hutcheon
Chalk River Laboratories
Chalk River, Ontario, Canada

ABSTRACT

The effect of microwave heating on solid state reactions is a key issue in materials processing. The microwave-assisted solid state reaction of alumina and anatase to form aluminum titanate was studied. The reaction was carried out in both air and nitrogen atmospheres. It was found that aluminum titanate can be synthesized at 1150°C in air and at 1050°C in nitrogen. Dielectric studies show a three-fold increase in the dielectric constant when processed in nitrogen as opposed to air, indicating the evolution of defects. Comparison of the dielectric measurement data shows the onset of the solid state reaction is enhanced when nitrogen processing is employed. XRD analysis of the resulting materials shows that anatase is converted to rutile before the Al_2TiO_5 reaction occurs. Additionally, Al_2TiO_5 was synthesized from an alumina-defect rutile route in nitrogen. The Al_2TiO_5 was nucleated at 600°C and was about 70% reacted by 900°C.

INTRODUCTION

Microwave processing has produced enhancements during sintering [1-3] when compared with conventional sintering. It has been reported in some cases that the activation energies for sintering are lowered by microwave processing [2] resulting in faster material sintering rates. The reasons for the apparent reduction of activation energies are the subject of debate. Recently microwave synthesis has been attempted on zinc aluminate spinel from zinc oxide and alumina [4] and aluminum titanate made from alumina and rutile [5,6]. In each case the reactions in question showed a reduction in the temperatures and times necessary to synthesize each phase.

Unlike conventional sintering or synthesis, microwave synthesis depends largely upon the ability of one or more phases in the reactant mixture to couple with the applied field. This may enhance reaction rates due to the fact that heating occurs throughout the reactant mixture, rather than by heat conduction from the surface to the core of the conventionally processed material. In this work, attempts were made to enhance the microwave absorption of a reactant mixture by introducing defects, in the form of oxygen vacancies, in one of the phases. The incorporation of defects into the crystal structure of a material can increase its dielectric loss [7] by introducing a mechanism for microwave absorption.

In this work Al_2TiO_5 was microwave synthesized from Al_2O_3 and TiO_2 phases in both air and nitrogen environments. In the latter case, the nitrogen gas was not used as a reactant in the system, as in the case of another process described elsewhere [8] in these proceedings, but rather to prevent or slow the oxidation of defect titania during heating and reaction. Al_2TiO_5 is an interesting material because of its low thermal expansion properties [9,10] due its ability to form microcracks.

EXPERIMENTAL

Alumina-titania (anatase) system: Equimolar mixtures of alumina (lot A16SG, Alcoa Inc), with a 0.2μm particle size, and anatase (Cerac, Inc), with a <5μm particle size, were wet-milled for 16 hours. The resulting powder was dried by microwave initially and then conventionally. Pellets, weighing approximately 12 grams, were pressed uniaxially to increase particle contact. The pellets had diameters of 1.13 inches and heights of about 0.3 inches for all microwave and conventional experiments.

Alumina-defect titania (rutile) system: Anatase was fired to 1100°C under forming gas (5% hydrogen-95% argon, Linde Inc.) for 3 hours to give oxygen-defective rutile. After firing, defect rutile was the main phase detected by X-ray diffraction (XRD). The defect powder was dry-milled with alumina overnight. Pellets, having the same dimension described earlier, were uniaxially pressed from the resulting powder mixture. The pellets were processed in a nitrogen (Linde, Inc.) atmosphere for these experiments.

Equipment: The microwave equipment, conventional drying oven and supporting materials used in this work are described elsewhere [11]. Microwave temperature measurements were done by K and C-type thermocouples [11] (Omega, Inc.). The temperatures at the various soak times reported in this work were maintained by the controller [11] cycling the microwave power within soak-tolerance limits (10-15°C of the setpoint temperature). The conventional furnace used for reduction of rutile and high temperature aluminum titanate synthesis was a Rapid Temperature tube furnace (CM Furnaces, Inc.).

Characterization: Phase analysis was done by XRD. The instrument used was a Phillips APD model 3600 using copper K-alpha radiation. Scans were done from 5 to 80 degrees 2-theta. Dielectric measurements were made at microwave frequencies over a temperature range of 25°C-1400°C under nitrogen and air. The heat rate was 20°C per minute.

High temperature X-ray diffraction (HTXRD) analysis was done on the anatase-alumina reaction to determine the temperatures for the anatase-rutile transition as well as to determine the temperatures necessary for conventional aluminum titanate production. The sample was heated on a platinum stage at various temperature intervals. At the end of an interval cycle a diffraction pattern was obtained while the sample remains at the final temperature of that

interval. These experiments were done in air, helium and under vacuum as needed by the specifics and limitations of the instrumentation.

RESULTS AND DISCUSSION

Analysis of the Anatase-Alumina Reaction

HTXRD data showed that aluminum titanate is not synthesized conventionally from anatase, but rather from the rutile phase. Anatase is converted to rutile between 900-1000°C in air. The conversion occurs near 800°C when the experiment is done in inert atmospheres or in a vacuum.

Al_2TiO_5 synthesis was not observed by 1200°C when heated in air (the instrument could not be operated above 1200°C in air). In an inert atmosphere or in a vacuum, the reaction began between 1300-1400°C and was not complete by 1500°C, which is the maximum working temperature achievable by the HTXRD.

Aluminum Titanate Synthesis in Air

Tables 1 and 2 summarize the XRD findings of attempts to synthesize Al_2TiO_5 from α-alumina and anatase by microwave and conventional synthesis respectively. Both tables show the relative intensities for representative peaks of the reactant and product phases.

Table 1. XRD Analysis of the Microwave Synthesis of Al_2TiO_5 in Air

Temp-Time	α-Al_2O_3	TiO_2 (anatase)	TiO_2 (rutile)	Al_2TiO_5
1050°C-30 min	71	---	100	---
1150°C-30 min	42	---	100	(1)
1250°C-30 min	26	---	100	38
1300°C-15 min	19	---	96	100
1300°C-30 min	4	---	36	100
1300°C-45 min	---	---	11	100

Table 2. XRD Analysis of the Conventional Synthesis of Al_2TiO_5 in Air

Temp-Time	α-Al_2O_3	TiO_2 (anatase)	TiO_2 (rutile)	Al_2TiO_5
1300°C-30 min	31	---	100	---
1300°C-45 min	32	---	100	6
1400°C-45 min	8	---	86	100

Table 1 shows that Al_2TiO_5 begins to nucleate at 1150°C using microwave heating. After 15 minutes at 1300°C the microwave reaction is near completion with minor amounts of alumina and rutile phases remaining. Synthesis by conventional methods does not show any product formation until 1300°C. These findings are consistent with work done by Ph. Boch, et al. [5,6], whereby alumina was reacted with rutile-phase titania. In that work aluminum titanate was nucleated near 1210°C by microwave and near 1300°C conventionally.

Aluminum Titanate Synthesis in Nitrogen

Tables 3 and 4 show the XRD results from Al_2TiO_5 synthesis from alumina and anatase in nitrogen by microwave (Table 3) and conventional (Table 4) methods. The relative peak intensities were used to compare the composition data at each temperature and time interval.

Table 3. XRD Analysis of the Microwave Synthesis of Al_2TiO_5 in Nitrogen

Temp-Time	α-Al_2O_3	TiO_2 (anatase)	TiO_2 (rutile)	Al_2TiO_5
950°C-15 min	52	7	100	---
950°C-30 min	46	5	100	---
1050°C-15 min	55	---	100	1
1050°C-30 min	49	---	100	3
1100°C-15 min	57	---	100	23
1100°C-30 min	37	---	100	34

Table 4. XRD Analysis of the Conventional Synthesis of Al_2TiO_5 in Nitrogen

Temp-Time	α-Al_2O_3	TiO_2 (anatase)	TiO_2 (rutile)	Al_2TiO_5
1150°C-30 min	47	3	98	---
1250°C-30 min	41	---	100	---
1300°C-30 min	35	---	100	(1)
1400°C-30 min	18	---	93	100

From the data obtained it appears that in an atmosphere that promotes oxygen defects and reduction of titania, the product phase can be synthesized at lower temperatures. This can further be observed by the dielectric measurements[12,13]done at microwave frequencies in both air and nitrogen atmospheres using the same starting reactants, i.e. alumina and anatase. The dielectric measurements [13] quantify the "real-part" (ε') and the "imaginary or loss part"(ε'') of the dielectric constant at 2460 MHz as temperature increases. The green and final densities of the sample measured in Figure 1 were 1.82 and 2.08g/cm^3 respectively.

From the data shown in Figure 1, the reactants heated in air up to 1400°C show an increase in ε' and ε'' starting at about 250°C. This rise peaks at 600°C and is possibly due to water evolving from the test sample. The increase in ε' and ε'' starting at 800°C may be consistent with the phase change of anatase to rutile, as shown by HTXRD. The sharp rise in ε' and ε'' starting at about 1160°C is thought to be the nucleation of Al_2TiO_5, which corresponds to XRD results obtained on the reaction (Table 1). Hence, in this case, dielectric measurements can be used as a technique to follow the reaction in the microwave region.

Figure 2 is the dielectric measurement data ($\varepsilon'$and ε'') for the alumina-anatase reaction in a nitrogen atmosphere. The green and final densities of the test sample were 1.81 and 2.06g/cm^3 respectively. The data can be interpreted analogously with Figure 1, with respect to phase transitions; however, what is important to note is that the onset of the Al_2TiO_5 reaction is at about 1050°C. This information is consistent with XRD results obtained (Table 3). Conventionally (Table 4), the results are almost analogous with Table 2.

Nitrogen processing lowered the onset of reaction about 110°C as opposed to air processing. Additionally, the increase in ε'' starting at 250°C is greater than in the air measurement case. A value of 1.2 is observed for the nitrogen case, while 0.3 is observed for the air case. In the air measurement case, this peak had dropped by 800°C, after which the peak for the phase transition of anatase to rutile begins. However, in the nitrogen measurement case, the 250°C loss-peak decreased to a minimum by 1000°C and probably includes the phase transition. This may indicate a decrease in the temperature necessary to convert anatase to rutile, which is consistent with HTXRD data. The HTXRD (in inert atmosphere) data showed the phase transition occurred about 100°C lower than in air.

The increase in magnitude likely indicates an increase in electronic conductivity. Conductivity (dielectric loss) continues to increase as temperature rises, indicating an increase in the defect concentration in the nitrogen system.

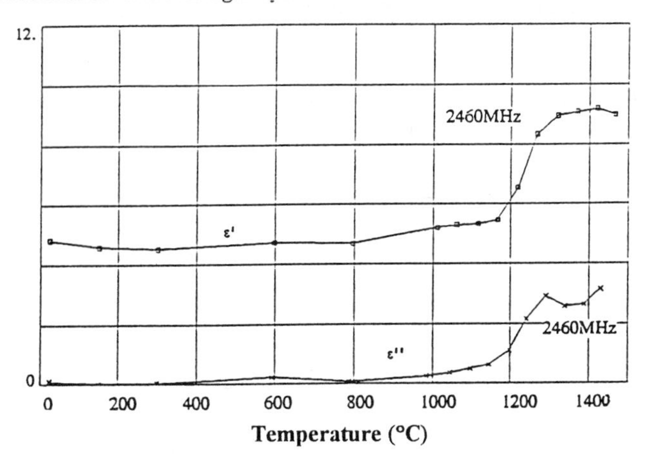

Figure 1. Measurement of ε'(-□-) and ε" (-x-) at 2460 MHz during the Alumina-Anatase Reaction in Air with Conventional Heating

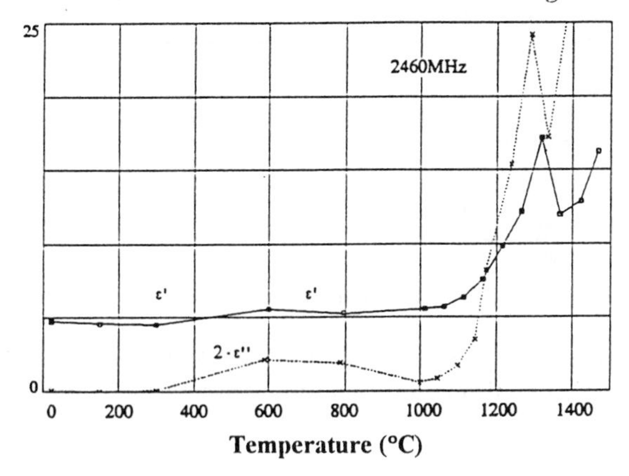

Figure 2. Measurement of ε'(-□-) and ε" (-x-) at 2460 MHz during the Alumina-Anatase Reaction in Nitrogen with Conventional Heating

Table 5, the XRD results of the microwave synthesis of Al_2TiO_5 from alumina and defect rutile, show that Al_2TiO_5 can be nucleated at 600°C. By 750°C the reaction is about 60% completed. At 900°C the reaction may be about 70% completed. This would indicate the possibility of very fast nucleation of the aluminum titanate phase. As the product phase becomes the major phase, the reaction is probably limited by diffusion processes through the product interfacial layer [14-16]. Additionally, the defect rutile-alumina mixture was dry-milled instead of wet-milled prior to processing. Hence the homogeneity of the reaction mixture is questionable. This would lead to pockets of alumina and titania that are not in good intimate contact with the other phase. Currently work is underway to better quantify the reaction amounts and increase particle contact.

Table 5. XRD Analysis of the Microwave Synthesis of Al_2TiO_5 from Al_2O_3 and Defect TiO_2 in Nitrogen

Temp-Time	α-Al_2O_3	TiO_2 (rutile)	Al_2TiO_5
600°C-15 min	44	100	3
750°C-10 min	40	48	100
900°C-15 min	49	43	100
1050°C-15 min	26	21	100

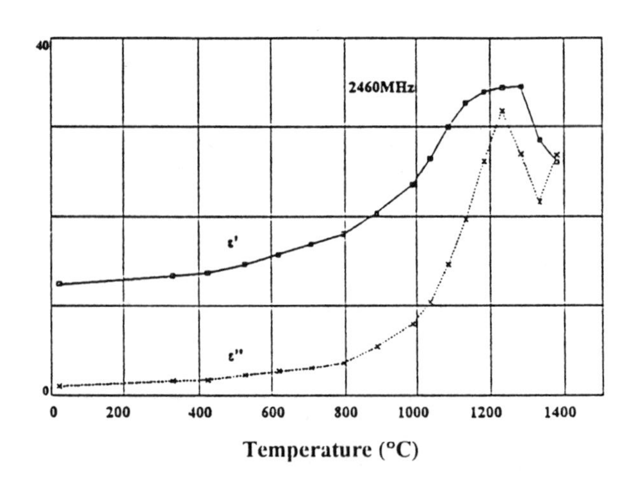

Figure 3. Measurement of ε'(-□-) and ε" (-x-) at 2460 MHz during the Alumina-Defect TiO_2 Reaction in Nitrogen with Conventional Heating

Dielectric measurements (Figure 3) done on the alumina-defect rutile system show overall increases in ε' and ε" above the case of alumina-anatase in nitrogen synthesis. The green and final densities of the test sample were 2.09 and 2.87g/cm^3 respectively. What is interesting to note is the constant rise in conductivity (ε") as the mixture heats. From the dielectric studies it is difficult to pinpoint the nucleation of Al_2TiO_5; however, the area where changes in the slopes of ε' and ε" occur (starting at about 430°C followed by a brief plateau at 600°C) might be the nucleation. This coincides with the XRD data.

In the actual microwave chamber, the reactants can be heated at about 0.78kw (60% of the total power) without the use of secondary coupling. Heating rates have been found to be near 200°C per minute in some experiments after the temperature reaches about 400°C. In the case of the alumina-anatase nitrogen synthesis, SiC rods were needed as secondary couplers and at no point in the reaction did heating rates reach 200°C/ per minute. These facts are likely attributable to the generation of defects and subsequent defect synthesis.

CONCLUSION

Aluminum titanate has been microwave synthesized from α-Al_2O_3 and TiO_2 (anatase and rutile phases) in air and nitrogen. It was found that before the product phase was formed the anatase conversion to rutile occurred. During this work no anatase phase was found when Al_2TiO_5 was detected by XRD analysis.

It was also found that Al_2TiO_5 could be microwave synthesized from anatase in a nitrogen atmosphere at about 1050°C. In air, the microwave reaction was found to initiate at 1160°C. This would tend to indicate that the microwave field is coupling with the defects that are generated in rutile by nitrogen processing. This can be observed from the dielectric measurements.

Aluminum titanate can be microwave synthesized from α-Al_2O_3 and defect rutile. The reaction is initiated near 600°C after a soak time of 15 minutes. Conventionally the synthesis does not occur until temperatures near 1300°C are achieved. The XRD data shows the microwave reaction nucleates quickly but growth is slowed as temperature increases. This is possibly due to difficulties encountered by the reactants as they attempt to diffuse across the product interface layer. Due to the poor mixing of the reactants in the defect rutile synthesis from dry milling, future work will involve the investigation of wet-mixing. Attempts will be made to synthesize Al_2TiO_5 in a reducing atmosphere. Slowing the oxidation of the defect rutile could result in increased microwave coupling and heightened reaction rates. Attempts are also being made to characterize the microstructure and to quantify defect concentration. Currently there is no information on the relation of defect concentration to the synthesis reaction. Such information might be vital to determining the reaction mechanisms and rates.

ACKNOWLEDGEMENTS

We thank Brian Lynch and Myles Brostrom for the XRD and the HTXRD analyses, respectively.

REFERENCES

(1) Sutton, W. H., "Microwave Processing of Ceramic Materials", *Ceram. Bull.*, 68[2], pp 376-86 (1989)

(2) Janney, M. A., Kimrey, H. D., " Diffusion-Controlled Processes in Microwave-Fired Oxide Ceramics" Mat. Res. Soc. Symp. Proc., 189, pp 215-27 (1991)

(3) Roy, R., Yang, L. J. and Komarneni, S., "Controlled Microwave Sintering and Melting of Gels", *Ceram. Bull.*, 63[3], p 459 (1984)

(4) Ahmad, I. and Clark, D. E., "Effect of Microwave Heating on Solid State Reactions of Ceramics", **Microwaves: Theory and Application in Materials Processing**, *Ceram. Trans.*, 21, pp 605-12 (1991)
(5) Boch, Ph., Lequeux, N. and Piluso, P., Mat. Res. Soc. Symp. Proc., 269, pp 211-16 (1992)
(6) Piluso, P., Lequeux, N. and Boch, P., *Ann Chim. Fr.*, 18, pp 353-60 (1993)
(7) Katz, J. D., "Microwave Enhanced Diffusion?", **Microwaves: Theory and Application in Materials Processing** , *Ceram. Trans.*, 21, pp 95-105 (1991)
(8) Mathis, M. D., Agrawal, D. K., Roy, R. and Plovnick, R. H., "Microwave Synthesis of SiAlONs", this volume
(9) Thomas, H. A. J. and Stevens, R., "Aluminum Titanate- A Literature Review: Part 2 Engineering Properties and Thermal Stability", *Br. Ceram. Trans. J.*, 88, pp 184-190 (1989)
(10) Thomas, H. A. J. and Stevens, R., "Aluminum Titanate- A Literature Review: Part 1 Microcracking Phenomena", *Br. Ceram. Trans. J.*, 88, pp 144-151 (1989)
(11) Mathis, M. D., Dewan, H. D., Agrawal, D. K., Roy, R. and Plovnick, R. H., "Microwave Processing of Ni-Al_2O_3 Composites", **Microwaves: Theory and Application in Materials Processing II**, *Ceram. Trans.*, 36, pp 431-438 (1993)
(12) Hutcheon, R. M., De Jong, M. S., Adams, F., Wood, G., McGregor, J. and Smith, B., "A System for Rapid Measurements of RF and Microwave Properties up to 1400°C", *J. Micro. Power Elec. Energy*, 27[2], pp 93-102 (1992)
(13) Hutcheon, R. M., De Jong, M. S., Adams, F. P., Lucuta, P. G., McGregor, J. E. and Bahen, L., "RF and Microwave Dielectric Measurements to 1400°C and Dielectric Loss Mechanisms", Mat. Res. Soc. Symp. Proc., 269, pp 541-51 (1992)
(14) Freudenberg, B. and Mocellin, A., "Aluminum Titanate Formation by Solid State Reaction of Coarse Al_2O_3 and TiO_2 Powders", *J. Am. Ceram. Soc.*, 71[1], pp 22-28 (1988)
(15) Freudenberg, B. and Mocellin, A., "Aluminum Titanate Formation by Solid State Reaction of Fine Al_2O_3 and TiO_2 Powders", *J. Am. Ceram. Soc.*, 70[1], pp 33-38 (1987)
(16) Lee, S. C. and Wahlbeck, P. G., "Chemical Kinetics and Mechanism for the Formation Reaction of Al_2TiO_5 from Alumina and Titania", *High Temp. Sci.*, 21, pp 27-40 (1986)

MICROWAVE SYNTHESIS OF NON-OXIDE CERAMIC POWDERS

JGP Binner[†], NA Hassine[†] and TE Cross[‡]
† Department of Materials Engineering and Materials Design
‡ Department of Electrical and Electronic Engineering
The University of Nottingham, University Park, Nottingham, UK

ABSTRACT

The synthesis of titanium carbide (TiC) and tantalum carbide (TaC) via microwave carbothermal reduction of the oxides has been examined. The titanium system suffered problems with isomorphism between a suboxide and the carbide crystal structure which prevented full conversion. This problem was avoided in the case of the tantalum carbide however. Much faster reaction rates were also observed in both cases. This phenomenon has been tentatively explained by a three fold increase in the pre-exponential factor for the titanium carbide system.

INTRODUCTION

The use of microwave energy has been attracting significant attention in recent years as an alternative method for processing ceramic materials. One of the reasons for this is the potential for reductions in manufacturing costs due to shorter processing times and consequent energy savings [1]. Although it is known that many oxides such as titania are almost transparent to microwaves at room temperature and hence cannot be heated readily [2], microwaves will heat carbon powders almost instantaneously provided that the particle size is similar to the depth of penetration [3]. This yields the possibility of heating intimate mixes of TiO_2 and carbon black extremely fast and volumetrically with a higher energy conversion efficiency than can be achieved via conventional furnace heating. The primary aim of the research reported here was thus to investigate the suitability of using microwave heating for the synthesis of two refractory metal carbide powders; titanium carbide, TiC, and tantalum carbide, TaC.

EXPERIMENTAL

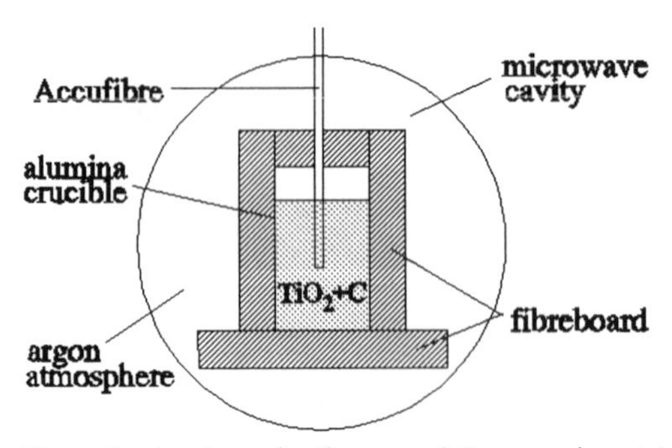

Figure 1. A schematic diagram of the experimental arrangement for heating oxide/carbon black powder mixtures using microwave energy. (Not to scale)

Stoichiometric mixes of TiO_2 + C black and Ta_2O_5 + C black powders weighing 180 g were heated in a 2.45 GHz multimode controlled atmosphere microwave furnace under a range of holding temperature and time conditions. The furnace arrangement is shown schematically in figure 1. Microwave power was smoothly controllable up to 5 kW; typically the power level was varied between 1 and 3 kW during each heat treatment cycle to obtain the desired temperature-time profile. The alumina reaction crucible was insulated within the microwave cavity using low dielectric loss alumina fibreboard to minimise heat losses. A high purity argon atmosphere was maintained throughout the duration of each experiment. Temperatures were measured using an Accufibre optical fibre thermometer buried deep within the powder mix. Some of the experiments were repeated using a conventional controlled atmosphere electrical resistance furnace.

The powders were weighed before and after heating to enable the extent of reaction to be determined. Powder products were characterised by a range of techniques including X-ray diffraction (XRD) to determine the phase composition, and particle size analysis and scanning electron microscopy to study particle size and particle size distribution. In addition, the total and free carbon content, the oxygen and nitrogen content could be monitored when required. Further experimental details can be found in reference [4].

RESULTS AND DISCUSSION

Figure 2 overleaf shows the effect of holding temperature on the formation of TiC, including particle size and oxygen content, using microwave heating. The curve is approximately exponential in form as would be expected for a diffusion-based phenomenon. However, it will also be noted that particle size increases substantially, if more slowly, with temperature and that, despite the use of a high-purity argon atmosphere, the oxygen content is relatively high even at the higher

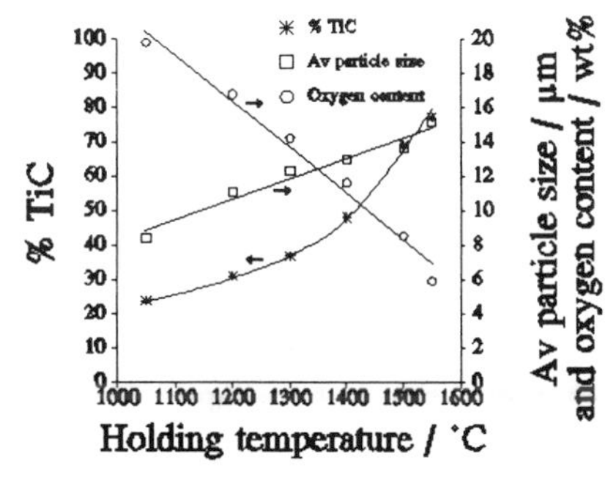

Figure 2. TiC formation and particle size as a function of holding temperature for microwave synthesis. Holding time 1 hour for all experiments.

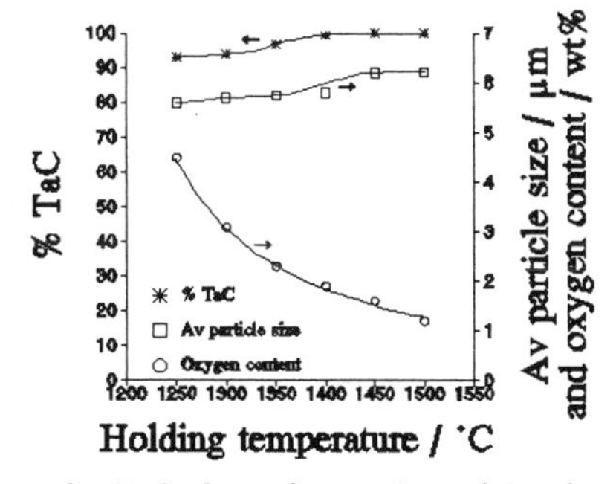

Figure 3. TaC formation and particle size as a function of holding temperature for microwave synthesis. Holding time 1 hour for all experiments.

temperatures. These phenomena may be explained by the fact that a pure TiC phase was never achieved as the final powder product. Rather the product was an oxycarbide based on solid solutions between the isomorphous compounds TiO and TiC as shown by the x-ray diffraction data presented in table 1.

As seen from figure 3, much higher yields of TaC were obtained than for the TiC production. At all holding temperatures greater than about 1400°C, complete conversion to TaC was achieved. Chemical and x-ray diffraction analysis showed no evidence to support the formation of any intermediate phases or lower oxides. In addition, the TaC powder showed a much lower degree of oxygen contamination than was achieved with the TiC and grain growth was also negligible.

Table I. Phase composition of the powders formed from 1 hour at the temperatures given.

Phases present	Holding temp/°C
TiO_2, Ti_4O_7, $Ti(O_{0.8},C_{0.2})$	1050
$Ti(O_{0.7},C_{0.3})$	1200
$Ti(O_{0.5},C_{0.4})$	1300
$Ti(O_{0.4},C_{0.5})$	1400
$Ti(O_{0.3},C_{0.7})$	1500
$Ti(O_{0.2},C_{0.8})$	1550

A significant feature of note in both figures 2 and 3 is the initiation of carbothermal reduction at temperatures much lower than expected. Industrial production of both titanium carbide and tantalum carbide is carried out in inductively heated vacuum furnaces using intimate powder mixes of the appropriate oxide and carbon. These are reacted at approximately 2200°C with a typical heating cycle of 24 hours for TiC powders and 1750°C with holding times of at least six hours for TaC powders [5]. Even under laboratory conditions, there is no evidence in the open literature which suggests that conventional heating can induce any reaction between TiO_2 and C black at a temperature as low as 1050°C. The lowest temperature reported was that observed by Licko et al [6] who suggested that the reduction of TiO_2 began at 1170°C. During the conventional heating experiments performed by the present authors, reduction started at approximately 1200°C. This apparent enhancement in the chemical reaction is a common feature associated with the use of microwaves for processing of materials [7].

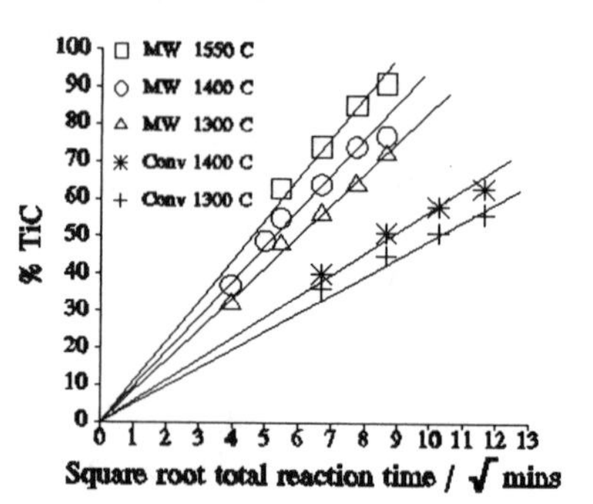

Figure 4. % TiC formed versus reaction time for microwave and conventionally heated carbothermal reductions.

Figure 4 provides plots of % TiC formed versus total reaction time for both the microwave and conventionally heated TiO_2 + C mixtures. The total reaction time, t_{react}, that the samples had undergone was assumed to be given by:

$$t_{react} = t_{hold} + (\frac{t_{heat}}{2}) + (\frac{t_{cool}}{2}) \qquad (1)$$

where:

t_{hold} = holding time at temperature

t_{heat} = time spent during heating between the minimum temperature for any reaction to occur and the holding temperature

t_{cool} = time spent during cooling between the holding temperature and the minimum temperature for any reaction to occur

The results show that both sets of data satisfy a parabolic model indicating that diffusion is the rate controlling step in the reaction.

The feature of greatest interest in figure 4 is the further evidence for faster diffusion rates using microwave heating compared with conventional heating. This is demonstrated more quantitatively in table 2, which presents the reaction (diffusion) rate constant, k, from the parabolic model:

$$degree\ of\ reaction = \sqrt{k t_{react}} \tag{2}$$

Table II. Reaction rates achieved by microwave and conventional heating

Holding temp /°C	Heating method	Reaction rate constant, k /min^{-1}	k_{mw} / k_{conv}
1300	Microwave	15.8	3.3
1300	Conventional	4.8	
1400	Microwave	21.8	3.4
1400	Conventional	6.5	
1550	Microwave	28.4	-

It can be seen that the diffusion rates using microwaves are more than treble those achieved using conventional heating. If it is assumed that this is due to enhanced mobility of the atoms then we can investigate the origins of this behaviour by considering the Arrhenius equation:

$$k = A\exp\left(-\frac{Q}{RT}\right) \tag{3}$$

where the terms have their usual meaning.

Plotting ln k versus 1/T yields the activation energies Q for microwave and conventional heating, see figure 5. These are 56±3 kJmol^{-1} and 63±3 kJmol^{-1} respectively; an apparent 10% reduction in activation energy with microwaves. However, given the small number of experimental points considered and the errors associated with the above values, it seems not unreasonable to assume that the activation energies are actually the same. On this basis, and using an intermediate value of 60 kJmol^{-1}, calculations [8] yield values for the pre-exponential factor of approximately 1650 min^{-1} and 500 min^{-1} for the microwave and conventional heating cases respectively. Thus the use of microwaves has resulted in an apparent increase in the pre-exponential factor by a factor of 3.3. These results support those of Katz et al [7] who found a factor of 3 increase in

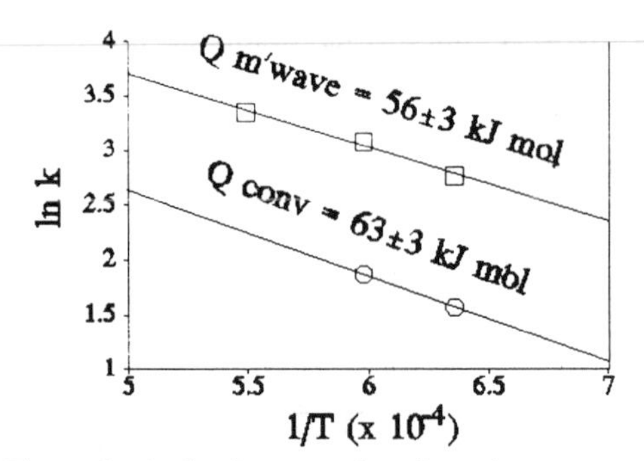

Figure 5. Activation energies for microwave and conventionally heated carbothermal reductions.

the preexponential factor and no reduction in the activation energy for interdiffusivity in the Al_2O_3-Cr_2O_3 system when using microwaves.

An alternative explanation could be the well known difficulties in measuring temperature in a microwave environment [9]. To investigate this let us assume that our microwave temperature measurements are in fact in error and that $Q_{m'wave} = Q_{conv} = 60$ kJmol^{-1} and $A_{m'wave} = A_{conv} = 500$ min^{-1} over the temperature range 1300-1550°C. Calculations based on this assumption [8] show that a temperature inaccuracy of 500°C - 700°C would be required to account for the trebling of the reaction rate when microwaves are used. It is believed that a variation of the magnitude required would most certainly have been detected.

A further possibility is that sparking occurred between the carbon particles yielding very high local temperatures. Since the sparks would be expected to generate similar temperatures, such a phenomenon would not be expected to yield the three clearly defined slopes for microwave heating seen in figure 4.

The Arrhenius pre-exponential factor, A, is dependent on a small number of materials' factors by an equation of the general form [10]:

$$A = \gamma \lambda^2 \Gamma \tag{4}$$

where:
γ is a geometric factor which includes the number of nearest-neighbour jump sites
λ is the distance between adjacent lattice planes, ie the jump distance, and
Γ is the jump frequency which is given by [10]:

$$\Gamma = \nu \exp\left(\frac{-Q}{kT}\right) \tag{5}$$

where:
ν is the natural vibration frequency of the atoms, and
Q is the activation energy.

γ, λ and Q appear to be unaffected, hence it would seem that the microwave field is somehow affecting ν. For bulk crystals a value for ν which is often quoted is 10^{13} s^{-1}. This makes a change in ν difficult to understand since the microwave frequencies used are approximately 10^9 s^{-1} and hence resonance effects would be unexpected. Nevertheless, the possibility of some change in ν cannot be discounted at this stage, especially for phenomena relying on diffusion along grain boundaries or across surfaces where frequency factors are less well defined.

CONCLUSIONS

During the microwave carbothermal reduction of TiO_2, the reaction never progressed beyond the oxycarbide $Ti(O_{0.8},C_{0.2})$ under the maximum conditions used of 1550°C with a holding time of 60 minutes. This limitation is ascribed to the isomorphism of the TiO suboxide with TiC. The net effect was that commercial grade TiC could not be produced.

Conversely, TaC was found to form directly from Ta_2O_5 without the formation of any intermediate phases. This led to i) a product with minimal oxidation contamination and ii) the ability to achieve complete conversion under conditions substantially less demanding than in current commercial practice, viz 60 minutes at 1500°C versus at least six hours at 1750°C.

The reaction between TiO_2 and carbon black was found to be diffusion controlled with both microwave and conventional heating data satisfying the parabolic model $x = \sqrt{kt}$. Reaction rates achieved by microwave heating were more than treble those obtained by conventional heating. These results might be explained by a 3.3-fold increase in the Arrhenius pre-exponential factor, A, with no change in the activation energy. The pre-exponential factor is dependent on the vibration frequency of the atoms at the reaction interface and hence it could be postulated that this might be being affected by the microwave field. Attempts to explain the increased reaction rates in terms of faulty temperature measurement were unsuccessful.

Finally, the results suggest that microwave processing may be a technically viable option for producing carbides by carbothermal reduction provided that the carbide is not isomorphous with a suboxide. Faster production rates and the ability to use lower temperatures may offer economic attractions as well.

ACKNOWLEDGEMENTS

The authors would like to acknowledge the financial contribution made to this project by the London & Scandinavian Metallurgical Company, the Department of Trade and Industry and the Engineering and Physical Science Research Council all in the UK.

REFERENCES

1. W.H. Sutton, MRS Bulletin, **18** [11] 22-24, 1993.
2. S.L. McGill, J.W. Walkiewicz and G.A. Smynes, in *Microwave Processing of Materials*, Material Research Society Symposium Proceedings **124**, ed. WH Sutton, MH Brooks and IJ Chabinsky, Materials Research Society, Pittsburgh, pp. 247-252 (1988).
3. C.P. Lorenson, M.C.L. Patterson, G. Risto and R. Kimber, in *Microwave Processing of Materials III*, Material Research Society Symposium Proceedings **269**, ed. RL Beatty, WH Sutton and MF Iskander, Materials Research Society, Pittsburgh, pp. 129-135 (1992).
4. N.A. Hassine, J.G.P. Binner and T.E. Cross, accepted for publication in Int J of Refrac & Hard Mat.
5. D. Wardle, London & Scandinavian Metallurgical Co Ltd Internal Report.
6. T. Licko, V. Figusch and J. Puchyova, J. Eur. Ceram. Soc. **5**, 257-265 (1989).
7. J.D. Katz, R.D. Blake and V.M. Kenkre, in *Microwaves: Theory and Application in Materials Processing*, Ceramic Transactions **21**, ed DE Clark, FD Gac and WH Sutton, American Ceramic Society, Westerville, pp. 95-98 (1991).
8. J.G.P. Binner, N.A. Hassine and T.E. Cross, accepted for publication in J Mat Sci.
9. A.C. Metaxas and J.G.P. Binner, in *Advanced Ceramic Processing and Technology, vol I*, ed Binner JGP, Noyes Publications, New Jersey, pp. 285-367 (1990).
10. W.D. Kingery, H.K. Bowen and D.R. Uhlman, *Introduction to Ceramics*, 2nd edition, John Wiley & Sons Inc., New York, chapt. 7 (1976).

Current Issues and Future Activities

OBSTACLES AND REMEDIES IN MICROWAVE PROCESSING RESEARCH AND COMMERCIALIZATION - AN OPEN FORUM

Wayne R. Tinga and Steven J. Oda*
Dept. of Electrical Engineering, Univ. of Alberta, Edmonton, AB, CANADA T6G 2G7
*36 Eleventh Street, Etobicoke, Ontario, CANADA M8V 3G3

Over 75 material scientists, engineers, physicists and equipment manufacturers vocalized their views on the above topic. Encouraging spontaneous discussion was the aim of this Open Forum, chaired by W.R. Tinga, with S.J. Oda acting as recording secretary. The Open Forum format, once again, proved to be a useful and interesting medium to get people to communicate their own unique perspectives on the various issues raised. Related issues were addressed at a 1993 forum [1].

LIST OF QUESTIONS POSED

1. In which process areas can research and development (R&D) significantly advance the use of microwave energy?

2. What are the best niche application areas?

3. Which industries are adversely affected by new regulatory limits?

4. What kind of research and development is most needed?

5. How can we be better knowledge brokers?

6. Scale-up: What are the significant parameters?

7. How can material processors be attracted to a meeting?

8. What are the user's needs?

9. What are the real problems in industrial processing?

Not all of these questions were answered. Many of the answers given indicated the complexity and difficulty of forging the proper linkages between researchers, equipments suppliers and the end users. Nonetheless, it appears that part of the necessary process is an open-ended discussion like this.

IN WHICH PROCESS AREAS CAN RESEARCH AND DEVELOPMENT SIGNIFICANTLY ADVANCE THE USE OF MICROWAVE ENERGY

No simple nor profound answers came to the fore. One requirement for the probable adoption of high technology equipment in low technology industries is that it solves existing problems not solvable by more conventional means, e.g., dumping a large amount of energy into a product in a very short time to achieve a desired result. Significant financial investment may be needed beyond the normal feasibility study stage. Such investment is warranted when there is demonstrated need for the use of a specific technology and when a market's time window is long enough to ensure the product is still relevant when production finally starts.

Applying this metric to various waste remediation problems, makes microwave processing look favorable. Creation of an entirely new and superior Parallam® wood product, according to MacMillan Bloedel, Ltd., was only possible through the selective use of many megawatts of microwave power along with many other process and design innovations. Interestingly, the process development took over 10 years, was very costly, but was based on fundamental R&D done both in university labs and in the company's own R&D labs. Two large plants are being built to accomodate demand for this unique microwave processed Parallam® product. Microwave energy use in the process constitutes only about 10% of the total process energy requirement.

WHAT ARE THE BEST NICHES APPLICATION AREAS?

Certainly those areas where there are products with a high value added component, because of processing, are more likely candidates since higher process costs are offset by the high market value of the end product. Microwave processed ceramic toolbits, a possible niche area, show improved characteristics and can be manufactured in commercially significant quantities according to Sherritt Gordon/Westaim of Alberta, Canada. In this case, binder burnout, quoted as a problem by some researchers, was no hindrance to the success of this process. In the opinion of others, binder burnout, a time-consuming operation in ceramic prototyping, is thought to be a niche area itself where microwaves may be able to reduce the burnout time if it can be done uniformly for complex products.

It may be beneficial to turn the question around and to ask: "What processes need significant improvement?" Consider, especially, ecology-sensitive processes, where conventional methods can not meet the stricter environmental constraints. Don't try to convert more exotic processes that already work well. Someone suggested it might be a good idea to have researchers and equipment developers take time to walk through industrial process plants to try and identify specific processing problems and subsequent solution opportunities.

On the other hand, moving an engineered solution to a real problem to commercialization is not a one-step process. Team effort is needed to supply knowledge about the materials, microwave interaction, microwave hardware design and process control, among others. Other critical facets of a real processing solution require the involvement of the end-user in the development process. These certainly can't be done overnight and require a significant investment of time and money. Venture capitalists will take risks to provide money if a good business plan is presented. However, researchers usually are not very good at making up such a plan. Clearly, team effort is needed, on either a small or large scale, to enhance successful application of a new solution to a new or old problem. Research must be "mission oriented," that is, looking for a solution to a real tangible problem. EA Technologies in the U.K., and MacMillan Bloedel Ltd in Canada, have found success in making microwave energy part of the solution to a vexing conventional problem or to the creation of an entirely new product.

Unfortunately, constraints imposed by proprietary demands, from the process innovator or the end user, can stifle a public teamwork approach forcing instead an in-house formation of an expert team. In either case, university researchers can become involved if they are proactive in seeking and pursuing practical linkages with end-users and equipment manufacturers.

SCALE-UP: WHAT ARE THE SIGNIFICANT PARAMETERS?

From an engineering point of view, mechanical microwave structures are very scaleable. Yet, from an energy deposition point of view, scaling from a small sample study to a large volume process is by no means straightforward. Microwave heat deposition is a non-linear process with respect to the fundamental electromagnetic field behavior. Power deposited in a material is proportional to material properties which vary nonlinearly with distance, temperature, composition and density, to name just a few parameters. Instabilities, at least in the microwave heating of small samples such as alumina and silicon nitride, appear to confound many in the quest to scale-up to larger process quantities. Despite this difficulty, it is possible to judiciously scale-up a process with respect to its production volume. Examples of such successes were mentioned earlier. One, however, needs to ensure that these apparent, scale-up limiting phenomena are not caused by fundamental limits of the process physics.

Nonetheless, in practice, the non-uniformity of a batch process may often be overcome by space and time-averaging in a continuously fed microwave applicator. Using traveling wave fields in place of standing wave fields may greatly change the dynamic process behavior in a given situation. In fact, continuous, conveyor-type, microwave systems use a combination of traveling and standing waves and are often easier to control than batch type systems. Distributed coupling structures can enhance energy deposition uniformity. Conversely, suppression of microwave leakage from continuous systems can be a real engineering challenge.

Electromagnetic interaction with materials depends very strongly on the wavelength or its inverse, the frequency. One practical example of this is the widespread use of radar to detect large aircraft, small raindrops, water vapor, or speeding cars. All these radars operate on the sample principle of wave reflection, but all operate at different frequencies. Detectable objects are those which have a characteristic size comparable to the radar's wavelength. Application of this scaling principle, along with equipment availability constraints, caused MacMillan Bloedel to select 915 MHz as the preferred frequency of operation in their Parallam® process. Similarly, it is this principle which makes the variable frequency power source a useful tool in creating greater product heating uniformity.

Finally, continuous processing may yield benefits not readily obvious in the batch testing stage, e.g., improved uniformity and better energy utilization because of increased applicator filling factor and dynamic microwave to material coupling changes. Conversely, replacing part of a conventional process sequence by a faster microwave process can cause throughput incompatibility problems.

HOW CAN WE BE BETTER KNOWLEDGE BROKERS?

Another way of phrasing this question is: How can material processors be attracted to a meeting of technical knowledge brokers? It was generally agreed that round-table discussions like this should involve equipment people, researchers and materials processors. It was pointed out that many users are not interested in technology as such. Replacing conventional working systems, even though problematic or limiting in some ways, with a "Hi-Tech" microwave system will typically meet with considerable resistance from the processor's side. It may be more appropriate to talk to them about "New Techniques For Process Optimization and Control." Investigators should ask: "What part of the conventional system needs improvement?" For example, all conventional ceramic processing systems need kiln furniture, similar to casketing, which makes up 30 to 40% of the load. Elimination of the kiln furniture requirement would, therefore, be a highly desirable goal. Benefits for the end users must be identified to attract potential processors while avoiding unnecessary technical jargon or details which cloud the issues.

CONCLUSIONS

Undoubtedly, this forum helped to identify some of the problems and possible remedies in making microwave processing technology a more attractive alternative to modern material processing problems. It is also clear that this requires a continuing and ongoing effort and dialogue between researchers, equipment makers and end users.

REFERENCES

1. W.R. Tinga and W.H. Sutton in, **Microwaves: Theory and Application in Materials Processing II**, Ceramic Transactions, Vol. 36 (D.E. Clark, W.R. Tinga and J.R. Laia, Jr., eds), American Ceramic Society, Westerville, OH, pp. 45-49 (1993).

WORLD CONGRESS ON MICROWAVE PROCESSING

David A. Lewis
IBM, T.J. Watson Research Center,
P.O. Box 218,
Yorktown Heights, NY 10598

INTRODUCTION

Over the past decade, interest in the utilization of microwave energy to process materials has increased substantially, following a lull in the late 1970's and early 1980's. In response to this increased attention, and the highly multidisciplinary nature of the technology, a number of societies have sponsored conferences relating to microwave processing. Despite the increasingly high interest in the technology, there have been very few successful commercializations of the microwave processing.

Two years ago a group of people who had been responsible for organizing symposia on microwave processing in different technology areas decided to attempt to co-ordinate the conferences, and assist planning to attempt to voluntarily prevent conferences "competing" with each other. It is important to note that this group was not intended to "dominate" the field and prevent others from organizing conferences, if they so chose. The group was named the Microwave Working Group and its primary goal was designated as to assist in co-ordinating and/or promoting new symposia in the area of micrwoave technologies. One of the consensus decisions was to hold a "World Congress", with the principle aim of bringing together the experts in the field with potential customers to increase communications and hopefully better address the problems associated with the utilization of the technology while informing customers of the potential and the state of the art of the technology.

The Congress will be held from **January 5 - 9, 1997** at the **Hilton Hotel, Walt Disney Village**, Orlando Florida.

The Congress is being sponsored by EPRI (Electric Power Research Institute) and is endorsed by a wide range of societies, including American Ceramics Society (ACerS), American Chemical Society, Polymer Division (ACS-POLYMER), Association of Microwave Power in Europe for Research and Education (AMPERE), Institute of Electromagnetic Wave Application - Japan (IEAJ), International Microwave Power Institute (IMPI), Materials Research Society (MRS) and National Institute of Ceramic Engineers (NICE). It is anticipated that additional societies will agree to endorse this Congress as time proceeds.

The Congress is being deliberately directed to a wider audience than other conferences in this area to achieve the aim of increasing technology transfer for this technology. The audience being targeted through the technical program consists of

(i) scientists and technologists experienced with microwave/RF technologies;
(ii) microwave equipment vendors;
(iii) representatives of the Electric Utilities who regularly communicate with their customers concerning alternate technologies;
(iv) companies with potential interest in utilizing microwave technologies.

ORGANIZATION

The World Congress is being co-chaired by Will Sutton (United Technologies) and Bernie Krieger (Cober Electronics) under the auspices of the Microwave Working Group. Most of these people have active roles in the Congress, in addition to being the overseeing committee. The organizing committee includes

J. Binner	D. Lewis**
I. Campisi	R. Metaxas
D. Clark	S. Oda
G. Eckhart	J. Osepchuk
J. Gerling	R. Silberglitt
M. Iskander	C. Shibata
D.L. Johnson	W. Sutton*
J. Katz	W. Tinga
H. Kimrey	R. Schiffmann
B. Krieger*	

* Congress Co-Chairs ** Technical Program Chair

PROGRAM

Since we are reaching out to a new, different audience, in addition to attracting the best microwave scientists and engineers in the world, the Congress will be a combination of highly technical presentations, applications to interest the users and the utilities representatives, an exhibition for the vendors and open discussion to facilitate bridging the "knowledge gap" between these groups. Additionally, there will be sufficient time to meet informally and initiate discussions of the technology.

To overcome the technology gap between the utilities group and the technologists, a series of "Overview of the Technology" presentations will be made to provide an abbreviated overview / summary of each of 5 - 6 technology areas, which will be directed to the "non-expert" in a particular field. These are expected to be of interest to both the expert and non-expert alike.

"Overview of the Technology" Series

This series - each taking one session (90 minutes) - is targets the utilities representatives and non-experts in the specific topic and will provide an advanced shortcourse level of difficulty describing the current shortcomings, state of the art and directions of the technology.

Some of the areas which will be included are

(i) Technology trends, reliability, capital cost and financing opportunities
(ii) Comparison of electrotechnologies alternatives
(iii) Ceramics
(iv) Polymers and composites
(v) Radio frequency processing
(vi) Plasma processing
(vii) Waste remediation

As an example of the content of this series, the Polymers and Composites session may include topics such as (i) what are polymers and why do people want to use them; (ii) how are these materials processed today; (iii) why would people want to change the processing technology; (iv) what are the problems and advantages of using microwave radiation to process these materials; (v) case studies of successful applications - why were they successful and what can be learned from this; (vi) how much is it going to cost to make the transfer to this technology and what is the business case; etc. The other technology areas would have an appropriate breakdown in topics suitable to those areas.

The aim is to arm the attendees of these presentations with sufficient knowledge of the particular part of the technology to enable them to (i) participate and understand the remaining Congress; (ii) talk knowledgeably with their customers about where the technology is and help them be most effective in contributing to their customers problems and simultaneously promote microwave technology, if the application is appropriate and (iii) expand their knowledge into other microwave processing areas.

There may be additional topics added as necessary and the titles will also change.

Technical Program

The technical portion of the program will focus on highly technical presentations ranging from basic science to applications within technology areas. Due to the limited number of oral sessions available, the poster session(s) will take on a very important role. Posters are very good for generating discussion and the dissemination of ideas and in some ways a good learning medium for those not as experienced in the field (eg the Utilities reps). We are expecting a large number of posters.

The technology areas which will be addressed include

Ceramics/joining
Polymer/chemistry/composites
Waste
RF
Plasma
Modelling
Dielectric Properties
Systems Design
Emerging Technologies

The topics names are for informational purposes only and will likely change according to the symposium organizers desires.

SUMMARY

The World Congress on Microwave Processing intends to bring together the users and potential users of microwave technologies in a manner to improve the transfer of ideas and technology.

On behalf of the organizers, we hope to see you in Orlando in 1997.

For more information, contact

David Lewis	OR	Bill Wisecup
Technical Program Chair		Congress Executive Director
IBM Research		W.L. Associates
PO. Box 218		7519 Ridge Rd
Yorktown Heights, NY 10598		Frederick, MD 21702
Phone (914)945-2074		(301)663-1915
FAX (914)945-4033		(301)371-8955
e-mail:DLEWIS@watson.ibm.com		75230.1222@compuserve.com

Keyword and Author Index